The Routledge Companion to Smart Cities

The Routledge Companion to Smart Cities explores the question of what it means for a city to be 'smart', raises some of the tensions emerging in smart city developments and considers the implications for future ways of inhabiting and understanding the urban condition. The volume draws together a critical and cross-disciplinary overview of the emerging topic of smart cities and explores it from a range of theoretical and empirical viewpoints.

This timely book brings together key thinkers and projects from a wide range of fields and perspectives into one volume to provide a valuable resource that would enable the reader to take their own critical position within the topic. To situate the topic of the smart city for the reader and establish key concepts, the volume sets out the various interpretations and aspects of what constitutes and defines smart cities. It investigates and considers the range of factors that shape the characteristics of smart cities and draws together different disciplinary perspectives. The consideration of what shapes the smart city is explored through discussing three broad 'parts' – issues of governance, the nature of urban development and how visions are realised – and includes chapters that draw on empirical studies to frame the discussion with an understanding not just of the nature of the smart city but also how it is studied, understood and reflected upon.

The Companion will appeal to academics and advanced undergraduates and postgraduates from across many disciplines, including Urban Studies, Geography, Urban Planning, Sociology and Architecture, by providing state of the art reviews of key themes by leading scholars in the field, arranged under clearly themed sections.

Katharine S. Willis is Associate Professor (Reader) in the School of Architecture, Design and Environment at the University of Plymouth, UK.

Alessandro Aurigi is Professor of Urban Design at the University of Plymouth, UK.

The Routledge Companion to Smart Cities

Edited by Katharine S. Willis and Alessandro Aurigi

LONDON AND NEW YORK

First published 2020
by Routledge
4 Park Square, Milton Park, Abingdon, Oxon OX14 4RN
605 Third Avenue, New York, NY 10017

First issued in paperback 2023

Routledge is an imprint of the Taylor & Francis Group, an informa business

British Library Cataloguing-in-Publication Data
A catalogue record for this book is available from the British Library

Library of Congress Cataloging-in-Publication Data
Names: Willis, Katharine S., editor. | Aurigi, Alessandro, editor. | Routledge (Firm)
Title: The Routledge companion to smart cities / edited by Katharine S. Willis and Alessandro Aurigi.
Description: New York : Routledge, 2020. | Includes bibliographical references and index.
Identifiers: LCCN 2019044046 (print) | LCCN 2019044047 (ebook)
Subjects: LCSH: Smart cities. | Cities and towns. | City planning. | Municipal government.
Classification: LCC TD159.4 .R68 2020 (print) | LCC TD159.4 (ebook) | DDC 307.1/216–dc23
LC record available at https://lccn.loc.gov/2019044046
LC ebook record available at https://lccn.loc.gov/2019044047

ISBN: 978-1-03-257004-4 (pbk)
ISBN: 978-1-138-03667-3 (hbk)
ISBN: 978-1-315-17838-7 (ebk)

DOI: 10.4324/9781315178387

Typeset in Bembo
by Swales & Willis, Exeter, Devon, UK

Publisher's Note
The publisher has gone to great lengths to ensure the quality of this reprint but points out that some imperfections in the original copies may be apparent.

Contents

Contents

Contents

Figures

Tables

Contributors

Alessandro Aurigi is Professor of Urban Design and Associate Dean at the University of Plymouth. Alex has previously worked at Newcastle University as Head of the Architecture Department and at the Bartlett School of Planning and CASA, University College London. His research focuses on the relationships between our digital society and the ways we conceive, design, and manage urban space to enhance and support place quality. Alex has published *Digital and Smart Cities* (Routledge, 2018, with Katharine S. Willis), the multi-disciplinary book *Augmented Urban Spaces* (Ashgate, 2008, edited with Fiorella De Cindio), and *Making the Digital City* (Ashgate, 2016). He is member of the Peer Review College of the Arts and Humanities Research Council (UK).

Michael Batty is Emeritus Professor of Planning at the Centre for Advanced Spatial Analysis, Faculty of the Built Environment, UCL, London. His research group is working on simulating long-term structural change and dynamics in cities as well as visualisation and urban analytics for smart cities, and he has published many books and articles in this area. His book *Cities and Complexity* (MIT Press, 2005) won the Alonso Prize of the Regional Science Association in 2010. His most recent books are *The New Science of Cities* (MIT Press, 2013) and the edited volumes *Virtual Geographic Environments* (ESRI Press, 2011) and *Agent Based Models of Geographical Systems* (Springer, 2012). He is editor of the journal *Environment and Planning B: Planning and Design*. He is a Fellow of the British Academy (FBA) and a Fellow of the Royal Society (FRS) and was awarded the CBS in the 2004 Queen's Birthday Honours list. Most recently he was the 2013 recipient of the Laureat Prix International de Geographie Vautrin Lud and in 2015 he received the Founders Gold Medal of the Royal Geographical Society.

Matthijs Bouw is a Dutch architect and founder of One Architecture (est. 1995), which focuses on urban design and resilient architecture. He currently serves as the Rockefeller Urban Resilience Fellow for Penn Design at the University of Pennsylvania. Mr Bouw has been a Guest Professor at, a.o., TU Delft, Berlage Institute, TU Graz, University of Kentucky College of Design and Sci-Arc, and was Professor I.V. of 'Gebaeudelehre und Grundlagen des Entwerfens' at the RWTH Aachen. A leading voice on resilient design, he has published several dozen papers and given numerous talks to both students and professionals on incorporating resiliency into design practice. Bouw's own practice is known for its unique approach in which programmatic, financial, technical and organisational issues are addressed, communicated and resolved through design. Bouw has been a pioneer in the use of design as a tool for collaboration, for instance through the development of 'Design Studios' as an instrument to support the Netherlands' Ministry of Infrastructure and the Environment with its long-term planning, and in community-development projects.

Christopher T. Boyko is a 50th Anniversary Lecturer in Design at Lancaster University. He explores critical connections between urban design and behaviour, informing both design practice and policy. He has undertaken research on the relationship between wellbeing, sharing and neighbourhoods for the Liveable Cities project; urban density and decision-making processes on Urban Futures; and urban design decision-making process and sustainability for VivaCity2020.

Glenda Amayo Caldwell is a Senior Lecturer in Architecture, School of Design, Creative Industries Faculty at the Queensland University of Technology (QUT). Embracing transdisciplinary approaches from architecture, interaction design, human computer interaction and design robotics, Glenda's research places design at the forefront of robotic research in design-led manufacturing. She is the Associate Director of the QUT Design Lab and the author of numerous publications in the areas of media architecture, urban informatics and design robotics.

Federico Caprotti is an Associate Professor in human geography at the University of Exeter. He is an urban geographer with an interest in urban futures, and his work has largely focused on eco-city and smart city projects in China and Europe. He has recently led the SMART-ECO multi-year, multi-partner research consortium on these topics, funded by several European research agencies and by the Chinese NSFC. From 2018–2020, Federico is a Fellow of the Alan Turing Institute, where he is developing his research in the area of platform urbanism.

Claudio Coletta is a Research Manager in the Urban Studies Institute at the University of Antwerp. His research is interdisciplinary, drawing on the fields of organisation studies, science and technology studies and urban studies. He is particularly interested in the relationship between digital technology and urban and social phenomena, especially working practices, organising processes and innovation.

Rachel Cooper OBE is Distinguished Professor of Design Management and Policy at Lancaster University. Her research interests cover: design thinking, design management, design policy, design for wellbeing and socially responsible design. She is President of the Design Research Society and a Visiting Professor and Council member of the Royal College of Art.

Claire Coulton has worked at Lancaster University for the last six years, as Project Coordinator and Researcher for the Liveable Cities project, working on wellbeing, sharing and future cities. She also is Project Coordinator for the PETRAS project, exploring critical issues related to the Internet of Things. Along with Chris Boyko, Claire is co-editor of the *Liveable Cities* Little Book series (Lancaster University).

Tim Davies is a researcher and consultant exploring the creation of open data infrastructures to address public policy problems. He is co-editor of *The State of Open Data: Histories and Horizons* (African Minds, 2019), and led the Open Data in Developing Countries research network from 2012–2014. He has an MSc in Social Science of the Internet from the Oxford Internet Institute, and has been a Fellow at the Harvard Berkman Centre for Internet and Society. He is co-author of a number of open data standards, including the Open Contracting Data Standard (OCDS) and Beneficial Ownership Data Standard (BODS).

Michiel de Lange is an Assistant Professor in New Media Studies, Department of Media and Culture Studies, Utrecht University, Netherlands; co-founder of The Mobile City (www.themobilecity.nl), a platform for the study of new media and urbanism; and co-founder of the (urban interfaces) research platform at Utrecht University. His research interests concern how (mobile) media technologies shape urban culture and vice versa. Currently, he is Co-lead of the current NWO funded project, Designing for Controversies in Responsible Smart Cities. See http://blog.bijt.org/

Martijn de Waal is a Professor at the research group of Play & Civic Media at the Amsterdam University of Applied Sciences. He is co-author (with José van Dijck and Thomas Poell) of *The Platform Society: Public Values in a Connective World* (Oxford University Press, 2018). In 2014 he wrote *The City as Interface. How Digital Media Are Changing the City* (Nai010 Publishers, 2014), a book on the relation between digital media and the urban public sphere. He was the Project Lead for The Hackable City research project. With Michiel de Lange, in 2007 he founded TheMobileCity.nl, an international think tank and research network on new media and urban culture. Together they edited *The Hackable City: Digital Media and Collaborative City-Making in the Network Society* (Springer, 2019).

Giulia Desogus is a Researcher in the Department of Civil, Environmental and Architecture Engineering at University of Calgari, Italy. Her research is focused on issues of urban planning and landscape planning, urban ecology and urban metabolism processes. She graduated in Architecture (Politecnico di Torino and Hosei University Tokyo 2014). Since then, she has actively continued post-graduate research, holding a Master's in Strategic Environmental Management in University of Padova (2017) and two Research Fellowships (2016, 2017) at the DICAAR, University of Cagliari.

Fábio Duarte is a Research Scientist at MIT Senseable City Lab, where he leads projects including Roboat (a fleet of autonomous boats for Amsterdam) and Underworlds (using robots to monitor public health through the sewage system). He is the author of *Unplugging the City* (2018).

Nick Dunn is Executive Director of Imagination Lancaster at Lancaster University, where he is Professor of Urban Design. He leads research on the future of cities and urbanism, responding to the contemporary city through experimentation and writing on the nature of urban space. His latest book, *Dark Matters: A Manifesto for the Nocturnal City* (Zero, 2016), invites us to rethink urban environments at night.

Carlos Estrada-Grajales is a Research Fellow in the School of Civil Engineering and Built Environment in the Science and Engineering Faculty at Queensland University of Technology (QUT). Carlos is a sessional academic at the School of Design in the Creative Industries Faculty. His research explores a diversity of topics, including digital technologies and media platforms, activism and political participation, citizen engagement, and multi-stakeholder governance practices.

Marcus Foth is Professor of Urban Informatics in the QUT Design Lab, Brisbane, Australia, and an Honorary Professor in the School of Communication and Culture at Aarhus University, Denmark. His research brings together people, place and technology with a focus on smart cities, community engagement, media architecture, internet studies, ubiquitous computing and sustainability. Professor Foth founded the Urban Informatics Research Lab in 2006 and the QUT Design Lab in 2016.

Chiara Garau is Assistant Professor in Urban and Regional Planning in the Department of Civil, Environmental and Architecture Engineering (DICAR) at University of Calgari, Italy. Her main research interests are focused on issues of urban planning, landscape planning, cultural heritage, planning systems and participatory processes. A referee of international scientific journals, she is author of over 26 scientific publications, including monographs, conference proceedings, articles in books, national and international journals. She is the Principal Investigator on the project GHOST (Governing the Smart City: A Governance-Centred Approach To Smart Urbanism) funded by MIUR (Italian Ministry of Education, Universities and Research).

Mark Graham is a Professor at Oxford University, a Visiting Researcher at the Wissenschaftszentrum Berlin für Sozialforschung (WZB) and the Director of the Fairwork Foundation. His work focuses on digital geographies and how they both reflect and reproduce digital inequalities. His full list of publications is available at www.markgraham.space

Robert G. Hollands is a Professor of Sociology at Newcastle University, UK. His work spans youth and urban studies and he is the co-author of the highly acclaimed book *Urban Nightscapes* (Routledge, 2003). He is also known for his critiques of the smart city idea, and his most recent work has been on the alternative and egalitarian arts, exemplified by his Leverhulme Trust funded project 'Alternative Creative Spaces and Urban Cultural Movements'.

Andrew Hudson-Smith is Chair of Digital Urban Systems at the Bartlett Centre for Advanced Spatial Analysis (CASA). He is a member of the Mayor of London Smart London Board and Visiting Professor, University of Plymouth, School of Art, Design and Architecture. His research is focused on sensing and visualising the built environment.

Stephan Hügel is a Research Fellow at Trinity College Dublin. Previously a doctoral researcher at UCL CASA, he is interested in urban cybernetics, 20th-century urban technology projects – particularly those of the 1980s and 1990s – and urban data visualisation. His research also explores conceptions of citizenship and agency in relation to the smart city.

Lenna Johnsen is a Master in City Planning student at MIT and a Research Assistant at MIT Senseable City Lab.

Rob Kitchin is a Professor in the Department of Geography and Maynooth University Social Sciences Institute. His research focuses on the relationship between society, space and digital technologies. He was principal investigator of the Programmable City project and is a co-PI of the Building City Dashboards project.

Francisco Klauser is Professor in Political Geography at the University of Neuchâtel, Switzerland. His work explores the socio-spatial implications, power and surveillance issues arising from the digitisation of present-day life. Main research topics include video surveillance, megaevent security and predictive policing, smart cities, civil drones and big data in agriculture.

Max Kortlander is a Public Researcher at Waag. He holds bachelor's degrees in International Relations and Spanish from Bucknell University and a master's degree in Media Studies from the University of Amsterdam. As a researcher, writer and editor, he has had the opportunity to work with interesting people from all over the world – novelists, nutritionists, filmmakers, NGOs and others who brought him on board to tell their stories. His research interests include co-creation, peacemaking and digital identity.

Frank Kresin is Dean of the Faculty of Digital Media and Creative Industries at the Amsterdam University of Applied Science. He is also Associate Fellow of DesignLab at the University of Twente, Research Fellow at Waag Society and Board Member of V2, Tetem, CREA and The Mobile City. Frank was trained in cinematography and artificial intelligence, and worked previously at the University of Amsterdam, the Dutch Digital University Consortium, Waag Society and the University of Twente, and served as a board member for the Dutch Chapter of the Internet Society. He is interested in citizen science, trans-disciplinary innovation and responsible design.

Taibat Lawanson is Associate Professor of Urban Planning at the University of Lagos, Nigeria, where she leads the Pro-Poor Development and Urban Management research cluster. She holds a PhD in Urban and Regional Planning from the Federal University of Technology, Akure. Her research interests traverse African urbanisation dynamics, governance, spatial justice, formal–informal linkages (infrastructure, systems and people) and understanding the agency of the urban poor. She serves on the board of directors of the Lagos Studies Association, and is international corresponding editor of *Urban Studies* journal. She can be reached at tlawanson@unilag.edu.ng

Charles Leleux is a researcher at CRISP – the Centre for Research into Information Surveillance and Privacy, located at the University of Stirling in Scotland. He is an expert on local government studies and a researcher on the ESRC-funded SmartGov project.

Aale Luusua is an architect whose research revolves around the themes of urban technologies and democratic participation. An important current focus is on the role of artificial intelligence in city-making and urban life. Methodologically, Dr Luusua has utilised and developed participatory methods for the design and research of technologies in public urban places and spaces.

Justien Marseille is a Senior Lecturer at the Rotterdam University of applied science and founder of the Future Institute, a micro organisation aiming to create better understanding in future thinking and forecasting techniques as well as insights in possible futures. Her background is in communication science and she has an MA from the University of Amsterdam.

Shannon Mattern is a Professor at the New School for Social Research. Her writing and teaching focus on media spaces, media infrastructures and spatial epistemologies. She has written books on libraries, maps and histories of urban intelligence, and she contributes a regular long-form column about urban data and mediated infrastructures to *Places Journal*. You can find her at wordsinspace.net.

Gavin McArdle is an Assistant Professor in the UCD School of Computer Science and a Collaborator with CeADAR – Centre for Applied Data Analytics. He was previously a Research Fellow at IBM. His research focus includes geovisual analysis, smart city technology and urban dynamics.

Peta Mitchell is Associate Professor in QUT's Digital Media Research Centre. Her research focuses on digital geographies, location awareness and mobile media, algorithmic culture and network contagion. Peta is author of *Cartographic Strategies of Postmodernity* (Routledge, 2008) and *Contagious Metaphor* (Bloomsbury Academic, 2012) and co-author of *Imagined Landscapes: Geovisualizing Australian Spatial Narratives* (Indiana UP, 2016).

Ingrid Mulder is an Associate Professor at Delft University of Technology. With a background in policy and organisation sciences (MA, University of Tilburg) and behavioral sciences (PhD, University of Twente), her ongoing research on design for social transformations interestingly combines strategic design with diffuse design while addressing the interplay between top-down policy and bottom-up participatory innovation.

Christian Nold is a Research Associate in the Extreme Citizen Science group at University College London. He created large-scale public art projects such as 'Bio Mapping' and experimental currencies in Holland and Finland. He has written: Mobile Vulgus (Bookworks, 2001), *Emotional Cartography: Technologies of the Self* (CC, 2009), *The Internet of People for a Post-Oil World* (Architectural League of New York, 2011) and *Autopsy of an Island Currency* (Pixelache, 2014).

Nancy Odendaal is Associate Professor in City and Regional Planning at the University of Cape Town. Her research focuses on three interconnected areas of enquiry: infrastructure development, technology innovation and socio-spatial change in cities. She has published extensively on smart cities, with her research focused on the interface between new technologies and marginalised spaces. She is the co-chair of the Global Planning Education Association Network (GPEAN) and incoming chair of the AAPS. Nancy is a member of the editorial boards of the *International Journal of E-Planning Research and Urban Geography*. Together with other African academics, she co-edited a volume entitled *Planning and the Case Study Method in Africa* (2014).

Dietmar Offenhuber is Associate Professor at Northeastern University in the fields of Art + Design and Public Policy. He holds a PhD in Urban Planning from MIT and master's degrees from the MIT Media Lab and TU Vienna. His research focuses on the relationship between design, technology and governance. Dietmar is the author of the award-winning monograph *Waste is Information* (MIT Press, 2017), works as an advisor to the United Nations and has published books on the subjects of urban data, accountability technologies and urban informatics.

Timo Ojala is a full Professor of Computer Science and Engineering, and the Director of the Center for Ubiquitous Computing at the University of Oulu. For 15 years his multidisciplinary research group focused on studying future ubiquitous computing systems at authentic urban settings. In recent years, his research has shifted towards large-scale immersive 3D virtual models interacted with VR headsets and 3D web.

Till Paasche was a Postdoctoral Fellow at the University of Neuchâtel, Switzerland during the research for this book chapter. He was then Associate Professor in Geography at the University of Soran, Kurdistan, and later moved to work outside academia.

Serena Pollastri is a Lecturer in Design at ImaginationLancaster at Lancaster University. Through a mix of design practice and research she explores processes and artefacts for articulating pluralistic, polyphonic and multispecies pasts, presents and futures. Serena teaches at the Lancaster University/Beijing Jiaotong University double degree programme in Media Art and Design.

Matti Pouke is a computer scientist whose research has been focusing on 3D graphics, mirror world applications as well as engineering methods for transforming human activities into virtual environments. Currently Dr Pouke is especially focusing on research regarding perceptual factors in graphics and audio, and their effect on virtual reality experiences.

Carlo Ratti is founder of the CRA Architecture Firm (Turin and New York) and Director of the Senseable City Lab at MIT, Boston. Ratti has co-authored over 500 publications and holds several patents. His work has been exhibited worldwide at venues such as the Venice Biennale, the Design Museum in Barcelona, the Science Museum in London and The Museum of Modern Art in New York. Two of his projects – the Digital Water Pavilion and the Copenhagen Wheel – have been included by TIME Magazine in the list of the 'Best Inventions of the Year'.

Gillian Rose is Professor of Human Geography in the School of Geography and the Environment at the University of Oxford. She led the ESRC-funded project Smart Cities in the Making: Learning from Milton Keynes.

Flora Roumpani is a research fellow at the British Library and The Alan Turing Institute and a visiting lecturer in the Royal College of Art in the Environmental Architecture MSc. Her interests cover urban-regional planning, urban modelling and simulations, spatial analysis and visualisation. She is also a registered architect in Greece and the UK.

Joe Shaw is a PhD candidate at the Oxford Internet Institute where his research is focused on the adoption of digital platforms and data analytics by the real estate market.

Mark Shepard is an Associate Professor of Architecture and Media Study at the University at Buffalo. His research investigates the entanglements of contemporary technologies and urban life. He is an editor of the Situated Technologies Pamphlets Series and editor of *Sentient City: Ubiquitous Computing, Architecture and the Future of Urban Space* (Architectural League of New York and MIT Press, 2011).

Ola Söderström is Professor in Social and Cultural Geography at the University of Neuchâtel, Switzerland. His recent work analyses transnational logics of urban development, on the one hand, and urban geographies of mental health, on the other. Main research topics include urban policy mobilities, smart cities, visual urban studies and biosocial processes in mental health.

Olamide Udoma-Ejorh is an urban activist researcher, writer and artist. She is currently involved in governance and social issues within the urban environment of Lagos, Nigeria. She is an advocate for sustainable transportation and social engagement within street spaces. She is Executive Director at Lagos Urban Network, Trustee at Open House Lagos and the Editor-in-Chief of the *Lost in Lagos Magazine*.

Judith Veenkamp is Lead of the Smart Citizens Lab at Waag. With her passion for new technologies and data solutions for societal issues, she develops strategies, concepts and shapes projects to create innovative solutions together with citizens and stakeholders. She is responsible for project development and coordination of ongoing projects on air quality, sound pollution, gamma radiation and water quality. She has experience as Communication Advisor on the edge of governance, policy and citizenship and earlier studied political science, conflict studies and human rights.

C. William R. Webster is Professor of Public Policy and Management at the University of Stirling, Scotland. He is a Director of CRISP – the Centre for Research into Information, Surveillance and Privacy. William is an expert on the governance of surveillance, and is Co-Editor-in-Chief of Information Polity and Co-Chair of the EGPA Permanent Study Group on eGov.

Edward Wigley was a Research Associate on the ESRC-funded project Smart Cities in the Making: Learning from Milton Keynes based at the Open University, and was a Teaching Fellow in Cultural Geography at University of Reading. He currently works at the Open University.

Katharine S. Willis is Associate Professor (Reader) in the School of Architecture, Design and Environment at University of Plymouth, UK. Her research interests include smart cities and the role of space, place and inclusion. She has published widely on the topic, including *Digital and Smart Cities* (Routledge, 2018, with Alessandro Aurigi) and *Netspaces: Space and Place in a Networked World* (Routledge, 2014). She is programme leader for the Masters in Smart Urban Futures at University of Plymouth, UK.

Johanna Ylipulli is a cultural anthropologist studying cultural and social implications of new digital technologies in urban contexts. For the last ten years, her work has focused on empirical research of digital innovations, scrutinising their design as well as appropriation processes and use. In addition to ethnography, Dr Ylipulli has specialised in participatory approaches to design and creative and speculative methods. She works at the Department of Computer Science, Aalto University.

Paola Zamperlin is a Lecturer in the Applied Geography Lab, Department of History, Archaeology, Geography, Art & Performance Arts (SAGAS), University of Firenze, Italy.

Acknowledgements

We are grateful to the opportunity provided by the AHRC Whose Right to the Smart City International Research Network for discussion with colleagues from UK, India and Brazil during the preparation of the book.

We would particularly like to thank the Routledge editorial team for their guidance, support and patience: Andrew Mould, Commissioning Editor and Egle Zigaite, Editorial Assistant.

My thanks to those who have helped with inspiring conversations at various stages during the writing of the book: Rob Kitchin, Ayona Datta, Nancy Odendaal, Lorena Melgaco, Satyarupa Shekhar, Ana Paula Baltazahar, Martijn de Waal, Michael de Lange, Ava Fatah, Saskia Sassen, Aale Luusua and E Leal.

1

Introduction

Katharine S. Willis and Alessandro Aurigi

Introduction

Cities have always been infused by technologies; in fact the urban condition is inherently underpinned by technological processes, interactions and practices. But the relationship of the digital and technical in society and the city is both changeful and evolving (Graham, 2004). The 'smart city' is defined by the emergence of new ways in which material urban systems are interconnected through information and data, changes in the processes through which cities are monitored, managed and analysed and a shift in how citizens participate, interact with the city and inhabit its spaces. This raises questions as to the future governance of cities and the role of interconnected data, people, places and urban systems which makes the challenge of understanding, designing and reflecting on smart cities an important new field to be investigated. To address this challenge, the volume aims to answer the question of what it means for a city to be 'smart', raise some of the tensions emerging in smart city developments and consider the implications for future ways of inhabiting and understanding the urban condition.

The key feature of this volume is that it draws on perspectives from the field of urban studies, architecture, urban design and urban planning. This recognises that the smart city agenda can be seen as part of a legacy of urban studies, urban design and planning thinking as well as being informed by critical thinking from the social sciences and more development-oriented enquiry from fields like computing science and interaction design. It sits within a growing body of edited books that address the smart city from a range of critical perspectives and draw together multidisciplinary empirical research on the topic (Cardullo *et al.*, 2019; Deakin and Wear, 2012; Karvonen *et al.*, 2019; Kitchin *et al.*, 2019; Marvin *et al.*, 2016). To situate the topic of the smart city for the reader, the volume sets out the various interpretations and aspects of what constitutes and defines smart cities in order to frame the topic and establish key concepts. It investigates and considers the range of factors that shape the characteristics of smart cities and draw together different disciplinary perspectives. The consideration of what shapes the smart city will be explored through discussing three broad 'parts': issues of governance, the nature of urban development and how visions are realised, and includes chapters that draw on empirical studies to frame the discussion with an understanding not just of the nature of the smart city but how it is studied, understood and reflected upon.

Overall the book situates the topic as capturing the landscape of the discussion by drawing together a range of disciplinary approaches and discussions and aims to provide a resource to enable readers to take their own critical position within the field of smart cities discourse.

The smart city in context

It is important to recognise that, whilst widely used, the term 'smart cities' is inherently ambiguous and used to describe and characterise a wide range of urban technological systems, strategies and also agendas. The label 'smart city' was first used in the early stages of the noughties, but it was not until around 2006 that it started to be widely accepted (Kitchin *et al.*, 2019; Willis and Aurigi, 2017, p. 2), and is often institutionally led as part of city or industry development or investment strategies. The term smart city is used across commercial, city and academic fields to characterise cities where technology is both embedded within the city in the form of sensors and other monitoring infrastructure and also the devices and platforms that enable people and often commercial or city governments to 'manage' this data in a large scale and 'real-time' way (Cocchia, 2014; Willis and Aurigi, 2017). According to Marvin, Luque-Ayala and McFarlane, the smart city opens up 'a new language of "smartness"' that is reshaping debates around contemporary cities (Marvin *et al.*, 2016, p. 2). In a smart city, computing power moves beyond wired or wireless infrastructure such as the broadband networks of the ubiquitous city, and pervades everyday objects and systems of the city, from parking sensors, to pollution monitoring to pedestrian footfall. William Mitchell, one of the authors to provide the first accessible introduction to the links between the city and technology as they emerged in the nineties (Mitchell, 1995, 2000, 2004), also provides an early overview of some core concepts and technologies that make up a smart city which sees cities 'fast transforming into artificial ecosystems of interconnected, interdependent intelligent digital organisms', which he argues is a 'fundamentally new technological condition confronting architects and product designers in the twenty-first century' (Mitchell, 2006). Mitchell captures some of the basic underlying features of what is termed smart city: a systemisation of city services and infrastructures together with an embedding of technological sensing and monitoring software and hardware into the fabric of the city. Although cities have always inherently been reflexive in nature, they are shaped by the people that inhabit them and their practices, cultures and infrastructures (Sassen 2013). The degree to which smart cities include a level of technological systemisation can be seen to shift the balance of this reflexivity (Crang and Graham, 2007; Shepard, 2011). This is where the challenge of understanding, designing and reflecting on smart cities becomes an important new field to be investigated. The terminology of smart city is still evolving, and it is expected that the term itself will soon be superseded by the new label with new agendas, interests and technological references and dependencies.

Introduction to the structure of the book

The volume is divided into three parts: governance, development and visions, each of which each is also divided into two sub-sections. The aim of the three parts of the book is to frame different perspectives to read and interpret smart urbanism. These underpin different approaches to the smart city agenda in a series of contexts and projects. Each part contributes a critical introduction to a core approaches, drawing on a key text which is introduced with a contextual commentary. The volume also draws on a broad range of different methodological approaches that are used and applied to design, study and analyse smart cities. This ranges from methods from data driven tools from the field of urban science and urban informatics to the more socially constructed, participatory approaches from the social sciences. These chapters include clearly described examples or case studies that demonstrate how the method is used in a specific context and to investigate the nature of inclusivity, diversity and participation.

Part I Smart city governance

Urban governance, data and participatory infrastructure

In the last decade the issue of governance in cities has become more complex as the nature of data flows have themselves become more integrated and city infrastructures, processes and social practices. The fundamental problem with the technocratic approach in the emergence of smart cities is that they tend to operate on a techno-deterministic logic that prioritises market-led solutions for urban development based on a promise of optimisation and efficiency of resources. Authors such as Aurigi and De Cindio (2008), Kitchin (2015; Kitchin and Perng, 2016), Marvin *et al.* (2016), Rose (2015) and Sassen (2012) have critiqued this approach since it not only reinforces a universalising view of urban development, but also masks the social tensions, issues and roles of its citizens in its construction and the role of the city itself. Haklay highlights how the failure of digital technologies to solve urban challenges is 'linked to the strategy where technology is used to disenfranchise, fails to enable local knowledge, and black boxes devices and technical infrastructure' (Haklay, 2013). The increasing role of software, data and artificial intelligence (AI) in city services and processes has implications for the governance of cities because software is embedded in often subtle and invisible ways and it produces data-driven outcomes that are not analogous with the material, physical and social life of the city. There is also the underlying issue of not only ownership of the software that 'manages' the city, but also consequently management and control (Luque-Ayala and Marvin, 2015; Vanolo, 2013). The high complexity of such data-driven systems means that they rarely, if ever, are developed and managed through city governance mechanisms, but controlled by private sector IT companies such as IBM, Cisco and Siemens who have vested interests well beyond those of the city itself. This leads to new forms of governmentality that rely on generating and monitoring systematic information about individuals which makes the systems and apparatus of governance more panoptical in nature (Kitchin, 2011, p. 949).

Key to this is understanding the role of data within new modes of participatory governance and what Kitchin has highlighted as the multitude ethical issues in smart cities governance. Databases and data analytics are not neutral, technical means of assembling and making sense of data but instead are socio-technical in nature, shaped by philosophical ideas and technical means (2016). In fact, the proper consideration of ethical considerations within smart city projects may prove to be one of the defining points in the development of projects on the ground, as this points to significantly more participatory and open modes of governance than are currently being implemented.

Chapters in this section are as follows:

- A city is not a computer
- Bias in urban research: from tools to environments
- Urban science: a short primer
- Defining smart cities: high and low frequency cities, big data and urban theory
- Digital information and the right to the city
- Shaping participatory public data infrastructure in the smart city: open data standards and the turn to transparency.

Governing, inclusion and smart citizens

Smart cities implement new socio-technical processes that require critical reflection around what constitutes urban management and control. They lead to new thinking about how technology affects cities where one of the key factors is that the digital is not simply about technology, computers or networks or code but about how people use, interact and behave with technology. It is through inhabitation and a context in their lives that the digital becomes meaningful. This is not simply because technology can establish different patterns of social relations and ways of living but because they can act as a potential reorganisation of social relations. This also includes patterns of inclusion and exclusion that emerge through these social relations constructed through technologies, and in particular a discussion on how those that lack digital skills and access to equipment may become excluded as cities become increasingly digital (Cardullo *et al.*, 2019). In this context, there is an increasing body of critical analysis that looks beyond celebrative and top-down approaches to include more diverse thinking about inclusion and 'citizenship' within smart city initiatives (Datta, 2018; Gabrys, 2014; Rabari and Storper, 2015; Vanolo, 2013). These document how smart city agendas rarely address issues of social differences in already-existing cities and March and Ribera-Fumaz (2014, p. 826) highlight the corresponding need to respond to the question of 'whose smartness and whose cities?' This particularly includes cases of smart city projects in the Global South, and the various ways in which new urban technologies are used, negotiated and even subverted by citizens. In this context, a number of authors have drawn on empirical evidence for how smart city projects arguably lead to the exacerbation of existing urban historical, material and social inequalities (Odendaal, 2011; Sadoway and Shekhar, 2014; Vanolo, 2016; Wiig, 2016).

These premium and highly connected networked infrastructures often ignore less-favoured and intervening places, enabling connectivity to operate 'selectively, linking valuable segments and discarding used up, or irrelevant, locales and people' (Castells, 1998, p. 390). Recent work has revealed that many so-called smart technologies do not empower citizens to become active players in their cities (de Lange and de Waal, 2019). This can particularly be seen in the promotion and development of private tech-led smart city initiatives, which are typically underpinned by a focus on highly connected, highly urbanised global cities with a highly skilled workforce (Hollands, 2014). But as Hollands points out, this has the potential to lead to social polarisation and 'the smart/creative city can become not only more economically polarized, but also socially, culturally and spatially divided by the growing contrast between incoming knowledge and creative workers, and the unskilled and IT illiterate sections of the local poorer population' (Hollands, 2008, p. 312). Therefore, the politics and power networks that underlie smart cities are important to address in order to establish how certain groups may benefit and others, often marginalised groups, may be excluded from any benefits of smart city initiatives. Vanolo highlights how this leads to patterns of exclusion since there is 'little room for the technologically illiterate, the poor and, in general, those who are marginalised from the smart city discourse' (2013, p. 893). Although this work is growing, Cardullo and Kitchin highlight that there is still work to do since 'despite the re-orientation towards creating "smart citizens" to date there has been little critical conceptual scrutiny as to how citizens are imagined and engaged by different smart city technologies' (2018, p. 5). Addressing the realities of how to enable smart citizenship in projects that offer more than just tokenistic participation is one of the key challenges for smart city projects, and there are significant implications for this in terms of who participates and how marginalised groups can be given a right to the smart city (Willis, 2019).

Chapters in this section are as follows:

- Towards an agenda of place, local agency-based and inclusive smart urbanism
- Governmentality and urban control
- How smart is smart city Lagos?
- Smart citizens in Amsterdam: an alternative to the smart city
- Governing technology-based urbanism: technocratic governance or progressive planning?

Part II Smart city development

Creative, smart or sustainable?

The smart city agenda is often marketed as a form of urban innovation project, which develops a range of information and communication technology (ICT) infrastructures to support learning and creativity at an urban scale, and draws heavily on an innovation and knowledge approach linked with ICTs (Hollands, 2014). Key to this is a broader, post-industrial shift from economies that rely on manufacturing of goods to those that operate through capitalising knowledge and information led by innovation (Komninos, 2008). Due to its focus on innovation systems, the smart city agenda gives implicit priority to competitiveness and economic growth. In other words, it is how ICTs, in conjunction with human and social capital and wider economic policy, are used to leverage growth and manage urban development that makes a city smart (Caragliu et al., 2011). Often running in parallel to the smart city agenda, the 'knowledge economy' underpins the development of smart city programmes, although associated economic models of what are termed the 'creative economy' and the 'learning economy' are also commonly used. The use of terms such as clever, smart, skilful, creative, networked, connected and competitive are seen as integral to the characterisation of knowledge-based urban development. This can particularly be seen in the promotion and development of smart city initiatives, which are typically underpinned by a focus on highly connected, highly urbanised global cities with a highly skilled workforce. Therefore the politics and power networks that underlie smart cities are important to address in order to establish how certain groups may benefit and others enhance innovation, learning, knowledge and problem solving (Hollands, 2008, p. 305).

In parallel to the link made between ICT investment and economic growth, the smart city is also part of an agenda that sees technological innovation as key to addressing the global challenge of sustainability (Joss et al., 2019). This links to the economic or resource concept of optimisation, where one of the central concepts is that in a world where resources are scarce, seeking solutions that enable a city to be more efficient in its use of resources can lead to sustainable urban development. More developed economies are seen as growing as a result of the more intelligent use of resources to produce greater value, rather than through the addition of new resources (Berkhout and Hertin, 2001). Smart city development plans are seen as key to underpinning alternative economic models where economic value is created primarily through the manipulation of ideas (the knowledge economy), rather than the exploitation of energy and materials. ICTs can contribute to a long-standing structural change in the economy away from materials-intensive activity and towards more service-based and information-intensive activities (Caragliu et al., 2011).

Chapters in this section are as follows:

- Will the real smart city please stand up?
- Smart to green: smart eco-cities in the green economy

- Towards ethical legibility: an inclusive view of waste technologies
- Stand up please, the real sustainable smart city.

Citizen science and co-production

There is a need to ensure that human and environmental values are taken into account in the design and implementation of governance processes and platforms that will influence the way cities operate and are governed (Haklay, 2013). One of the key ways that this can be achieved is through enabling people to participate actively in the way that smart cities are developed. For example, Cardullo and Kitchin identify that 'the normative challenge to creating truly "citizen-centric" smart cities will be to re-imagine the role citizens are to play in their conception, development and governance' (2018, p. 20). Arnstein's 'ladder of participation' has been adopted by a number of authors as a model to critically understand the actual role of citizenship in smart city projects (Cardullo and Kitchin, 2018; Shelton et al., 2015). This has identified that, in the same way as there are different levels of participation, from manipulation and therapy (which are non-participation) to informing and placating (tokenistic), all the way to partnership and citizen control, there are multiple degrees of public engagement in smart cities. For example Haklay distinguishes between four levels of citizen science, from citizens as sensors at the lowest level, to participatory science and extreme collaborative science (2012). The challenge is often to distinguish between the claims of projects around participation and the 'actually existing' reality of smart citizenship in practice (Shelton et al., 2015).

The discussions around citizenship and participation are fundamentally important for forms of engagement and involvement that can be invented and controlled by the people (Mclaren and Agyeman, 2015). This takes a model of participation, or sharing data that is termed 'co-production' whereby 'citizens perform the role of partner rather than customer in the delivery of public services' (Linders, 2012, p. 446). This sees new forms of sharing, enabled by technological devices and platforms (Willis and Aurigi, 2017), that work by enabling citizens to create, adapt and exploit data (Cowley, 2010) and can create new ways in which citizens participate in the governance of the city. A sharing cities approach focuses on bringing local people together through shared activities and cooperation for the benefit of the city and includes initiatives such as carsharing, community currencies, cohousing, hackerspaces, time-banks and tool or kitchen libraries. For example, civic apps developed by citizens, civic organisations and commercial companies (Desouza and Bhagwatwar, 2012) have become widespread and typically create some form of two-way interaction where citizens contribute to commenting on or providing data on public services usually offered by the city such as crime prevention, rubbish collection, public transportation and pollution reduction.

These experiments also initiate new ways of collaboration between citizens and researchers, and between entrepreneurs and city officials and can be seen in cities such Amsterdam, Eindhoven, Aarhus, London, Santander or Barcelona (Brynskov et al., 2018). One of the ways that this model of participation is played out in smart cities is the role of what are termed 'living labs' or urban 'test beds', where the city itself are treated as living laboratories; that is, as sources of data and as test beds to validate the science and test the practical interventions produced (Evans et al., 2016; Laurent and Pontille, 2019). Halpern and colleagues have critiqued the model of the test bed model since 'the logic of the test bed, past data are always used to produce the future' (2013, p. 164) and argue that we need to 'begin to design with less authority and greater interest in the space of society and culture that is produced in the interstices between what is human, machine, and more than human' (Halpern et al., 2013, p. 164). There is still much to be done to understand how co-production, citizen science and other people-centred models of participation in urban test beds and living lab-type platforms can produce a better model of engagement in smart cities that moves up and across Arnstein's ladder of participation.

Chapters in this section are as follows:

- Sharing in smart cities: what are we missing out on?
- Taxonomy of environmental sensing in smart cities
- Co-creating sociable smart city futures.

Part III Smart city visions

Urban planning, city models and smart storytelling

Smart labelling and rebranding of creative and eco city projects follow the genealogy of technology and urban planning that deals in utopian visions, where historicity is abolished and the vision of a technologically enabled bright future is all pervasive (Anthopoulos, 2017). This is part of a much longer history of urban visioning around future cities that adopt utopian ideals to provide a rationale for technical intervention (Carey, 1999). In fact, much of the smart city rhetoric is characterised by a focus on nominal futures. This is exemplified in the bright future promised by IBM's smarter cities model which does not suggest a revolution in urban morphology but a 'reformist optimization through data, monitoring, interconnectedness and automatic steering mechanisms' (Söderström et al., 2014, p. 317). As Söderström and colleagues reveal, the technique of imagining futures appropriates forms of storytelling to contextualise and to lend a reality to a speculative technology (2014). The smart city rhetoric that markets technology as a revolutionary approach to solving complex urban social and spatial problems is increasingly being shown to be disingenuous. This is evidenced by the failures of many over-hyped smart city projects that have yet to be realised at the scale planned, and that the reality of what has been built is homogeneous and bland urban space (Cugurullo, 2013; Joss et al., 2019). One of the ways to speculate on possible futures is to build models or scenarios that enable the future city to be visualised or 'made real' (Rose, 2018). Increasingly virtual models are being developed that enable the smart city to be experienced as a 'mirror world' (Gelerntner, 1991) that scale up to mirror cities.

The counter approach to techno-deterministic utopian visions is to enable a more relational set of practices for collaborative citymaking as well as affordances of systems for innovation, adaptation and social change. This opens up new ways of thinking and designing for how citizens participate and act in the public spaces of the city. It also introduces different models of placemaking and how people can interact and seek to bring resources back into the public domain, through a citizen-based ownership of the city (de Lange and de Waal, 2013).

Chapters in this section are as follows:

- Smart cities as corporate storytelling
- Will the real smart city please make itself visible?
- From hybrid spaces to 'imagination cities': a speculative approach to virtual reality
- The museum in the smart city: the role of cultural institutions in co-creating urban imaginaries.

Cities and placemaking

At the scale of the built environment, the smart city promises a new model of integrated urban design. This is set in the context of profound shifts in the balance between production and consumption to mutuality: from professional amateur to wisdom of the crowd, from do-it-yourself culture to the hacker ethic (Botsman and Rogers, 2010). Central to this is the question how collaborative principles and participatory ethics from online culture can be ported to the urban realm in order to coordinate collective action and help solve some of the urgent complex issues that cities are facing.

It also introduces different models of placemaking (Aurigi, 2012) and how people can interact and shape the city, that 'start with the neighbourhood and not with the technology' (McFarlane and Söderström, 2017, p. 321). The smart infrastructures and citizen sensing networks enable new modes of communication and feedback both in terms of people-to-people but also people-to-city (Gabrys, 2014). Similarly, when participatory aspects of social networks are coupled with highly mobile urban citizens then this creates opportunities for new types of social organisation and collaborative decisionmaking (Cowley, 2010; Linders, 2012). Crowdsourcing urban services is also a new model of citizen interaction that can operate at a city scale. Exploiting the model of swarming, the new tools of the 'sharing economy' (Mclaren and Agyeman, 2015) include shared ownership platforms (Shaheen, 2011), makerspaces and fablabs (Niaros *et al.*, 2017), crowdfunding (Carè *et al.*, 2018) and platform coops (Scholz, 2016). These platforms work on models of sharing as a new paradigm of distribution and ownership of resources, and include people sharing transportation modes, public space, information and new services. The common thread in these concepts is that technologies need to serve and work for people and communities first in terms of their design and deployment, but also in relation to setting local civic and infrastructural priorities. This addresses a gap in the approaches to smart cities that has failed to value the importance of urban community insights, as well as a recognition of civic and third sector organisations, social enterprises, cooperatives and places such as libraries and community centres. These need to 'draw lessons from urban planning traditions that emphasize deep and meaningful civic engagement or community control in questions about local urban planning and design' (Sadoway and Shekhar, 2014). So, we will need to engage 'smart' in novel ways, driven not by the adoption of whatever technology is trending to produce global solutions, but by leveraging of local resources, people and wisdom (Willis and Aurigi, 2017) in placemaking and urban design practices in real cities and with real people.

Chapters in this section are as follows:

- The hackable city: a model for collaborative citymaking
- Designing the city as a place or a product? How space is marginalised in the smart city
- Self-monitoring, analysis and reporting technologies: smart cities and real-time data
- Reimagining urban infrastructure through design and experimentation: autonomous boat technology in the canals of Amsterdam
- The death and life of smart cities.

Summary

Smart city rhetoric can be polarising: on the one hand it presents a technocratic paradigm of homogenous and globalised smart driven urban development, whilst on the other it focuses on the positive societal impacts of connected and shared smart networks in the context of the urban condition. Yet, even where the social impacts of smart cities are considered, discourses are usually based on a fairly superficial and 'tokenistic' community participation where the social benefits tend to privilege digitally literate, highly educated (i.e. 'intelligent'), highly skilled young people (Leontidou, 2015, p. 84) rather than on effective participation (Cardullo and Kitchin, 2018; Willis, 2019). To counter this, academics have tried to highlight and prioritise the role of citizen participation in the governance and making of the smart city, and the importance of recognising the ethical implications in order to solve real problems and to localised projects that work with specific places and communities. Central to this is recognising that many of the urban problems currently being defined in corporate-led smart city projects are often disingenuous and derive from a flawed techno-deterministic model of optimisation of resources which are deemed to require a 'spatial fix' (Martin *et al.*, 2019). Alternatives draw on new peer-to-peer economic models and participatory urban planning techniques, approaches such as 'hackable', sharing and open source cities are recognised as having the potential to enable people to become active in shaping their urban environment to collaboratively address shared urban issues (de Lange and de Waal, 2013) and also importantly recognising the value of the range of Global South and non-Western approaches to technological development.

The essays in this volume bring together a diverse range of perspectives from fields such as geography, computer science, urban studies, urban planning, design and sociology to address not just critical readings but also different methodological approaches to address three broad areas of smart city governance, development and visions. One of the features of the volume is that it draws on voices from outside academia and from academics who have had sustained or collaborative projects with private or third sector partners. That bring insights into how, in practice, questions of citizenship, co-production and governance are answered in the actually existing smart city. Whilst smart cities hold a focus on certain types of digital technologies and infrastructures, they also more importantly require a bridging of disciplinary perspectives and empirical fields that will become increasingly important as smartness and societal challenges such as the climate emergency become increasingly entwined. This leads to asking more inclusive, diverse and sustainable questions of the smart city.

References

Anthopoulos, L. (2017) 'Smart utopia vs smart reality: Learning by experience from 10 smart city cases'. *Cities*, 63, pp. 128–148.

Aurigi, A. (2012) 'Reflections towards an agenda for urban-designing the digital city'. *Urban design international*, 18 (2), pp. 131–144.

Aurigi, A. & De Cindio, F. (2008) *Augmented Urban Spaces: Articulating the Physical and Electronic City.* Aldershot: Ashgate.

Berkhout, F. & Hertin, J. (2001) *Impacts of information and communication technologies on environmental sustainability: Speculations and evidence.* www.oecd.org/sti/inno/1897156.pdf: OECD. Available.

Botsman, R. & Rogers, R. (2010) *What's Mine is Yours: How Collaborative Consumption is Changing the Way we Live.* New York: HarperCollins.

Brynskov, M., Heijnen, A., Balestrini, M. & Raetzsch, C. (2018) 'Experimentation at scale: Challenges for making urban informatics work'. *Smart and sustainable built environment*, 7 (1), pp. 150–163.

Caragliu, A., Del Bo, C. & Nijkamp, P. (2011) 'Smart cities in Europe'. *Journal of urban technology*, 18 (2), pp. 65–82.

Cardullo, P., Feliciantonio, C. D. & Kitchin, R. (eds.) (2019) *The Right to the Smart City.* Bingley: Emerald Publishing.

Cardullo, P. & Kitchin, R. (2018) 'Being a 'citizen' in the smart city: Up and down the scaffold of smart citizen participation in Dublin, Ireland'. *GeoJournal*, 84 (1), pp. 1–13.

Carè, S., Trotta, A., Carè, R. & Rizzello, A. (2018) 'Crowdfunding for the development of smart cities'. *Business horizons*, 61 (4), pp. 501–509.

Carey, J. (1999) *The Faber Book of Utopias.* London: Faber.

Castells, M. (1998) *The Information Age: Economy, Society And Culture, Vol.3, End of Millennium.* ed. Castells, M. Malden: Blackwell Publishers.

Cocchia, A. (2014) 'Smart and Digital City: A Systematic Literature Review', In Dameri, R.P. and Rosenthal-Sabroux, C. (eds) *Smart City: How to Create Public and Economic Value with High Technology in Urban Space.* Switzerland: Springer, pp. 13–43.

Cowley, J. E. (2010) 'Planning in the age of Facebook: The role of social networking in planning processes'. *GeoJournal*, 75 (5), pp. 407–420.

Crang, M. & Graham, S. (2007) 'Sentient cities: Ambient intelligence and the politics of urban space'. *Information, communication society*, 10 (6), pp. 789–817.

Cugurullo, F. (2013) 'How to build a sandcastle: An analysis of the genesis and development of Masdar city'. *Journal of urban technology*, 20 (1), pp. 23–37.

Datta, A. (2018) 'The digital turn in postcolonial urbanism: Smart citizenship in the making of India's 100 smart cities'. *Transactions of institute of British geographers*, 0, pp. 1–15.

de Lange, M. & de Waal, M. (2013) 'Owning the city: New media and citizen engagement in urban design'. *First monday*, 18 (11). Online.

de Lange, M. & de Waal, M. (eds.) (2019) *The Hackable City: Digital Media and Collaborative City Making in the Network Society*. Singapore: Springer.

Deakin, M. & Wear, H. A. (eds.) (2012) *From Intelligent to Smart Cities*. Abingdon: Routledge.

Desouza, K. C. & Bhagwatwar, A. (2012) 'Citizen apps to solve complex urban problems'. *Journal of urban technology*, 19 (3), pp. 107–136.

Evans, J., Karvonen, A. & Raven, R. (eds.) (2016) *The Experimental City*. London: Routledge.

Gabrys, J. (2014) 'Programming environments: Environmentality and citizen sensing in the smart city'. *Environment and planning D: Society and space*, 32 (1), pp. 30–48.

Gelerntner, D. (1991) *Mirror Worlds*. Oxford: Oxford University Press.

Graham, S. (ed.) (2004) *The Cybercities Reader*. London: Routledge.

Haklay, M. (2012) 'Citizen science and Volunteered Geographic Information: Overview and Typology of Participation', In Sui, D., Elwood, S., and Goodchild, M. (eds) *Crowdsourcing Geographic Knowledge*. Dordrecht: Springer Netherlands, pp. 105–122.

Haklay, M. (2013) 'Beyond Quantification: A Role for Citizen Science and Community Science in a Smart City', In Campkin, B. and Ross, R. (eds) *UCL Urban Laboratory Pamphleteer*. London: UCL.

Halpern, O., LeCavalier, J., Calvillo, N. & Pietsch, W. (2013) 'Test Bed as Urban Epistemology', In Marvin, S., Luque-Ayala, A. and McFarlane, C. (eds) *Smart Urbanism: Utopian Vision or False Dawn?* Abingdon: Routledge, pp. 146–168.

Hollands, R. G. (2008) 'Will the real smart city please stand up?: Intelligent, progressive or entrepreneurial?'. *City*, 12 (3), pp. 303–320.

Hollands, R. G. (2014) 'Critical interventions into the corporate smart city'. *Cambridge journal of regions, economy and society*, 8 (1), pp 61–77.

Joss, S., Sengers, F., Schraaven, D., Caprotti, F. & Dayot, Y. (2019) 'The smart city as global discourse: Storylines and critical junctures across 27 cities'. *Journal of urban technology*, 26 (1), pp. 3–34.

Karvonen, A., Cugurullo, F. & Caprotti, F. (eds.) (2019) *Inside Smart Cities: Place, Politics and Urban Innovation*. London: Routledge.

Kitchin, R. (2011) 'The programmable city'. *Environment and planning B*, 38 (6), pp. 945–951.

Kitchin, R. (2015) 'Making sense of smart cities: Addressing present shortcomings'. *Cambridge journal of regions, economy and society*, 8, pp. 131–136.

Kitchin, R. (2016) 'The ethics of smart cities and urban science'. *Philosophical transactions of the royal society A: Mathematical, physical and engineering sciences*, https://doi.org/10.1098/rsta.2016.0115.

Kitchin, R., Coletta, C., Evans, L. & Heaphy, L. (2019) 'Creating Smart Cities', In Kitchin, R., Coletta, C., Evans, L. & Heaphy, L. (eds) *Creating Smart Cities*. Abingdon: Routledge, pp. 1–18.

Kitchin, R. & Perng, S.-Y. (eds.) (2016) *Code and the City*. Abingdon: Routledge.

Komninos, N. (2008) *Intelligent Cities and Globalisation of Innovation Networks*. London and New York: Routledge.

Laurent, B. & Pontille, D. (2019) 'Towards a Study of City Experiments', In Kitchin, R., Cardullo, P. and Feliciantonio, C. D. (eds.) *The Right to the Smart City*. Bingley: Emerald, pp. 90–103.

Leontidou, L. (2015) 'Smart cities'of the debt crisis: Grassrots creativity in mediterranean Europe'. *The Greek review of social research*, 144 (144), pp. 69–101.

Linders, D. (2012) 'From e-government to we-government: Defining a typology for citizen coproduction in the age of social media'. *Government information quarterly*, 29 (4), pp. 446–454.

Luque-Ayala, A. & Marvin, S. (2015) 'Developing a critical understanding of smart urbanism?'. *Urban studies*, 52 (12), pp. 105–2116.

March, H. & Ribera-Fumaz, R. (2014) 'Smart contradictions: The politics of making Barcelona a Self-sufficient city'. *European urban and regional studies*, 23 (4), pp. 816–830.

Martin, C., Evans, J., Karvonen, A., Paskaleva, K., Yang, D. & Linjordet, T. (2019) 'Smart-sustainability: A new urban fix?'. *Sustainable cities and society*, 45, pp. 640–648.

Marvin, S., Luque-Ayala, A. & Mcfarlane, C. (eds.) (2016) *Smart Urbanism: Utopian Vision or False Dawn?*. Abingdon: Routledge.

McFarlane, C. & Söderström, O. (2017) 'On alternative smart cities: From a technology-intensive to a knowledge-intensive smart urbanism'. *City*, 21 (3-4), pp. 312–328.

Mclaren, D. & Agyeman, J. (2015) *Sharing Cities: A Case for Truly Smart and Sustainable Cities*. Cambridge, MA: MIT Press.

Mitchell, W. (2000) *E-Topia*. Cambridge, MA: MIT Press.

Mitchell, W. (2006) 'Smart city 2020', *Metropolis*, April.

Mitchell, W. J. (1995) *City of Bits*. Cambridge, MA: MIT Press.

Mitchell, W. J. (2004) *Me++: The Cyborg Self and the Networked City*. Cambridge, MA: MIT Press.

Niaros, V., Kostakis, V. & Drechsler, W. (2017) 'Making (in) the smart city: The emergence of makerspaces'. *Telematics and informatics*, 34 (7), pp. 1143–1152.

Odendaal, N. (2011) 'The spaces between: ICT and marginalization in the South African city'. *Proceedings of the 5th International Conference on Communities and Technologies*. Brisbane and Australia: ACM, pp 150–158.

Rabari, C. & Storper, M. (2015) 'The digital skin of cities: Urban theory and research in the age of the sensored and metered city, ubiquitous computing and big data'. *Cambridge journal of regions, economy and society*, 8 (1), pp. 27–42.

Rose, G. (2015) 'Smart cities and why they need a lot more social scientists to get involved'. Available at https://visualmethodculture.wordpress.com/2015/03/20/smart-cities-and-why-they-need-a-lot-more-social-scientists-to-get-involved/: 2017.

Rose, G. (2018) 'Smart Urban: Imaginary, Interiority, Intelligence', In Lindner, C. and Meissner, M. (eds.) *The Routledge Handbook of Urban Imaginaries*. Basingstoke: Routledge, pp. 105–112.

Sadoway, D. & Shekhar, S. (2014) '(Re)prioritizing citizens in smart cities governance: Examples of smart citizenship from urban India'. *The journal of community informatics*, 10 (3). Online.

Sassen, S. (2012) 'Urbanising Technology', In Burdett, R. and Rode, P. (eds.) *The Electric City Newspaper*. http://ec2012.lsecities.net/newspaper/: LSE Cities. 12-14 Available at http://ec2012.lsecities.net/newspaper/

Sassen, S. (2013) 'Does the city have speech?'. *Public culture*, 25(2), pp. 209–221, 2013. Available at SSRN: https://ssrn.com/abstract=2846094.

Scholz, T. (2016) *Platform Cooperativism Challenging the Corporate Sharing Economy*. New York: Rosa Luxemborg Stiftung.

Shaheen, S. (2011) 'Hangzhou public bicycle: Understanding early adoption and behavioral response to bikesharing in Hangzhou, China'. *Transportation research record*, 2247, pp. 34–41.

Shelton, T., Zook, M. & Wiig, A. (2015) 'The actually existing smart city'. *Cambridge journal of regions, economy and society*, 8 (1), pp. 13–25.

Shepard, M. (ed.) (2011) *Sentient City: Ubiquitous Computing, Architecture, and the Future of Urban Space*. Cambridge, MA: MIT Press.

Söderström, O., Paasche, T. & Klauser, F. (2014) 'Smart cities as corporate storytelling'. *City*, 18 (3), pp. 307–320.

Vanolo, A. (2013) 'Smartmentality: The smart city as disciplinary strategy'. *Urban studies*, 51, pp. 883–898.

Vanolo, A. (2016) 'Is there anybody out there? The place and role of citizens in tomorrow's smart cities'. *Futures*, 82, pp. 26–36.

Wiig, A. (2016) 'The empty rhetoric of the smart city: From digital inclusion to economic promotion in Philadelphia'. *Urban geography*, 37 (4), pp. 535–553.

Willis, K. (2019) 'Whose Right to the Smart City?', In Kitchin, R., Cardullo, P. and Feliciantonio, C. D. (eds.) *The Right to the Smart City*. Bingley: Emerald Publishing, pp. 27–41.

Willis, K. S. & Aurigi, A. (2017) *Digital and Smart Cities*. London: Routledge.

Part I
Smart city governance

Section 1
Urban governance, data and participatory infrastructure

2

A city is not a computer

Shannon Mattern

Editor's introduction

"We want to build a city" is the strapline on a Y Combinator Research project web page titled "New Cities". Meanwhile Sidewalk Labs, an Alphabet company, sets out its challenge to respond to the following question – "What would a city look like if you started from scratch in the internet era—if you built a city 'from the internet up?'".[1] These tech visions set out the city as the ultimate Silicon Valley "start up". In this chapter, Mattern builds a case for challenging the rhetoric of this sort of computational thinking on cities.[2] Drawing on the marketing strategies of tech funders such as Y Combinator and Alphabet that position a city as the next challenge for the technological optimization model, she highlights the flaws in taking this approach to city-making. The broader argument draws on multiple readings which have revealed the problems with seeking to address complex socio-economic urban challenges with programmable solutions (Gabrys, 2014; Greenfield, 2013; Kitchin, 2014; Marvin et al., 2016; Vanolo, 2016). This ranges from the infrastructural, where a city embedded with networked sensors is reduced to a complex dataset down to the scale of the everyday where city inhabitants and their interactions are considered as users of technology with little or no agency. Mattern argues in this chapter that this focus on treating cities as complex problems to be solved by code means that we have lost a "critical perspective on how urban data become meaningful spatial information or translate into place-based knowledge".

Central to Mattern's discussion is the need to recognize the shortcomings in computational and data-driven models that presume an objectivity in urban data and fail to recognize or accommodate the range of critical and more importantly ethical decisions to the machine (Mattern, 2016, 2017). From IBM's original strategic move into the Smarter Cities arena to the more recent Sidewalk Labs' Toronto Waterfront project, technology companies have failed to adequately acknowledge and address the fact that technology is not neutral and is culturally, socially, economically and politically situated. By conveniently sidestepping the political implications of a city governed by data, schemes such as Sidewalk

Labs' Toronto Waterfront have failed to give adequate space to how the project might impact the relationship between local government and residents. Mattern makes a powerful case for demonstrating that treating the city as a problem to be solved through optimization is inherently a rewriting of any urban governance model, and therefore needs to be scrutinized on this level and not in terms of key performance indicators and productivity gains.

The dream of informatic urbanism is one underpinned by urban science approaches and propagated by tech companies, whilst the broader challenge is what role urban planners, designers and city inhabitants have in a city "designed from the internet up". Fundamentally, the city as a computer project is one where data analysts, marketeers and software engineers are the new planners and designers. As Mattern argues below, the processes of city-making are more complicated than simply rewriting code, and is challenged by "technologists (and political actors) who speak as if they could reduce urban planning to algorithms".

In the current urgent crisis of climate breakdown, it's widely acknowledged that we need to think radically about our cities. Treating the city as a computer may create a slick and palatable solution to city marketing programmes and looks compelling as a strapline on a website. Yet, as Mattern demonstrates, urban intelligence is more than data capture, feedback and processing, it is a much broader kind of knowledge that lives within bodies, minds and communities. This chapter thoughtfully and comprehensively captures the range of issues that combine to form a rejection of the technocractic, data driven vision of problem-solving in cities.

References

Doctoroff, D. (2016) 'Reimagining Cities from the Internet Up'. In Jaffe, E., Doctoroff, D. L. and Quirk, V. (eds) *Sidewalk Talk*. https://medium.com/sidewalk-talk/: 2019. Available at: https://medium.com/sidewalk-talk/reimagining-cities-from-the-internet-up-5923d6be63ba.

Gabrys, J. (2014) 'Programming Environments: Environmentality and Citizen Sensing in the Smart City'. *Environment and Planning D: Society and Space*, 32 (1), pp. 30–48. https://doi.org/10.1068/d16812.

Gibson, W. (1995) *Neuromancer*. London: Harper Voyager.

Greenfield, A. (2013) *Against the Smart City*. New York: Do Projects.

Kitchin, R. (2014) 'The Real-time City? Big Data and Smart Urbanism'. *GeoJournal*, 79 (1), pp. 1–14.

Marvin, S., Luque-Ayala, A. & Mcfarlane, C. (eds.) (2016) *Smart Urbanism - Utopian Vision or False Dawn?* Abingdon: Routledge.

Mattern, S. (2016) 'Instrumental City: The View from Hudson Yards, circa 2019'. *Places Journal*, April 2016

Mattern, S. (2017) *Code + Clay, Data + Dirt: Five Thousand Years of Urban Media*. Minnesota and London: University of Minnesota Press.

Vanolo, A. (2016) 'Is There Anybody Out There? The Place and Role of Citizens in Tomorrow's Smart Cities'. *Futures*, 82, pp. 26–36.

Introduction

"What should a city optimize for?" Even in the age of peak Silicon Valley, that's a hard question to take seriously. (Hecklers on Twitter had a few ideas, like "fish tacos" and "pez dispensers.") Look past the sarcasm, though, and you'll find an ideology on the rise. The question was posed last summer[3] by Y Combinator—the formidable tech accelerator that has hatched a thousand startups, from AirBnB and Dropbox to robotic greenhouses and wine-by-the-glass delivery—as the entrepreneurs announced a new research agenda: building cities from scratch. *Wired's* verdict: "Not Actually Crazy" (Rhodes, 2016). Which is not to say wise. For every reasonable question Y Combinator asked—"How can cities help more of their residents be happy and reach their potential?"—there was a preposterous one: "How should we measure the effectiveness of a city (what are its KPIs)?" That's key performance indicators, for those not steeped in business intelligence jargon. There was hardly any mention of the urban designers, planners, and scholars who have been asking the big questions for centuries: "How do cities function, and how can they function better?"

Of course, it's possible that no city will be harmed in the making of this research. Half a year later, the public output of the New Cities project[4] consists of two blog posts, one announcing the program and the other reporting the first hire. Still, the rhetoric deserves close attention, because, frankly, in this new political age, all rhetoric demands scrutiny. At the highest levels of government, we see evidence and quantitative data manipulated or manufactured to justify reckless orders, disrupting not only "politics as usual," but also fundamental democratic principles. Much of the work in urban tech has the potential to play right into this new mode of governance.

Tech companies have come out forcefully against the Muslim travel ban, but where will they stand on subtler questions of social "optimization"? Autonomous vehicles and pervasive cameras and sensors are just the sort of disruptive technologies that an infrastructure-championing president might deem "tremendous." Donald Trump's chief strategist (who, years ago, ran the Biosphere 2 experiment into the ground) is also on the board of a data mining and analytics firm that seeks government contracts. Will the president start tweeting about how crime-ridden (and racialized) "inner cities" would be a whole lot better if they were run like computers?

It's a politically complicated environment, to say the least. Into the ring steps the first hire at New Cities: Ben Huh, founder of the meme-and-cat-pic empire Cheezburger. "There's no shortage of space to build new cities," he effervesced, in a post explaining his decision to join the Y Combinator project. "Technology can seed fertile starting conditions across nations and geographies." His goal for the six-month research position: to "create an open, repeatable system for rapid *cityforming* that maximize[s] human potential" (Huh, 2016). No pressure.

Meanwhile, Alphabet (née Google) is moving forward with plans to build its own optimized cities. Its urban-tech division, Sidewalk Labs, has already installed public WiFi kiosks on New York City streets: infrastructural nodes (known as "Links") that may someday exchange data with autonomous vehicles, public transit, and other urban systems.[5]

The company is also partnering with the U.S. Department of Transportation on efforts like the "Smart City Challenge," which awarded a US$50 million grant to Columbus, Ohio. Last June, on the same day Y Combinator announced its New Cities project, *The Guardian* published details of Alphabet's "Flow," the cloud software behind the mobility experiments in Columbus (Harris, 2016). Within months, partnerships were underway in 16 other cities (Davis, 2016).

Urban transportation is the first target for disruption, but it won't end there. Dan Doctoroff, the Michael Bloomberg associate who founded Sidewalk Labs, wonders, "What would a city look like if you started from scratch in the internet era—if you built a city 'from the internet up?'" In November, the company took another step in that direction, launching four new "labs" that will work on housing affordability, health care and social services, municipal processes, and community collaboration. The company plans to run pilot projects in select urban districts, then scale up. Announcing the expansion, Doctoroff recalled past "revolutions" in urban technologies:

> Looking at history, one can make the argument that the greatest periods of economic growth and productivity have occurred when we have integrated innovation into the physical environment, especially in cities. The steam engine, electricity grid, and automobile all fundamentally transformed urban life, but we haven't really seen much change in our cities since before World War II. If you compare pictures of cities from 1870 to 1940, it's like night and day. If you make the same comparison from 1940 to today, hardly anything has changed. Thus it's not surprising that, despite the rise of computers and the internet, growth has slowed and productivity increases are so low ... So our mission is to accelerate the process of urban innovation.
>
> *(2016, n.p.)*

While Doctoroff has been telling some version of this story since Sidewalk Labs launched in 2015, the timing of the new expansion, three weeks after the U.S. presidential election, alters the context. As everyone was watching the drama at Trump Tower, the world's largest searching-mapping-driving-advertising-information-organizing company was throwing its resources behind a "fourth revolution" in urban infrastructure.

Dreams of an informatic urbanism

Of course, major companies like Alphabet have already dramatically reshaped the cities where they are headquartered,[6] but they have not yet had the luxury of building on a blank slate. The idea of the "new city" certainly isn't new, and the model now emerging in the United States has precedents in Asian and Middle Eastern countries, where Cisco, Siemens, and IBM have partnered with real-estate developers and governments to build "smart cities" *tabula rasa*.

We don't know how these urban experiments will fare. Since they are in a constant state of development, always "versioning" toward an optimized model ever on the horizon, they are not easily evaluated or critiqued (Halpern et al., 2017). If you believe the marketing hype, though, we're on the cusp of an urban future in which embedded sensors, ubiquitous cameras and beacons, networked smartphones, and the operating systems that link them all together, will produce unprecedented efficiency, connectivity, and social harmony. We're transforming the idealized topology of the open web and Internet of Things into urban form.

Programmer and tech writer Paul McFedries explains this thinking:

> The city is a computer, the streetscape is the interface, you are the cursor, and your smartphone is the input device. This is the user-based, bottom-up version of the city-as-computer idea, but there's also a top-down version, which is systems-based. It looks

at urban systems such as transit, garbage, and water and wonders whether the city could be more efficient and better organized if these systems were "smart."

(2014, p.36)

While projects like Sidewalk Labs and Y Combinator's New Cities were conceived in an age of big data and cloud computing, they are rooted in earlier reveries. Ever since the internet was little more than a few linked nodes, urbanists, technologists, and sci-fi writers have envisioned cybercities and e-topias built "from the 'net up'" (Boyer, 1995; Castells, 1989; Gibson, 1995; Mitchell, 2000, 1995). Modernist designers and futurists saw morphological parallels between urban forms and circuit boards. Just as new modes of telecommunication have always reshaped physical terrains and political economies, new computational methods have informed urban planning, modeling, and administration (Graham and Marvin, 1996; Light, 2004; Vallianatos, 2015).

Modernity is good at renewing metaphors, from the city as machine, to the city as organism or ecology, to the city as cyborgian merger of the technological and the organic.[7] Our current paradigm, the *city as computer*, appeals because it frames the messiness of urban life as programmable and subject to rational order. Anthropologist Hannah Knox explains, "As technical solutions to social problems, information and communications technologies encapsulate the promise of order over disarray ... as a path to an emancipatory politics of modernity" (Knox, 2010, pp.187–188). And there are echoes of the pre-modern, too. The computational city draws power from an urban imaginary that goes back millennia, to the city as an apparatus for record-keeping and information management.

We've long conceived of our cities as knowledge repositories and data processors, and they've always functioned as such. Lewis Mumford observed that when the wandering rulers of the European Middle Ages settled in capital cities, they installed a "regiment of clerks and permanent officials" and established all manner of paperwork and policies (deeds, tax records, passports, fines, regulations), which necessitated a new urban apparatus, the office building, to house its bureaus and bureaucracy (Mumford, 1961, p.344). The classic example is the Uffizi (Offices) in Florence, designed by Giorgio Vasari in the mid-16th century, which provided an architectural template copied in cities around the world. "The repetitions and regimentations of the bureaucratic system"—the work of data processing, formatting, and storage—left a "deep mark," as Mumford put it, on the early modern city (Kittler, 1996, pp.721–722).

Yet the city's informational role began even earlier than that. Writing and urbanization developed concurrently in the ancient world, and those early scripts—on clay tablets, mudbrick walls, and landforms of various types—were used to record transactions, mark territory, celebrate ritual, and embed contextual information in landscape (Mattern, 2016b). Mumford described the city as a fundamentally communicative space, rich in information:

Through its concentration of physical and cultural power, the city heightened the tempo of human intercourse and translated its products into forms that could be stored and reproduced. Through its monuments, written records, and orderly habits of association, the city enlarged the scope of all human activities, extending them backwards and forwards in time. By means of its storage facilities (buildings, vaults, archives, monuments, tablets, books), the city became capable of transmitting a complex culture from generation to generation, for it marshaled together not only the physical means but the human agents needed to pass on and enlarge this heritage. That remains the greatest of the city's gifts.

As compared with the complex human order of the city, our present ingenious electronic mechanisms for storing and transmitting information are crude and limited.

(Mumford, 1961, p.569)

Mumford's city is an assemblage of media forms (vaults, archives, monuments, physical and electronic records, oral histories, lived cultural heritage), agents (architectures, institutions, media technologies, people), and functions (storage, processing, transmission, reproduction, contextualization, operationalization).[8] It is a large, complex, and varied epistemological and bureaucratic apparatus. It is an information processor, to be sure, but it is also more than that.

Were he alive today, Mumford would reject the creeping notion that the city is simply the internet writ large. He would remind us that the processes of city-making are more complicated than writing parameters for rapid spatial optimization. He would inject history and happenstance. *The city is not a computer.* This seems an obvious truth, but it is being challenged now (again) by technologists (and political actors) who speak as if they could reduce urban planning to algorithms (for more on the algorithm as a timely conceptual model, see Mazzotti, 2017).

Why should we care about debunking obviously false metaphors? It matters because the metaphors give rise to technical models, which inform design processes, which in turn shape knowledges and politics, not to mention material cities. The sites and systems where we locate the city's informational functions—the places where we see information-processing, storage, and transmission "happening" in the urban landscape—shape larger understandings of urban intelligence.

Informational ecologies of the city

The idea of the city as an information-processing machine has in recent years manifested as a cultural obsession with urban sites of data storage and transmission. Scholars, artists, and designers write books, conduct walking tours, and make maps of internet infrastructures. We take pleasure in pointing at nondescript buildings that hold thousands of whirring servers, at surveillance cameras, camouflaged antennae, and hovering drones. We declare: "the city's computation happens here" (Mattern, 2013, 2016d).[9]

Yet such work runs the risk of reifying and essentializing information, even depoliticizing it. When we treat data as a "given" (which is, in fact, the etymology of the word), we see it in the abstract, as an urban fixture like traffic or crowds. We need to shift our gaze and look at data in context, at the lifecycle of urban information, distributed within a varied ecology of urban sites and subjects who interact with it in multiple ways. We need to see data's human, institutional, and technological creators, its curators, its preservers, its owners and brokers, its "users," its hackers and critics. As Mumford understood, there is more than information *processing* going on here. Urban information is *made*, commodified, accessed, secreted, politicized, and operationalized.

But where? Can we point to the chips and drives, cables and warehouses—the specific urban architectures and infrastructures—where this expanded ecology of information management resides and operates? I've written about the challenges of reducing complicated technical and intellectual structures to their material, geographic manifestations, i.e., mapping "where the data live" (Mattern, 2016d) (see also Amoore, 2018). Yet such exercises can be useful in identifying points of entry to the larger system. It's not only the infrastructural object that matters; it's also the personnel and paperwork and protocols, the machines and management practices, the conduits and cultural variables that shape terrain within the larger ecology of urban information.

So the next time you're staring up at a Domain Awareness camera, ask how it got there, how it generates data—not only how the equipment operates technically, but also what information it claims to be harvesting, and through what methodology—and whose interests it serves. And don't let the totalizing idea of the *city as computer* blind you to the countless other forms of data and sites of intelligence-generation in the city: municipal agencies and departments, universities, hospitals, laboratories, corporations. Each of these sites has a distinctive orientation toward urban intelligence. Let us consider a few of the more public ones.

First, the municipal archive. Most cities today have archives that contains records of administrative activity, finances, land ownership and taxes, legislation and labor. The archives of ancient Mesopotamian and Egyptian cities held similar material, although historians debate whether ancient record-keeping practices served similar documentary functions (O'Toole, 2004). Archives ensure financial accountability, symbolically legitimize governing bodies and colonial rulers, and erase the heritage of previous regimes and conquered populations. They monumentalize a culture's historical consciousness and intellectual riches. In the modern age, they also support scholarship (Walsham, 2016). Thus, the "information" inherent in the archive resides not solely in the content of its documents, but also in their very existence, their provenance and organization (there's much to be learned about the ideals of a culture by examining its archival forms), and even in the archive's omissions and erasures (Stoler, 2010).

Of course, not all archives are ideologically equal. Community archives validate the personal histories and intellectual contributions of diverse publics. Meanwhile, law enforcement agencies and customs and immigration offices are networked with geographically distributed National Security Agency repositories and other federal black boxes. These archives are not of the same species, nor do they "process" "data" in the same fashion.

Practices and politics of curation and access have historically distinguished archives from another key site of urban information: libraries. Whereas archives collect unpublished materials and attend primarily to their preservation and security, libraries collect published materials and aim to make them intelligible and accessible to patrons. In practice, such distinctions are fuzzy and contested, especially today, as many archives seek to be more public-facing. Nevertheless, these two institutions embody different knowledge regimes and ideologies.

Modern libraries and librarians have sought to empower patrons to access information across platforms and formats, and to critically assess bias, privacy, and other issues under the rubric of "information literacy" (Mattern, 2016a). They build a critical framework around their resources, often in partnership with schools and universities. Further, libraries perform vital symbolic functions, embodying the city's commitment to its intellectual heritage (which may include heritage commandeered through imperial activities).

Similarly, the city's museums reflect its commitment to knowledge in embodied form, to its artifacts and material culture. Again, such institutions are open to ideological critique. Acquisition policies, display practices, and access protocols are immediate and tangible, and they reflect particular cultural and intellectual politics.

Just as important as the data stored and accessed on city servers, in archival boxes, on library shelves and museum walls are the forms of urban intelligence that cannot be easily contained, framed, and catalogued. We need to ask: What place-based "information" doesn't fit on a shelf or in a database? What are the non-textual, un-recordable forms of cultural memory? These questions are especially relevant for marginalized populations, indigenous cultures, and developing nations. Performance studies scholar Diana Taylor urges us to acknowledge ephemeral, performative forms of knowledge, such as dance, ritual, cooking, sports, and speech (Taylor, 2003). These forms cannot be reduced to "information," nor can they be "processed," stored, or transmitted via fiber-optic cable. Yet they are vital urban intelligences that live within bodies, minds, and communities.

Finally, consider data of the environmental, ambient, "immanent" kind. Malcolm McCullough has shown that our cities are full of fixed architectures, persistent terrains, and reliable environmental patterns that anchor all the unstructured data and image streams that float on top (McCullough, 2014, p.36, 42). What can we learn from the "nonsemantic information" inherent in shadows, wind, rust, in the signs of wear on a well-trodden staircase, the creaks of a battered bridge—all the indexical messages of our material environments? I'd argue that the intellectual value of this ambient, immanent information exceeds its function as stable ground for the city's digital flux. Environmental data are just as much figure as they are ground. They remind us of necessary truths: that urban intelligence comes in multiple forms, that it is produced within environmental as well as cultural contexts, that it is reshaped over the *longue durée* by elemental exposure and urban development, that it can be lost or forgotten. These data remind us to think on a climatic scale, a geologic scale, as opposed to the scale of financial markets, transit patterns, and news cycles.

Here's some geologic insight from T. S. Eliot's 1934 poem "The Rock":

> Where is the Life we have lost in living?
> Where is the wisdom we have lost in knowledge?
> Where is the knowledge we have lost in the information?
> *Eliot, T. S. (1934)*

Management theorist Russell Ackoff took Eliot's idea one step further, proposing the now famous (and widely debated) hierarchy: Data < Information < Knowledge < Wisdom (Sharma, 2008; Weinberger, 2010). Each level of processing implies an extraction of utility from the level before. Thus, contextualized or patterned data can be called information. Or, to quote philosopher and computer scientist Frederick Thompson, information is "a product that results from applying the processes of organization to the raw material of experience, much like steel is obtained from iron ore." Swapping the industrial metaphor for an artistic one, he writes, "data are to the scientist like the colors on the palette of the painter. It is by the artistry of his theories that we are informed. It is the organization that is the information".[10] Thompson's mixed metaphors suggest that there are multiple ways of turning data into information and knowledge into wisdom.

Yet the term "information processing," whether employed within computer science, cognitive psychology, or urban design, typically refers to *computational* methods. As Riccardo Manzotti explains, when neuroscientists adopt the metaphor of the *brain as computer*, they imply that information is "stuff" that's mentally "processed," which they know is not true in any real sense. The metaphor survives because it makes an irresistible claim about "how marvelously complex we are and how clever scientists have become" (Manzotti and Parks, 2016). Psychologist Robert Epstein laments that "some of the world's most influential thinkers have made grand predictions about humanity's future that depend on the validity of the metaphor" (Epstein, 2016). But the appeal of analogy is nothing new. Throughout history, the brain (like the city) has been subjected to bad metaphors derived from the technologies of the time. According to Epstein, we've imagined ourselves as lumps of clay infused with spirits, as hydraulic or electro-chemical systems, as automata. The *brain as computer* is just the latest link in a long chain of metaphors that powerfully shape scientific endeavor in their own images.

The *city as computer* model likewise conditions urban design, planning, policy, and administration—even residents' everyday experience—in ways that hinder the development of healthy, just, and resilient cities. Let's apply Manzotti's and Epstein's critiques at the city scale. We have seen that urban ecologies "process" data by means that are not strictly algorithmic, and that not all urban intelligences can be called "information." One can't "process" the local cultural effects of long-term weather patterns or derive insights from the generational evolution of a neighborhood without a degree of sensitivity that exceeds mere computation. Urban intelligence of this kind involves site-based experience, participant observation, sensory engagement. We need new models for thinking about cities that *do not compute*, and we need new terminology. In contemporary urban discourses, where "data" rhetoric is often frothy and fetishistic, we seem to have lost critical perspective on how urban data become meaningful spatial information or translate into place-based knowledge.

We need to expand our *repertoire* (to borrow a term from Diana Taylor) of urban intelligences, to draw upon the wisdom of information scientists and theorists, archivists, librarians, intellectual historians, cognitive scientists, philosophers, and others who think about the management of information and the production of knowledge (Foth et al., 2007). They can help us better understand the breadth of intelligences that are integrated within our cities, which would be greatly impoverished if they were to be rebuilt, or built anew, with computational logic as their prevailing epistemology.

We could also be better attuned to the lifecycles of urban information resources—to their creation, curation, provision, preservation, and destruction—and to the assemblages of urban sites and subjects that make up our cities' intellectual ecologies. "If we think of the city as a long-term construct, with more complex behaviors and processes of formation, feedback, and processing," architect Tom Verebes proposes, then we can imagine it as an organization, or even an organism, that can learn (Verebes, 2016). Urbanists and designers are already drawing on concepts and methods from artificial intelligence research: neural nets, cellular processes, evolutionary algorithms, mutation and evolution.[11] Perhaps quantum entanglement and other computer science breakthroughs could reshape the way we think about urban information, too. Yet we must be cautious to avoid translating this interdisciplinary intelligence into a new urban formalism.

Instead of more gratuitous parametric modeling, we need to think about urban epistemologies that embrace memory and history; that recognize spatial intelligence as sensory and experiential; that consider other species' ways of knowing; that appreciate the wisdom of local crowds and communities; that acknowledge the information embedded in the city's facades, flora, statuary, and stairways; that aim to integrate forms of distributed cognition paralleling our brains' own distributed cognitive processes.

We must also recognize the shortcomings in models that presume the objectivity of urban data and conveniently delegate critical, often ethical decisions to the machine. We, humans, *make* urban information by various means: through sensory experience, through long-term exposure to a place, and, yes, by systematically filtering data. It's essential to make space in our cities for those diverse methods of knowledge production. And we have to grapple with the political and ethical implications of our methods and models, embedded in all acts of planning and design. *City-making* is always, simultaneously, an enactment of *city-knowing*—which cannot be reduced to computation.

Acknowledgments

This chapter originally appeared in: Mattern, S. (2017) 'A City Is Not a Computer'. *Places Journal*, February, and is reprinted with permission.

Notes

1 Doctoroff (2016) cited in https://medium.com/sidewalk-talk/reimagining-cities-from-the-internet-up-5923d6be63ba
2 Originally published in *Places Journal* in 2017.
3 See Cheung andAltman (2016) The post drew responses on Twitter from designer and urbanist Fred Scharmen ("fish tacos") and visual journalist Erik Reyna ("pez dispensers"), among others.
4 https://cities.ycr.org/
5 Sidewalk Labs is a key investor in Intersection, the "municipal media company" that is a partner in LinkNYC. See Brown, E. (2016), Mattern, S. (2016c), Lessin, (2016), and Weinberg (2016).
6 See Susie Cagle, "Why One Silicon Valley City Said 'No' to Google," *Next City*, May 11, 2015; Sean Hollister, "Welcome to Googletown," *The Verge*, February 26, 2014; Chris Morris-Lent,"How Amazon Swallowed Seattle," *Gawker*, August 18, 2015.
7 Some argue that the city-as-machine has a much deeper history, as evidenced by use of grid layouts, linear patterns, and regular geometric forms since ancient times, and by the use of standardized patterns for colonial urban development. See, for instance, Kevin Lynch, *Good City Form* (Cambridge, MA: MIT Press, 1981): 81–88. See also Matthew Gandy, "Cyborg Urbanization: Complexity and Monstrosity in the Contemporary City," *International Journal of Urban and Regional Research* 29: (March 2005): 26–49, https://doi.org/10.1111/j.1468–2427.2005.00568.x; Peter Nientied, "Metaphor and Urban Studies: A Crossover, Theory and a Case Study of SS Rotterdam," *City, Territory and Architecture* 3:21 (2016), https://doi.org/10.1186/s40410-016-0051-z; William Solesbury, "How Metaphors Help Us Understand Cities," *Geography* 99:3 (Autumn 2014): 139–42; Tom Verebes (2016).
8 Marcus Foth's conception of "urban informatics" is similarly capacious: it encompasses "the collection, classification, storage, retrieval, and dissemination of recorded knowledge," either (1) in a city or (2) "of, relating to, characteristic of, or constituting a city." See Foth, M. (ed.) *Handbook of Research on Urban Informatics: The Practice and Promise of the Real-Time City* (Hershey, PA: Information Science Reference, 2009): xxiii. Such a definition acknowledges a wide variety of informational functions, contents, and contexts. Yet his focus on recorded knowledge, and on informatics' reputation as a "science" of data processing, still limits our understanding of the city's epistemological functions.
9 For prominent examples, see Andrew Blum, *Tubes: A Journey to the Center of the Internet* (New York: HarperCollins, 2012), and the work of Ingrid Burrington and Mél Hogan.
10 Quoted in Marcia J. Bates, "Information," in Marcia J. Bates, Mary Niles Maac, eds., Encyclopedia of Library and Information Sciences, 3rd ed. (New York: CRC Press, 2010): 2347–2360, available online https://pages.gseis.ucla.edu/faculty/bates/articles/information.html". See also Rafael Capurro and Birger Hjørland, "The Concept of Information," in Blaise Cronin (ed.), *The Annual Review of Information Science and Technology*, Vol 37 (2003): 343–411.
11 See, for instance, the work of Michael Batty.

References

Amoore, L. (2018) 'Cloud Geographies: Computing, Data, Sovereignty'. *Progress in Human Geography*, 42 (1), pp. 4–24.
Boyer, M. C. (1995) *CyberCities: Visual Perception in the Age of Electronic Communication*. New York: Princeton Architectural Press.
Brown, E. (2016) 'Alphabet's Next Big Thing: Building a 'Smart' City'. *Wall Street Journal*. www.wsj.com/articles/alphabets-next-big-thing-building-a-smart-city-1461688156
Castells, M. (1989) *The Informational City: Information Technology, Economic Restructuring, and the Urban–Regional Process*. Oxford: Blackwell.

Cheung, A. & Altman, S. (2016) 'New Cities'. https://blog.ycombinator.com/new-cities/: Y Combinator Blog.

Davis, S. L. (2016) '16 Cities Join T4 America's Smart Cities Collaborative to Tackle Urban Mobility Challenges Together'. http://t4america.org/2016/10/18/16-cities-join-t4americas-smart-cities-collaborative-to-tackle-urban-mobility-challenges-together/ Transportation for America.

Eliot, T.S. (1934) *The Rock*. New York: Harcourt, Brace and Company, p. 7.

Epstein, R. (2016) 'The Empty Brain'. *Aeon*, May 18. https://aeon.co/essays/your-brain-does-not-process-information-and-it-is-not-a-computer.

Foth, M., Odendaal, N. & Hearn, G. (2007) 'The View from Everywhere: Towards an Epistemology for Urbanites', 4th International Conference on Intellectual Capital, Knowledge Management and Organizational Learning. Cape Town, South Africa.

Graham, S. & Marvin, S. (1996) *Telecommunications and the City*. London: Routledge.

Halpern, O., Mitchell, R. & Geoghegan, B. D. (2017) 'The Smartness Mandate: Notes Toward a Critique'. *Grey Room*, 68, pp. 106–129.

Harris, M. (2016) 'Secretive Alphabet Division Funded by Google Aims to Fix Public Transit in US'. *The Guardian*, June 27, 2016. www.theguardian.com/technology/2016/jun/27/google-flow-sidewalk-labs-columbus-ohio-parking-transit.

Huh, B. (2016) 'Should I Pursue My Passion or Business?' https://medium.com/@benhuh/should-i-pursue-my-passion-or-business-76187b6b83fb: Medium.

Kittler, F. A. (1996) 'The City is a Medium'. *New Literary History*, 27 (4), pp. 717–729.

Knox, H. (2010) 'Cities and Organisation: The Information City and Urban Form'. *Culture and Organization*, 16 (3), pp. 185–195.

Lessin, J. E. (2016) 'Alphabet's Sidewalk Preps Proposal for Digital District'. *The Information*. www.theinformation.com/articles/sidewalk-labs-preps-proposal-for-digital-district

Light, J. (2004) *From Warfare to Welfare: Defense Intellectuals and Urban Problems in Cold War America*. Baltimore, MD: Johns Hopkins University Press.

Manzotti, R. & Parks, T. (2016) 'Does Information Smell?' *New York Review of Books*. www.nybooks.com/daily/2016/12/30/consciousness-does-information-smell/.

Mattern, S. (2013) 'Infrastructural Tourism'. *Places Journal*, July 2013.

Mattern, S. (2016a) 'Public In/Formation'. *Places Journal*, November 2016.

Mattern, S. (2016b) 'Of Mud, Media, and the Metropolis: Aggregating Histories of Writing and Urbanization'. *Cultural Politics*, 12 (3), pp. 310–331.

Mattern, S. (2016c) 'Instrumental City: The View from Hudson Yards, circa 2019'. *Places Journal*, April 2016.

Mattern, S. (2016d) 'Cloud and Field'. *Places Journal*, August 2016.

Mazzotti, M. (2017) 'Algorithmic Life'. *Los Angeles Review of Books*. https://lareviewofbooks.org/article/algorithmic-life/.

McCullough, M. (2014) *Ambient Commons: Attention in the Age of Embodied Information*. Cambridge, MA: MIT Press.

McFedries, P. (2014) 'The City as System [Technically Speaking]'. *IEEE Spectrum*, 51 (4), p. 36.

Mitchell, W. (2000) *E-Topia*. Cambridge, MA: MIT Press.

Mitchell, W. J. (1995) *City of Bits*. Cambridge, MA: MIT Press.

Mumford, L. (1961) *The City in History*. New York: Harcourt Brace.

O'Toole, J. J. (2004) 'Back to the Future: Ernst Posner's Archives in the Ancient World'. *The American Archivist*, 67, pp. 161–175.

Rhodes, M. (2016) 'Y Combinator's Plan to Build a New City? Not Actually Crazy'. *Wired*, www.wired.com/2016/07/y-combinators-plan-build-new-city-not-actually-crazy/.

Sharma, N. (2008) 'The Origin of the "Data Information Knowledge Wisdom" (DIKW) Hierarchy'. www.researchgate.net/publication/292335202_The_Origin_of_Data_Information_Knowledge_Wisdom_DIKW_Hierarchy.

Stoler, A. (2010) *Against the Archival Grain: Epistemic Anxieties and Colonial Common Sense*. Princeton, NJ: Princeton University Press.

Taylor, D. (2003) *The Archive and the Repertoire: Performing Cultural Memory in the Americas*. Durham, NC: Duke University Press.

Thompson, F. B. (1968) 'The Organization is the Information'. *American Documentation*, 19 (3), pp. 305–308.

Vallianatos, M. (2015) *Uncovering the Early History of 'Big Data' and 'Smart City' in Los Angeles*. https://boomcalifornia.com/2015/06/16/uncovering-the-early-history-of-big-data-and-the-smart-city-in-la/: Boom California.

Verebes, T. (2016) 'The Interactive Urban Model: Histories and Legacies Related to Prototyping the Twenty-First Century City'. *Frontiers in Digital Humanities*, 3. https://doi.org/10.3389/fdigh.2016.00001.

Walsham, A. (2016) 'The Social History of the Archive: Record-Keeping in Early Modern Europe'. *Past & Present*, 230 (Supplement 11), pp. 9–48.

Weinberg, C. (2016) 'Is Alphabet Going to Build a City?' *The Information*, www.theinformation.com.

Weinberger, D. (2010) 'The Problem with the Data-Information-Knowledge-Wisdom Hierarchy'. *Harvard Business Review*, https://hbr.org/2010/02/data-is-to-info-as-info-is-not.

3

Bias in urban research

From tools to environments

Mark Shepard

Introduction

What we see is influenced by how we see, which in turn is conditioned by the tools we use to see with. One could say these tools *bias* what we see. Yet while human bias is defined in terms of a preference, predisposition, prejudice or predilection for or against something or someone, instrument bias can occur for very different reasons. On the one hand, measuring tools can be improperly calibrated, leading to a scale producing inaccurate measurements for weight, for example. Alternately, instruments can be designed to weigh various aspects or qualities of the object under investigation differently, such as when an ultraviolet (UV) filter reduces transmission of specific wavelengths of light through a camera lens. Regardless of whether the bias in question is the result of error or intent, we can say the relation between the tools and objects of research is anything but neutral.

As we more frequently view cities through the data they generate, we often deploy algorithms as tools for insight. Methods involving big data and machine learning introduce forms of bias that are both inherited from human bias residing in the dataset itself, as well as generated by the way in which the algorithms operate on that data. Algorithmic bias can include the filtering of content for particular individuals based on a history of their online activity, as often occurs with social media. Alternately, where algorithms operate within decision-making contexts, bias can arise from how an algorithm generates outputs that discriminate against a protected population. Consider predictive policing platforms that have been shown to produce a feedback loop which often results in the allocation of more patrol cars to neighbourhoods with higher populations of racial minorities, irrespective of their "true" crime rate (Ensign *et al.*, 2018).

To the extent that our understanding of a city is shaped by the methods by which we apprehend it, so too is the city shaped by this understanding. The evolution of urban environments can be understood as an ontogenetic process, whereby the relation between the tools and objects of urban research and design is recursive and mutually reinforcing. Indeed, as neighbourhoods become instrumented with arrays of sensors, and their residents in turn generate ever-larger volumes of data as they go about their daily business, the tools themselves are beginning to merge with the environments on which they are reporting. This chapter traces this shift from observational tools to environments that observe in an attempt to examine the changing nature of bias in urban research and design and the subsequent implications for epistemologies of urban environments. Following the introduction of two pre-cinematic optical devices that embody radically different epistemological models, I proceed through a comparative analysis of two approaches to urban research that employ techniques of moving image analysis that contrast small data studies with big data analytics. I conclude by asking how recent developments in the quantification of urban life through smart city initiatives are altering not only how we conceive cities but also how we perceive their citizens.

Observational devices and their epistemic implications

Observational devices have long been used in the effort to represent urban space. Canaletto's use of the camera obscura to chart the urban landscape of 18th century Venice is well known (Figure 3.1). While the optical principles of the device were known for centuries, by the early 19th century, the camera obscura was recognized as the dominant model for observation. As art historian Jonathan Crary has noted, the device represented more than just the performance of optical principles; it also articulated an epistemology of the relation between observer and world (1990). The camera obscura posited a disembodied observer occupying the interior of a darkened enclosure into which the exterior world is projected by means of a tiny aperture in one wall as an inverted two-dimensional image. This model of observation served as an analogy for human vision: the aperture of the room was a corollary for the human eye, and the dark interior a metaphor for the mind to which the world is represented as image. In this model, the observing subject is constructed as both monocular and devoid of other senses. Such veracity of observation, already a conviction of Enlightenment thought, was firmly grounded in an empirical demonstration of the mechanical optics of vision in which the other senses are not to be trusted.

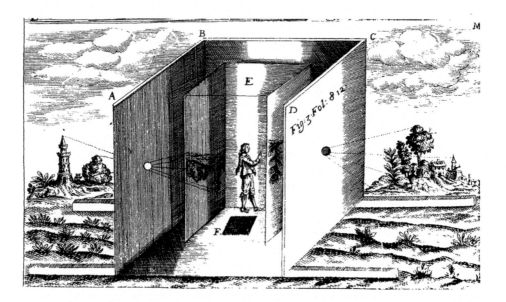

Figure 3.1   Camera obscura – Athanasius Kircher, 1646
Source: Public domain via Wikimedia Commons

By comparison, the stereoscope, itself a by-product of early 19th century advances in physiological optics, capitalized on the discovery that with binocular vision, each eye sees something slightly different due to the angular disparity existing between them (Figure 3.2). The production of depth in sight was subsequently understood to be related to the mind's ability to unite and reconcile two dissimilar images. The stereoscope was developed to reproduce this optical experience mechanically. Significantly, the device marks an intent not just to *represent* a given space, but to *simulate* its presence. What is sought is not merely a likeness, but a lucid tangibility. With the stereoscope, one is confronted not with a view of the world through an aperture or frame, but with the technical reconstitution of an already reproduced world fragmented into two non-identical models (Crary, 1990). Through the incorporation of the observing subject into the mechanics of the device, the stereoscopic image is produced. The body is immobilized and integrated with the apparatus. The subject becomes a participant in the production of a verisimilitude through a process of unifying and reconciling the experience of difference. The disjunction between an experience and its cause is reified, the "real" conflated with the "optical." Absent is the notion of a "point of view" in a Cartesian sense. There is, in the end, nothing out there.

These pre-cinematic optical devices present different modes of observation that condition how one engages with an environment and what can be known about that environment. Embedded within the historical and scientific contexts from which they emerged, they present contrasting models of the relation between an observer and world as mediated through an optical device. What's striking is not only how each constructs radically different observing subjects, but also how both illustrate divergent epistemological assumptions

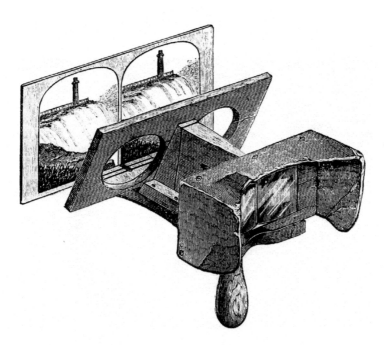

Figure 3.2 Stereoscope, 1861
Source: Public domain via Wikimedia Commons

underpinning the truth claims they articulate. The shift from an accurate likeness forming the basis of a truthful representation to a tangible presence that enacts the visual experience of a given space marks changing notions of the role and status of the body and its sensing capabilities in the production of knowledge: from a disembodied, monocular subject occupying the interior of the device to the integration of the observing subject into the mechanics of device itself. The radical differences between these devices highlight the role of the apparatus in constituting not only the parameters of what we know about the world, but also how we conceive our relationship to it, and ultimately how we construe who we are and our agency to act within it.

With the introduction of film at the close of the 19th century came the ability to capture movement and change in urban space over the course of time. The early city symphony films, for example, use the lens of the cinematic apparatus to record the rhythms of the industrial city. Walter Rutmann's *Berlin. Die Sinfonie der Großstadt* (1927) is a catalogue of urban movements that follows a linear progression from morning to night. Repetitive, cyclical operations of machines are juxtaposed with the actions of people over the course of the day. Urban life converges with the industrialized city into a tightly synchronized composition. The convergence of urban life with the mechanics of film is even more pronounced in the classic *The Man with the Movie Camera* (1929) by Dziga Vertov in which the mobilized camera itself becomes a protagonist in a series of scenes that depict life in the city (Figure 3.3). From dawn to dusk, citizens in Kiev, Kharkov, Moscow and Odessa are shown at work and play through their interactions with the machinery of modern life. The film culminates in a rapid montage that juxtaposes the aperture of the camera with a human eye: a dizzying fusion of observer and observational device. The superimposition of the observing subject into the observational device is complete.

Figure 3.3 Film still from *The Man with the Movie Camera*, Dziga Vertov, 1929
Source: Public domain via Wikimedia Commons

The quantification of vision

By the latter half of the 20th century, the role of the moving image had shifted from that of representing urban environments to serving as an explicit tool of empirical urban research. William Whyte employed time-lapse photography in the 1970s to study the interaction of people with and within urban space in Manhattan. In the film *The Social Life of Small Urban Spaces* (1979) (Figure 3.4), he presented the outcomes of his Street Life Project, a decade-long study of open public space and street life in New York City that had been commissioned by the New York City Planning Commission (Whyte, 1980). Whyte's research used direct observation as a method to focus on small-scale, street-level studies that examined human behaviour in public places. The time-lapse opening shot of the Seagram building's plaza over the course of a day correlates a moving patch of sun with areas of activity within the plaza. Within the filmic frame we see a clock, a sign of the empiricism underlying the researchers' aspirations toward the factual verification of a set of hypotheses.

Whyte mapped the micro-interactions between people and those between people and urban space in order to document patterns of use and activity over time. This street-level investigation incorporated both an observational device and a research methodology focused on urban amenities such as "sittable" space, street, sun, food, water, trees. The correlation between the path of the sun and the activity within a plaza, for example, is perhaps obvious, as Whyte remarks. Yet the rhetorical role of the mechanical apparatus is clear: the camera is understood as a transparent research tool, enabling the study of the role of movement and social interaction in urban space, as well as the use of moving images in spatial analysis. Designed to influence public policy on the design of urban plazas, Whyte's filmic observations and detailed analysis claimed the status of factual representations of how small urban spaces are used in New York City.

At a time when the rhetoric of the quantifiable is re-emerging as the primary driver of urban development, Whyte's project can be understood as a precedent to more recent initiatives in the commercial software industry that embrace empirical methods of observation in studying urban environments. Placemeter, for example, was a technology start-up founded in 2012 that used algorithms to extract data about urban life from video feeds and sensors that are distributed throughout

Figure 3.4 Film still from *Social Life of Small Urban Spaces*, William H. Wythe, (c) 1980

the city (Figures 3.5 and 3.6). Their product was a software platform that employed crowd-sourced, window-mounted smart phone cameras and machine vision algorithms to develop datasets on urban activity. Yet unlike Whyte's controlled research project, Placemeter leveraged video streams sourced from the public at large who had signed up with the service. Users streamed video data captured by a camera mounted on their window to Placemeter's servers, where the data was analysed. The results were subsequently accessed through an online dashboard.

Placemeter used crowd-sourced data to quantify movement in urban spaces. Through proprietary machine vision algorithms, the software first classified different kinds of moving objects appearing within the video frame: pedestrians, bicycles, motorcycles, vehicles and large vehicles. Subsequent analysis extrapolated various attributes about this activity, including volume of foot traffic, speed and dwell time of moving bodies, and the use of specific urban amenities, for example. Various "solutions" were offered for smart cities, transportation, retail, advertising and what the company's website terms "tactical urbanism." Applications included

> discovering crowded and under-used areas through looking at user flow data; analysing the use of specific design features (park benches, recycling bins, playground equipment); measuring the impact of special programming (concerts, farmers markets); determining the impact of temporary events (street closures, art installations).
>
> *(Placemeter.com, 2016)*

Figure 3.5 Placemeter smartphone cameras capturing pedestrian traffic
Source: Courtesy Lee Kim

Figure 3.6 Videostill of Placemeter demonstration video
Source: Courtesy Placemaker.com

The platform essentially applied the logic of website-analytics to the task of measuring urban activity, tailored to the needs and interests – the *biases* – of the transportation, retail and advertising industries.

While quantifying street-level activity such as foot traffic in front of a retail store has obvious implications for how real estate is valued and marketed, what is at stake when the system is deployed at an urban scale? In one such case, the city of Paris and Cisco have worked with Placemeter and other sensing platforms to test different urban planning models for the redevelopment of the Place de la Nation. As part of an initiative known as the €1 billion Paris Smart City 2020, the project was viewed as an experiment that would yield results that could be extrapolated across the entire city, from the Place de la Bastille to the Place des Fêtes, the Place Gambetta, the Place d'Italie, the Place de la Madeleine and the Place du Panthéon.

The Place de la Nation is a large, circular intersection in eastern Paris that is divided into a series of concentric traffic islands by broad streets. The guillotine that used to dominate its centre island has been replaced by a monument by Jules Dalou that commemorates the French Revolution. "Le triomphe de la République," as the monument is entitled, portrays a figure that symbolizes the Republic being held aloft by a lion-drawn chariot that is being led by the figure of Liberty, attended to by those of Labor and Justice, and followed by that of Abundance. Today Place de la Nation is a busy traffic circle devoid of pedestrians. The project for Place de la Nation integrated a range of technologies for sensing and acquiring data. A device called the "Breezometer" analysed the air quality from sensors located in the plaza. Fullness levels of waste containers were monitored by sensors embedded inside their shells. Anti-noise panels measured real-time sound levels. All collected data were reportedly fed to ParisData, an open data portal developed and maintained by the City of Paris. Information panels situated on-site were to provide passers-by with data visualizations about the project. Cisco deployed Placemeter to study the number of pedestrians and bikers, automobile traffic patterns and other activities occurring in the plaza. These data were intended to be combined with other data to study the effects of closing streets at certain points for a period of time, for example, or how widening bike lanes and moving benches, chairs and other amenities to different locations affects patterns of use and activity in the plaza.

Both Placemeter's platform and Whyte's method are based on empirical visual evidence recorded through time-lapse photography. Whyte's results, however, were presented in the form of charts and graphs that are now referred to as "small data." As Rob Kitchin and Tracey Lauriault note, prior to the emergence of big data, all data studies were essentially small data studies (Kitchin and Lauriault, 2015). Small data is common to social science research, and is usually produced through surveys, interviews and other qualitative research methods. Small data studies generally involve a targeted inquiry designed to answer specific research questions. They tend to be context-rich, in-depth investigations of a specific issue or set of issues. They are based on limited volume and variety of data collected at specific points in time. Placemeter and the associated technologies employed at the Place de la Nation, by contrast, are based on a distributed sensing model where data streams from multiple sources are aggregated and interpreted by machine learning algorithms. These data streams share attributes with what are called big data. As described by Kitchin (2014), big data are huge in volume, high in velocity, created in or near real-time, diverse in variety, structured and unstructured in nature, temporally and spatially referenced, exhaustive in scope, fine-grained in resolution, uniquely indexical in identification, relational in nature, flexible, extendable and scalable. Significantly, machine learning based on big data eschews an initial hypothesis in favour of pattern-recognition, aiming to reveal previously unknown correlations and other insights.

Whyte's research departed from a set of research questions focused on understanding the micro-interactions between people and an existing urban environment, around which a methodology for observation was developed. The Place de la Nation project, by contrast, deploys a suite of existing sensing devices as part of the design process to iteratively study how people behave in a specific urban space and test alternatives for its modification. That the project presupposes methods of quantification based on sensor data might lead one to paraphrase Maslow in saying that if the only instruments we have are sensors, we will treat everything as if it were data. In effect, the observational device in this case becomes the process of quantification and analysis of data, a process that generates and tests hypotheses about urban activity based on iterative analysis of a series of design proposals.

In contrast to Whyte's method which used time-lapse photography to test a set of hypotheses about how people inhabit urban space, Placemeter employed machine learning algorithms trained on big data in an attempt to derive hypotheses from patterns of movement and activity, quantifying the life of the street in terms of pre-established classifiers. Former editor-in-chief of *Wired* magazine Chris Anderson has described this new era of big data and machine learning as one of knowledge production characterized by "the end of theory" (Anderson, 2008). In other words, big data and machine learning enable an entirely new epistemology for making sense of the world; rather than testing a theory by gathering and analysing relevant data as Whyte did, this new approach seeks to gain insights "born from the data." Here, correlation supersedes causation.

Testbed neighbourhoods

While platforms such as Placemeter add big data and machine learning to the instruments available for the research and design of urban environments, other initiatives look toward the instrumentalization of the urban environment itself. New York City's Hudson Yards project is the largest private real estate project ever to be built in the United States. Over the next decade, the $20 billion project – which spans the seven blocks of 30th to 34th Street cordoned by 10th and 12th Avenues – will add 17 million square feet of commercial, residential and civic space. Reports claim that it will contain the nation's first "quantified community," a testbed for applied urban data science (Leber, 2016). Led by Constantine E. Kontokosta, a professor of urban informatics at the Center for Urban Science and Progress (CUSP), the Quantified Community, as he refers to it, aims to be a fully instrumented urban neighbourhood that uses an integrated, expandable sensor network to support the measurement, integration and analysis of neighbourhood conditions, activity and outcomes (Kontokosta, 2016).

The premise of the Quantified Community is drawn from the Quantified Self movement. People associated with this movement employ fitness activity trackers to monitor a range of health factors – from heart rate, to steps taken, floors climbed, calories burned, even sleep quality – and to produce representations of their progress toward self-identified goals that are shared and aggregated through online portals (Wolf, 2010). In another form of self-tracking, dieting apps enable us to record what we eat and track how many calories we have consumed, what proportion of those calories are from protein, carbohydrates or fat, and provide nutrition information about daily eating habits. Central to this movement is the idea that these personal monitoring devices can support behavioural change.

Scaling this paradigm to the neighbourhood, the Quantified Community posits that the continuous monitoring of the built environment, from its technical systems to the human activity within it, can be fed back into that environment so as to alter its future performance and the behaviour of its inhabitants. Kontokosta himself explicitly identifies his behaviourist intentions for the quantified community: "My focus is much more on understanding how the data influences behaviour, and using the type of information that's now available to really democratize the planning process much more" (Libby, 2016). His version of the Quantified Community aims to measure, model and predict a wide range of activity, including: pedestrian flows through traffic and transit points, open spaces and retail space; air quality both within individual buildings as well as across open spaces and surrounding public areas; health and activity levels of residents and workers using a proprietary mobile phone app; and solid waste with particular focus on increasing the recovery of recyclables and organic waste, energy production and usage throughout the project's lifecycle (NYU CUSP, 2016). The Quantified Community at Hudson Yards thus presents a test bed for future urban life, where urban intelligence – the "smartness" of the smart city – is rendered not as conscious, liberal or objective, but rather as performative (Figure 3.7).

If Vertov's film presented the collapse of distance and distinction between observer and observational device, the Quantified Community renders citizens themselves as sensors alongside those embedded within buildings and associated infrastructural systems. The observing subject is replaced by a suite of algorithms that mine, aggregate and extrapolate from these data in search of patterns of activity and behaviour. Here, people become not just residents of the neighbourhood, but also consumers of its amenities at the same time as they generate data about these activities. As Whyte and his researchers observed, the physical environment shapes our behaviour within it. In the Quantified Community, then, data collected about that behaviour from a diverse and unevenly distributed set of sources are fed back into that environment in the form of modulations made to it designed to alter that behaviour.

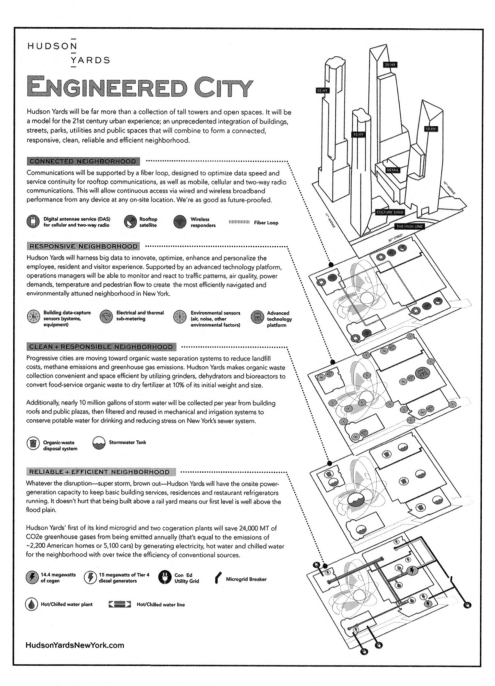

HUDSON YARDS

ENGINEERED CITY

Hudson Yards will be far more than a collection of tall towers and open spaces. It will be a model for the 21st century urban experience; an unprecedented integration of buildings, streets, parks, utilities and public spaces that will combine to form a connected, responsive, clean, reliable and efficient neighborhood.

CONNECTED NEIGHBORHOOD

Communications will be supported by a fiber loop, designed to optimize data speed and service continuity for rooftop communications, as well as mobile, cellular and two-way radio communications. This will allow continuous access via wired and wireless broadband performance from any device at any on-site location. We're as good as future-proofed.

- Digital antennae service (DAS) for cellular and two-way radio
- Rooftop satellite
- Wireless responders
- Fiber Loop

RESPONSIVE NEIGHBORHOOD

Hudson Yards will harness big data to innovate, optimize, enhance and personalize the employee, resident and visitor experience. Supported by an advanced technology platform, operations managers will be able to monitor and react to traffic patterns, air quality, power demands, temperature and pedestrian flow to create the most efficiently navigated and environmentally attuned neighborhood in New York.

- Building data-capture sensors (systems, equipment)
- Electrical and thermal sub-metering
- Environmental sensors (air, noise, other environmental factors)
- Advanced technology platform

CLEAN + RESPONSIBLE NEIGHBORHOOD

Progressive cities are moving toward organic waste separation systems to reduce landfill costs, methane emissions and greenhouse gas emissions. Hudson Yards makes organic waste collection convenient and space efficient by utilizing grinders, dehydrators and bioreactors to convert food-service organic waste to dry fertilizer at 10% of its initial weight and size.

Additionally, nearly 10 million gallons of storm water will be collected per year from building roofs and public plazas, then filtered and reused in mechanical and irrigation systems to conserve potable water for drinking and reducing stress on New York's sewer system.

- Organic-waste disposal system
- Stormwater Tank

RELIABLE + EFFICIENT NEIGHBORHOOD

Whatever the disruption—super storm, brown out—Hudson Yards will have the onsite power-generation capacity to keep basic building services, residences and restaurant refrigerators running. It doesn't hurt that being built above a rail yard means our first level is well above the flood plain.

Hudson Yards' first of its kind microgrid and two cogeneration plants will save 24,000 MT of CO_2e greenhouse gases from being emitted annually (that's equal to the emissions of ~2,200 American homes or 5,100 cars) by generating electricity, hot water and chilled water for the neighborhood with over twice the efficiency of conventional sources.

- 14.4 megawatts of cogen
- 15 megawatts of Tier 4 diesel generators
- Con Ed Utility Grid
- Microgrid Breaker
- Hot/Chilled water plant
- Hot/Chilled water line

HudsonYardsNewYork.com

Figure 3.7 Hudson Yards: Engineered City

Source: Courtesy Related Companies

The critiques of behaviourist explanations of human activity and their implications for urban design are well known. See for example, Skinner (1971) and critiques by Chomsky (1971) and Koestler (1968). What is perhaps less evident is what happens to those aspects of urban life that are not easily measured in a Quantified Community. Not everyone will choose to opt-in to a proprietary app designed to measure the health and activity of the community, and not all choices or decisions about how we inhabit or otherwise occupy urban space are reducible to quantifiable data points. What we measure is limited by the instruments we have available. If the behaviour of these citizen-sensors is taken to indicate levels of engagement with each other and with their neighbourhood, these indications are inevitably biased by the instruments that make those actions visible.

Whyte's studies departed from a series of focused research questions about micro-interactions of people in public spaces and the role the built environment plays in supporting or hindering these interactions. By contrast, the Quantified Community would appear to depart from a suite of technical capabilities for quantifying behaviour of people, the environment and infrastructural systems with an eye toward increasing the optimization and efficiency of each. In this shift from observational tools to environments that observe, both the city and its citizens merge into populations of human and non-human actors and actants that comprise not individual bodies but rather patterns of activity and behaviour iteratively mined, clustered and interpreted by algorithmic processes. This test-bed world of big data and machine learning – where correlation supersedes causation –"is a probabilistic one where few things are certain and most are only probable," as Halpern et al. write (2013, p.294). It presents an urban epistemology not concerned with documenting facts, representing spaces or developing representative models, but rather evolving models that are in and of themselves territories.

Conclusion

"The trouble with modern theories of behaviourism," Hannah Arendt wrote, "is not that they are wrong but that they could become true, that they actually are the best possible conceptualization of certain obvious trends in modern society" (1958). The trend toward the quantifiable, measurable and accountable in urban design would appear to reflect a return to what Brenner and Schmid describe as "technoscientific urbanism," where sensing space and analysing behavioural data become the dominant methods for empirically driven urban design aimed at finding solutions for perennial urban problems. The neo-positivist, neo-naturalist revival of post-war systems thinking at the core of smart city developments such as Hudson Yards not only reinforces this view of cities and urban life as universally replicable, but also as depoliticized subjects to be more optimally and efficiently managed (Brenner and Schmid, 2015).

Biasing the quantifiable in this way, however, only makes sense when data are engendered with the capability of being "true." Yet, as Daniel Rosenberg (2013) reminds us, there is no truth in data, and the use of the word "data" in the English language has been intertwined in conflicting ways with related concepts of "fact" and "evidence" since its emergence. At the beginning of the 18th century, "data" referred to either "principles accepted as a basis of argument or to facts gleaned from scripture that were unavailable to questioning." By the end of the century, the word more commonly referred to "facts in evidence determined by experiment, experience, or collection" (Rosenberg, 2013, p.33). While this shift from understanding data as the rhetorical premise of an argument to the result of an empirical investigation laid the groundwork for mid-20th century claims of scientific veracity in urban design and planning, today, as Rosenberg suggests, "[i]t may be that the data we collect and transmit has no relation to truth or reality whatsoever beyond the reality that data helps us to construct" (2013, p.37).

As tools for urban research and design merge with the very environments they aim to both study and project, new urban territories emerge that are populated more by statistical imaginaries derived from aggregate data than by communities of embodied citizens. If the camera obscura presented an epistemology that objectified the world through optical principles for an isolated, interiorized, monocular subject, the stereoscope employed principles of physiology to engage a disembodied observer in the co-production of a verisimilitude of the world. Vertov followed by collapsing the distinctions between an observing subject and an observational device entirely, presenting the world itself as a purely cinematic construct. The territories constructed by algorithms discussed above, then, represent an urban epistemology that dispenses with the very idea of observing a world altogether, positing instead "insights" born from data potentially bearing no relation to an observable truth or reality.

From this perspective, unbiased urban research is an oxymoron. The question becomes: what bias do we want to bring to tomorrow's cities? The first step in responding to this question is to dispense with the notion that any given method will result in an empirically "true" observation, and subsequently to foreground the selection of the tools and methods by which urban space is analysed and projected as a rhetorical one. For this we will not only need better tools and methods than are currently available, but also a willingness to think critically and reflexively about the epistemological implications of their application to the shaping of urban environments, and in turn, of urban life.

References

Anderson, C. (2008) The End of Theory: The Data Deluge Makes the Scientific Method Obsolete. *Wired*. [online] Accessible at www.wired.com/2008/06/pb-theory/ [Accessed 28 May 2019].

Arendt, H. (1958). *The Human Condition*. Chicago: University of Chicago Press.

Brenner, N. and Schmid, C. (2015) Towards a New Epistemology of the Urban? *City*, 19(2-3), pp. 151–182.

Chomsky, N. (1971) The Case Against B. F. Skinner. *The New York Review of Books*, 30 December.

Crary, J. (1990). *Techniques of the Observer: On Vision and Modernity in the Nineteenth Century*. Cambridge: MIT.

Ensign, D., Friedler, S. A., Neville, S., Scheidegger, C., and Venkatasubramanian, S. (2018). "Runaway Feedback Loops in Predictive Policing." In *Proceedings of the 1st Conference on Fairness, Accountability and Transparency*. New York: PLMR, vol. 81, pp. 160–171. [online] Available at http://proceedings.mlr.press/v81/ [Accessed 28 May 2019].

Halpern, O., LeCavalier, J., Calvillo, N. and Pietsch, W. (2013) Test-Bed Urbanism. *Public Culture*, 25, pp. 272–306.

Kitchin, R. (2014) Big Data, New Epistemologies and Paradigm Shifts. *Big Data and Society*, 1(1), pp. 1–12.

Kitchin, R. and Lauriault, T. P. (2015) Small Data in the Era of Big Data. *GeoJournal*, 80(4), pp. 463–475.

Koestler, A. (1968). *The Ghost in the Machine*. New York: Macmillan.

Kontokosta, C. E. (2016) The Quantified Community and Neighborhood Labs: A Framework for Computational Urban Science and Civic Technology Innovation. *Journal of Urban Technology*, 23(4), pp. 67–84.

Leber, J. (2016) Beyond the Quantified Self: The World's Largest Quantified Community. Co.Exist. [online] Accessible at www.fastcoexist.com/3029255/beyond-the-quantified-self-the-worlds-largest-quantified-community [Accessed 14 August 2016].

Libby, B. (2016). Quantifying the Livable City. *Citylab*. [online] Accessible at www.citylab.com/tech/2014/10/quantifying-the-livable-city/381657/ [Accessed 14 August 2016].

NYU CUSP (NYU Center for Urban Science and Progress (2016). NYU CUSP, Related Companies, And Oxford Properties Group Team Up to Create "First Quantified Community" in the United States at Hudson Yards. [online] Accessible at http://cusp.nyu.edu/press-release/nyu-cusp-related-companies-oxford-properties-group-team-create-first-quantified-community-united-states-hudson-yards/ [Accessed 14 August 2016].

Placemeter.com (2016). Quantify the World with Placemeter. [online] Available at www.placemeter.com [Accessed 14 August 2016].

Rosenberg, D. (2013). Data Before the Fact. In Gitelman, Lisa. *"Raw Data" Is an Oxymoron*. Cambridge: MIT Press.

Skinner, B. F. (1971). *Beyond Freedom and Dignity*. New York: Knopf.

Whyte, W. H. (1980). *The Social Life of Small Urban Spaces*. Washington, DC: The Conservation Foundation.

Wolf, G. (2010). The Quantified Self. *TED*. [video] Accessible at www.ted.com/talks/gary_wolf_the_ quantified_self [Accessed 28 May 2019].

4

Urban science

Prospect and critique

Rob Kitchin

Introducing urban science

Urban science practices and promotes an interdisciplinary scientific and computational approach to city systems and the processes of urbanization. It uses statistical analysis and data analytics – including machine learning, data mining, visual analytics, modelling, and simulation – to identify causal relationships and predict how city systems work. This is in contrast to urban studies, which uses both quantitative and qualitative methods and adopts a more contextual approach with respect to politics, culture, policy, and history. While urban studies generally conceives of cities as constellations of places, with analysis usually based upon fairly static empirical data (small samples, generated at specific places and times), urban science views cities as systems (or a system of systems) with analysis utilizing urban big data (massive samples generated on a continuous basis). Typically, urban science seeks to map and model urban *dynamics* – patterns of flow, urban processes, and system interactions. The aim is to determine urban 'laws', conduct real-time analysis of systems, produce new theoretical insights, develop a synoptic and integrative science of cities, and to translate the knowledge produced into practical application, including urban design and planning, city management, and economic development.

Urban science builds on a longer history of quantitative social science, including quantitative geography, geographic information science, urban and transportation modelling, social physics, urban and regional economics, urban cybernetics, social ecology, and location theory, that have sought to explain and model urban processes and the functioning of city systems (Batty 2013a). However, many academics and industry analysts now practising and promoting urban science have little grounding in this history, with their training being rooted in the fields of data and information science, computer science, physics, and engineering. Moreover, they have little knowledge of the deep history of research in urban studies and allied disciplines of sociology, geography, anthropology, economics, history, architecture, and planning. Urban science researchers have been attracted to investigating urban processes by the massive volumes of urban big data now being generated, the call for science to tackle global challenges such as rapid urbanization, sustainability, and climate action, along with associated research funding streams, and the wider promotion by industry and governments for the creation of 'smart cities' (Townsend 2013, 2015; O'Sullivan and Manson 2015). These drivers have also led to the creation of a number of large, interdisciplinary urban science research centres across the globe (see Batty 2013b; Townsend 2015). Industry and government are often external partners or stakeholders in these centres and their projects, providing data, funding, and other in-kind contributions, and are often direct beneficiaries of the research in terms of new knowledge, intellectual property, products, and networks.

For many of those practising urban science, the approach is promoted as a paradigm-shifting endeavour – where urban science promises to provide a more integrative and insightful understanding of cities than urban studies that will transform how urban policy making and planning is undertaken, and will become the dominant approach for urban research. Indeed, Solecki et al. (2013) argue that urban studies, rooted in a social sciences rather than computational, scientific tradition, has failed to deliver knowledge that effectively solves city issues and is inappropriate for delivering solutions for the major challenges ahead as urbanization continues apace. This unsuitability is due to its disciplinary fragmentation, a panoply of approaches that create disparate viewpoints, and its focus on cities as places and on the symptoms of urban problems. Instead, Solecki et al. (2013) call for an urban science that focuses on urban processes and systems, and underlying causes and potential solutions (not place and symptoms), and shares a common epistemological approach underpinned by a scientific method. They propose three basic goals for urban science:

(1) To detail the basic components of urbanization across scales;
(2) To identify the universal laws of city-building;
(3) To find relationships between urbanization and other aspects of Earth's systems.

Only urban science, Solecki et al. (2013) contend, can produce 'a theory of urbanization with fundamental and unique components that can withstand scientific scrutiny' and 'lead to systemic solutions that address the whole rather than separate components' (p. 14)

Further, because urban science conducts analyses and builds models based on urban big data, it is posited that it offers the potential for urban knowledge that has greater breadth, depth, scale, and timeliness, and is inherently longitudinal, in contrast to that derived from urban studies (Batty *et al.* 2012). Big data have fundamentally different properties to traditional 'small' datasets, being generated in real-time, exhaustive in scope, and having fine resolution (Kitchin 2014). For example, rather than data being derived from a travel survey with a handful of city dwellers during a specific time period at particular locations (i.e., sampled 'small data'), transport big data consist of a continual survey of every traveller, for example, collecting *all* the tap-ins and tap-outs of travel cards, or using automatic number plate recognition-enabled cameras to track *all* vehicles, or using sensors to monitor the mobile phone MAC (media access control) addresses to track *all* pedestrians with a phone. It thus becomes possible to determine detailed patterns of travel across times of the day, days of the week, and seasons, and to do this for all nodes on the network (e.g., junctions, bus stops, sensor locations), and to make predictions about future system performance under different conditions. As a consequence, data from such systems have the potential to produce a highly granular, longitudinal, whole system understanding of a city system and enable it to be managed in real time.

Over the past couple of decades, this transformation from slow and sampled 'small' data to fast and exhaustive 'big' data has been enabled by the roll-out of a raft of new networked, digital technologies embedded and integrated into the fabric of urban environments and infrastructures. Such technologies include digital cameras, sensors, transponders, meters, actuators, and GPS that monitor various phenomena and continually send data to an array of control and management systems, such as city operating systems, centralized control rooms, intelligent transport systems, logistics management systems, smart energy grids, and building management systems. In addition, a multitude of smartphone apps and sharing economy platforms generate a range of real-time location, movement, and activity data. The result is a deluge of real-time, fine-grained, contextual, and actionable data which are routinely generated about cities and their citizens upon which urban science can be practised (Koonin and Holland 2013). However, such data are largely closed in terms of access given that they are mostly generated by privately owned systems, meaning that practising urban science often requires developing access rights with companies and states.

The remainder of the chapter sets out urban science's relationship to urban informatics and explains its epistemology. It then summarizes criticism of urban science with respect to epistemology, instrumental rationality, data issues, and ethics.

Urban science and its relationship to urban informatics

There seems to be some confusion in the literature as to the relationship between urban science and urban informatics, a term that pre-dates urban science and is used to describe a form of academic enterprise concerning the generation, management, processing, analysis, and utilization of urban data (Foth 2009). It is worth teasing out the overlap and differences here in order to make clear the nature of urban science. While Batty (2013b) frames urban science within a larger domain of urban informatics, Townsend (2015) positions urban informatics as sub-branch of urban science, and Kitchin (2016) has them as separate but complementary fields that often intersect. This confusion is due to how urban informatics has been conceived.

For Batty (2013b: 3) urban informatics is the 'application of computers to the functioning of cities' and 'the ways in which computers are being embedded into cities'. Here, urban science is one way, within the broader remit of urban informatics, in which computation is being utilized to understand the functioning of cities and in turn informs how computation is used to manage and control urban systems. For Foth (2009) urban informatics is an interdisciplinary enterprise that includes three broad communities: social (e.g., media studies, communication studies, cultural studies, sociology); urban (e.g., urban studies, geography, planning, architecture), and the technical (e.g., computer science, data science, electronic engineering, human–computer interaction). From this perspective, urban informatics is primarily concerned with the development of informational tools and management systems for controlling and communicating urban processes, understanding human interactions with such systems, and studying the relationship between people, place, and digital technology (Foth 2009), rather than being centrally concerned with urban modelling, statistical analysis, simulation and prediction, and finding 'urban laws'. Moreover, urban informatics given its wider body of constituent practitioners is less likely to be positivistic in nature or follow the 'scientific method'.

Many of the new urban research centres detailed by Batty (2013b) and Townsend (2015) conduct both urban science and urban informatics research. For example, the Centre for Advanced Spatial Analysis (CASA) in University College London undertakes a range of applied and fundamental geospatial research focused on modelling and simulating cities, including creating 3D virtual models and city dashboards. The Centre for Urban Science and Progress (CUSP) at New York University seeks to use a scientific approach to develop data-driven solutions for explaining and tackling urban problems and offers a Masters programme in Urban Science and Informatics.

Urban science and its epistemologies

Urban science is broadly rooted in a positivistic tradition that has sought to apply scientific principles and methods, drawn from the natural, hard, and computing sciences, to social phenomena in order to explain them. The aim is to statistically test relationships between variables or build models as a means to produce and verify laws that explain and predict how systems work. Central to this endeavour is the objective collection of data through common and standardized methods of observation (that can be replicated) and the formulation of theories which can be tested and verified. In general, a realist epistemology is adopted that supposes the existence of an external reality which operates independently of an observer and which can be objectively and accurately measured, tracked, statistically analysed, modelled, and visualized to reveal the world as it actually is (Kitchin *et al.* 2015). In other words, it is held that urban data can be abstracted from the world in neutral, value-free, and objective ways and are understood to be essential in nature. That is, data are representative of that which is being measured, faithfully capturing its essence and are independent of the measuring process (though it is acknowledged that there might be data quality issues related to error, bias, calibration, etc.). These data, when analysed in similarly objective ways through statistical analysis, modelling, and simulation, reveal deep insights about cities that can be used to reshape urban policy and enhance urban infrastructures (though it is appreciated that there might be constraints and limitations due to the methodology employed). While cybernetic modelling approaches recognize the complexity and emergent qualities of city systems, such systems are still understood in machinic terms and are largely closed and bounded in nature.

Three epistemological variations of urban science are practised (Kitchin 2014). The first is a traditional, hypothesis-driven, deductive scientific method, with questions and approach guided by established theory. The second is a form of inductive empiricism in which it is argued that, by employing data analytics, urban big data can speak for themselves free of theory or human bias or framing (Kitchin 2014, 2016). Such an approach is best exemplified by Anderson (2008: n.p.) who argues that 'the data deluge makes the scientific method obsolete' and that '[c]orrelation supersedes causation, and science can advance even without coherent models, unified theories, or really any mechanistic explanation at all'. In other words, rather than being guided by theory, the data can be wrangled through hundreds of algorithms to discover the most salient factors with regards to a particular phenomenon. Such an approach has gained some traction in data science and within industry research. The third is data-driven science that seeks to hold to the tenets of the scientific method, but generate hypotheses and insights 'born from the data' rather than 'born from the theory' (Kelling *et al.* 2009). It uses guided knowledge discovery techniques to mine data to identify potential hypotheses, before a traditional deductive approach is employed to test their validity. This approach is more common because it rejects the idea of the 'end of theory' and maintains scientific values; extracts additional, valuable insights that traditional knowledge-driven science would fail to generate; and it produces more holistic and extensive models and theories of entire complex systems rather than elements of them (Kelling *et al.* 2009; Miller 2010).

In many cases, these approaches have been realized through applied research that uses city environments as 'living laboratories'; that is, as testbeds to validate the science and test the practical interventions produced (see Evans *et al.* 2016). For example, CASA uses London as its laboratory and CUSP uses New York, working with public and private stakeholders to tackle the real-world problems they have identified. Indeed, much urban science research is highly empirically grounded and applied in nature, with extensive collaboration between scientists, city administrations/state agencies, and industry partners. The potential benefits to each party are clear – academic gain access to key datasets, companies gain access to intellectual insight, and city administrations/state agencies gain access to potential interventions and solutions. The Smart Dublin initiative in Ireland, for example, offers university researchers and companies (often working in partnership) the opportunity to work with four local authorities and within a couple of designated urban testbed areas to experiment with new technologies, generate urban big data, and practice urban science. At the start of 2019 there were 109 active projects – not all of which were urban science orientated – aimed at understanding and solving a diverse set of urban problems.

Criticism of urban science

While urban science has expanded rapidly in the last decade, it is far from producing a paradigmatic shift in urban research and has been subject to critique from urban studies scholars and others (Crampton *et al.* 2013; Kitchin 2016; Mattern 2013). This critique is multi-pronged, with much of it mirroring that of positivistic social sciences and geographic information sciences in previous decades (see Crampton 2010; Kitchin 2015). In these earlier 'theory wars' urban and spatial science were roundly criticized for being too reductionist, mechanistic, essentialist, and deterministic, collapsing diverse individuals and complex, multi-dimensional social structures and relationships to abstract data points and universal formulae and laws. Moreover, rather than being epistemologically objective, neutral, and value-free, it was demonstrated that such science was framed and situated within power-geometries of knowledge and practice and often served particular interests (Pickles 1994). In addition, they also wilfully ignored the metaphysical aspects of human life and the role of politics, ideology, social structures, capital, and culture in shaping urban relations, governance, and development (Kitchin 2016).

Consequently, scientific approaches to cities have been critiqued as being rather naïve and narrow in perspective, producing overly simplified explanations and models, and a limited and limiting understanding of how cities work (foreclosing what kinds of questions can be asked and how they can be answered) and how urban issues can be tackled. They promote an instrumental rationality that posits that cities can be effectively steered and managed through scientific insights and technical instruments and that urban issues can be solved through a range of technical solutions (Kitchin *et al.* 2015; Mattern 2013). Urban science, it is argued, has thus far failed to recognize that cities are complex, multi-faceted, contingent, relational systems, full of contestation and wicked problems that are not easily captured or steered, and that urban issues are often best solved through political, civic society, fiscal, policy, and legal interventions rather than technical fixes and technocratic forms of governance (Kitchin *et al.* 2015). Indeed, critique of the first wave of cybernetic approaches to cities in the later 1960s and 1970s demonstrated that they produced knowledge and policy interventions that not only failed to live up to their promises but did much damage to city operations (Flood 2011; Townsend 2013). For example, New York's adoption of the RAND Corporation's cybernetic model for the redistribution of fire stations contributed to the destruction caused by fires that blighted the city in the 1970s (Flood 2011).

While advocates of computational social and urban science counter that the availability of big data and data analytics address some of the criticisms of earlier forms – especially those of reductionism and universalism by providing more finely grained, sensitive, and nuanced analysis that can take account of context and contingency (Kitchin 2014) – many concerns undoubtedly still hold for present forms of urban science (Kitchin 2016; Wyly 2014). For example, Batty (2013a, 2013b) notes that, despite drawing on complexity theory and advances in data analytics, urban science is still failing to provide detailed explanations of cities and their processes. He argues that there is often a naivety amongst those who do not have a background in urban thinking and policy with respect to framing cities and devising solutions, overly focusing on technology and engineering interventions and failing to heed lessons from the long history of urban policy and planning. As he notes, there are no easy solutions to the intractable problems of cities, and urban science will produce no silver bullets, though that is not to say that it will not produce useful insights. He also cautions against the search for universal laws, arguing that urban systems are too large, complex, fluid, and diverse, instead promoting a more tempered approach of understanding individual systems and recognizing, rather than dismissing, the value in other approaches to understanding cities.

It is also the case that scientific analysis is heavily dependent on data quality and contextual information. While urban big data undoubtedly provide numerous opportunities to examine particular systems and issues, they also have a number of limitations. For example, with respect to urban transportation data, while the datasets are rich in volume, they often have limited demographic context – we might know the journeys, but not who took them or why (Batty 2013b). In many cases, the data are being repurposed having been generated by commercial entities for their specific needs but not scientific research. There are thus questions concerning the extent to which repurposed big data provide adequate, rigorous, and reliable surrogates for more targeted, sampled data and how representative such data are of phenomena and populations (Struijs *et al.* 2014). Moreover, big data might seek to be exhaustive, but as with all data they are both a representation and a sample. For example, social media data only relate to those who use a service and are stratified by social class and age, and also include many anonymous and bot accounts. What big data are captured by a system is shaped by: the field of view/sampling frame (where data capture devices are deployed and their settings/parameters); the technology and platform used (different surveys, sensors, lens, textual prompts, and layouts all produce variances and biases in what data are generated); the context in which data are generated (unfolding events mean data are always situated with respect to circumstance); the data ontology employed (how the data are calibrated and classified); and the regulatory environment with respect to privacy, data protection, and security (Kitchin 2014).

Further, much big data have little methodological transparency concerning how they were produced and processed (especially those generated by companies); few metadata with respect to relevance, credibility, timeliness, accessibility, interpretability, coherence, and veracity (accuracy, fidelity, including details of uncertainty, error, bias, reliability, and calibration); and minimal documentation concerning the provenance and lineage of a dataset. And yet it is generally acknowledged that big data can be full of dirty, gamed, and faked data, as well as data being absent (Kitchin 2014). While some might argue that 'more trumps better' and that big data does not need the same standards of data quality, veracity, and lineage because the exhaustive nature of the dataset removes sampling biases and compensates for any errors or gaps or inconsistencies in the data (Mayer-Schonberger and Cukier 2013), it is still the case that garbage-data-in produces garbage-analysis-out.

Urban science and its use of urban big data also raise a number of ethical questions that so far have received little consideration. Since much urban big data are exhaustive and indexical, they raise concerns with respect to privacy, dataveillance and geosurveillance, social sorting, and anticipatory governance (Graham 2005; Kitchin 2016). Many cities are saturated with remote controllable digital CCTV cameras whose footage is increasingly analysed using facial, gait, and automatic number plate recognition software using machine vision algorithms, enabling individuals to be tracked (Kitchin 2016). Smartphones and their apps continuously communicate their location and share with third parties in order to create user profiles and practice targeted location-based advertising (Leszczynski 2018). In a number of cities, for example London and Chicago, sensor networks have been deployed across street infrastructure such as bins and lampposts to capture and track unique phone identifiers such as MAC addresses (Kitchin 2016). Big data analysis can thus reveal highly detailed patterns of spatial behaviour from which other insights, such as mode of travel, activity, lifestyle, co-travellers, can be inferred. The consequence is that individual privacy is eroded with people no longer lost in the crowd and it becomes possible to produce and predict detailed individual and place profiles. These profiles can be used to socially sort and redline populations or to socially sort places with respect to policy interventions. For example, a number of US police forces are now using predictive analytics rooted in urban science research to anticipate the location of future crimes and to direct police officers to increase patrols in those areas and to try and identify potential criminals (Jefferson 2018).

Smart city technologies, the data they generate, and the urban analytics applied to them thus have significant direct and indirect impact on people's everyday lives. Few of those whose data has fed into creating predictive profiles imagined that their data were going to be repurposed to social sort or regulate or control them, or nudge them towards certain behaviours. Generally, these studies – both in universities and industry R&D labs – circumvent notice and consent issues, as well as Institutional Research Boards ethics procedures, by anonymizing and aggregating the data. Nonetheless, the research being undertaken can have effects on those who are unwittingly participating by feeding back into the formulation and delivery of services. In other cases, studies ignore ethical procedures altogether, arguing that data in the public domain (e.g., social media data) are open to carte blanche analysis or that they are entitled to experiment on their own systems without user consent (Kitchin 2016).

To address some of these concerns, some have suggested reconceptualizing cities within urban science and reframing its epistemology (Kitchin 2016). With respect to the first, rather than being cast as bounded, knowable and manageable systems that can be captured, modelled, steered, and controlled in mechanical, linear ways, it is suggested cities need to be understood as fluid, open, complex, multi-level, contingent, and relational systems that are full of culture, politics, competing interests, and wicked problems (Kitchin 2016). With regards to the latter, it is proposed to shift the epistemology towards those employed in critical Geographic Information Science and radical statistics. These approaches employ quantitative techniques, inferential statistics, modelling, simulation, and visual analytics whilst being mindful and open with respect to their shortcomings, drawing on critical social theory to frame how the research is conducted, how sense is made of the findings, and how the knowledge employed (Kitchin 2014; Kitchin et al. 2015). Here, it is recognized that there is an inherent politics pervading the datasets analysed, the research conducted, and the interpretations made. Moreover, such a reframing does not foreclose complementing computational social science with 'small data' studies that provide additional and amplifying insights (Crampton et al. 2013). In addition, researchers – whether in the public or private domain – need to consider the ethical implications of their work and the uses to which their research is being deployed. Beyond complying with relevant laws and institutional review board (IRB) requirements, urban science practitioners should have a duty of care to citizens not to expose them to harm through its analysis (Kitchin 2016).

Conclusion

Building on earlier rounds of quantitative social science, geocomputation and natural science research, and extending them through the use of new data analytics to extract insights from urban big data, urban science has grown rapidly over the past decade. With the trend in creating smart cities, the on-going growth in the production of urban big data, and large-scale investment in urban science research, this expansion is likely to continue for some time. It is unlikely, however, that urban science will become a new paradigm, producing an integrative approach that replaces the diverse philosophical traditions within urban studies. This is because urban studies continues to produce useful and insightful research and the inherent weaknesses in the epistemology of urban science. Instead, urban science will provide a complementary approach to urban studies and urban informatics and its epistemology is likely to shift and fracture in the same manner as GIScience.

While one could argue that a better approach to city development might be achieved through abandoning urban science and smart cities, others suggest that they are recast in political and epistemological terms. In other words, rather than advocating against the smart city and urban science per se, it is argued that how they are presently conceived and practiced is transformed to be more contextual, relational, and contingent in orientation. Whether such recasting occurs or not, because urban science potentially provides technical, computational solutions to urban problems it will continue to be seen as a viable and profitable means of making sense of cities and creating new products through the analysis of urban big data. In turn, it will continue to influence the development of urban policy and planning and the rollout of smart city initiatives for the foreseeable future.

Acknowledgements

The research for this chapter was funded by an ERC Advanced Investigator award (ERC-2012-AdG 323636-SOFTCITY) and a Science Foundation Ireland grant, Building City Dashboards (15/IA/3090). The chapter in part draws on Kitchin (2016). Thanks to Claudio Coletta, Paolo Cardullo, Liam Heaphy, and Sung-Yueh Perng for comments on the initial draft.

References

Anderson, C. (2008) The end of theory: The data deluge makes the scientific method obsolete. *Wired*, 23rd June. See www.wired.com/2008/06/pb-theory/ (last accessed 2 January 2017).

Batty, M. (2013a) *The New Science of Cities*. Cambridge, MA: MIT Press.

Batty, M. (2013b) *Urban Informatics and Big Data*. London: CASA, University College London. www.spatialcomplexity.info/files/2015/07/Urban-Informatics-and-Big-Data.pdf last accessed 2 Jan 2017.

Batty, M., Axhausen, K.W., Giannotti, F., Pozdnoukhov, A., Bazzani, A., Wachowicz, M., Ouzounis, G. and Portugali, Y. (2012) Smart cities of the future. *European Physical Journal Special Topics* 214: 481–518.

Crampton, J.W. (2010) *Mapping: A Critical Introduction to Cartography and GIS*. Malden, MA: Wiley-Blackwell.

Crampton, J.W., Graham, M., Poorthuis, A., Shelton, T., Stephens, M. Wilson, M.W. and Zook, M. (2013) Beyond the Geotag: Situating 'big data' and leveraging the potential of the geoweb. *Cartography and Geographic Information Science* 40(2): 130–139.

Evans, J., Karvonen, A. and Raven, R. (eds) (2016) *The Experimental City*. London: Routledge.

Flood, J. (2011) *The Fires: How a Computer Formula, Big Ideas, and the Best of Intentions Burned Down New York City and Determined the Future of Cities*. New York: Riverhead.

Foth, M. (2009) Preface. In Foth, M. (ed.) *Handbook of Research on Urban Informatics: The Practice and Promise of the Real-Time City*, Hershey, PA: Information Science Reference, pp. xxviii–xxxi.

Graham, S.D.N. (2005) Software-sorted geographies. *Progress in Human Geography* 29(5): 562–580.

Jefferson, B.J. (2018) Predictable policing: Predictive crime mapping and geographies of policing and race. *Annals of the American Association of Geographers* 108(1): 1–16.

Kelling, S., Hochachka, W., Fink, D., Riedewald, M., Caruana, R., Ballard, G. and Hooker, G. (2009) Data-intensive science: A new paradigm for biodiversity studies. *BioScience* 59(7): 613–620.

Kitchin, R. (2014) *The Data Revolution: Big Data, Open Data, Data Infrastructures and Their Consequences.* London: Sage.

Kitchin, R. (2015) Positivistic geography. In Aitken, S. and Valentine, G. (eds) *Approaches to Human Geography*, 2nd ed. London: Sage, pp. 23–34.

Kitchin, R. (2016) The ethics of smart cities and urban science. *Philosophical Transactions A* 374(2083): 1–15.

Kitchin, R., Lauriault, T.P. and McArdle, G. (2015) Knowing and governing cities through urban indicators, city benchmarking & real-time dashboards. *Regional Studies, Regional Science* 2: 1–28.

Koonin, S.E. and Holland, M.J. (2013) The value of big data for urban science. In Lanes, J., Stodden, V., Bender, S. and Nissenbaum, H. (eds) *Privacy, Big Data, and the Public Good: Frameworks for Engagement*, Cambridge: Cambridge University Press, pp. 137–152.

Leszczynski, A. (2018) Geoprivacy. In Kitchin, R., Lauriault, T. and Wilson, M. (eds) *Understanding Spatial Media*, London: Sage, pp. 239–248.

Mattern, S. (2013) Methodolatry and the art of measure: The new wave of urban data science. *Design Observer: Places.* 5 November. https://placesjournal.org/article/methodolatry-and-the-art-of-measure/ (last accessed 2 Jan 2017).

Mayer-Schonberger, V. and Cukier, K. (2013) *Big Data: A Revolution that will Change How We Live, Work and Think.* London: John Murray.

Miller, H.J. (2010) The data avalanche is here. Shouldn't we be digging? *Journal of Regional Science* 50(1): 181–201.

O'Sullivan, D. and Manson, S.M. (2015) Do physicists have geography envy? And what can geographers learn from it? *Annals of the Association of American Geographers* 105(4): 704–722.

Pickles, J. (ed.) (1994) *Ground Truth: The Social Implications of Geographic Information Systems.* New York: Guildford Press.

Solecki, W., Seto, K.C. and Marcotullio, P.J. (2013) It's time for an urbanization science. *Environment: Science and Policy for Sustainable Development* 55(1): 12–17.

Struijs, P., Braaksma, B. and Daas, P.J.H. (2014) Official statistics and big data. *Big Data & Society* 1(1): 1–6.

Townsend, A. (2013) *Smart Cities: Big Data, Civic Hackers, and the Quest for a New Utopia.* New York: W.W. Norton & Co.

Townsend, A. (2015) Making sense of the new science of cities. Rudin Center for Transportation Policy & Management and Data and Society Research Institute, New York University. www.citiesofdata.org/wp-content/uploads/2015/04/Making-Sense-of- the-New-Science-of-Cities-FINAL-2015.7.7.pdf (last accessed 2 Jan 2017).

Wyly, E. (2014) Automated (post)positivism. *Urban Geography* 35(5): 669–690.

5

Defining smart cities

High and low frequency cities, big data and urban theory

Michael Batty

Introduction

Smart cities define the latest stage in the digital revolution which began with the invention of computers in the middle of the last century. Successive waves of ever smaller and more powerful computers have marked this evolution with computers and their communications first spreading out from their origins in scientific laboratories to business transactions processing in dedicated computer centres. As computers were scaled down from mainframes to minis in the 1960s and 1970s and with the advent of the microprocessor in 1971, they became personal and individualistic in the 1980s, being used as the primary devices to connect to the growing internet in the 1990s. By the millennium, mobile phones were becoming widespread, morphing into hand-held devices such as smart phones and these have predominated during the last decade. Currently (mid-2019), more than 35% of the world's population of 7.5 million persons have access to a smart phone and to all intents and purposes, by the year 2030, at least 75% will be able to connect with one another in this way.

This evolution has been governed by Moore's law, the regularity observed in the miniaturisation of electronic circuitry that began with the invention of the transistor in 1948 at Bell Labs. It was first articulated by Gordon Moore (1965), one of the pioneers of the integrated circuit in one of the first and still the world's most powerful semiconductor company, Intel. Moore observed that computer processing power was doubling every 18 months, increasing in speed in the same way, reducing in cost by one half and scaling down in size by the same order over the same period. This period doubling has continued ever since and it shows little sign of stopping, notwithstanding limits posed by the speed of light and other physical constraints. The miniaturisation generated by this evolution reached the point in the last decade where computers were small enough to be embedded into small objects. Combined with sensors which could be deployed extensively within the natural and built environment, they could be used to generate massive volumes of data, second by second and scaled over many locations in real time. Although similar data had been collected quite modestly for many decades using analogue devices, this digital scaling down has given rise to the era of 'big data' which is critical to the development of the smart city movement. Indeed smart cities and big data – and of course the analytics that has developed to make sense of all this – constitute the domain that we will define and describe in this chapter.

There are two quite different perspectives on smart cities. The first and most obvious, some might say superficial, is the actual embedding of hardware and software into the built environment. This represents the focus of the computer industry in all its forms, for the city and its public spaces now constitute a new marketplace for the buying and selling of computers and communications. We will outline the machinery of the smart city which computers offer in the next section but it is important to contrast this with a second perspective which is the impact that computer and communications are having on the social and economic nexus of the city, on the way computers are changing the way cities form and function and on our own behaviour – impacts that are likely to be substantial. This second perspective is a consequence of the first, while the way our behaviour is being changed also affects what we consider appropriate in terms of the hardware and software that we decide constitutes the smart city. To an extent, we might consider the first perspective centred around the impact of computation on the physicality of the city, while the second concentrates on how new technologies are impacting on social and economic structure which is more abstract, but both are interwoven with each other in terms of the smart cities movement. In fact, one of the dilemmas of smart cities is that many groups consider one or the other as the only perspective and often do not relate the two, as we do so in this chapter.

In the rest of this chapter, we will introduce these two perspectives and then focus on a related distinction between the high frequency and low frequency city. Change in cities over very fine time intervals, seconds, minutes, hours, days is high frequency in contrast to changes over years, decades, centuries and so on which is low frequency and there is some correlation between both the physical and the more abstract with respect to these changes. We will then focus on examples of high and low frequencies and suggest that a new, much more extensive science is required to understand the city, based on complexity, simulation and new forms of data mining and perhaps artificial intelligence (AI) or at least machine learning. Our argument slowly turns from more prosaic and realistic tools and methods to softer, more open ideas and speculations and these serve to link these arguments to practical tools to enable us to understand, predict and design future cities (Batty, 2018).

Embedded computation

In automating the city, the most highly constrained behaviours and procedures associated with how the population reacts to routine functions such as movement, the use of energy, everyday marketing such as retailing, repetitive production processes and so on, have always represented the activities that are most likely to be mechanised. Indeed many of these functions were the subject of extensive automation during the first and second industrial revolutions that led to steam power, mechanised routines and electricity. For example, once the automobile was invented, control of the road system using traffic lights, loop counters for assessing traffic volumes, and even control centres based on augmenting human interaction with TV support became routine, coinciding but not yet part of the emerging digital revolution in the mid- to late 20th century. The same kinds of functions emerged for routine marketing and for energy usage with the deployment of cash registers, lighting in cities, more automated waste management and so on. It was fairly obvious that once computers scaled down to the point where they could replace mechanical devices which were intrinsically less reliable and required more maintenance, all these functions would be subject to digital control.

Only since the millennium have sensors linked to digital computers been widely installed, almost simultaneously for card payments in transit, retailing and energy use, with digital replacing analogue in a relatively seamless fashion. With digital sensing, however, vast streams of data monitoring and recording of the operations and usage of these systems have become available. This is 'big data' in the terminology of the current digital revolution, big in the sense that it is voluminous (with its actual volume actually only being determined once the sensors are switched off) and rapid in its delivery – second by second or at finer intervals or more periodic in terms of minute by minute. Much depends on the systems put in place, although raw data can be delivered at a precise instant notwithstanding the fact that it might be delivered to users or to its archives at much less frequent intervals. For example, pollution data in London are made available to the public at large every hour, although it is captured at a much finer temporal interval. In fact digital control, sensing and related services are being introduced into many routine domains within the city and it is virtually impossible to catalogue all of these. Our general view of cities as being entities that are managed, controlled and planned from the top down – which was the collective wisdom of how the city was organised in the mid-20th century, has all but disappeared as the idea of cities evolving from the bottom up has taken root under the banner of complexity theory (Batty, 2005). In such contexts, new information technologies are being introduced to transport, energy, marketing, finance, health care, education – the list is as long as we can count the many functions that define cities in everyday life – and this is why we can never say that one city is any smarter than another. Many of these technologies are invisible or at least, not immediately visible without special scrutiny; they are introduced from the bottom up and thus intrinsically without any central coordination.

Moreover, these new technologies are being introduced without any grand plan, although we can divide them into two kinds. First, there are fixed sensors that can be operated automatically or activated by human touch or sense. That is, sensors that simply record natural and human events that are always switched on, such as pollution monitors and sensors that are only activated by human touch, such as the recording of payments whenever someone makes a tap or swipes a credit card for payment. The second type of sensor is now much more ubiquitous and this is the device that we carry around with us all the time – the smart phone which was introduced in 2007 and then took the world by storm. In fact, much greater volumes of data are being captured using smart phones, by ourselves activating them or their being used as passive sensors. This is a kind of crowd sourcing and much data is being collected this way, some of it being captured by the providers of the services that such phones can activate but a lot being initiated by phone users themselves in social media and related activities.

In another sense, the hard and soft infrastructure of the smart city in terms of computation itself and the communications equipment that is needed to keep our networks up and running, constitutes another physical domain. This also includes the software that enables data to be communicated and computation to be initiated and all of this hardware and software does have physical presence in the great proliferation of fibre optics and data centres which are increasingly visible and use an increasing amount of the world's electricity. To an extent, all this physical plant is relatively unobtrusive, much of it buried underground and built like cities from the ground up. Like the internet itself in terms of the flow of data and information, the network hardware is constructed in smaller pieces or if constituting some global network, is simply one of many, some of which are linked together organisationally but many of which are separate from one another, only linked at key locations for routing and switching (Blum, 2013). To an extent, the provision of all this hardware and software does not appear to be making a major difference to the form of the city but the fact that it is relatively invisible still opens up the possibility that combined with globalisation and changes in human behaviour occasioned by our usage of this technology, there may be profound changes to the physical and spatial structure of cities as this century progresses.

Changing spatial behaviours

If one wishes to understand and predict the physical infrastructure of the smart city within the wider information society, it is essential to abstract and explain the ways this infrastructure is acting on urban populations and the way these populations are organising and reorganising their activities in the city. This constitutes our second focus, involving the way these new infrastructures are impacting on current behaviours, disrupting them, adding dramatically new forms of network to the city, and enabling cities to easily connect up globally, thus facilitating world-wide specialisation as well as new waves of global urbanisation and migration. The hardware and software that we noted in the previous section generates many new kinds of electronic network whose diversity is best seen not with respect to the hardware of the network but in terms of the multiple functions that these networks allow and enable: email to social media to accessing web resources are typical of these new forms of communication that have become significant in the last 30 years.

The smart city is thus based on multiple networks that connect very different kinds of computer together for complex purposes. Prior to the telegraph and the telephone, networks were operated largely by the physical labour of signalling or at best by the middle of the 19th century, by fast transportation, usually stage coach. The telegraph invented round about 1840, the telephone in the late 1880s and television in the 1920s all provided passive networks of communication where users could communicate but not transform or manipulate information. The big breakthrough did not happen until computers were linked to networks that enabled users not only to communicate data and information but manipulate it through enabling computation at a distance. In essence, the computer was an interactive device in stark contrast to earlier information technologies that were largely passive.

To an extent, it is communication *and* computation that defines the smart city and it is connectivity between 'things' and individuals but with individuals enabling the manipulation of 'things' using these new technologies. The Internet of Things (IoT), coined by the Kevin Ashton of Massachusetts Institute of Technology's Auto-ID Labs in 1999, but having a much longer history in terms of ubiquitous computing (see Weiser, 1991), has been popularised by several writers, in particular in the science fiction writer Bruce Sterling's (2005) prescient essays about the potential for communicating between any kind of object within which sensors, computers and their networks might be embedded and linked. It is however the internet that has made all this possible, not only the network of networks in terms of its hardware and software but the very evolution of the net from the bottom up. No one planned it, no one could ever have planned it, but it could have been thwarted had its genesis not been in the public domain, ironically as part of the US cold war effort to share computer resources. By the time the net was truly global it was almost impossible for private capital to take it over and charge for its use. Had this not happened, much of what is now contained within the smart city, through the way we access information using Google and email and social media, would not have occurred, or if it had, it would be in a very different form from the kind of information society that we have today.

The internet revolutionised email and web services but it is also making possible a new kind of economy in which the biggest players exploit the power of the net which is largely costless to access for most users, at least through direct use. What is fast emerging are platforms built around the fact that the internet is free where companies use our ability to connect with anyone, in any place and at any time to exploit this access, thus enabling many different kinds of populations to buy, sell, browse, access, educate, travel – any of a myriad of activities that define the way we interact with one another in modern societies. These platforms are sites/companies/digital forums such as Facebook, Tencent (WeChat), Google, Amazon, Twitter, Alibaba, Baidu, Weibo and so on, as well as the newer more disruptive ones such as Airbnb, Uber and Didi. Their implications are best summed up in Tom Goodwin's immortal opening lines from his article in *TechCrunch* in 2015 when he said:

> Uber, the world's largest taxi company, owns no vehicles. Facebook, the world's most popular media owner, creates no content. Alibaba, the most valuable retailer, has no inventory. And Airbnb, the world's largest accommodation provider, owns no real estate. Something interesting is happening.

The platforms are still emerging very rapidly as we are able to capture more and more information on our devices, have enormous potential to disrupt existing activities in cities and in the economy. Uber is killing some taxi and public transportation services, especially those that serve underprivileged populations that require public subsidies, and Airbnb is taking stock from the housing market that would otherwise serve long term residents who need to rent. Combined with software that is now only accessible on the web and requires extensive password security, it is an open question as to whether or not productivity is beginning to decline because of extensive details needed to simply log on, never mind navigate. During the last 10 years since the Great Recession, productivity in the Western world (and we suspect in China and other parts of Asia Pacific) appears to have been dropping and some of this must be due to the kinds of security that are required to access much information technology. Banks are a good case in point. It is now hard to access bank accounts other than through the web and the security this requires takes a significant proportion of time to initiate and activate. This is a type of disruption from previous practices, discriminating of course against the elderly, the very young and those not able to swing along with the technologies which are getting ever more complex and convoluted. The productivity paradox however is deep-seated and in terms of the smart city seems to conflict with the very idea of introducing automated technologies; we simply flag this as a pointer to the future of smart cities, but we must watch carefully for this area needs substantial research and reflection, just as questions of privacy and confidentiality in all these data services need to be explored with respect to regulation (Goldin, Koutroumpis, Lafond, Rochowicz, and Winkler, 2019).

In 1997, Frances Cairncross popularised the term *The Death of Distance* in her article and book of the same name, articulating what had been clear during much of the industrial revolution. As new movement technologies – railways, automobiles and planes – were invented thus enabling people to travel faster, longer and at lower cost, the role of distance was transformed. Cities were able to get bigger, to grow beyond one million which had been the constraint until the early 19th century, while the non-physical information technologies of the telegraph, telephone and television lowered the friction of distance, enabling people to communicate globally. The invention of the web thus appeared to be the culmination of this process, with there being little need for physical contact if everything could be manipulated digitally. Of course, this was a caricature but the implications were clear. This, Cairncross (1997) argued, represented the 'death of distance', culminating in nearly two centuries of urban growth that might see everyone enriching their activities even further by working, shopping, playing etc. from home. This has not happened over the last 30 years and is unlikely to; big cities have tended to agglomerate even faster with more, not less, face-to-face contact occurring. However what is happening is that urban form is being increasingly disconnected from function and in terms of the smart city, the implications of this have not really been mapped out. In fact, the impact of the death of distance on the hardware and software being introduced into the city, which we noted in the previous section, has barely been considered.

With the majority of the population now connected through hand-held devices, the proliferation of software apps has been dramatic. These range from simple routine tools that enable improved services to much bigger, more fully fledged software applications that can be used on the move to access what once were applications that required a mainframe, a mini or a desktop. In fact a good deal of new software is itself being embedded into the built environment or if not embedded as such, is available on the person through mobile devices which have become essential for many routine activities. The next great wave of change will involve the move to a truly cashless society and the platform companies such as Facebook are already beginning to predict their own development of such media. The development of software *in situ* – code space as Kitchin and Dodge (2011) have called it – is also proceeding apace as the systems to both design and maintain the built, natural and social environment in which we all live are being infected with software which eventually becomes essential for those originally non-digital systems to continue functioning. This is the idea of the 'digital twin', a digital version of the system which begins to merge into the system of which it is a model. Building information models are a case in point and much of the smart city will eventually be composed of such twins, for these are no more nor less than apps that scale to entire areas from individual applications. To elaborate these developments, however, we need to change tack slightly and examine both theory and analytics that enables us to make sense of the contemporary city, particularly the smart city that has introduced many new features into our daily lives that require explanation and understanding if we are to produce effective and equitable designs for the city of the future.

High and low frequency cities

The smart city movement has thrown onto the agenda in stark relief the notion of the 24-hour city. Up until the end of the last century, most of our theories about cities were rooted in thinking of the city as a static entity, largely in an equilibrium that we considered would help us in thinking about how cities might change to a new equilibrium over timespans that were measured in decades or generations. The master plan was the mechanism that would accomplish this through explicit top down planning. The idea of theorising about how we might organise ourselves in the 24 hour city at high frequencies rather than low, although implicitly considered through routine management, was largely absent from our arsenal of theories and methods. Lack of temporal data was as much a cause of this dilemma as any lack of thinking about how cities changed over all spans of time from the shortest to the longest. The development of scaled-down sensors, the vast proliferation of network technologies and tiny computers has changed all this. In the last 30 years, many systems to monitor what happens in cities over very fine time scales have been put in place.

The other change has been in the way we think about urban dynamics. No longer is the idea of the city in equilibrium credible. Cities are manifestly systems in disequilibrium, always, and thus equilibrium is a simplification too far. In fact cities are often thought about as far-from-equilibrium, which is the concept adopted by complexity science in which systems like cities evolve from millions of individual and group actions, all originating from the bottom up (Batty, 2017). If cities evolve in this way, then their dynamics are on all scales, from the smallest to the largest and in this sense, the high frequency merges into the low frequency city. However most of our understanding based on formal and systematic models is very strongly physical in focus. Basically we still think of cities as being forms that follow function, despite the fact that this mantra was only ever useful as an analogy in emergence of the modern movement in the late 19th century and it is now very wide of the mark. Most of our analytics is thus based on methods of spatial analysis which are statistically based on regularities in land use, activities and transportation in cities, on demographic and employment location, while more formal models explaining how housing markets are structured are still used to think about the way populations locate and the way residential development takes place. These ideas come from social physics and urban economics but some of them are increasingly passé.

The changes that have made many of these theories, established from the middle of the last century, somewhat redundant and obsolete, relate to the fact that cities have become ever more complex over the last 200 years and the increase in their complexity shows no sign of stopping. The smart city movement thus sits on top of all the evident complexity that had emerged physically, involving the decentralisation of jobs, agglomeration of financial functions in central business districts, urban sprawl, continuing migration and globalisation. Our tool box of techniques and models which constitute urban analytics is now beginning to include methods that deal with the 24 hour, high frequency city, for we now have data that enable us to find patterns at fine temporal scales; the heterogeneity that is uncovered when we move from aggregate to disaggregate, and from equilibrium to disequilibrating change can now be disentangled using a variety of multivariate techniques that are being continually refined as we learn more about urban dynamics across these scales.

The wide range of tools that are becoming available to enable us to make sense of cities build on methods that have been established over many years certainly since the 1960s but to an extent, from the late 19th century when location theory and social physics began to emerge. From these theoretical ideas about how cities are structured in terms of the location of activities and land uses and the ways in which transportation patterns become established came a flurry of computer models beginning in the 1950s in the United States, where the focus was very largely on transportation systems which were then being developed to cater for the automobile and the construction of the interstate highway system. These computer models met with only partial success in that theory was limited, data were sparse and incomplete, computational power was never enough and the focus on using such models to address contemporary urban policy problems was minimal. Nevertheless, new generations of model were built around different approaches emphasising dynamics, land development and disaggregate behaviour as well as non-equilibrium perspectives. By the time the smart city emerged in the early 21st century, the arsenal of techniques included land use transportation interaction (LUTI), agent-based (ABM), cellular automata (CA) and micro-simulation models, along with many well developed approaches to spatial analysis built around appropriate applications of spatial statistics and econometrics.

However, much of this heritage is in the spirit of building models that are deductive and predictive, whereas the emergence of big data from highly routinised fine scale temporal processes requires tools to extract patterns from such data and models built on these patterns which tend to be pragmatic and hence much less reliant on theory than previous generations of techniques. So far, despite considerable hype about data mining, deep learning and the power of neural nets in building predictive models for pattern recognition and prediction, there are few examples of how these tools can be used in urban analytics for the smart city. In the next section, we will recount experiences with exploring big data sets in cities, particularly in transit, where potentially there are some interesting and insightful patterns to be mined but our ability to explain these without powerful theory is quite limited. There is little doubt that highly routinised processes such as those that pertain to traffic movements of various kinds from cars to trains, might be automated using patterns that pertain within highly constrained situations, and examples of traffic control already exist from previous times when analogue systems were first developed. The promise of autonomous vehicles depends on our assumptions that such regularities can be extracted from massive amounts of data pertaining to the driver experience and the surrounding environment, and although the tests so far appear promising, coping with very complex environments will be a real challenge.

These kinds of automation are of the simplest kind but when it comes to much more complex behavioural structures in cities involving housing markets, the location of financial services, the provision of transport infrastructure, questions of heritage and preservation and a whole range of issues pertaining to location and interaction in cities, the kinds of analytics we require are not very different from those that pre-date the smart city. As we have noted, these theories and tools require substantial work if they are to be used intelligently to manage and plan the city over any time scale, and the fact that cities are becoming more complex and our focus has extended to embrace the much higher frequency city has simply increased the challenge of developing good science for a much more profound understanding than anything hitherto. Our theories, which in the past pertained to the low frequency city, need to be extended to the high frequency and this requires a continuum of data from the shortest temporal scales to the longest and from the finest spatial scales to the coarsest. This is a continuing dilemma that shows little sign of abating and perhaps it heralds an era when we need to accept that we will need to keep running to stand still, which may well be an intrinsic characteristic of post-industrial, digital cities and the society therein.

Big data in the smart city

A crude definition of big data is 'anything that will not fit in an Excel spreadsheet'. As these sheets have reached a million rows, then it might be argued that if you had a million items of data with a much more modest number of attributes, say 256 or 1024, then you would find it difficult to pack these into such a sheet anyway, but the point is that data greater than such volumes require special software and tools to deal with it in terms of storage, retrieval and analysis. For example, in my research group,[1] we have a dataset from Transport for London consisting of all the tap-ins and tap-outs ('taps') for all passengers using the Oyster card – the Radiofrequency Identification (RFID) card (or equivalent) that were used for payment on all public transport in Greater London – that is nearly one billion taps (rows if you like) for three months from July to September in 2012 (Reades, Zhong, Manley, Milton, and Batty, 2016). This is big data by our standards; it is comprehensive and accurate, although such data are often incomplete in that sometimes the tap-ins and tap-outs are not captured by the system, if the barriers are left open occasionally (as they are in some stations) and if the traveller does not need to validate their payment due to their having a free card or season ticket.

Such data in principle then give all movements between origin and destination stations on the rail network where both tap-ins and tap-outs are required but the actual paths between these stations – the flows of traffic – need to be constructed using various shortest route algorithms which are by no means perfect in that travellers do not always follow their shortest route for many reasons involving cognition, navigation, travel preferences and so on. Nevertheless the matrix of trips is as complete as possible, notwithstanding that external data, for example, from questionnaire surveys might be used to enhance it. However, integrating this dataset into the set of data needed to examine the wider pattern of travel is highly problematic. Origins and destinations at stations for example are only part of the story in that what we need to explore overall movements are the trips that are made before the traveller enters the station and the trips when the traveller exits it. In short we need to 'add value' to the data, synthesising it with all the other movements that the traveller in question makes. In the absence of such data, the problem is incomplete in that the data are limited. In fact, it is hard to know how such data might be used in understanding the overall pattern of trips by rail, and when the problem is scaled to all trips, these problems become ever more severe. This does not mean that insights cannot be generated from the analysis of such data even though it is partial data. We can explore such data with respect to disruptions to the system from which we can extract how people behave if a rail line goes down due to stalled trains or signal failures, working out the cascades onto other lines and out of the system. Because there is no tap-out on a bus, only a tap-in, there are limits to exploring multimodal travel on rail and bus. We can however examine the pattern of travel with respect to the very limited set of attributes in the data – such as the kind of card: child, elderly and so on – noting spatial variations in travel demand and tying this to other datasets such as that involving geo-demographics from the Population Census and such databases. This can give some insights into patterns in the data but the extent of analysis is still quite limited.

If you think of these kinds of problems writ-large across the city for all the big datasets that are now being collected, the scale of the challenge becomes quite evident. Big data generated routinely in real time is only as good as the attributes that are tagged in such recording and very often there is no common key, other than locational or temporal, that enables data to be stitched together. Even if data can be linked with a locational referent or temporal index, this does not necessarily enable any cross-classification, and the datasets thus have limited applicability. When one turns to other data that are potentially big, such as mobile phone records which it is suggested correlate with travel patterns and from which it appears that travel patterns might be extracted, then the problems are even greater. Mobile phone calls are tagged to cell towers, and even on the assumption that such calls are tagged to the nearest tower (which is not necessarily the case), a very brave assumption is required that such data are correlated with actual traffic patterns. As with transit smart card data, much of the analysis of such data is exploratory and although there are interesting features and patterns that might be extracted, it is still an open question as to how useful such data might be for transport planning.

Despite all the hype about the smart city and the generation of big data from networks of sensors that are likely to be installed everywhere, none of this has resolved the basic problem that faces us in our understanding of cities and the means we have to predict and design their future. These are age-old problems compounded by the fact that cities are getting more complex, of which the smart cities movement and the embedding of technologies into the city are part and parcel but only one part of this increasing complexity. The notion proposed by Anderson (2008) that we no longer need theory and that all will be revealed in big data is incredulous and it is more than 10 years since he wrote this. During that time, there has been very little progress with respect to the development of new ideas coming from pattern recognition in big data. Most of what has been developed is highly routinised information technologies being embedded into what were once entirely mechanical systems such as cars, traffic lights and so on. One can never say that eventually new theory and analysis will not emerge from big data, for if we know anything, we know that the future in unknown and unpredictable, but perhaps even more so in an era when the speed at which inventions and innovations are taking place is continuously accelerating (Kelly, 2016).

Conclusion

The smart cities movement, and we call it a movement because it is so wide in scope, is being used as a label to cover a variety of opinions and ideologies about the city. These range from the provision of new information technologies being embedded into cities by the private sector that supply computer networks and software all the way to those who consider the major issues in cities to be problems of sustainability, regeneration, improving the quality of the environment and many of the traditional concerns of urbanists, planners and politicians who have a concern for creating a better life. In this chapter, we have argued that the smart city must be seen in terms of urban dynamics that cover all temporal and spatial scales, not just the 24 hour city that is dominated by big data, sensing and the generation of real time data but by the low frequency city.

Our argument that the tools we have been developing for the last half century need to be reinvigorated rather than replaced with inductive pattern seeking in the manner of machine learning and data mining, does not imply that there is no place for these new approaches but that they must complement and supplement rather than substitute for our current arsenal of techniques. The chapters in this book show how diverse the range of theories, ideologies, methods, models and tools to tackle the problems of the contemporary city are. What is required is a much more focused effort on adapting these tools to new circumstances, and this chapter has sketched some of the new directions in which this quest might proceed.

Note

1 Centre for Advanced Spatial Analysis (CASA) at University College London (UCL).

References

Anderson, C. (2008) The end of theory: The data deluge makes the scientific method obsolete, *Wired*, 23rd June, www.wired.com/2008/06/pb-theory/

Batty, M. (2005) *Cities and Complexity: Understanding Cities with Cellular Automata, Agent-Based Models, and Fractals*, MIT Press, Cambridge, MA.

Batty, M. (2017) Cities in Disequilibrium, In Johnson, J., Nowak, A., Ormerod, P., Rosewell, B., and Zhang, Y.-C. (eds) *Non-Equilibrium Social Science and Policy: Introduction and Essays on New and Changing Paradigms in Socio-Economic Thinking*, Springer, New York, 81–96.

Batty, M. (2018) *Inventing Future Cities*, MIT Press, Cambridge, MA.

Blum, A. (2013) *Tubes: Behind the Scenes at the Internet*, Penguin Books, New York.

Cairncross, F. (1997) *The Death of Distance: How the Communications Revolution Is Changing Our Lives*, Harvard Business School Press, Cambridge, MA.

Goldin, I., Koutroumpis, P., Lafond, F., Rochowicz, N., and Winkler, J. (2019) Why is productivity slowing down? www.oxfordmartin.ox.ac.uk/downloads/academic/201809_ProductivityParadox.pdf

Goodwin, T. (2015) The battle is for the customer interface, *Techcrunch*, 3rd March, https://techcrunch.com/

Kelly, K. (2016) *The Inevitable: Understanding the 12 Technological Forces That Will Shape Our Future*, Penguin Books and Random House, New York.

Kitchin, R., and Dodge, M. (2011) *Code/Space: Software and Everyday Life*, MIT Press, Cambridge, MA.

Moore, G. (1965) Cramming more components onto integrated circuits, *Electronics*, 38, 8, April 19, 114–117.

Reades, J., Zhong, C., Manley, E., Milton, R., and Batty, M. (2016) Finding pearls in London's oysters, *Built Environment*, 42, 3, 365–381.

Sterling, B. (2005) *Shaping Things*, MIT Press, Cambridge, MA.

Weiser, M. (1991) The computer for the 21st century, *Scientific American*, 265, 94–104.

6

Digital information and the right to the city

Joe Shaw and Mark Graham

The urbanization of information

Should we feed all the data for a given problem to a computer? Why not? Because the machine only uses data based on questions that can be answered with a yes or a no. And the computer itself only responds with a yes or a no. Moreover, can anyone claim that all the data have been assembled? Who is going to legitimate this use of totality? Who is going to demonstrate that the "language of the city", to the extent that it is a language, coincides with ALGOL, Syntol, or FORTRAN, the languages of machines, and that this translation is not a betrayal? Doesn't the machine risk becoming an instrument in the hands of pressure groups and politicians? Isn't it already a weapon for those in power and those who serve them?

(Lefebvre, 2003: 59)

In prioritizing urban space as the object of political struggle, Henri Lefebvre conceived of a *right to the city* as a broad and ambitious transformation of political life. Amongst other things, this demanded citizens grasp a renewed access and self-management of resources, surplus production and the urban core (Harvey, 2012; Lefebvre, 1968) – revolutionary demands that recognize the potential and value of city life to both citizens and capital alike, and in the most political sense (Purcell, 2002). Importantly, Lefebvre also called for a complementary *right to information* that would assist in facilitating a withering away of the state and superseding *metro-work-sleep* with a more egalitarian and fulfilling urban society (Lefebvre, 2003). He argued that such concepts would help dispense with the 'urban problematic' as produced and ideologically sustained by the forces of capitalism (Lefebvre, 2003).

Given that the world's urban population now has more access to information than ever before, and yet urban injustice persists en masse, we contend that information's complement to a right to the city is now a more complex aspect of political struggle than Lefebvre could perhaps realize at the time. This may have been due to the relative power of state actors at the time of his writing (Lefebvre, 2014a: 810, 1991: 285), or because of the less imaginable fact that citizens would one day not simple consume digital information but also come to produce it themselves – and in huge quantities (Lefebvre, 2014b).

So despite Lefebvre's often visionary and prescient concerns with statistical models, visualization logics and even early e-commerce (Lefebvre, 1991: 41; 2014a: 824), he did not quite foresee a reality where these technologies would run riot within powerful, affordable and geo-located computers that count our steps from our wrists, or monitor our sleep from our bedsides. Such ubiquity of digital information technology within the urban environment is very different from that of Lefebvre's century; and so in this chapter we argue that the *right to the city* now depends upon a better reading of today's critical phase in urbanization as a period where the city is increasingly reproduced through this digital information (Shaw and Graham, 2017a).

Since Lefebvre's death,[1] other scholars have begun to address the geographies of these dynamics. For example, Dodge and Kitchin (2011) have focused on the ways that computer code can shape how spaces are brought into being; and Graham et al. (2013) have discussed how digital information augments spatial experiences of the urban environment as a hybrid, densely layered and heavily mediated process. While none of these authors explicitly draw on Lefebvre, such conceptualizations of the spatiality of code and content serve as a useful starting point to begin thinking about the problematic entanglements between digital information and a Lefebvrian understanding of the *production of space* (Lefebvre, 1991: 26). Or, where each society's space is continuously produced through a triad of *lived* spatial practices; *perceived* experiences; and most crucially here, those *abstract* representations of space that are *conceived* in maps, drawings, schematics, ideas and information.

Such abstract space dominates our world today in a variety of ways. In particular, it plays a role in the production of space through concepts, techniques and technologies that range from naturalized ideas of neighbourhoods to real-time data visualizations, models, path-finding algorithms and now even Pokémon (Madden, 2014: 479; Kitchin *et al.*, 2015; Hern, 2016). None of which may seem *real* as such, but all of which can take a 'deeply, troublingly real' role in the (re)production of material space (Merrifield, 2015). And whilst the interactions between all three types of space come to embody a society's total social relations of reproduction, it is the role of this abstract space today that seems most relevant to digital information. Since from smartphone applications to GPS devices, Uber, Wikipedia and TripAdvisor, the code and content relating to the buildings and spaces of our cities can now seem as important as their bricks and mortar. And consequently, the power afforded by Lefebvre to traditional actors of urban power via representational technologies – developers, planners, landlords – is now rivalled by the rise of new corporate informational monopolies and platforms.

In this respect, much as the city has been conceptualized as the correlate of the road (Deleuze and Guattari, 1997), we argue that today it should be thought of as the correlate of the optic fibre. Urban society is now materially produced as a function of digitally networked informational circulation – a point defined by entries and exits. And with a focus on cities as key points of this information circulation, the wider urbanization process has now assembled information and communication technologies (ICTs) and people as a productive force that is both powerfully creative and planetary in scale (Brenner, 2014; Lefebvre, 2003: 173). To develop Lefebvre's right to the city in this context, we argue that the city must be read as unequivocally informational (Mattern, 2017), and with a renewed attention to these flows of digital information. Flows which are produced and mediated by a technology that further saturates the urban environment and yet which also retains the city as a primary site of experimentation and focus (Luque-Ayala and Marvin, 2016: 194; Tranos, 2013).

From a perspective of spatial justice and a right to the city, a key task of this reading is to critically examine the power relations around conduits of digital information as it becomes urban: the urbanization of information. In other words, just as the urbanization of water is a processional notion which can uncover 'stories about the city's structure and development', so can a reading of the city's 'political, social, and economic conduits' through which information flows also 'carry the potential for an improved, more just, and more equitable right to the city' (Swyngedouw, 2004: 4). The urbanization of information is now just as relevant to questions of spatial justice and the city as those which surround other historical infrastructures and commodities.

The conceptual division between a right to information and the right to the city is therefore problematic inasmuch as ICTs have become an integral part of everyday urban life. Digital information produces space and the urban environment; it is produced by all and circulates as a commodity which can be accumulated; and we have become increasingly dependent upon it as such. This raises many questions: what occurs when information becomes the urban? What spatial processes typify the reproduction of the digital-informational city? And how is this relevant to spatial justice and the right to the city? To begin to answer these questions we will broadly consider just one particular informational monopoly, Alphabet Inc.'s Google.

Google's right to the city

We argue that Google currently has the power to curate the right to the city through its dominance of digital urban information and therefore also of abstract space. Whilst they are certainly not the only actor that is relevant to this discussion,[2] it is Google's unrivalled position as the world's number one search engine (Andriole, 2018; European Commission, 2013; Jackson, 2015), as well as their many connected services across other domains including maps and mobile connectivity, which means that they occupy an especially powerful and privileged position in the *production* of urban space as described by Lefebvre (1991).[3] Their more recent entrance as an internet service provider (ISP) (e.g. Google Fiber), as well as a business and unelected city planning consultant supposedly concerned with today's 'big urban problems' (e.g. Google Digital Garage Bus, Sidewalk Labs), only compounds this power. But across all domains of activity, it is their massive aggregate power in digital spatial representation that can influence where people go, how and when they get there, what they do, the geography and characteristics of economic or social and political activities, and especially the way in which some parts of the city are made (in)visible. In the city they are now an ultimate 'infrastructural journalist of the everyday' (Luque-Ayala and Marvin, 2016); and their right to the city is also a power to choose how a city is reduced to information and to control the manner in which it is translated into knowledge and re-introduced to material everyday reality (Lefebvre, 1991: 137, 230). Therefore, everyday urban reality is being increasingly reproduced, read and experienced as the space of Google – an increasingly measured and quantifiable form of abstract space.

Despite the undeniable utility of Google's wide array of services – perhaps as an 'arena of practical actions' (Lefebvre, 1991: 288) – it must be understood that this abstract space has very real and unexpected (and often undesirable) consequences for the city. For example, as Google reduces the city to information, this process includes decisions which are (opaquely, seemingly magically) made on our behalf: which voices, bodies, gestures and paths to include and which to exclude, including who or which spaces are reproduced as 'outsider' or 'other' – or which 'databodies' to map to which 'codespaces' (Mattern, 2018). And despite their claimed objectivity (Google, 2015), the various interfaces of Google also ensure that not every place is seen the same, and not everyone sees the same place (Graham and Zook,

2013), and particular attention should be paid to the fact that such digital information is already produced, consumed and accumulated to reflect starkly uneven patterns (Graham *et al.*, 2015). Or, that there is often more digital content available about small European countries like Belgium than there is about the continents of South America or Africa – these data presences and absences reveal that digital geographies can be at least as uneven as their economic counterparts. In practical terms, this may mean that some classes of urban society will be better equipped to promote their vision of place over others (Zukin *et al.*, 2015); and as with other large digital platforms, this will undoubtedly come to leave its own particular mark of spatial-informational injustice on history.

Therefore, Google's right to the city concerns a power to reproduce and amplify a range of pre-existing spatial inequalities, and to potentially create new urban divides through this dominance of spatial information technologies. Less ubiquitous technologies that have warranted similar critical examination might include Microsoft's 'Avoid the Ghetto' patent or 'Pot Hole Apps' (Matyszczyk, 2012). In these examples, it is easy to hypothesize that poorer urban areas might become worse-off and those that are adequately digitally connected thrive in clean, safe and well-maintained streets. In Google's more-encompassing case, the outcome of these processes could include reproducing existing segregation through serving Hebrew and Arabic users with different search results (Graham and Zook, 2013); the definition of city attributes via semantic data (Ford and Graham, 2015, 2016); the making or breaking of small independent businesses (Poulsen, 2014); or where the disappearance of the word 'Chinatown' from a particular neighbourhood in Google Maps becomes a real moment of victory or defeat for groups battling over a particular urban core or *centrality* (Lefebvre, 2003; Tosoni and Tarantino, 2013).

By extension of this point, we argue that this filtering of material-social space can also alter its potential for political import (Dikeç, 2012). This is because any material manifestation of socio-spatial segregation[4] will reconfigure and nullify spatial politics as something which relies on encounters, possibilities and ruptures between different groups and individuals (e.g. Rancière, 1998): 'In order for space to have political import, it has to be associated in some way with change in the established order of things, leading to new distributions, relations, connections and disconnections' (Dikeç, 2012: 675). For Lefebvre this is described in a similar manner, and the possibility of encounters between diverse groups and individuals is itself a pathway to a right to the city, not to mention one of the great attractions of city life itself (Jacobs, 1961). Such a power to nullify dissensus exemplifies Lefebvre's theses surrounding the homogenizing tendencies of abstract space and the reductive nature of digital information (Lefebvre, 1991: 287, 2003: 59, 2014a: 811) – albeit both combined in a powerful new technology that thus represents a threat to urban democracy in its most proper sense (Rancière, 1998; Swyngedouw, 2011). So just as TripAdvisor or Yelp might have become a tyranny for the world of restaurateurs,[5] the 'neutral' and 'objective' algorithms of Google Maps can themselves become a hegemonic order of consensus for the broader urban population, spelling a death-knell for the enunciation of dissent.

Therefore, the regulatory processes behind Google – including algorithms, code and systems of data governance – now take a powerful role in the spatial processes of urban politics. They may privilege the spatial projects of some subjects or groups, and also play a role in the curation or foreclosing of political dissent for others. The increasingly opaque and complex nature of these algorithms might therefore be said to represent an ideological structure of power which is capable of producing space in much the same way that other regimes of regulatory power can (Mager, 2014; Ruppert, 2013). Hence, the (dis)ability to action change within the representational regime of Google becomes a (in)capacity to engage in the production of urban space itself.

Thinking about these examples also aptly demonstrate how digital information complicates Lefebvre's right to the city as a right to *centrality*,[6] since the assertion that 'centrality is always possible' is rendered moot by the ubiquity of Google's digital information (Lefebvre, 2003: 130): how can anyone begin to approach *habiting* within the urban periphery when Google is already quantifying where, when and how they go there (in real-time) – all with a view towards advertising these facts as such? Thus, over time, the flows of goods, services and people can be both monitored and re-aligned towards new cores of informational affluence – and away from informational peripheries – based on the algorithmic interpretation of gigantic databases controlled by a single monopolistic corporation. Lefebvre's (2003: 81) question of how to dwell in this quantifiable abstract space has never been more important, and the application of Google's technology to a range of domains ranging from transport to real estate investment (Shaw and Graham, 2017a, 2017b) all speak to a power 'to control space and even to produce it' (Lefebvre quoted in Elden, 2004: 84). Hence the landlord, developer and planning actors of Lefebvre's day are now complemented by a new type of actor in the production of urban space – one that endlessly operates upon urban social space through the medium of digital information. For example – and with the historical precedent of blockbusting[7] in mind (Hirsch, 2015) – Google now has the capacity to act as a type of algorithmic blockbuster for almost any major global city. But instead of slowly intervening in valuable urban socio-economic processes by distributing information on paper leaflets, the instant aggregate informational redirection of pedestrian flows and intentions on one side of the street can now impact those on the other – the 'beaten path' of urban consumption is reproduced by Google (Poulsen, 2014). Or in other words, Google's ability to mediate and redefine centrality is also a power to control flows of urban information as a productive and profitable force.

Therefore, Google has become a dominant force in the informational reproduction of urban space for the vast majority of cities. Particularly in the global North, it is Google that now occupies a type of informational right to the city (Shaw and Graham, 2017a), and it will be Google that can increasingly control a city's surplus production or best further their own vision and ideology of how it might develop. The current urban form is typified by an assemblage and distribution of digital information over which Google can preside – a position supported by the dominant ideology of an abstract and quantifiable space that is filled with connected digital devices and subjects. Just as the agora became the shopping mall in Lefebvre's reading of the city, now the people, objects and social relations within the shopping mall must all be digitally connected, quantified and informationally productive (Lefebvre, 2003: 9). This is the city of Google, but can we change it?

(Mis)understanding the right to the city

Lefebvre was optimistic that technology could improve everyday life in urban society. But he was also concerned by how powerful actors might control the discourse around any such potential in order to 'mask other, less obvious motives' (Lefebvre, 2003: 143). In terms of motives, Google is an information broker and advertising company with a $110 billion annual turnover in 2017 (Fiegerman, 2018). Their immediate and irrefutable interests concern profits, growth and a return on investment for shareholders. The corporation wholly dominates the global internet advertising arena, and their detailed (but relatively risk-free) capture of social processes and lived space is how they profit as an intermediary between production and consumption (Lefebvre, 2003: 101). And, when combined with powerful geospatial information from individual users, their vast resource of urban information can be used to create innovative and profitable advertising products like 'radius bidding' – where advertising strategies can be based upon a synthesis between an individual's profile data and exact real-time location (Google, 2014). As the corporation becomes better at producing such accumulated knowledge (Lefebvre, 1991: 137), Lefebvre would argue that such abstract space will be increasingly re-integrated into social practice and material production, resulting in an increasingly commodified urban space and an economy that valorizes certain relationships along the exact same lines (Lefebvre, 1991: 56). Or, that the truth regime of Google is fully integrated with the performances of normal daily life, giving us all occasion to raise the ultimate existential urban question: 'It's *not* on Google?!'

Fully unmasking these dynamics and their implications is crucial to developing a counter-strategy for the right to the city (Lefebvre, 2003: 144) – what are the commodity interests, politics and infrastructures of the urbanization of information? These facts are often masked, where as part of a broader infrastructure, the seemingly benign spatial representations provided by Google's services may often be politically overlooked due to the gradual banalization of their delivery through desktop computers, tablets, smart phones or other domestic things. All of which become commodified points-of-service where mediating technology disappears beneath clean white interfaces, communicates with mobile phone masts disguised as trees, or travels through cables buried deep beneath the road (Kaika and Swyngedouw, 2000). So just as water emerges from the tap (via dams, reservoirs and monitoring systems), the processes and operations enabled by Google mask an enormous process of operations and social relations from the gaze of the regular user.

Nevertheless, digital (and geospatial) urban information is being produced and exchanged as a valuable commodity, and so the most important basic questions still remain: 'Who produces information? How? For whom? And who consumes it?' (Lefebvre, 2014a: 813–816). In a period where the grand claims and cultural capital of Silicon Valley have resulted in a new 'information ideology' (Lefebvre, 2014a: 818) – and with it re-defined social understandings of risk, labour and value (Irani, 2015; Neff, 2012) – these are therefore often difficult questions to ask. But to transcend the fantasies behind these narratives and uncover the political economy of digital information is now crucial to locating the right to the city; whether amidst the operation of Google or indeed any large smart city provider.

However just as the political-economic narrative behind ICTs has shifted from a Keynesian concern with ameliorating the effects of capitalism, to talk of flexible market regulation and a prioritization for the technological ingenuity and innovation of forces like Silicon Valley (Ampuja and Koivisto, 2014), so too has the complicit position of the user as an information producer become apparent. The arena of practical actions provided by Google is a powerful invitation to individuals around the world to engage in the production of Google's valuable hoards of urban data. As a collective, city dwellers using Google have therefore become a significantly culpable audience who reproduce and enact the lived ideology of informational capitalism with each click, swipe or even 'gesture of intent' (e.g. Google's Soli Chip): 'The ideological superstructure and the economic base meet with, and feed, each other in every singly Google query', and so 'consent is reached by way of creating win-win situations that make individuals play by the rules of capitalism' (Mager, 2014: 32).

Assessing accountability for power in this networked context is difficult. If the use of Google is a 'decision for those at the base' (in everyday life) (Lefebvre, 2014a: 824), then why does it seem like there is such little opportunity for actions of resistance? We suggest that a better account of Google's power instead focuses on a situation whereby control is achieved precisely through such possible individual actions: Google's power draws on a self-governance where individual autonomy and action is what lies at the heart of disciplinary control (McNay, 2009: 56; Barnett *et al.*, 2008). 'Openness', 'democracy', 'user-feedback' and 'participation' are all part of the mode of governance for informational capitalism – it is precisely the invitation to behave as an individual that reproduces this power in the hands of an elite. Consenting to the fantasy that we can 'spread the love' if we 'put our cities on the map!' (Google, 2016) – or attempt to technically overcome the uneven urbanization of information by engaging with Google more to correct the situation – is no true path to the citizen's right to the city.

Taken together, the highly valuable nature of digital information as a commodity and the difficulty in mobilizing political acts of disagreement (yet still engaging with digital technology) present enormous challenges for the right to the city. Attempts to regulate Google's activities by states and international cooperation might be one way to tame their particular form of information-centric capitalism (and regulation *is* important too), but it remains the case that large digital platforms in general depend on mantra such as 'The More We Connect The Better It Gets' (Facebook). This critically calls into question initiatives to redistribute digital technology more evenly, whether through the magical promises of free laptops, new submarine cables or hot air balloon Wi-Fi (for example, Google's Project Loon). In other words, justice in informational distribution is not the default ideal order in the face of informational injustice (Badiou quoted in Dikeç, 2002: 96); and the right to the city cannot simply appear as a tamed form of data-driven capitalism. There is something that is always fundamentally in conflict with a citizen's right to the city as long as they continue to engage with a corporate monopoly like Google; and this reflects criticism of other misunderstandings around the right to the city as either post-political or liberalized to the point of being stripped of meaning (Purcell, 2002). Only immediately seeking to supersede Google entirely is an ambition equal to *our* right to the city. This right to the city demands a strategy to fully 'escape the quantifiable' and to discover an alternative ICT-based 'path to the possible' (Lefebvre, 2003: 185, 6): working for a Google-free city.

Working for a Google-free city

The political utility of a concept is not just in providing better understanding of reality, it also needs to enable an experimentation with reality that reveals new possibilities and openings (Merrifield, 2011). In this respect, the original right to the city has been misunderstood as an impotent and post-political directive that does not require that we get rid of the city of actors like Google, and that they can still somehow 'improve life in cities for everyone' (Google Sidewalk Labs, 2016; Souza, 2010). A present failure to traverse the ideologies of Silicon Valley, 'start-up culture' and wider technological determinism has resulted in a situation where disagreement with Google's particular infrastructures and assemblages of spatial production is extremely difficult; complicated by the alienation of a system predicated on individual action and characterized by a lack of effective collective action. What might benefit the urban dweller's right to the city amidst this scenario? Where might we find a 'right to the smart city' (CAG, 2016) and disagree with this commodification and simulation of social space through the medium of digital information?

A straightforward reading of Lefebvre suggests that the technology needs first to be somehow re-appropriated (Elden, 2004: 152); and small acts of such re-appropriation or resistance have long been abundant around digital technology and information. For example, tools which have been developed to enable users to destroy or devalue their own data (e.g. Web 2.0 Suicide Machine) could yet be applied to the city and urban space. Perhaps revitalized notions of *détournement* (Debord, 1967) need to be re-applied in the present; and perhaps urban dwellers could critique processes of data-driven gentrification by fabricating several hundred non-existent artisanal cup-cake bakeries in the South Bronx on Google Maps (alongside their fantastically gushing reviews on Yelp). Or they might produce a hoard of bots to troll the conspicuous consumption of Central London (Dwyer and Shaw, 2017). All acts which *might* mischievously disrupt or disagree with the 'ideo-logic' of abstract substitution or visualization (Lefebvre, 2003: 183) to espouse a spontaneous politics of flux and vitality (Lash, 2005), and when pursued collectively can demonstrate (perhaps like the Anonymous collective) 'the importance of art, expression, autonomy, and creation through unalienated labour' (Coleman, 2014: 270).

The problem here however is that these acts usually require significant skills, are resisted by Google and other corporate platforms, and take a lot of work to bring into being – time and resources which might only be available to the privileged classes mentioned above; many of whom currently labour continuously to directly increase the power of Google themselves. It is also the case that these acts are themselves often predicated on a logic that prioritizes the individual and the continued simulation of societal relations. In other words, we cannot all join Anonymous on the 4chan forum (not least because we are mostly 'newfags'), and second, these actions often have little more political motivation than 'for the lulz'.[8] This often risks committing little more than appropriation for appropriation's sake, and would arguably produce a city that is ultimately no more egalitarian than the city of Google – and perhaps where all sorts of 'moralfaggotry' would be unwelcome (Coleman, 2014: 62). This is where Lefebvre's questions of how we might begin to replace *habitat* with *habiting* arises: how do we replace the ideology of habitat ('*Your data-driven life online!*') with a practice of actually *living with* the digital information we produce and the tools we produce it with?

Beyond such isolated instances of re-appropriation and participation, Lefebvre also focused on a sustained practice of *autogestion* (Lefebvre, 2003: 150): the self-management of technologies, resources and surplus. To this end, there has been a resurgence in reference to the value of commons-based digital platforms or 'working class ICTs' (Fuchs, 2014). Such ideas have helped re-contextualize ICTs within a class struggle that is yet to be rendered obsolete by the grand promises of these technologies (Dyer-Witheford, 2015: 9). And deploying such traditional systems in a new setting – for example, in Uber driver unions or municipal Airbnb-style coops – might help re-imagine Google's functions as a collectively owned and operated technology for the good of the city. Hence a focus on the labour relations around digital urban information remains important.

Therefore, commons-based platforms like Wikipedia continue to merit attention as relatively open mechanisms for encouraging a broad-base of participation in the creation of a diverse range of geographic information. They release all core platform data freely to the public and have transparent mechanisms for resolving conflicts about how places, people and processes should be represented. This still takes work (Ostrom, 1990), but the skills of the privileged few are perhaps better directed at these collective projects that foster the emergence of cooperative urban platforms over corporate ones like Google. An informational right to the city can therefore begin to be located in projects like Public Code (http://smart cities.publiccode.net), OpenStreetMap and the work being done towards the end of a cooperative cloud (e.g. webarchitects.coop). Use of, contribution to and support of these diverse projects is vital if we are to ultimately solidify and produce an urban commons around any particular city as conceived by its density of relations across all of Lefebvre's perceived, lived and abstract space.

Any notion of living well in the city also invokes Lefebvre's cry for the right to the *oeuvre* (Lefebvre, 1996: 65): the demand that technologies should be employed as much for human works of joy (*oeuvres*) as for outright profits. Discovering such a deliberated and joyful production of information also undoubtedly requires *more work*, but we (and others) contend that a less-alienated informational experience of the city will ultimately be the better one (Purcell, 2017). Or, that the mountaineer's slow but self-mastered ascent is a relatively worthy pursuit in light of the predetermined speed of the ski-lift rider (Lefebvre quoted in Elden, 2004: 133). To this end, many worthwhile productions of informational *oeuvres* already exist: in 'sousveillance' or 'Citizen Science' projects that address an array of social, political and environmental concerns (e.g. Public Lab, SPLASSH, the Air Quality Egg) (Cohn, 2008; Mann *et al.*, 2002); or in projects like Dewey Maps that build upon globally maintained open source software with locally focused interests that might 'bring together practical information to *live well* locally without breaking the bank'.

Furthermore, whereas platforms like Wikipedia might be criticized for their growing professionalism, many of these smaller projects foster the powerful enjoyment of labour by the *amateur* (Merrifield, 2015). And, such triumphs might even arrive in the success of those smaller, more locally embedded businesses. For example, Carteiro Amigo ('Friendly Mailman'), a novel and much-needed addressing system and postal service in Rocinha, a Rio *favela*: 'Google came by here last month. They asked if they could take a photo of our map. I said: "No way." Let them do their own' (Mier, 2014). Such local initiatives demonstrate that not everyone is willing to be another passive user in the battle to control the legibility of the city – and some are still *enjoying* the fight. In each case, the *autogestion* of urban information has been realized as a worthwhile activity which can contribute to a never-ending struggle towards greater empowerment and democracy for all urban citizens. This is working for a Google-free city.

In this light it should be considered that the early days of the internet were almost everywhere an *oeuvre* (dominated by skilled, empowered, fun-loving and explorative community of creative amateurs and geeks) compared with today's electronic drudgery (dominated by rules and IT professionals; populated by individuals who are largely un-skilled and disempowered; often experienced as a boring necessity that is nevertheless constantly advertised as exhilarating – if only we believe the latest superfast broadband adverts!). Similarly, the problem is that our city's information and abstract space is now being managed for us as a part of someone else's *oeuvre*: an *oeuvre* of the venture capitalist, the bureaucrat or the Smart City evangelist. And often, as an *oeuvre* that belongs to Google.

Therefore, a *right to the city* must employ all of the above – re-appropriation and participation as part of a sustained *autogestion* – as well as to behold and recover the digital information *oeuvre* as a worthwhile pursuit. We must actively work towards enjoying the practice of producing and managing our urban information. It will never be enough to expect Google to provide you with a joy of your own.

Conclusion

The concept of an informational right to the city is useful in understanding the power of an informational monopoly like Google. Their control over a newly ubiquitous form of digital abstract space enables them to reproduce and control urban space itself. In this capacity, they have now joined – and in some cases, perhaps even superseded – the ranks of urban planners, developers and landlords from Lefebvre's era in terms of their power over the city, as well as its many joys and problems. Similarly, this power is also masked by a newly dominant ideology of Google as technology serving the 'general interest' of the city (Lefebvre, 2014a: 253): we can 'spread the love' if we 'put our cities on the[*ir*] map!' (Google, 2016). As such forces begin to re-shape the city, these powers and discourses merit critical attention through the lens of Lefebvre, and his theory on the production of space provides a strong starting point.

However, the dependence of these new technologies of abstract space upon vast flows of digital information also demonstrates that Lefebvre's original separation of a right to information and a right to the city is problematic. Concatenating the two has helped us to re-think both the relationship of information to a right to the city, and the way that such technology complicates and challenges Lefebvre's original thesis on urbanization, power and space. Urban citizens are no longer simply passive information consumers in a time of a powerful regulatory state. Instead, through the user's complicit and near-constant production of urban information, Google can increasingly control urban centralities and political representations; homogenize urban space; embed abstract advertising products in material space; prioritize and valorize some urban subjects or relations over others; harness surplus production through technological innovation; and dominate the digital process of reducing concrete social practices to abstract information. Yet, we live in a world where even the poorest can possess a Google-ready smartphone, producing Google-ready data. What role will the more passive users and information producers take in the urbanization of information?

In trying to capture an appropriately political understanding of what an informational *right to the city* might entail, this consideration of Google's right to the city has highlighted that traversing and unmasking the fantasies and ideologies around their operation is crucial to resisting such power. This also follows for a host of other corporate urban platforms and smart city projects. All can threaten the urban dweller's capacity to politically disagree with their particular forms of abstract space and operation; and finding ways to disagree is the embodiment of potential for a better urban condition. This disagreement will take work, but particular paths to the right to the city today include the laborious (but potentially joyful) pursuit of *autogestion* through cooperative and commons-based models. Or, the production of new, non-corporate and democratic platforms for our digital urban information. If we want cities that are built on models of openness, participation, representativeness, inclusiveness, collective freedoms and the simply joy that comes with the playfulness of space, then Henri Lefebvre would have likely said that ultimately it has to be Google-free.

Notes

1 Just 38 days before the creation of the first webpage.
2 For example, Facebook, Strava, Tripadvisor and Foursquare all possess some similar types of powers, functions and traits as those which will be discussed here in relation to Google.
3 Google most obviously produces urban space through its Google Maps platform. Google Maps brings together road maps, satellite imagery, a variety of urban place information, traffic data, routing algorithms, and user-generated content in order to provide an interface to the city. It is used for anything from finding reviews of pizza restaurants to museum opening hours to figuring out which bus to take home.
4 For example, as a spatial 'Googlearchy' (Hindman, 2008) or 'filter bubble' (Pariser, 2011).
5 Not only might TripAdvisor make or break small hospitality businesses, but the bimodal tendency of such review systems to award a place either one or five stars is in itself a tyranny that naturally reduces nuances and complexities of value to simple pass/fail outcomes (Shaw, 2015). The internet does not produce many three-stars-out-of-five reviews.
6 Urban centrality is a key concept for Lefebvre (2003), and seemingly suggests access and participation to/within the city's core(s) and central resources. However, Andy Merrifield (2011) argues that it is also an existential notion as well as a geographical one. This seems to represent another instance where Lefebvre's more singular notion of power struggles with contemporary urbanization; and we suggest that Google's power is very much a phenomenon that renders Lefebvre's account of power somewhat archaic.

7 The term 'blockbusting' refers to the surreptitious practices of US real estate agents in the mid-20th century, which included a range of informational tactics to encourage fearful white residents/families to sell their homes quickly for a low price, and then sell them on to black residents/families at a profit (Hirsch, 2015). For example, through leaflet distribution, false greeting letters from 'new' black neighbours, using white 'proxy buyers', or the hiring of agent provocateurs from racial minorities.

8 Usually meaning laughter at someone else's expense – where decisions as to *whose* expense are often just as opaque or arbitrary as Google's algorithms.

References

Ampuja, M., & Koivisto, J. (2014). 'From "Post-Industrial" to 'Network Society" and Beyond'. *tripleC*. 12: 447–463.

Andriole, S. (2018). 'Apple, Google, Microsoft, Amazon And Facebook Own Huge Market Shares = Technology Oligarchy'. *Forbes*. www.forbes.com/sites/steveandriole/2018/09/26/apple-google-microsoft-amazon-and-facebook-own-huge-market-shares-technology-oligarchy

Barnett, C., Clarke, N., Cloke, P., & Malpass, A. (2008). 'The Elusive Subjects of Neo-Liberalism'. *Cultural Studies*, 22/5: 624–653. DOI: 10.1080/09502380802245902

Brenner, N. (2014). *Implosions/Explosions: Towards a Study of Planetary Urbanization*. Berlin, BE: JOVIS Verlag.

CAG. (2016). *(Re)Prioritizing Citizenship: Setting a new agenda for Smart Cities Governance*. Chennai: CAG. https://socialsmartcities.files.wordpress.com/2016/12/whosesmartcity_chennai-workshop-report-2016.pdf

Cohn, J.P. (2008). 'Citizen Science: Can Volunteers Do Real Research?' *BioScience*, 58/3: 192–197. DOI: 10.1641/B580303

Coleman, G. (2014). *Hacker, Hoaxer, Whistleblower, Spy: The Many Faces of Anonymous*. London: Verso.

Debord, G. (1967). *La Société du Spectacle*. Paris: Buchet/Chastel.

Deleuze, G., & Guattari, F. (1997). 'City/State'. In Leach N. (ed.) *Rethinking Architecture*, London: Routledge: 296–299.

Dikeç, M. (2002). 'Police, Politics, and The Right to The City'. *GeoJournal*, 2–3: 91–98.

Dikeç, M. (2012). 'Space as a Mode Of Political Thinking'. *Geoforum*, 43/4: 669–676. DOI: 10.1016/j.geoforum.2012.01.008

Dwyer, A., & Shaw, J. (2017). 'Place-Faking: Fermenting Resistance Through Digital Productions of Space'. The Museum of Contemporary Commodities: creative propositions and provocations on the heritages of data-trade-place-value. Presented at the RGS-IBG Annual Conference, London, UK.

Dyer-Witheford, N. (2015). *Cyber-proletariat*. London: University of Chicago Press.

Elden, S. (2004). *Understanding Henri Lefebvre: Theory and the Possible*. Continuum studies in philosophy. London: Continuum.

European Commission. (2013). Press Release – Commission Seeks Feedback on Commitments Offered by Google to Address Competition Concerns – Questions and Answers. Brussels: European Commission. http://europa.eu/rapid/press-release_MEMO-13-383_en.htm

Fiegerman, S. (2018). 'Google posts its first $100 billion year'. *CNNMoney*. https://money.cnn.com/2018/02/01/technology/google-earnings/index.html

Ford, H., & Graham, M. (2015). 'Semantic Cities: Coded Geopolitics and the Rise of the Semantic Web'. In Kitchin R. & Perng S.-Y. (eds) *Code and the City*, London: Routledge: 1–37.

Ford, H., & Graham, M. (2016). 'Provencance, Power and Place: Linked Data and Opaque Digital Geographies'. *Environment and Planning D*, DOI: 10.1017/CBO9781107415324.004

Fuchs, C. (2014). *Digital Labour and Karl Marx*. London: Routledge.

Google. (2014). 'Understanding Consumers' Local Search Behavior'. http://think.storage.googleapis.com/docs/how-advertisers-can-extend-their-relevance-with-search_research-studies.pdf

Google. (2015). 'Ten Things We Know To Be True'. www.google.com/about/philosophy.html

Google. (2016). 'Get Your Business Online'. Google. http://gybo.com/

Google Sidewalk Labs. (2016). 'Sidewalk Labs'. www.sidewalkinc.com

Graham, M., Sabbata, S., & Zook, M. (2015). 'Towards a Study of Information Geographies: Immutable Augmentations and a Mapping of the Geographies of Information'. *Geo: Geography and Environment*, 2/1: 88.

Graham, M., & Zook, M. (2013). 'Augmented Realities and Uneven Geographies: Exploring the Geolinguistic Contours of the Web'. *Environment and Planning A*, 45/1: 77–99. DOI: 10.1068/a44674

Graham, M., Zook, M., & Boulton, A. (2013). 'Augmented Reality in Urban Places: Contested Content and the Duplicity of Code'. *Transactions of the Institute of British Geographers*, 38/3: 464–479. DOI: 10.1111/j.1475-5661.2012.00539.x

Harvey, D. (2012). *Rebel Cities: From the Right to the City to the Urban Revolution*. London: Verso.

Hern, A. (2016). Pokémon Go: Who owns the virtual space around your home? *The Guardian*. https://www.theguardian.com/technology/2016/jul/13/pokemon-virtual-space-home

Hindman, M. (2008). *The Myth of Digital Democracy*. New Jersey: Princeton University Press.

Hirsch, A. (2015). 'Blockbusting'. www.encyclopedia.chicagohistory.org/pages/147.html

Irani, L. (2015). 'Hackathons and the Making of Entrepreneurial Citizenship'. *Science, Technology & Human Values*, 40/5: 799–824. DOI: 10.1177/0162243915578486

Jackson, G. (2015). 'Datawatch: Global search market share'. *FT Blogs*. Retrieved September 1, 2015, from http://blogs.ft.com/ftdata/2015/04/17/datawatch-global-search-market-share/

Jacobs, J. (1961). *The Death and Life of Great American Cities*. Vintage book New York: Random House.

Kaika, M., & Swyngedouw, E. (2000). 'Fetishizing the Modern City: The Phantasmagoria of Urban Technological Networks'. *International Journal of Urban and Regional Research*, 24/1: 120–138.

Kitchin, R., & Dodge, M. (2011). *Code/space: Software and Everyday Life*. University Press Scholarship Online. Cambridge, MA: MIT Press.

Kitchin, R., Lauriault, T.P., & McArdle, G. (2015). 'Knowing and Governing Cities Through Urban Indicators, City Benchmarking and Real-time Dashboards'. *Regional Studies, Regional Science*, 2/1: 6–28. DOI: 10.1080/21681376.2014.983149

Lash, S. (2005). 'Lebenssoziologie: Georg Simmel in the Information Age'. *Theory, Culture & Society*, 22/3: 1–23. DOI: 10.1177/0263276405053717

Lefebvre, H. (1968). *Le droit à la ville*. Société et Urbanisme. Paris: Anthropos.

Lefebvre, H. (1991). *The Production of Space*. Oxford: Blackwell.

Lefebvre, H. (1996). *Writings on Cities*. Oxford: Blackwell.

Lefebvre, H. (2003). *The Urban Revolution*. Minneapolis: University of Minnesota Press.

Lefebvre, H. (2014a). *Critique of Everyday Life*. Single-vol. London: Verso.

Lefebvre, H. (2014b). 'Dissolving City, Planetary Metamorphosis'. *Environment and Planning D: Society and Space*, 32/2: 203–205. DOI: 10.1068/d3202tra

Luque-Ayala, A., & Marvin, S. (2016). 'The Maintenance of Urban Circulation: An Operational Logic of Infrastructural Control'. *Environment and Planning D: Society and Space*, 34/2: 191–208. DOI: 10.1177/0263775815611422

Madden, D. J. (2014). 'Neighborhood as Spatial Project: Making the Urban Order on the Downtown Brooklyn Waterfront'. *International Journal of Urban and Regional Research*, 38/2: 471–497. DOI: 10.1111/1468-2427.12068

Mager, A. (2014). 'Defining Algorithmic Ideology: Using Ideology Critique to Scrutinize Corporate Search Engines'. *tripleC*, 12/1: 28–39.

Mann, S., Nolan, J., & Wellman, B. (2002). 'Sousveillance: Inventing and Using Wearable Computing Devices for Data Collection in Surveillance Environments'. *Surveillance & Society*, 1/3: 331–355.

Mattern, S. (2017). *Code and Clay, Data and Dirt: Five Thousand Years of Urban Media*. Minneapolis: University of Minnesota Press.

Mattern, S. (2018). 'Databodies in Codespace'. *Places Journal*, from https://placesjournal.org/article/databodies-in-codespace/

Matyszczyk, C. (2012). 'The joy of Microsoft's "avoid ghetto" GPS patent'. *CNET*. Retrieved May 20, 2007, from www.cnet.com/uk/news/the-joy-of-microsofts-avoid-ghetto-gps-patent

McNay, L. (2009). 'Self as Enterprise: Dilemmas of Control and Resistance in Foucault's The Birth of Biopolitics'. *Theory, Culture & Society*, 26/6: 55–77. DOI: 10.1177/0263276409347697

Merrifield, A. (2011). 'The Right to the City and Beyond'. *City*, 15/3–4: 473–481. DOI: 10.1080/13604813.2011.595116

Merrifield, A. (2015). 'Future Shock'. http://antipodefoundation.org/2015/03/18/future-shock

Mier, B. (2014). 'How One Startup Mapped Brazil's Confusing favelas'. *Motherboard*, from http://motherboard.vice.com/read/the-mailman-mapping-brazils-largest-favela-by-hand

Neff, G. (2012). *Venture Labor: Work and the Burden of Risk in Innovative Industries*. University Press Scholarship Online. Cambridge, MA: MIT Press.

Ostrom, E. (1990). *Governing the Commons*. Cambridge: Cambridge University Press.

Pariser, E. (2011). *The Filter Bubble: What the Internet is Hiding from You*. London: Viking.

Poulsen, K. (2014). 'How Google Map Hackers Can Destroy a Business at Will'. *Wired*. www.wired.com/2014/07/hacking-google-maps

Purcell, M. (2002). 'Excavating Lefebvre: The right to the city and its urban politics of the inhabitant'. *GeoJournal*, 2–3: 99–108.

Purcell, M. (2017). 'The City Is Ours (If We Decide It Is)'. In Shaw, J. & Graham, M. (eds) *Our Digital Rights to the City*. London: Meatspace Press: 30–33.

Rancière, J. (1998). *Disagreement: Politics and Philosophy*. Minneapolis: University of Minnesota Press.

Ruppert, E.S. (2013). 'Rights to Public Space: Regulatory Reconfigurations of Liberty'. *Urban Geography*, 27/3: 271–292. DOI: 10.2747/0272-3638.27.3.271

Shaw, J., & Graham, M. (2017a). 'An Informational Right to the City? Code, Content, Control, and the Urbanization of Information: An Informational Right to the City?'. *Antipode*, 49/4: 907–927. DOI: 10.1111/anti.12312

Shaw, J., & Graham M. (eds) (2017b). *Our Digital Rights to the City*. London: Meatspace Press.

Shaw, Z. (2015). 'A Few Interesting Ratings System Observations'. From http://zedshaw.com/archive/a-few-interesting-ratings-system-observations/

Souza, M.L.D. (2010). 'Which Right to Which City? In Defence of Political-Strategic Clarity'. *Interface*, 2/1: 315–333.

Swyngedouw, E. (2004). *Social Power and the Urbanization of Water*. Oxford SRC, GoogleScholar: OUP.

Swyngedouw, E. (2011). *Designing the Post-Political City and the Insurgent Polis*. London: Bedford.

Tosoni, S., & Tarantino, M. (2013). 'Space, Translations and Media'. *First Monday*, 18(11–4), from http://journals.uic.edu/ojs/index.php/fm/article/view/4956/3788

Tranos, E. (2013). *The Geography of the Internet : Cities, Regions and Internet Infrastructure in Europe*. New horizons in regional science. Cheltenham: Edward Elgar.

Zukin, S., Lindeman, S., & Hurson, L. (2015). 'The Omnivore's Neighborhood? Online Restaurant Reviews, Race, and Gentrification'. *Journal of Consumer Culture*, 0/0: 1–21. DOI: 10.1177/1469540515611203

Shaping participatory public data infrastructure in the smart city

Open data standards and the turn to transparency

Tim Davies

Introduction

One of the earliest expressions of the modern open data movement evolved out of the CitiStat analytical dashboard in Washington DC. In 2007, seeking to allow actors outside the administration work with city data, the DC chief technology officer launched data.dc. gov, a data portal offering direct access to machine-readable government datasets (Tauberer, 2014, Ch. 1). Over the following decade, the idea that governments should publish their data holdings as machine-readable, freely accessible and openly licensed data has taken hold across the world at national and sub-national levels (Davies *et al.*, 2019). Open data ideas have found particular traction in urban areas, initially connecting with a culture and practice of civic hacking (Landry, 2019), and open data has also acted as a component of a number of smart cities programmes, both rhetorically and substantively. Scholars have been particularly interested in the potential uses of open data to support democratic engagement and collaborative models of governance in the smart city (Bartenberger & Grubmüller-régent, 2014; Goldsmith & Crawford, 2014). Yet open data communities have also been the source of a number of critical perspectives on the smart city: questioning centralisation and corporate control of urban data infrastructure, and challenging the presentation of a rationalised urban domain filled with consumers and service recipients, rather than a rich urban environment of diverse citizens, political struggles and lives only partially digitised (e.g. Sadoway & Shekhar, 2014).

One source of this dual role of open data, in both enabling and opposing smart city narratives, can be traced to the complex origins of the open data movement, which brought together both public sector information businesses and civil society activists, united by a common cause in gaining access to government data, but ultimately motivated by divergent long-term goals (Gonzalez-Zapata & Heeks, 2015; Gray, 2014). These unusual allies ranged from those seeking transparency, accountability and new forms of civic engagement (Davies, 2010; Huber & Maier-rabler, 2012; Kassen, 2013; Sieber & Johnson, 2015), to those looking to develop new business models and promote the idea of 'government as a platform' outsourcing many more aspects of service provision to the private sector (Gurin, 2014).

In this chapter, my goal is to explore open data-related strategies for re-asserting the role of citizens within the smart city. Such strategies are able draw in particular on the political narrative of transparency, and the role of technical standards in delivering transparency. I will outline how these two components can be used not only to secure access to data from government, but also to open up two-way communication channels between citizens, states and private providers. I will argue that a focus on opening up the data infrastructures of the smart city not only offers the opportunity to make processes of governance more visible and open to scrutiny, but it also creates a space for debate over the collection, management and use of data within governance. This can give citizens an opportunity to shape the data infrastructures that do so much to shape the operation of smart cities and of the modern data-driven policy environment.

The chapter proceeds in four parts, the first three unpacking different aspects of the title, and the forth offers a model for thinking about the relationship between transparency, open data and standards in the future development of inclusive and participatory data practice in the smart city.

Part 1: Participatory public data infrastructure

Data infrastructure

Infrastructures provide the shared set of physical and organisational arrangements upon which everyday life is built. The notion of infrastructure is central to conventional imaginations of the smart city. Fibre-optic cables, wireless access points, cameras, control systems and sensors embedded in just about anything, constitute the digital infrastructure that feed into new, more automated, organisational processes. These, in turn, direct the operation of existing physical infrastructures for transportation, the distribution of water and power, and the provision of city services. However, between the physical and the organisational lies another form of infrastructure: data and information infrastructure.

Although in the literature the term 'information infrastructure' is often used to cover both data and information, I use the two terms separately here to draw attention to an important analytical distinction. The General Definition of Information (GDI) describes information as 'data + meaning' (Floridi, 2004). Information, as the basis for human decision-making, requires data that is filtered, organised and contextualised. Data, by contrast, is, in its purest form, decontextualised: with each individual aspect of a phenomena encoded as a distinct data point, open to be re-assembled and represented as information, but also open to a range of different representations and forms of analysis. It is this re-interpretability that gives digital data its particular value. Of course, in practice a digital dataset rarely encodes all the possible variables that describe a phenomenon; instead, certain features of the world are selected for encoding and others discarded. Even with growing data storage and processing capacity, the need for this explicit or implicit selection is not avoided.

It is by being rendered as structured data that signals from the myriad sensors of the smart city, or the submissions by hundreds of citizens through reporting portals, are turned into management information and fed into human- or machine-based decision-making, and back into the actions of actuators (Dunleavy et al., 2006) within the city. Seen as a set of physical or digital artefacts, the data infrastructure of a city involves ETL (Extract, Transform, Load) processes, APIs (Application Programming Interfaces), databases and data warehouses, stored queries and dashboards, schema, codelists and standards. Seen as part of a wider 'data assemblage' (Kitchin & Lauriault, 2014) this data infrastructure also involves various processes of data entry and management (Denis & Goëta, 2017; Goeta & Davies, 2016), of design, analysis and use, as well relationships to other external datasets, systems and standards. Dodds and Wells capture this by defining data infrastructure to incorporate not only data assets, such as datasets, identifiers and registers, but also the organisations and organisational processes used to provide access to those assets (Dodds & Wells, 2019)

It is, however, often very hard to 'see' data infrastructure. By their very natures, infrastructures move into the background, often only 'visible upon breakdown' (Star & Ruhleder, 1996). For example, you may only really pay attention to the shape and structure of the road network when your planned route is blocked. It takes a process of 'infrastructural inversion' to bring information infrastructures into view (Bowker & Star, 2000), deliberately foregrounding what has been so far the background. I will argue shortly that 'transparency' as a policy performs much the same function as 'breakdown' in making the contours of infrastructure more visible. In taking something created with one set of use-cases in mind, and placing it in front of a range of alternative use-cases, transparency allows the affordances and limitations of a data infrastructure to be more fully scrutinised. Such critical scrutiny can then feed into shaping the future development of that infrastructure. But before developing that argument further, I will first outline the extent of 'public data infrastructure' and the different ways in which we might understand the idea of a 'participatory public data infrastructure'.

Public data infrastructure

A city may have a wealth of public data yet without a coherent public data infrastructure. In *The Responsive City*, Goldsmith and Crawford describe the status quo for many as 'The century-old framework of local government – centralized, compartmentalized bureaucracies that jealously guard information' (Goldsmith & Crawford, 2014). In such a city, datasets may exist, but they are disconnected. For example, one department may maintain detailed digital maps showing the land the city is responsible for maintaining and city assets, whilst another manages contracts with outsourced maintenance crews, who work from entirely separate maps. Even so, such a city may have come to publish some data online in an open data portal in response to transparency edicts, but each dataset exists as an island, published using different formats and structures, without any attention to interoperability

It is against this background that initiatives to construct public data infrastructure have sought to introduce shared technology, standards and practices that provide access to a more coherent collection of data generated by, and focusing on, the public tasks of government. For example, in 2012, Denmark launched their 'Basic Data' programme, looking to consolidate the management of geographic, address, property and business data across government, and to provide common approaches to data management, update and distribution (Government of Denmark, 2012). In the European Union, the INSPIRE Directive and programme has been driving creation of a shared 'Spatial Data Infrastructure' since 2007, providing reference frameworks, interoperability rules and data sharing processes (Bartha & Kocsis, 2011). And more recently, the UK government launched a 'Registers programme' (Miller & Roe, 2018) to create centralised reference lists and identifiers of everything from the names of countries to a list of government departments, framed as part of building governments digital infrastructure. At the city level, increasing adoption of smart city technology, particularly when cities choose to work with a range of vendors, is likely to drive greater development of urban data infrastructure and frameworks for city data interoperability.

The creation of these data infrastructures can clearly have significant benefits for both citizens and government. For example, instead of citizens having to share the same information with multiple services, often in subtly different ways, through a functioning data infrastructure governments can pick up and share information between services, and could provide a more joined up experience of interacting with the state. By sharing common codelists, registers and datasets, agencies can end the duplication of effort and increase their intelligence, drawing more effectively on the data that the state has collected.

However, at the same time, these data infrastructures tend to have a particularly centralising effect. Whereas a single agency maintaining their own dataset has the freedom to add in data fields or to restructure their working processes in order to meet a particular local need, when that data is managed as part of a centralised infrastructure, their ability to influence change in the way data is managed will be constrained both by the technical design and the institutional and funding arrangements of the data infrastructure (Miller & Roe, 2018). A more responsive government is not only about better intelligence at the centre, it is also about autonomy at the edges, and this is something that data infrastructures need to be explicitly designed to enable, and something that they are generally not oriented towards.

An analogy may be useful to illustrate the tensions present in how data infrastructure is designed. In the eighteenth century, Britain underwent an infrastructural transformation, with the development of a national network of roads and highways (Guldi, 2012). As metaled roads spread out across the country there were debates over whether to use local materials, which were easy to maintain with local knowledge, or to apply a centralised 'tarmacadam' standard to all roads. There were questions of how the network should balance the needs of the majority, with road access for those on the fringes of the country, and how the infrastructure should be funded. The shape, structure and design of this public infrastructure were highly contested, and the choices made over its design have had profound social consequences. Whilst Guldi uses the road network as an analogy for debates over modern internet infrastructures, it can also illuminate questions around equally intangible public data infrastructure.

If you build roads to connect the largest cities but leave out a smaller town, the relative access of people in that town to services, trade and wider society is diminished. In the same way, if your data infrastructure lacks the categories to describe the needs of a particular population, their needs are less likely to be met. Yet, a town might equally reject the idea of being connected directly to the road network, feeling that this might see its uniqueness and character eroded, much like some groups may also want to resist their categorisation and integration in the data infrastructure in ways that restrict their ability to self-define and develop autonomous solutions in the face of centralised data systems that are necessarily reductive (c.f. Rainie et al., 2019).

Another debate that was central to the development of early transport infrastructure also resonates today for data infrastructure. That is the question of ownership, access and the public or private provision of infrastructure. Increasingly our nominally public data infrastructures may rely upon stocks and flows of data that are not publicly owned. In the United Kingdom, for example, the Postal Address File, which is the basis of any addressing service, was one of the assets transferred to the private sector when Royal Mail was sold off (Hope, 2013; Pollock & Lämmerhirt, 2019). The UK's national mapping agency, the Ordnance Survey, retains ownership and management of the Unique Property Reference Number (UPRN), a central part of the data infrastructure for local public service delivery, yet access to this is heavily restricted, and complex agreements govern the ability of even the public sector to use it (Ordnance Survey, 2016). Historically, authorities have faced major challenges in relation to 'derived data' from Ordnance Survey datasets, where the use of proprietary mapping products as a base layer when generating local records contaminates those local datasets with intellectual property rights of the proprietary dataset, and restricts who they can be shared with (Yates *et al.*, 2018). Whilst open data advocacy has secured substantially increased access to many publicly owned datasets in recent years, when the datasets the state is using are privately owned in the first place, and only licensed to the state, the potential scope for public re-use and scrutiny of the data, and scrutiny of the policy made on the basis of it, is substantially limited.

In the case of smart cities this concern is likely to be particularly significant. Take transit data for example. In 2015 the City of Boston, Massachusetts, agreed a deal with Uber to allow access to data from the data-rich transportation firm to support urban planning and to identify approaches to regulation. Whilst the data shared revealed some information about travel times, the limited granularity rendered it practically useless for planning purposes. Boston turned to senate regulations to try to secure improved access to granular data (Vaccaro, 2016). Yet, even if the city does get improved access to data about movements via Uber and Lyft in the city, the ability of citizens to get involved in the conversations about policy from that data may be substantially limited by continued access restrictions on the data. With the introduction of privately owned sensors, networks and processes central to many smart city models, the extent to which the data infrastructure for public tasks cease to have the properties that we will shortly see are essential to a 'participatory' public data infrastructure is a question worth addressing.

Participatory public data infrastructure

Even if we assume that the growth of urban data infrastructures in the smart city is almost inevitable, the shape that data infrastructures will take is far from predetermined. For many decades, community activists, advocates and civic campaigners have sought to ensure that citizens have the right to shape their cities, creating a vibrant field of participatory practice. Participation practitioners have sought to empower residents in decisions over the physical architecture of the city, the design of public services and the distribution of infrastructure such as roads, bus services, schools and hospitals (see https://participedia.net/ for examples and methods). Participatory design of the data systems of the city has been less common. Yet, with the design of data infrastructure affecting the policy options open to the city, and affecting the distribution of power and autonomy within the city, it becomes vital to explore what it may take to create a participatory public data infrastructure.

For Gray (Gray, 2015; Gray & Davies, 2015; Gray & Venturini, 2015) who has done much to develop the concept of a participatory data infrastructure, it is crucial to secure wider public awareness and engagement with questions of how data are collected, structured and published. I suggest that in thinking about the participation of citizens in public data, there are three critical aspects:

- Participation in data use
- Participation in data production
- Participation in data design

These are different in kind, but any case of participation may also be different in degree. Arnstein's ladder of participation (1969) offers a useful analytical tool to understand that within any form of citizen engagement with data the extent of participation can range from tokenism through to full, shared decision making (Figure 7.1). And, as for all participation projects, any assessment of participation is not complete without an answer to the question 'who is participating?', and a consideration of how inclusive or exclusive the participation opportunities are.

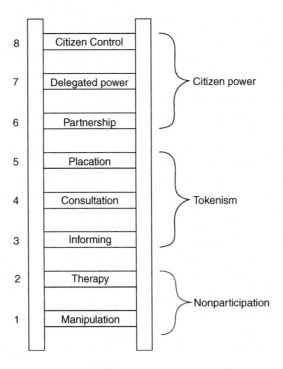

Figure 7.1 Arnstein's ladder of participation
Source: Arnstein (1969)

A data infrastructure assessed across all categories at the bottom-level 'non-participation' rung of Arnstein's ladder might record data 'about' citizens, without their active consent or involvement. It might exclude them from access to the data itself, and then uses the data to set rules, 'deliver' services and enact policies over which citizens have no influence in either their design or delivery. In such cases, the data infrastructure treats the citizen as an object, not an agent. Notably, this description might not be far off from the role that some fear is envisaged for citizens within a number of contemporary smart city visions (Krivý, 2016; Vanolo, 2016).

By contrast, when a data infrastructure is designed to enable citizens to participate in data use they are able to work with public data to engage in both service delivery and policy influence. Enabling this form of participation was a major focus for the early civic open data movement (Landry, 2019), drawing on ideas of co-production, and government-as-a-platform, to enable partnerships or citizen-controlled initiatives that made use of data to develop new solutions to local problems. In a more political sense, participation in data use can also address persistent patterns of information inequality between policy makers and citizens affected by policies, removing at least some of the power imbalances commonly at work when policies are discussed (Davies, 2010). In short, if the data relating to a city's population distribution and movement, electricity use, water connections, sanitation services and funding availability are shared, such that policy maker and citizen are working from the same data, then a data infrastructure can, in theory, act as an enabler of more meaningful participation, whether through new technology enabled channels or existing democratic fora.

Practical experience, however, often tells a different story. Instead of shared analysis of agreed data, it is not uncommon when engaging diverse citizen groups in seeking to use public data to find the process dominated by discussion of data gaps and ways in which the data itself needs to be improved. In some cases, the way data are being used might be uncontested, but the input might turn out to be misrepresenting the lived reality of citizens. This highlights the importance of citizen participation not only in data use, but also in dataset production. Simply having data collected from citizens does not make a data infrastructure participatory. The fact that sensors tracked my movement around an urban area does not make me an active participant in collecting data. By contrast, when citizens come together to collect new datasets, such as the water and air quality datasets generated by sensors from Public Lab (https://publiclab. org/wiki/nonprofit-initiatives), and when they are able to feed this into a shared corpus of data used by the state, there is much more genuine participation taking place. Similarly, the use of voluntary contributed data on Open Street Map, or submissions to issue-tracking platforms like FixMyStreet, can constitute a degree of participation in producing a public data infrastructure.

The opportunities for citizens are clearly positive when they use public data to shape policy and practice, and to contribute to the data on which policy and operational decisions are made. However, we should note that many participatory data projects, whether concerned with data use or with data production, are both patchy in their coverage and hard to sustain. They may offer an add-on to the public data infrastructure, but leave the core substantially untouched, not least because they generally rely on voluntary labour. This can also lead to substantial biases in terms of who gets to participate, what gets described and who benefits. Numerous studies have shown how contributions to crowd-sourced data platforms are often dominated by men, or by more affluent populations (Escher, 2011; Pak, 2017; Stephens, 2013), and thus tend to reflect their needs. There are also a limited number of topics that can attract a critical mass of user contributions adequate to address gaps in the state's own data infrastructure. For example, whilst crowdsourcing detailed urban cycle route data works in some cities, capturing detailed data on publicly accessible toilets with baby-changing facilities may be more difficult.

If then, participation is to have a sustainable impact on both policy and practice, it is important to consider how citizens can also be involved in shaping the core of public data infrastructure itself. This involves looking at the existing state-supported data collection activities that create public data infrastructure, and exploring whether or not choices over which data are collected, and how data are encoded, serve a broad public interest. If data collection and management practices are to allow the maximum range of choice and democratic freedom in policy making and implementation, then they cannot simply be taken as a given, but must themselves be subject to some degree of democratic oversight. This grounds a view of a participatory data infrastructure as one that enables citizens (and groups working on their behalf) not only to use and contribute data, but also to engage in discussions over data design.

The idea that communities and citizens should be involved in the design of data collection is not a new one. The history of public statistics owes a lot to the work of voluntary social reformers focused on health and social welfare in the eighteenth and nineteenth centuries, who initiated data collection to influence policy and then advocated for government to improve or take up on-going data collection (Dumpawar, 2015). In many countries, the design of the census, and of other government surveys, has long been a source of political contention (e.g. Thompson, 2010). Yet, with the vast expansion of connected data infrastructures, which rapidly become embedded, brittle and hard to change (Bowker & Star,

2000), we face a particular moment at which increased attention is needed to the place of citizens in shaping public data infrastructures. Seemingly technical choices, currently left to experts or to commercial providers, may set the long-term shape of public data infrastructures with substantial impact on the future of our urban environments.

It is the current expansion of data infrastructure that makes this a critical moment. Ribes and Baker (2007), in writing about the participation of social scientists in shaping research data infrastructures, underscore the importance of timing. They describe the limited window during which an infrastructure may be flexible enough to allow substantial insights from social science to be integrated into its development. As smart city agendas unfold, there may be a similarly limited window within which to establish data structures and systems that avoid foreclosing future political choices and that provide foundations for participatory public data infrastructure. Two critical tools to open and exploit this window come in the form of transparency policy and open data standard initiatives.

Part 2: Transparency and data standards

Although the movement for open data developed as a 'big tent', with different interests finding common ground around a limited technical definition of open data (Weinstein & Goldstein, 2012), arguments linking open data and transparency have played a particularly strong rhetorical role (Gray, 2014). Transparency was a central part of the framing of Obama's critical Open Government Directive in 2009, and transparency was in the foreground during the launch of data.gov.uk in the wake of a major political expenses scandal (Halonen, 2012). Transparency remains a core element of the narrative for many urban open data programmes.

It is also worth noting that transparency has also become an important resource in the regulatory toolbox of governments where governments mandate disclosure of information by private sector parties, for example, requiring publication of food safety inspection scores on restaurant windows and online (Fung et al., 2007). As Fung et al. argue in *Full Disclosure*, governments have turned to these forms of targeted transparency as a way of requiring that certain information (including from the private sector) is placed in the public domain, with the goal of disciplining markets or influencing the operation of marketised public services by improving the availability of information upon which citizens will make choices (Fung et al., 2007).

Thinking of open data as a transparency tool has two important consequences. Firstly, it draws a connection between open data as a form of proactive transparency, and pre-existing forms of reactive transparency, available through Right to Information laws. Sweden's Freedom of the Press law from 1766 was the first to establish a legal right to information (Michener, 2011), but it not until the middle of the last century that 'right to know' statutes started to appear elsewhere around the world. Today, over 100 countries have Right to Information laws in place, giving citizens a right to request documents held by the state. Increasingly, new or revised Right to Information laws recognise that transparency can be realised not only through access to documents, but also through providing access to datasets. Although there have been some tensions between document-based Right to Information communities, and more digitally oriented open data communities (Fumega, 2015; Janssen, 2012), there is a growing recognition of the complementary roles of reactive and proactive transparency through documents and data. When open data are not only provided by grant of government, but when citizens right to data is asserted, a key component of the right to the smart city is also enabled.

Secondly, linking open data and transparency provides an onwards link to accountability. Transparency policy is generally not only about access to information but also about asking an actor to give account for their actions. Full transparency is only realised when information is received, processed and understood (Heald, 2006; Larsson, 1998). This creates space for citizens to not only access information, but also to demand that it is made comprehensible and usable as part of public scrutiny of decision making. This challenges 'data dumping' approaches to open data that rely on simply uploading some existing dataset to open data portals, and instead places the focus on the creation and publication of new more legible public datasets that respond to citizen demand.

It is unrealistic, however, to imagine that each city government will negotiate the shape, structure and contents of each open dataset to be disclosed with each different citizen who wants access. Rather, cities may be encouraged to adopt particular domain-specific transparency standards that can act to align their data publication with citizen needs. This makes standards themselves an important site for participation in the creation of urban data infrastructure.

Part 3: Standards

Data standards in the smart city come in many forms: from low-level specifications that describe the protocols through which sensor networks exchange packets of data, through to data schemas that set out the columns, fields, identifiers and business rules to use when structuring data for publication. In general, it is this latter form of standard that is brought into focus by transparency efforts. Transparency data standards function on the boundary between the internal data infrastructure of the state and the public realm.

The term standard itself has overlapping meaning within different communities of practice (Russell, 2014). For a technical community, a data standard specifies the 'how' of expressing certain data, describing the schema through which data may be represented. For a policy community, a data standard may be more about the 'what' of disclosure, setting out normative guidance on the particular information that should be provided. In practice, these levels of specification can interact, as the presence of a technical standard-as-schema offers the ability to automatically validate whether the particular information demanded by a standard-as-content-requirement has been provided. Standards may furthermore support the analysis of data in third-party tools that help render the data more directly legible as information for citizens and other stakeholders.

Although early open data advocacy focused on the idea of 'raw data now' (Davies & Frank, 2013), envisaging a one-way route from data inside government to re-use of that data by outside actors, as explored above the opening up of data can reveal the limitations of existing data infrastructure. Far from being a simple template into which data from internal systems is directly exported, the introduction of standards for publication of open data can lead to a reconfiguration of data practices within the state (Goeta & Davies, 2016), particularly when those standards are used to define what it means to be transparent in relation to a particular topic. This makes the design of standards, and the involvement of citizens in those design processes, an area of critical importance. Three brief examples may illustrate this point.

GTFS: defining transit transparency

GTFS stands for General Transit Feed Specification. It was originally developed in 2005 through collaboration between Google and staff at the City of Portland (McHugh, 2013). It has gone on to become the de-facto standard for disclosure of public transit schedules, providing the data to drive thousands of urban transit apps and underpinning smart city visions of more efficient urban mobility. Both private transit providers and government officials have had to develop new working practices to turn schedules into GTFS format, and intermediary data infrastructures have been developed to bring together data from multiple existing datasets (Colpaert & Rojas Melendez, 2019; Goeta & Davies, 2016). However, although if you were to ask a city government if their public transport data were 'transparent' and they might point to their GTFS feeds, in practice these data only describe part of the transit story. For examples, fare information, transit performance or the accessibility of bus services is unlikely to be featured within the default GTFS datasets created. The data are tailored to real-time planning, but not to long-term transport planning. The 'official' version of the standard also lacks ways to describe informal transport models, particularly prevalent in the developing world, rendering the transport used by a large proportion of the population effectively invisible, unless governments adopt the independently maintained GTFS-flex extensions[1] (Colpaert & Rojas Melendez, 2019). The presence of a mechanism for independent extensions to GTFS does point, however, to a space for participatory engagement to at least partially reshape this component of smart city infrastructure. GTFS operates a loose governance model, in which a group of standard users is able to suggest and specify new elements that others are encouraged to use (https://developers.google.com/transit/gtfs/guides/changes-overview). Yet, unless these new elements feed into mainstream applications, or citizens campaign for their governments to publish data against them, they may remain theoretical rather than actual elements of government transparency.

OCDS: participation in design and adoption

When the Open Contracting Data Standard (OCDS) was developed in 2013, insights from GTFS adoption led the designers (the author of this article included) to include not only a technical specification, but also a set of data validation tools intended to support citizens to engage with their governments over the depth and breadth of information published about public procurement. OCDS emerged as the technical complement to a set of 'open contacting principles' (www.open-contracting.org/implement/global-principles/) that called on governments to publish detailed information about public tenders, contract awards and contracts signed with private parties (http://standard.open-contracting.org/). This information is particularly relevant in the context of the smart city, as smart city initiatives invoke Public Private Partnership (PPP) arrangements, license private vendors to collect data across the urban environment, or contract out delivery of public services. Fully implemented in a smart city, OCDS would allow citizens to understand and analyse this web of contractual relationships.

The development of OCDS followed an iterative process, starting from a review of supply-side data availability from government, and then mapping out a set of use-cases for transparent data on public procurement. By going back and forth between supply and demand, and consulting with potential data providers and users, the standard developed a set of data elements that struck a balance between the information citizens want and the data governments could theoretically produce. For example, a number of data use-cases called for geo-location of contracts, allowing public contracts and spending to be mapped by area. However, few, if any, e-procurement systems provide data fields to record location. As a result, the standard has specified location elements, but has not made these mandatory, providing space for citizens to engage with their own local governments to push for changes to upstream systems so that location information can be provided in future. This built on learning from Craviero et al.'s experience mapping public budget data in Sao Paulo, Brazil (Craveiro, 2013).

The governance and implementation model of OCDS provides two sites for participation public data infrastructure building. Firstly, the open governance process of the standard itself invites feedback from data users and publishers, and subjects proposed changes to a peer-review process that invites representatives from government, civil society and private sector to approve or reject changes. Secondly, when any city sets out to adopt OCDS, implementation guidance encourages a dialogue with local civil society to identify priority use-cases for the data, and to use these to guide the prioritisation of changes to the internal data infrastructures that generate the data and the development of external data infrastructures for data analysis.

Air quality: in need of standardisation

Transparency data standards can facilitate not only citizen access to city data, but can also be the means by which citizens are engaged in the co-production of public data. Air pollution is an increasingly critical issue for cities to address, causing millions on premature deaths worldwide every year (Landrigan, 2017). Yet, as the Open Data Institute describe, 'we are still struggling to "see" air pollution in our everyday lives' (Fawcett, 2016). They report the case of decision-making around a new runway at Heathrow Airport, where policy makers were presented with data from just 14 'official' NO2 sensors. By contrast, a network of citizen sensors provided much more granular information, and information from citizen's gardens and households offered a contrasting account from to that presented with official sensor data.

Mapping data from official government air quality sensors in the UK reveals just how limited their coverage is and backs up the Open Data Institute's calls for a collaborative, or participatory, data infrastructure. Fawcett described how:

> Our current data infrastructure for air quality is fragmented. Projects each have their own goals and ambitions. Their sensor networks and data feeds often sit in silos, separated by technical choices, organizational ambition and disputes over data quality and sensor placement. The concerns might be valid, but they stand in the way of their common purpose, their common goals.
>
> *(2016)*

He concludes, 'We need to commit to providing real-time open data using open standards.'

This is a call for transparency targeted at both public and private actors: inviting those with access to sensors to allow re-use of their data, and to render it comparable with other data through use of common standards. The future design of such initiatives will need to carefully balance public and private interests, and to consider issues of privacy (Scassa, 2019), centralised or decentralised architectures of data exchange, and how to incentivise contributions from different stakeholders to a data commons, as well as the interaction between voluntary standardisation and data sharing imposed by regulation or via contracts. These are not only technical decisions, but decisions with substantial consequences for the kinds of public debates over air quality that can take place, and the way policy responses might be designed.

Part 4: Conclusion: transparency, standards and the shaping of infrastructure

In the preceding sections I have introduced the concept of open data and explained its broad based origins. I have suggested that the opening of city data supports a process of infrastructural inversion, in which the particular features of the underlying data infrastructure that produce it are brought into view and subject to debate. I have then argued that these data infrastructures are an important site for civic participation, not only through the use of data, but also through citizen engagement in the (re-)design of the processes through which data are collected, structured and published. Without such participation, I argue, citizens become the object, rather than the subject, of the smart city. I have then outlined how the adoption of data standards to facilitate re-use of data creates a particular point for interventions that have the potential to shape the public data infrastructure that may result from transparency-driven open data initiatives.

Figure 7.2 offers a simplified schematic representation of the relationships at play, where processes of working with state data highlight limitations and surface new requirements, which can feed into standards and onwards into the internal data infrastructures of the state. The dotted line shows the boundary between data management 'inside' the state and the open engagement with data by citizens and other stakeholders in the lower part of the diagram.

It is important to note, however, that the 'virtuous cycle' described here, in which citizens are afforded space to participate in shaping public data infrastructures through data use and standardisation processes is by no means inevitable.

Firstly, a smart city initiative may lack a transparency-driven open data component. As Landry suggests (2019), although 'In some ways, open data is a golden thread running through modern urban development work' over recent years 'excitement for citizen–government collaboration based on a foundation of open data has often waned, and it is not clear how many cities have truly embedded a culture of openness through data into their organisational DNA'. Without both proactive and reactive open publication of datasets and an active citizenry requesting, exploring and engaging with data, opportunities to shape the data infrastructure of the smart city may be limited.

Secondly, there are competing pressures on a cities data infrastructure. Proprietary vendors may offer large 'integrated systems' in place of standards-based interoperable components. And even where standards are adopted, if these standards are proprietary or driven primarily by the

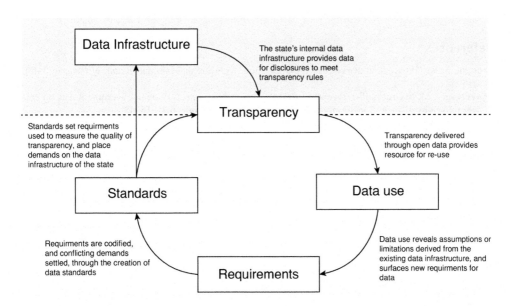

Figure 7.2 Mapping the relationship between transparency, data use, standards and data infrastructure

business requirements of commercial players in the smart city marketplace rather than by the concerns of citizens, infrastructures may be shaped to prioritise private, rather than public, interests. And even when citizens are engaged, if those citizens represent only a limited sub-set of the urban population, then the data infrastructures that emerge are likely to the tailored to their particular interests and needs of these niche groups. Just as good participation work involves pro-active outreach to under-represented communities, the use of participatory spaces enabled by open data will generally require careful attention to be paid to who participates. Without this, gender and socio-economic biases often inherent within the technical communities are likely to come to the fore (Brandusescu & Nwakanma, 2019).

In short then, transparency, open data and standardisation are not automatic generators of participatory public data infrastructure. Instead, they offer a strategic space for pro-public interventions, and they call for new forms of practice from civic actors seeking to make sure the smart city is one which maximises, rather than restricts, the democratic freedoms of its residents. In closing then, let me offer six questions to support strategic engagement with the data infrastructure of your (smart) city:

(1) What information is pro-actively published, or can be demanded, as a result of transparency and right to information policies?
(2) What does the structure of the data reveal about the process/project it relates to?
(3) What standards might be used to publish such data?
(4) Do these standards provide the data that I, or other citizens, need to be empowered in relevant ways to this process/project?
(5) Are these open standards? Whose needs were they designed to serve?
(6) Can I influence these standards? Can I afford not to?

Note

1 https://github.com/MobilityData/gtfs-flex.

References

Arnstein, S. R. (1969). A ladder of citizen participation. *Journal of the American Institute of Planners*, 34(5), 216–224.

Bartenberger, M. & Grubmüller-régent, V. (2014). The enabling effects of open government data on collaborative governance in smart city contexts. *Journal of E-Democracy*, 6(1), 36–48.

Bartha, G. & Kocsis, S. (2011). Standardization of geographic data: The European inspire directive. *European Journal of Geography*, 2(2), 79–89.

Bowker, G. C. & Star, S. L. (2000). *Sorting things out: Classification and its consequences*. Cambridge, MA: MIT Press.

Brandusescu, A. & Nwakanma, N. (2019). Issues in open data: Gender equity. In T. Davies, S. Walker, M. Rubinstein, & F. Perini (eds), *The state of open data: Histories and horizons* (pp. 287–299). Cape Town and Ottawa: African Minds and International Development Research Centre. www.stateofopendata.od4d.net/

Colpaert, P. & Rojas Melendez, J. A. (2019). Open data and transportation. In T. Davies, S. Walker, M. Rubinstein, & F. Perini (eds), *The state of open data: Histories and horizons* (pp. 215–224). Cape Town and Ottawa: African Minds and International Development Research Centre. http://stateofopendata.od4d.net

Craveiro, G. da S., Santana, M. T. De, & Alburquerque, J. P. de. (2013). Assessing open government budgetary data in Brazil. In *ICDS 2013*. Retrieved from www.gpopai.usp.br/IMG/pdf/Craveiro-ICDS2013.pdf

Davies, T. (2010). Open data, democracy and public sector reform: A look at open government data use from data.gov.uk. Practical Participation. September 29. Retrieved from www.opendataimpacts.net/report/

Davies, T., Walker, S., Rubinstein, M., & Perini, F. (eds) (2019). *The state of open data: Histories and horizons*. Cape Town and Ottawa: African Minds and International Development Research Centre. https://stateofopendata.od4d.net/

Davies, T. & Frank, M., 'There's no such thing as raw data': Exploring the socio-technical life of a government dataset, In *Proceedings of the 5th annual ACM web science conference*, pp. 75–78.

Denis, J. & Goëta, S. (2017). Rawification and the careful generation of open government data. *Social Studies of Science*, 47(5), 604–629.

Dodds, L. & Wells, P. (2019). Issues in open data: Data infrastructure. In T. Davies,S. Walker, M. Rubinstein, & F. Perini (eds), *The state of open data: Histories and horizons* (pp. 260–273). Cape Town and Ottawa: African Minds and International Development Research Centre. www.stateofopendata.od4d.net/

Dumpawar, S. (2015). *Open government data intermediaries: Mediating data to drive changes in the built environment*. Cambridge, MA: MIT Press.

Dunleavy, P., Margetts, H., Bastow, S., & Tinkler, J. (2006). *Digital era governance: IT corporations, the state, and E-government*. New York: Oxford University Press.

Escher, T. (2011). Writetothem. com: Analysis of users and usage for UK citizens online democracy. UK Citizens Online Democracy.

Fawcett, J. (2016) How to build the data infrastructure to tackle urban air pollution. https://oldsite.theodi.org/blog/how-to-build-the-data-infrastructure-to-tackle-urban-air-pollution

Floridi, L. (2004). Information. In L. Floridi (ed.), *The Blackwell guide to the philosophy of computing and information*. Wiley-Blackwell.

Fumega, S. (2015). Understanding two mechanisms for accessing government information and data around the world. World Wide Web Foundation, Washington, DC, July. https://idl-bnc-idrc.dspacedirect.org/handle/10625/56286

Fung, A., Graham, M., & Weil, D. (2007). *Full disclosure: The perils and promise of transparency* (1st ed.). Cambridge: Cambridge University Press.

Goeta, S. & Davies, T. (2016). The daily shaping of state transparency: Standards, machine-readability and the configuration of open government data policies. *Science & technology studies*, 29(4).

Goldsmith, S. & Crawford, S. (2014). *The responsive city*. John Wiley & Sons.

Gonzalez-Zapata, F. & Heeks, R. (2015). The multiple meanings of open government data: Understanding different stakeholders and their perspectives. *Government Information Quarterly*, 32(4), 441–452.

Government of Denmark. (2012). *Good basic data for everyone: A driver for growth and efficiency* (October). Copenhage: Government of Denmark.

Gray, J. (2014). *Towards a genealogy of open data*. Rochester, NY: Social Science Research Network 3 September. https://papers.ssrn.com/abstract=2605828

Gray, J. (2015). Democratising the data revolution: A discussion paper. Open Knowledge International.

Gray, J. & Davies, T. (2015). Fighting phantom firms in the UK: From opening up datasets to reshaping data infrastructures?. Working paper presented at the Open Data Research Symposium at the 3rd International Open Government Data Conference in Ottawa, on May 27th 2015.

Guldi, J. (2012). *Roads to power: Britain invents the infrastructure state*. Harvard University Press.

Gurin, J. (2014). Open Data Now. Open Data Now. McGraw-Hill Education.

Halonen, A. (2012). Being open about data analysis of the UK open data policies and applicability of open data. Report published by the Finish institute. www.fininst.uk/wp-content/uploads/2017/09/being-open-about-data.pdf

Heald, D. (2006). Varieties of transparency. *Proceedings of the British Academy*, 135, 25–43.

Hope, C. (2013). Everyone's postcodes to be privatised in Royal Mail flotation, despite objections from Sir Tim Berners-Lee. *The Telegraph*, 19 April. www.telegraph.co.uk/news/uknews/royal-mail/9994741/Everyones-postcodes-to-be-privatised-in-Royal-Mail-flotation-despite-objections-from-Sir-Tim-Berners-Lee.html

Huber, S. & Maier-rabler, U. (2012). The changing roles of citizens, civil society and public authority in Open Government. XXIInd World Congress of Political Science. Retrieved from http://paperroom.ipsa.org/papers/paper_11202.pdf

Janssen, K. (2012). Open government data and the right to Information: Opportunities and obstacles. *The Journal of Community Informatics*, 8(2).

Kassen, M. (2013). A promising phenomenon of open data: A case study of the Chicago open data project. *Government Information Quarterly*. doi:10.1016/j.giq.2013.05.012

Kitchin, R. & Lauriault, T. P. (2014). Towards critical data studies : Charting and unpacking data assemblages and their work. In J. Eckert, A. Shears, & J. Thatcher (eds), *Geoweb and big data*. University of Nebraska Press. Retrieved from http://papers.ssrn.com/sol3/papers.cfm?abstract_id=2474112

Krivý, M. (2016). Towards a critique of cybernetic urbanism: The Smart City and the Society of Control. *Planning Theory*. doi:10.1177/1473095216645631

Landrigan, P. J. (2017). Air pollution and health, *The Lancet Public Health*, 2(1). e4–5. doi:10.1016/S2468-2667(16)30023-8

Landry, J. (2019). Open data and urban development. In T. Davies, S. Walker, M. Rubinstein, & F. Perini (eds), *The state of open data: Histories and horizons*. Cape Town and Ottawa: African Minds and International Development Research Centre. www.stateofopendata.od4d.net/

Larsson, T. (1998). How open can a government be? The Swedish experience. Openness and Transparency in the European Union. www.worldcat.org/title/openness-and-transparency-in-the-european-union/oclc/39554644

McHugh, B. (2013). Pioneering open data standards: The GTFS story. In B. Goldstein & L. Dyson (eds), *Beyond transparency: Open data and the future of civic innovation* (pp. 125–136). San Francisco: Code for America Press. http://beyondtransparency.org/pdf/BeyondTransparency.pdf#page=136

Michener, G. (2011). FOI laws around the world. *Journal of Democracy*, 22(2), 145–159.

Miller, D. & Roe, S. (2018). Registers and collaboration: Making lists we can trust. The Open Data Institute. https://theodi.org/article/registers-and-collaboration-making-lists-we-can-trust-report/

Ordnance Survey (2016). 'Policy statement – UPRN', Ordnance Survey. www.ordnancesurvey.co.uk/about/governance/policies/addressbase-uprn.html [accessed 28 June 2019].

Pak, B., Alvin, C., & Andrew Vande, M. (2017). FixMyStreet Brussels: Socio-demographic inequality in crowdsourced civic participation. *Journal of Urban Technology*, 24(2), 65–87. doi:10.1080/10630732.2016.1270047

Pollock, R. & Lämmerhirt, D. (2019). Open data around the world: European Union. In T. Davies, S. Walker, M. Rubinstein, & F. Perini (Eds), *The state of open data: Histories and horizons* (pp. 465–484). Cape Town and Ottawa: African Minds and International Development Research Centre. http://stateofopendata.od4d.net

Rainie, S.C., Kukutai, T., Walter, M., Figueroa-Rodríguez, O.L., Walker, J., & Axelsson, P. (2019). Issues in open data: Indigenous data sovereignty. In T. Davies, S. Walker, M. Rubinstein, & F. Perini (eds), *The state of open data: Histories and horizons* (pp. 300–319). Cape Town and Ottawa: African Minds and International Development Research Centre. http://stateofopendata.od4d.net

Ribes, D. & Baker, K. (2007). Modes of social science engagement in community infrastructure design. *Proceedings of the 3rd Communities and Technologies Conference*, 107–130.

Russell, A. L. (2014). *Open standards and the digital age: History, ideology and networks*. New York: Cambridge University Press.

Sadoway, D. & Shekhar, S. (2014). (Re)Prioritizing citizens in smart cities governance: Examples of smart citizenship from urban india. *Journal of Community Informatics*, 10(3). http://ci-journal.org/index.php/ciej/article/view/1179

Scassa, T. (2019). Issues in open data: Privacy. In T. Davies, S. Walker, M. Rubinstein, & F. Perini (eds), *The state of open data: Histories and horizons* (pp. 339–350). Cape Town and Ottawa: African Minds and International Development Research Centre. www.stateofopendata.od4d.net/

Sieber, R. E. & Johnson, P. A. (2015). Civic open data at a crossroads: Dominant models and current challenges. *Government Information Quarterly*, 32, 308–315. doi:10.1016/j.giq.2015.05.003

Star, S. L. & Ruhleder, K. (1996). Steps toward an Ecology of Infrastructure: Design and access for large information spaces. *Information Systems Research*, 17, 111–134.

Stephens, M. (2013). Gender and the GeoWeb: Divisions in the production of user-generated cartographic information. *GeoJournal*, 78(6), 981–996. doi:10.1007/s10708-013-9492-z

Tauberer, J. (2014). *Open government data: The book* (2nd ed.) https://opengovdata.io/

Thompson, D. (2010). 'The Politics of the census: Lessons from abroad', *Canadian Public Policy*. doi:10.3138/cp.36.3.377

US government (2009). Open Government Directive. Washington DC. Retrieved from www.whitehouse.gov/open/documents/open-government-directive

Vaccaro (2016) Boston wants better data from Uber, and is taking a roundabout route to try and get it. www.boston.com/news/business/2016/06/28/uber-data-boston-wants

Vanolo, A. (2016). Is there anybody out there? The place and role of citizens in tomorrow's smart cities. *Futures*, 82, 26–36. doi:10.1016/j.futures.2016.05.010

Weinstein, J. & Goldstein, J. (2012). The benefits of a big tent: Opening up government in developing countries. *UCLA Law Review Discourse*, 38, 38–48.

Yates, D., Keller, J., Wilson, R. & Dodds, L. (2018). The UK's geospatial data infrastructure: Challenges and Opportunities. The Open Data Institute. https://theodi.org/wp-content/uploads/2018/11/2018-11-ODI-Geospatial-data-infrastructure-paper.pdf/

Section 2
Governing, inclusion and smart citizens

8

Towards an agenda of place, local agency-based and inclusive smart urbanism

Nancy Odendaal and Alessandro Aurigi

Introduction

The relationship between smart cities and urban space is currently dominated by glossy representations that share a visual narrative devoid of context and the messiness that accompanies contemporary urbanity. The smart city imagination display an odd 'placelessness' that bares little relation to the urban experience. In many ways it perpetuates the theoretically worn-out but still dominant dualism of cyber/smart urbanism as separate and alternative to the ongoing evolution of city spaces. The spectre of the smart urban fantasy represents a visual narrative that represents the 'city in a box' or the 'mirror world' – a version of the future as an ideal-state of connection, speed and gloss. These images hardly change in their travels across the globe, whether a reformulation of what Kigali in Rwanda should look like or a representation of future life in Hanoi, or Qatar. This decontextualised visual language represents the city in an orderly, 'tidy', idealised future: free of pollution, informality, unplanned neighbourhoods and crime. The implication is that the old can be replaced with the new; the messy present can be subsumed by digitally enabled nirvana. But, the chapter argues, the production of urban (smart) space is not just an add-on 'solution' and product of rational expertise and technical skill. It is above all the embodiment of specific visions of society, and of its economic development, which inform the shaping of spaces, facilities and services. The smart city appears then more as a remediation of existing trajectories, than any type of revolutionary stance

The tendency to engage in the techno-romance of idealised futures (Coyne 2001) is not an uncommon feature of city imaginings. Utopian visions have been part of urban history since the first conceptions of the city. City planning practice since the early 1900s is replete with idealised visions of the perfect city. In early digital city debates, techno-determinism is discernible in the many projections from the late 1990s that envisaged a frictionless urban environment where space no longer matters. Yet decades later it has become apparent (perhaps increasingly so) that both the inconveniences and opportunities stemming from spatial distance and proximity continue to be a factor in cities and that whilst much innovation exists, life continues as normal in cities across the world.

What informs the conceptualisation of this chapter is a tension between two perspectives through which the smart city concept can be looked at. The first of these is that the notion of augmented urban space encourages us to look at how regeneration and its digitally enhanced aspects can end up being part of the same thread, informed by the same preconceptions, rather than the latter innovating and breaking away from the former. The second is the growing awareness of the ongoing city making that happens outside the realm of traditional urban planning and design practices. The theme of 'informality' is relevant here and relates to the examples in this chapter. Rather than seeing it as an issue of the urban global South, a feature of disorder and illegality, contemporary thinking suggests a decentring of the concept, departing from the formal–informal dichotomy and recognising the 'unofficial' and 'informal' exists everywhere in the everyday (Marx and Kelling 2019). The examples referred to in this chapter show how technology plays an interchangeable role with livelihood strategies of those living in marginalised conditions in the global South, as informal urbanisation has become part of the everyday. We therefore argue that learning from these experiences is not only relevant to the developing world, but resonates globally. From a place-based perspective, we explore the notion of interstices: placemaking practices that define less developed parts of cities outside the realm of traditional urban planning and design (Phelps and Silva 2018).

By considering these two qualities of contemporary urban change we explore an alternative conceptualisation of the smart city that engages the present and the everyday, that emphasises the importance of place and argues that innovation is iterative and often to be found in unexpected places. The future is continuously being made and remade, by its very nature a conversation between technological appropriation and local innovation. Rather than seeing the 'new' (digital) as replacing the 'old' (analogic) in a linear temporality, we see the interaction between the two as reciprocal and dynamic. Our aim is to argue for a situated reading of the high technology-augmented city that recognises the importance of the often more informal but crucial urban interstices that give us clues into the markers of contemporary urbanity. We argue for an approach that embraces current literature on the urban everyday through a relational lens as constituting a research agenda that embraces the continuous remediation of multiple, existing future trajectories.

Redesigned urban futures and missed opportunities

Recent urban debates have highlighted the problematic relationship between the smart city imagination and entrepreneurialism (Das 2015; Datta 2015), whilst earlier 'digital city' explorations sought to understand the relationship between city governance and smart technologies (Aurigi 2005; Odendaal 2003). The question is whether the smart city is a game-changer when it comes to conceiving and delivering innovative approaches to urban development and regeneration, as it is currently being portrayed by industry hype.

When Robert Hollands wrote his 2008 seminal paper titled 'Will the Real Smart City Please Stand Up?' he mainly referred to the concept of the smart or intelligent city as the leveraging on improvements in civic information and communication technology (ICT) infrastructures for either place marketing, or towards supporting gentrified forms of economic development. Such a model of an entrepreneurial city falls foul of environmental and social sustainability, and citizens' participation, yet many smart discourses claim to address and solve these very problems. Smart rhetoric tends to frame those problems as the product of 'a "sick city" permeated by a series of pathologies' (Söderström, Paasche and Klauser 2014, 316) mainly because of inefficiencies in its current systems. Prevalent smart city propositions therefore stem from an 'intense anti-urbanism' (Dear 1995, 31) that is a typical generator of utopian planning (Choay 1997, referred to in Söderström, Paasche and Klauser 2014, 316), but they do this without exploring any new ideas and models of inhabitation and social equality. As Söderström, Paasche and Klauser (2014) argue in their revealing analysis of IBM's Smarter Cities' discourse: '"smarter cities" is a mild utopianism: it promises efficiency rather than paradise on earth' (316). In other words, the 'smart' paradigm is following logics of optimisation of current placemaking practices, and the improvement of specific lifestyles and desirable models of development that underpin these, rather than posit any alternative futures.

Placemaking for the middle class

Such an approach has very significant spatial and social connotations, and supports specific agendas rather than being limited to adding some, allegedly neutral, functional virtual layer to cities. Globally, this is uncannily ubiquitous, but as Watson (2013) argues, in the African context, it is closely tied to real estate multinational firms' investment strategies in what has become known as the 'last frontier' for property investment. Representing a 'fantasy urbanism' of sorts that seemingly seeks to overcome the messy urban realities of informality and urban poverty, the impact of the implementation of these plans is profound. Land speculation can lead to dispossession and relocation, especially in regeneration processes. The digital fantasy can easily translate into an analogic nightmare for some as market exclusion sees well-located locations become available to the digital (and other) elites only. As shown in Datta and Das's ongoing work on the smart city programme in India, there are important questions to be asked regarding citizenship and exclusion (Datta 2015, 2017; Willis 2019). Not only do smart city interventions have little contextual relevance but they also have impacts on livelihoods in inequitable conditions. The largely infrastructure-led approach to the implementation of smart city programmes runs the risk of perpetuating inequality, at worst, but misses an opportunity to use technology to enhance livelihoods, at best. The relationship to informality, hence a significant aspect of 'context', for example, is largely unexplored, yet the footloose nature of technology enables an intimate relationship between livelihoods and smart appropriation (Odendaal 2014). Thus, the relationship between smart city and responsive placemaking has transformative potential, yet has recently become a code for particularly reductive trends of area-based regeneration and master planning.

Area-based approaches carry potential for equitable regeneration by approaching urban development and redevelopment from a holistic viewpoint, inclusive of the need of the whole population living in a place (Urban Settlements Working Group 2019). Conversely, contemporary smart city design narrative reinforces a very narrow imagination with regards to lifestyles and livelihoods. Perhaps its most glaring exclusionary focus is in urban regeneration aimed at shaping cities for a very specific social class. The tension between utilising the creative city discourse as a 'worlding device' and place-based responses, is well captured by Nkula-Wenz as she explores Cape Town's ongoing flirtation with entrepreneurial governance and its exclusionary logics (2018). The new citizen imagined as part of smart city narratives appears to be a version of that described in Alvin Toffler in *The Third Wave* (1981): someone mainly involved in the service economy, who could produce and work just from about anywhere. This person was therefore highly mobile and unbound to any specific place, whether by choice or not. Discourses surrounding creative cities (Landry 2012) that underpin liveability indices such as the AIA Principles for Liveable Communities (www.aia. org/) share these characteristics with recent conceptions of the smart city: an utopian spatial imagination that assumes blanket upward social mobility. And the socially mobile and exigent middle classes express themselves through consumption – be it the purchase of goods, of experience, the fruition of culture and entertainment. When urban living and citizenry are framed in this way, it is natural for urban space to become a commodity.

Regeneration: missed opportunities

Given such constructed needs of the hegemonic middle class as the ideal group of city dwellers, an area-based approach to regeneration driven by those needs enables urban and suburban areas – often in potentially prime and attractive locations – to be isolated and safely designated for renewal and investment. As ubiquitous in the South as they are in the North, often labelled in the US and UK as Business Improvement Districts (BIDs), in South Africa as City Improvement Districts (CIDs), they are based on introducing private control of entire urban precincts and their 'public' spaces:

> When it comes to the BIDs pyramid, which is geared towards creating the optimum trading environment, the first layer on which the whole structure depends is the creation of a clean and safe environment, so just as man needs to breathe and eat to survive, these parts of the city need to be clean and safe. The next layer is 'transport and access', the level up is 'marketing and branding of the area' and the apex is the creation of a 'memorable experience for visitors'.
>
> *(Minton 2009, 43)*

The emphasis on 'visitors' of the BID, rather than on citizens, is very revealing. On the one hand the emergence of these large chunks of urban territory has raised issues of the lack of authenticity of 'an architecture of deception which … is almost purely semiotic, playing the game of grafted signification, theme-park building' (Sorkin 1992, 4). But perhaps more importantly, this implies a significant shift and narrowing of the focus on what the city is expected to be for. It reveals an obsession for specific, clean, acceptable and dignified activities which are associated with gentrification, 'jettisoning a physical view of the whole, sacrificing the idea of the city as the site of community and human connection' (Sorkin 1992, xiii). The problem with gentrification, however, is that it involves the collateral damage of removing/shifting and ignoring those parts of society unable to gentrify. Where regeneration projects talk of 'reclaiming' public space, this is of limited benefit to those who can afford it and results in a 'cleaning out' those that cannot (Minton 2009, 52). The outcome is that urban poverty is simply displaced, rather than 'solved' through regeneration (Madden 2013). When smart urbanism visions are similarly skewed, the danger is that they could further amplify those social divides by boosting the potential of exclusive areas and the lifestyles of already empowered individuals.

Smart gentrification as part of a global imagination

These exclusionary dynamics becomes particularly poignant in urban spaces of the global South, where middle-class enclaves contrast with extreme urban poverty. 'Ordinary cities' such as Johannesburg and Nairobi may appear to fall outside the world city dynamic, but are sites of functional networks that connect across geographic boundaries (Robinson 2006). Departing from a hierarchical world city classification enables an analysis that recognises that global economic connections are diverse and complex, where 'the city is increasingly a key articulator in a new, regional geography of centrality, dispersal, mobility and connectivity that expands not only to the rest of the continent but around the globe' (Mbembe and Nuttall 2004, 360). Their 'apparent structural irrelevance' (Robinson 2006, 99) deems them less connected to the world economic hierarchy of global finance and economic transfers. Cities such as Johannesburg (discussed in Robinson's work) and Lagos (see Gandy 2005) are traditionally seen as marginal to the networks of economic flows that underpin the global economy, yet are intimately connected into global networks through the everyday.

State efforts to attract foreign direct investment (FDI) through place marketing and event attraction (the hosting of the FIFA World Cup in 2010 in South Africa, for example) are integral to policy objectives that seek global visibility. Increasingly the smart city narrative is used as marketing language for new satellite cities and redevelopment areas. These skew infrastructural investment to the exclusion of city spaces that are in dire need of intervention, and/or lead to resettlement of residents. This also mirrors in the plans for erecting many new 'smart' towns in Asia or the Middle East, where the pretext of the advantage of building a new techno-utopia from scratch fits very well such highly selective regeneration approaches. The case of Songdo is perhaps one of the first to be examined in detail, with planning efforts to enable liveability and connectivity largely failing (Townsend 2013) and African versions, such as Eko Atlantic having very little to do with the rest of Lagos (Ajibade 2017). The contrast between these 'premium networked spaces' (Graham and Marvin 2001) and the emergent and informal urban spaces in many cities of the global South is particularly pronounced and makes this dynamic particularly insidious.

However, dynamics of systematic spatial exclusion and economic marginaliz-sation are not unique to the South. The term 'emergification', initially coined by *Financial Times* reporter Joseph Cotterill (2010) in the context of investing in emerging economies, can also be very effective to highlight the limits – and demise, to an extent – of the 'we are all middle class now' (Jones 2012, 140) tenet of the city and the smart city itself. There is evidence of various 'first World' economies and societies 'emergifying', as citizens are forced by adverse economic conditions, loss of jobs or considerable reduction of income, to make ends meet through considering and adopting less consumption-based and wasteful lifestyles, and go back to valuing social capital basics such as local collaboration, communes and more informal markets. A major 2012 survey of Italians – despite their country still being included in the club of the main economic global powers – revealed the 'collective perception that reality is changing in an irreversible way, and one which is strongly pejorative' (Masci 2012, translation by the author). Cloza (2012) also uses the term 'emergification' when talking about the Italian ever-growing informal economy, exactly to describe how 'We are becoming, or maybe going back to, being like emerging countries'. Similarly, Hadjimatheou (2012) describes how small-scale farming can become a desirable alternative for former city dwellers in Greece. In the meantime, economic commentators are warning that the failing to adjust of even larger sections of the population of Western countries is generating an ever-increasing rise in household and personal unsecured debt, representing another economic crisis 'time bomb' (Inman and Barr 2017).

The fact that entire sections of society and cities face or will face the task of reinventing their economies is certainly not new. What is noticeable is a growing awareness that this cannot and will not be done simply through a process of turning everybody into white-collar workers operating in the service industries through large corporations. An increased emphasis needs to be put, especially in those many towns – the vast majority of the World, in fact – that do not enjoy any privileged position in global trade and financial networks, on the facilitation of bottom-up, social enterprise. It becomes crucial to enable communities to leverage local resources and re-establish local markets, promoting cheap, collective mobility means, and fostering all sorts of lateral ideas such as, just to name one, urban farming. It seems necessary, in other words, to re-empower the growing number of people who have been let down by the mirage of consumption-fuelled wealth-for-all and indeed the consequent vision of a service and leisure-only city where everybody makes a living 'opening doors for each other' (Jones 2012, 54).

Yet, the smart city visual and marketing narrative is closely tied to an imagination of the smart citizen that bears very little resemblance to the livelihood struggles of the everyday (Datta 2015). This therefore leads to a homogenisation of citizens according to pre-constructed lifestyle images; where these lifestyles are increasingly hinged on the availability of disposable income and leisure time. Meanwhile, those lacking such lifestyles can become increasingly invisible and negligible. As noted before, a tourist-like profile seems to be the new expected normal for city users. Many urban regeneration initiatives therefore verge towards making the profile of the 'citizen' and that of the 'tourist' converge, as argued by Elizabeth Wilson: 'Not only is the tourist becoming perhaps the most important kind of inhabitant, but we all become tourists in our own cities' (Wilson 1995, 157).

The reassuring and business-friendly features of a service-rich, clean and safe environment, with good transport links and offering a 'memorable experience' to citizens who very much become visitors, are in fact largely remediated by smart city technology offerings, and celebrated by the related commercial literature. The prevalent, mainstream visions of smart urbanism involve an urgency to address the needs for cleaner and environmentally more sustainable towns, though with a marked focus on high middle-class expectations. Hitachi's Smart Cities website for instance warns about 'the growth of slums, air pollution, the difficulty of acquiring fresh drinking water, the treatment of waste water and sewage, energy supplies, traffic congestion, and waste disposal' (Hitachi website, last accessed Nov. 2013). It however presents a series of scenarios about 'Living in a Smart City' which range from 'Freedom to Work When and Where you Want' to 'Convenient Vehicle Use as Part of the Community' and 'Well-balanced Lifestyles in Tune with How People Live'. In one of these for instance a hypothetical housewife states:

> Because we live in the suburbs, my husband normally commutes to work by car. He drives an electric vehicle (EV) that fully recharges while we sleep … He saves time in the morning by checking his schedule while he is in the car.
>
> *(Hitachi website, last accessed Nov 2013)*

This short description makes some revealing assumptions, as it describes an ideal smart city dwelling situation as one of urban sprawl, possibly functional with forms of gentrified and exclusive out-of-town living. Being in the 'suburbs' seems in fact to be associated with an absence or deficiency of public transport, and a dependency on the motor car. The 'husband' going to work seems to have a highly variable schedule which needs to be checked in the morning and requires high levels of continuous connectivity, suggesting some form of executive employment.

Similar interpretations come from GSMA, the worldwide trade association of mobile tele-communication operators. Its 'Connected City' initiative and exhibition claims to address 'making homes and cars smarter, travel swifter, shopping easier and urban living safer and more environmentally friendly' (GSMA website, last accessed Nov. 2013). GSMA also keeps a 'Smart City Index' categorisation and ranking, which interestingly is based on indicators named 'Smart Mobile Services', 'Business, Economic and Mobile Cluster Impact', 'Smart Mobile Citizens' and

'Mobile Infrastructure' (GSMA website, last accessed Nov. 2013). Another notable example is the impressive Living PlanIT documentation on what has been defined as the blueprint for an 'Urban Operating System' (UOS). This also makes explicit reference to, and places great emphasis on, the importance for cities to foster knowledge economies, and how smart urbanism can be central to it, arguing for 'strategies to increase the sophistication of their populations to service and attract advanced industries' (Living PlanIT 2011, 4).

So, the smart city – or at least an interpretation of it which is being promoted by the tech industry and often embraced as an out-of-the-box solution by municipalities and governments – shares and remediates the aims of gentrification-keen urbanism, and indeed provides new ways to project a certain vision of city 'users' further. In these visions, the 'smart citizen' uses the city in a way that implies high levels of mobility, and the need for ubiquitous services to support such mobile and knowledge-based occupations. It is of course necessary for the smart citizen to be conversant with high and mobile technology, be able to afford hi-tech gadgetry and be willing to interact with advanced systems of data feeding and reporting. This also implies that such citizens are themselves highly mobile and potentially disloyal – they can move somewhere else easily – and hence behave as the paying customers of the city, and that the latter is driven to provide them with the control, services, safety and cleanliness they expect. Most proposed projects therefore do not question any of the typical high middle-class models of living. For instance, the motor car and its presence in the city is never particularly put in doubt or challenged, but is remediated by technologies that offer enhanced ways to use it and find parking spaces (see for instance Lamba 2013), hence making it appear more, rather than less, socially and environmentally acceptable.

These two overlapping aspects of creating territorial exclusion, or exclusive enclaves of regeneration, and constructing the 'smart' citizen as a high middle-class dweller and user of services and infrastructure-rich areas offering specific lifestyle advantages, add up to make emergent smart city models.

For instance, the Konza 'technopolis' outside Nairobi, Kenya, is described as Africa's 'first smart city' and promises a connected future for its citizens. Estimated to cost US$14 billion in infrastructure investment, this 'silicon savannah' (Anderson 2017) is aimed at building on the country's growing ICT sector but creating a hub for entrepreneurs. Yet many of these hubs exist, in the city of Nairobi, albeit in less glamorous surroundings, with some arguing that a critical mass does not yet exist for justifying a decentralised location for the sector (Anderson 2017). The future population is projected at 250,000, with the marketing language replete with descriptions of service delivery, improved municipal infrastructure and enhancements, and real-time monitoring systems (Burger 2017). Whilst portrayed as technological support to enhance the lives of people, not only does Konza bear little relation to Kenyan urbanism but is also located 37 miles from Nairobi. Despite claims of future sustainability, very little evidence suggests that the spatial concept deviates from a modernist, car-centred, mono-centric zoned living environment the literature argues against. But this is not a tension limited to new developments in the global South. The highly debated case of the Quayside project by Google's Sidewalks Labs in Toronto reveals similar pitfalls, in contrast with promises of sustainable development and affordability. Canada's prime minister has described the hi-tech waterfront neighbourhood as 'a step toward "smarter, greener, more inclusive cities", and "creating a new type of neighbourhood that puts people first"' (Murphy 2017). But the nature itself of the scheme seems to be encouraging a strong selection when it comes to prospective residents, resulting in a homogenous, high middle-class neighbourhood.

Austen (2017) has noted how 'Quayside's current plans promise housing for people of all income levels. But the only company so far committed to moving there is Google Canada, suggesting an influx of young, affluent workers'.

Missing the revolution?

There seems to be a convergence between social and spatial gentrification and the tenets and consequences – intended or not – of the production of digitally augmented cities. Hollands (2008) outlined how wide and complex the range of interpretations of the 'smart city' concept was. But he also remarked how its aims ended up being very much aligned with the type of gentrified urban economic development this chapter is reflecting on. Instead of representing a change of direction, an innovative view and approach, it can very much mirror and reinforce the trend. In a way this does not have to be particularly surprising. Although the frequent – and often marketing-based – hype about the role of high technologies in the city tends to present these as a game-changing, revolutionary factor, the point of view of Bolter and Grusin (1999, 182) arguing that cyberspace is a 'remediating' force which extends earlier media and spaces is an enlightening and particularly relevant one here.

Remediation is providing an explanation on how what is being portrayed as a potential technological revolution does not seem to be underpinned by any particular revolutionary idea pointing at a progressive model, either in socio-economic organisation or indeed civic design. So, Hollands' observations fundamentally still stand, despite an apparently ever-evolving and increasingly technologically sophisticated landscape. The smart city therefore can end up being a simple, digitally enhanced way of repackaging recent forms of commercial urbanism.

This is where the utopia of industry-driven smart city visions shows its limitations. Townsend still compares past models with present developments when he argues that Ebenezer Howard's Garden City 'was the Songdo of its day – network technology undergirded its daring break from the past. While Londoners choked on smoke from a million coal-fired furnaces, Howard's utopia would run on clean municipal electricity' (Townsend 2013, 95). This is understandable with a lens of 'break from the past through technology', but – beyond any wider considerations on its actual impacts on city design – it has to be acknowledged that Howard's vision embedded a social reform agenda and a brand new model to propose. It was at least based on a revolution of the concept of land ownership based on an urban 'commonwealth'. Major smart city developments like the often-cited Songdo or Masdar, however, present no major, critical alternative view of the principles of living, working and socialising – and governing all this – in geographically contiguous settlements.

The question at this point can be: how can this be steered – or maybe hacked – in more transformative, adaptive, socially sustainable and place-aware trajectories? There are two related tensions that surface with regards to the relationship between the fantasy smart city, which can manifest in area-based, redevelopment efforts but nevertheless embraces their principles, and the somehow neglected 'analogic' city through which people move and pursue their lives. The one relates to the contrast between the pre-designed and programmed spaces of consumption and prioritisation of the knowledge-based economy and the incremental and messy continuous unfolding of the 'real' city. The second refers to city as an imagined future of order, seamlessness and low-friction mobility that contrast the contingency and emergence of present urbanity. In the following section we explore the notion of 'interstices' as a means to address this divide, engaging and leveraging on the authenticity and diversity that urban contexts provide.

Getting dirty and real with the 'smart city'

Short of some form of urban revolution based on radical, major changes, a way to start making smart city efforts more relevant to the realities of urban existence mentioned so far is to move away from the spatially and functionally selective and prejudiced models of a 'pure', safe and efficient smart urban machine. This means shifting our gaze from the investment and discourse-dominating mainstream smart urban 'solutions' and engaging with the 'grit' of everyday diversity and tensions, which can reveal local wisdom and resources.

Leaving out the parts of urban space which are not functional to a certain regeneration trajectory by focusing interest and visibility towards those that do is close to what Ingold calls an urbanism of 'assembly':

> This distinction between the walk and the assembly is the key to my argument ... Once the trace of a continuous gesture, the line has been fragmented – under the sway of modernity – into a succession of points or dots. This fragmentation, as I shall explain, has taken place in the related fields of travel, where wayfaring is replaced by destination-oriented transport, mapping, where the drawn sketch is replaced by the route-plan, and textuality, where story-telling is replaced by the pre-composed plot. It has also transformed our understanding of place: once a knot tied from multiple and interlaced strands of movement and growth, it now figures as a node in a static network of connectors. To an ever-increasing extent, people in modern metropolitan societies find themselves in environments built as assemblies of connected elements.
>
> *(Ingold 2007, 75)*

This can be increasingly facilitated by smart city technologies, as they allow and encourage – under logics of efficiency and rationalisation of movement – point-to-point interest. The efficiency of digitally enhanced navigation – and even more so the push towards autonomous vehicles – replaces the serendipitous, inefficient appreciation of interstices and the not necessarily negative chance of getting lost and discovering something or someone, as argued by Shapiro (1995) at the dawn of cyberspace-related debates, and more recently by Foth (2016). There is a desired 'seamlessness' that aims to reduce friction of movement and decision-making into a designed optimised 'whole'.

This approach aligns with the ways the creation of BIDs is concerned with maximising the potential of urban space as a commercial environment. Spatially, but also digitally, this is fundamentally achieved in two ways: by providing a series of arrival points – shops, cafes, attractions and services, for instance – and efficiently directing people to them, regulating and selecting access. It also involves sanitising and commodifying as much as possible the route itself to these points of consumption, so that it anticipates, prepares and leads to them, as well as embedding some. Street furniture interventions in such places show an obsession with signage and the management of the visitor's experience to keep them on the right route, and this is often seen as the most obvious priority in regenerating an area to boost its commercial potential. The smart layer promises a frictionless experience whereby the smart citizen is continuously connected and served by the sentient presence of the screens apps and urban services that constitute the Internet of Things (IoT).

As for excluding certain categories of people, there can therefore be an effort to design out certain forms of urban space which are seen as undesirable and incoherent with the dominating view which 'has become fragmented into individualized pieces incompatible with the creation of a physical plan', and as a consequence 'we have no map of the city linking together the poorer neighbourhoods with the enclaves of the well-to-do. A plan or a map might draw us closer together and underscore our collective plight' (Boyer 1993, 112–113)

All of this can therefore hide and undermine the value and role of those parts of the city that are not functional to the view of it as a series of easy to access, safe and sanitised commodified destinations. These have been called the 'interstitial' spaces in the city, which 'represent what is left of resistance in big cities – resistance to normativity and regulation, to homogenisation and appropriation' (Nicolas-le Strat 2007, 314) – both in spatial and social terms. These spaces of resistance are not necessarily confined to event-driven social action (such as occupations or protests), and they are very much part of many people's 'everyday'. In the global South, these 'interstices' are what can define many urban spaces and extend to the use of technology. The informal urbanisation that typifies urban growth in cities in Africa relates to how people house themselves and pursue their livelihoods.

The embedded need to control movement in accordance with an idealised blueprint is in itself a Modernist planning legacy. The temptation to apply standardised formulas of technological-fix and rationalisation to the city is particularly relevant to contemporary smart interpretations. Fragments and interstices are messy interlopers with no place in designed smart futures. In the 1990s Rem Koolhaas spoke of the 'city's defiant persistence and apparent vigour' as representative of the ongoing agency of ordinary citizens who continuously make and remake their cities, 'disconcerting and (for architects) humiliating' (Koolhaas 1995, 28; Tonkiss 2012). In architecture and urbanism, the adoption of an 'international' style based on modernist rationalism has affected – and is still affecting – the shape of developments in many Asian and Middle-Eastern centres, the quality of those places and indeed their ability (or inability) to leverage local historical wisdom and be sustainable. Similarly, the 'smart' city is generally being proposed as an extension and upgrade of that very modernist concept of the city, where the rational 'machine' is made yet more efficient through the deployment of digital service and control, based on the centralised digestion and elaboration of 'big data'. As steel and glass skyscrapers might not necessarily have been the most environmentally sustainable choice for sub-tropical developing towns, the smart urban machine with its often standardised, prêt-à-porter solutions is unlikely to generate a universally socially sustainable urban growth trajectory.

This urgency is multidimensional. If on the one hand it is easier to conceive that as many 'everydays' exist as the different geo-social contexts generating them, on the other hand it is also important to notice how diversity is key within the same location. Growing social polarisation and the need for alternative socio-economic approaches are widening the gaps between a range of diverse contexts within the domains of countries, regions and cities, which smart city initiatives cannot treat as homogeneous.

How do we look at interstices in the smart city? We would argue this needs two main – and obviously overlapping – foci. First, letting the interstices highlight issues and provide place-based wisdom, rather than simply raw data for some algorithm. Second, using ubiquitous computing for enabling local agency.

Drawing from place and its socially embedded wisdom

Imprinting a material layer on the interactions of the everyday, especially with regards to new technologies, surfaces qualities of socio-technical interaction that deviate substantially from the mainstream imaginations of the smart city. As urban dwellers appropriate technology as an ongoing input into livelihoods, the manifestations of these moments take on spatial forms that are contextually embedded and intimately related to place. One relates to the spatial modalities of ICT use in many cities of the global South for example, where container telecentres, informal phone shops on side walks and the sale of airtime by street vendors at road intersections reveals a more flexible and 'real-time' transformation of urban space. Innovations are mediated by culture and social norms. A second dimension relates to local innovations that have spatial implications such as the use of MPesa, a Kenyan mobile money transfer innovation, that enables remittances between cities and remote urban areas (a core part of livelihood strategies of urban migrants) (Gikunda, Odilla and Njeru 2014).

For infrastructural projects in developing countries, the economist Tim Harford has noted that 'although the technical properties of the system may have been understood and improved, the human properties of the system have not been addressed at all' (Harford 2006, 227). The process of shaping the smart city needs therefore to bring people back into the equation not simply as 'users' or controlled consumers of urban system and space, but as active contributors to it, and creators of markets. Local appropriation of technology by low-income communities can facilitate sustaining and enhancing informal but vital economic systems and markets in cities, with positive effects for those communities (Odendaal 2006). The relationship between space and informality is important, since 'understanding the spatial manifestations of marginal livelihoods is important. They provide us with clues on the entry points for technology appropriation in urban space' (Odendaal 2013, 33). Communities can do much, and can be extremely creative and 'lateral' at how technology-enabled networks are formed and utilised. The ability of the smart city to provide support and facilitation for such bottom-up approaches and ideas can be key in ensuring that those ideas – and the communities able to generate them – succeed.

This therefore involves accepting that digitally augmented urban space cannot be treated as a homogenous territory, or indeed a corporate commodity. The progressive smart city relies on the power and relevance of context. Ignoring local values, culture, knowledge and indeed space can result into making the machine-space rigid, blunt and insensitive; as argued by Sassen (2011): 'What stands out is the extent to which these technologies have not been sufficiently "urbanised". That is, they have not been made to work within a particular context'.

Envisaging, fostering and supporting local solutions, the type of approaches which this chapter has argued are needed for cities and communities in a post-bubble era (and for the next bubble burst), inevitably requires leveraging on local resource, culture and wisdom, and indeed on the value of the characteristics of place themselves: 'Context would not just be an opportunity, but it would become one of the central generators of the digital intervention' (Aurigi 2013, 138). As traditional and vernacular architecture, often ignored by modernist development, can hold the key to a deeper understanding of place and ideas for a truly sustainable use of resources and inhabitation (for a review of this see Vellinga 2013), the city of smart space and objects will greatly benefit from keeping in touch with the richness and wisdom embedded in place.

Enhancing local agency in smart cities

The ubiquitous computing characterising smart urbanism visions means the fixtures and utilities of contemporary life, the everyday functions and infrastructural uses, are, 'augmented with computational capacities' (Dourish and Bell 2007, 414). The boundaries between private and public have become less certain. As technologies become increasingly mobile and pervasive, opportunities for surveillance increase, as do a 'real-time' interface between everyday decision-making and technology use. As we purchase goods at supermarkets (using credit cards), stop at traffic intersections (through traffic web cams), acquire books and music online and enter buildings (through electronic entry points) we leave 'bits' of ourselves; 'These technologies allow spaces to both remember and anticipate our lives' (Crang and Graham 2007, 789) and they becoming particularly poignant to those who live their urban lives 'on the move' (Simone 2004) in precarious circumstances.

This has implications for the experience of space and movement between places. Ubiquitous computing anticipates a spatial dimension where the digital and physical co-produce an experiential dimension typified by seamless flows of information and interaction. A hybrid space is possible at the interface between infrastructure and human experience (Dourish and Bell 2007) that does not distinguish between the material and human but recognises the relational qualities of this socio-technical exchange. Whilst traditional networked infrastructures are tangible and fixed, ubiquitous computing is pervasive, mobile and increasingly footloose due to wireless capacity. This means that it can be employed to shape and characterise different, localised versions of hybrid space, instead of simply offering a sanitised and homogenous experience. What emerges is a 'dance' between the digital and the physical, the social and the technical, in the ongoing production of space and creation of place, where code itself – far from being seen as neutral – purposefully embeds specific values, interpretations and expectations related to local culture and agency. Merging the past, present and future as part of a socio-technical continuum provides a platform for contextually appropriate city-making. In the following section we explore what conceptual parameters for such an approach could be considered by smart city-makers.

The case for a more inclusive research agenda for smart cities

The making of cities is the ongoing interacting agency of designers, planners and citizens together with the many other stakeholders that contribute to their shaping and management. Inclusive cities stem from primarily considering communities or citizens and how their needs are accommodated, rather than a mechanistic logic, based on geometries, functional macrozoning or abstract guidelines. The type of smart city represented by the highly codified representations of IT company-driven urbanism shares many characteristics with the civic models driven in the past and it is no small coincidence that many plans share codifications with New Urbanism. Contemporary thinking on the making of place advocates a more fluid and emergent engagement with context, whilst work on cities of the global South acknowledges qualities of emergence, contingency and informality.

The normative urgency in cities across the globe is a transition towards more sustainable and inclusive cities (see the New Urban Agenda adopted at Habitat III in 2016). Many aspects of smart monitoring and control seem to align with this. In lower income environments, however, this entails allowing space for low-technology innovation, labour-intensive implementation and decentralised management of services. Pieterse for instance argues for an approach in the global South that pursues optimal resource efficiency, ensures universal service coverage while articulating with economic multipliers (2014). Technically this requires analysis of materials flows that not only links back to industrial ecology and urban political economy (administrative regimes and governance) (UNEP 2013) but also crucially to livelihoods strategies.

Despite cities experiencing a growing pressure on, and unevenness of infrastructure, urban life continues to be reinvented at the margins. Social networks function as critical livelihood arteries in the ongoing survival strategies of the poor. These transactions cannot be defined in space or frozen in time. This terrain comprises a divergent range of intentions, communications and movements exchanged between multiplicities of actors making sense of their life worlds, and is often articulated in the informal economy, in makeshift housing and fleeting urban encounters. The potential exists for harnessing these strategies, building on the social capital created despite the absence of, or in addition to, the usual resources available for survival. Despite their lack of guaranteed permanence, these social networks provide the vehicles for information sharing, resource negotiations and support in precarious living conditions. They are also closely tied to context.

Developing a smart city that enables economic production demands an engagement with the human and cultural relationships, and the local human and natural resource strengths that define trade and exchange of goods and services. Simplistically assuming the predominance of the tertiary economy falls foul of engaging the realities of city spaces worldwide. In the way they tend to be portrayed by mainstream smart discourses, 'smart citizens' actually comprise a very small proportion of inhabitants in cities of the North and South. In many cases livelihoods entail mobility: between cities, between the rural and urban, and often across borders (Simone 2010). These circuits rarely coincide with the more formal spatial delineations of regeneration. The implications are twofold. The potential of existing services and technologies is not maximised in terms of how much it could facilitate employment and economic growth, and secondly, it may constrain a wider perspective on mobility and livelihoods. 'This is more than simply building new roads, rails, power lines, and telecommunications. It is more of a matter of constructing synergies between the physical, the institutional, the economic, and the informational' (Simone 2010, 29). What emerges is a need to gain insight into how networks are constructed and maintained and how that ties in with how technology is used. This calls for an approach that goes beyond a simple infrastructure viewpoint and more towards a networking one, and emphasises facilitation over simple provision of services or products. The articulations between circuits of exchange and technology access need to be enabled and understood in how they contribute to the making of place.

Thus, there is a need to consider cities as socio-technical systems that incorporate human ingenuity, reinvention at the margins and, more than anything, to recognise that urban change is iterative and experimental. Spaces for learning and creativity need to be enabled, and research efforts to uncover these and understand how they work and what city-makers can learn from them need to be a key enabler. Such micro-level socio-technical environments encompass small networks of actors that add new technologies to the agenda, promoting innovations and novel technological developments. This may reveal configurations of actors hitherto unexplored in studies of smart cities. How social learning from niches can be applied at the city scale to help reshape the existing infrastructure regime is a challenge that requires a multi-scalar perspective that is mindful of the connections necessary for survival (UNEP 2013: 14). The aim is to not only uncover the actors, institutions, technologies that affect beneficial change, but also the relations between them. It is through such constellations that agency emerges.

Conclusion

We argue for a smart city-making approach and a research and development agenda that untangles these relations in urban space. It requires 'thick description' of the everyday that leads to uncovering 'emergent groups, multiple lay-expert knowledge forms, programmes of action, valuation regimes, fluid topologies' (Farías 2011, 367). The emphasis on the empirical also enables deep contextual inquiry. Case studies that document the interstitial manifestations of the interface between digital and analog, as an ongoing unfolding of the urban everyday, enables inquiry into how contingent the smart city is. That in itself is political: regime change is about continuous questioning (Farías 2011); furthermore, as an 'empirical tracing of how it is that materials come to matter' (McFarlane 2011: 734). Technology innovation can be tracked from above and below. The learning that informs and emanates from the appropriation of technological artefacts could then be used more broadly, leading to a conceptualisation of the smart city as a living, breathing phenomenon, continuously being reinvented. It requires an engagement with social and spatial agency at the local scale. A conceptual frame that reveals hopeful engagement with the 'everyday' encounters between people and technology offers opportunities for emancipation; 'an engaged political project that asks evaluative questions about how urban technologies are socially appropriated, why and in whose favour?' (Coutard and Guy 2007, 731).

This chapter commenced with an account of the ways through which the smart city narrative overlaps with utopian ambitions often associated with exclusionary regeneration strategies. Smart infrastructure may be part of the problem as we continue to associate it with glossy futures that have nothing to do with the everyday city. Thus, a regime change is required, that goes beyond mechanistic visions of efficient design and placemaking, to understanding the emergent properties of street urbanism and how the poor get by. Much of this day-to-day invention speaks to the interface between the material and human. The availability of technology is generative – it leads to outcomes that are assimilated back into livelihoods – and how appropriation interfaces with livelihoods is part of the conversation that needs to be shared.

Rethinking the conceptual lens, that frames ideas surrounding smart cities, presents an opportunity for researching and understanding urban space from an agency perspective. This requires empirical openness, rethinking scale by zooming in on the hyperlocal, informal and non-prime dimension of place, what we have called the 'interstices', as well as setting aside usual assumptions regarding the relations between technology, the state and the economy. In this chapter we have sought to make the case that this is good for urban theory, and could be excellent for city-making.

References

Ajibade, I. (2017). Can a future city enhance urban resilience and sustainability? A political ecology analysis of Eko Atlantic city, Nigeria. *International Journal of Disaster Risk Reduction*, *26*, 85–92.

Anderson, M. (2017). Kenya's tech entrepreneurs shun Konza 'silicon savannah''', *The Guardian*, 5/01/2015, www.theguardian.com/global-development/2015/jan/05/kenya-technology-entrepreneurs-konza-silicon-savannah, last accessed 10/12/2017.

Aurigi, A. (2005). Competing urban visions and the shaping of the digital city. *Knowledge, Technology and Policy*, *18*(1), 12–26.

Aurigi, A. (2012). Reflections towards an agenda for urban-designing the digital city. *Urban Design International*, *18*, 131–144.

Austen, I. (2017). City of the future? Humans, not technology, are the challenge in Toronto, *The New York Times*, 29/12/17, www.nytimes.com/2017/12/29/world/canada/google-toronto-city-future.html, last accessed 5/4/2019.

Bolter, J.D. & Grusin, R. (1999). *Remediation: Understanding New Media*. Cambridge, MA: MIT Press.

Boyer, M.C. (1993). The city of illusion: New York's public places. In Knox, P. (ed.) *The Restless Urban Landscape*. Englewood Cliffs, NJ: Prentice Hall, pp. 111–126.

Burger, S. (2017). Kenya's Konza Techno City emerging as Africa's smart-city frontrunner, *Engineering News*, 28/4/2017, www.engineeringnews.co.za/article/kenya-building-konza-connected-smart-city-2017-04-28, last accessed 20/01/2018.

Cloza, G. (2012). The Truman show, *Bassa Finanza* 1/10/12, www.bassafinanza.com/archivio_newsletter/2012/Bassa_Finanza_n_44_Ottobre_2012.pdf, last accessed 16/10/13.

Cotterill, J. (2010). Emergification has arrived, the *Financial Times*' Alphaville blog, 26/10/10, http://ftalphaville.ft.com/2010/10/26/382091/emergification-has-arrived/, last accessed 18/10/12.

Coutard, O. & Guy, S. (2007). STS and the city: Politics and practices of hope. *Science, Technology and Human Values*, 32(6), 713–734.

Coyne, R. (2001). *Technoromanticism: Digital Narrative, Holism, and the Romance of the Real* Leonardo Cambridge, MA: MIT Press.

Crang, M.C., & Graham, S. (2007). Variable geographies of connection: Urban digital divides and the uses of information technology. *Urban Studies, 43*(13), 2551–2570.

Das, D. (2015). Hyderabad: Visioning, restructuring and making of a high-tech city. *Cities, 43*, 48–58.

Datta, A. (2015). New urban utopias of postcolonial India: 'Entrepreneurial urbanization' in Dholera smart city, Gujarat. *Dialogues in Human Geography, 5*(1), 3–22.

Dear, M. (1995). Prolegomena to a postmodern urbanism. In Healey, P., Cameron, S., Davoudi, S., Graham, S., Madani-Pour, A. (eds) *Managing Cities: The New Urban Context*. Chichester: Wiley, pp. 27–44.

Dourish, P. & Bell, G. (2007). The infrastructure of experience and the experience of infrastructure: Meaning and structure. *Environment and Planning B: Planning and Design, 34*(3), 414–430.

Farías, I. (2011). The politics of urban assemblages. *City, 15*(3–4), 2011.

Foth, M. (2016). Why we should design smart cities for getting lost, *The Conversation*, 7/8/16, http://thecon versation.com/why-we-should-design-smart-cities-for-getting-lost-56492, last accessed 4/12/17.

Gandy, M. (2005). Learning from Lagos. *New Left Review, 33*, 37.

Gikunda, R. M., Abura, G. O., & Njeru, S. G. (2014). Socio-economic effects of Mpesa adoption on the livelihoods of people in Bureti Sub County, Kenya. *International Journal of Academic Research in Business and Social Sciences, 4*(12), 348.

Graham, S. & Marvin, S. (2001). *Splintering Urbanism: Networked Infrastructures, Technological Mobilities and the Urban Condition*. London: Routledge.

GSMA website, Connected City, www.gsma.com/connectedliving/gsma-connected-city/, last accessed 14/11/13.

Hadjimatheou, C. (2012). Greeks go back to basics as recession bites, 20/8/12, BBC News World Radio and TV, www.bbc.co.uk/news/world-radio-and-tv-19289566, last accessed 21/10/13.

Harford, T. (2006). *The Undercover Economist*. London: Abacus.

Hitachi website, Smart cities – Living in a smart city, www.hitachi.com/products/smartcity/smart-life/index.html, last accessed 14/11/13.

Hollands, R.G. (2008). Will the real smart city please stand up? *City: Analysis of Urban Trends, Culture, Theory, Policy, Action, 12*(3), 303–320.

Ingold, T. (2007). *Lines: A Brief History*. Abingdon: Routledge.

Inman, P. & Barr, C. (2017). The UK debt crisis – in figures, *The Guardian*, www.theguardian.com/busi ness/2017/sep/18/uk-debt-crisis-credit-cards-car-loans, last accessed 23/11/17.

Jones, O. (2012). *Chavs: The Demonization of the Working Class*. London: Verso.

Koolhaas, Rem. (1995). Whatever happened to urbanism? *Design Quarterly, 164*, 28–31, www.jstor.org/stable/4091351

Lamba, N. (2013). Innovative parking plan could help clear Birmingham's traffic and skies, *Building a Smarter Planet Blog*, http://asmarterplanet.com/blog/2013/01/22902.html, last accessed 18/11/13.

Landry, C. (2012). *The Creative City: A Toolkit for Urban Innovators* (2nd ed.). London: Earthscan.

Living PlanIT. (2011). Cities in the cloud: A Living PlanIT introduction to future city technology, www.livingplanit.com/resources/Living_PlanIT_SA_Cities_in_the_Cloud_Whitepaper_Website_Edition_(2011-09-10-v01).pdf, last accessed 11/06/12.

Madden, D. (2013). Gentrification doesn't trickle down to help everyone, *The Guardian*, 10/10/13, www.theguardian.com/commentisfree/2013/oct/10/gentrification-not-urban-renaissance, last accessed 18/10/13.

Marx, C. & Kelling, E. (2019). Knowing urban informalities. *Urban Studies, 56*(3), 494–509.

Masci, R. (2012). Censis, L'Italia fra poverta' e ansia, *La Stampa*, 7/12/12, www.lastampa.it/2012/12/07/italia/cronache/censis-l-italia-fra-poverta-e-ansia-Ab4OkHGTEMRT53EAoa8rMO/pagina.html, last accessed 14/10/13.

Mbembe, A. & Nuttall, S. (2004). Writing the world from an African metropolis. *Public Culture, 16*(3), 347–372.

McFarlane, C. (2011). Encountering, describing and transforming urbanism. *City, 15*(6), 731–739.

Minton, A. (2009). *Ground Control: Fear and Happiness in the Twenty-First-Century City*. London: Penguin Books.

Murphy, C. (2017). Toronto must share the sidewalk equitably, *The Star*, 22/10/17, www.thestar.com/opin ion/commentary/2017/10/22/toronto-must-share-the-sidewalk-equitably.html, last accessed 19/6/19.

Nicolas-le Strat, P. (2007). Interstitial multiplicity. In Petcou, C., Petrescu, D., Marchand, N. (eds) *Urban/Act*. Montrouge: aaa – PEPRAV, pp. 314–318.

Nkula-Wenz, L. (2019). Worlding Cape Town by design: Encounters with creative cityness. *Environment and Planning A: Economy and Space*, *51*(3), 581–597.

Odendaal, N. (2003). Information and communication technology and local governance: Understanding the difference between cities in developed and emerging economies. *Computers, Environment and Urban Systems*, *27*, 585–607.

Odendaal, N. (2006). Towards the digital city in South Africa: Issues and constraints. *Journal of Urban Technology*, *13*(3), 29–48.

Odendaal, N. (2013). You have the presence of someone – The ubiquity of smart. In Hemment, D., Townsend, A. (eds) (2013) *Smart Citizens*. Manchester: FutureEverything Publications, pp. 31–34.

Odendaal, N. (2014). 'Space matters: The relational power of mobile technologies/O espaço importa: Poder relacional das tecnologias móveis'. *URBE: Brazilian Journal of Urban Management*, *6*(1), 33–45.

Phelps, N.A. & Silva, C. (2018). Mind the gaps! A research agenda for urban interstices. *Urban Studies*, *55*(6), 1203–1222.

Pieterse, E. (2014). Filling the void: An agenda for tackling African urbanization. In Parnell, S., Pieterse, E. (eds) *Africa's Urban Revolution*. New York: Zed Books, pp. 200–220.

Robinson, J. (2006). *Ordinary Cities: Between Modernity and Development*. Oxon: Routledge.

Sassen, S. (2011). Talking back to your intelligent city, http://whatmatters.mckinseydigital.com/cities/talking-back-to-your-intelligent-city, last accessed 28/04/13.

Shapiro, A.L. (1995). Street corners in cyberspace, *The Nation*, 3/7/95.

Simone, A. (2004). *For the City Yet to Come*. London: Duke University Press.

Simone, A. (2010). Infrastructure, real economies, and social transformation: Assembling the components for regional urban development in Africa. In Pieterse, E. (ed.) *Urbanization Imperatives for Africa: Transcending Policy Inertia*. Cape Town: African Centre for Cities, pp. 221–236.

Söderström, O., Paasche, T., & Klauser, F. (2014). Smart cities as corporate storytelling. *City: Analysis of Urban Trends, Culture, Theory, Policy, Action*, *18*(3), 307–320.

Sorkin, M. (1992). *Variations on a Theme Park: The New American City and the End of Public Space*. New York: The Noonday Press.

Toffler, A. (1981). *The Third Wave*. London: Pan Books.

Tonkiss, Fran. (2012). Informality and its discontents. In Angélil, M. & Hehl, R. (eds) *Informalize!: Essays on the Political Economy of Urban Form*. LSE Cities, 1. Berlin, Germany: Ruby Press, pp. 55–70. ISBN 9783981343663.

Townsend, A.M. (2013). *Smart Cities: Big Data, Civic Hackers, and the Quest for a New Utopia*. New York: W.W. Norton & Company.

UNEP. (2013). *City-Level Decoupling: Urban resource flows and the Governance of Infrastructure Transitions. Summary for Policy Makers*, Swilling, M., Robinson, B., Marvin, S., Hodson, M. (eds). Nairobi: UNEP.

Urban Settlements Working Group. (2019). *Area-Based Approaches in Urban Settings: Compendium of Case Studies*, Global Shelter Cluster, https://reliefweb.int/sites/reliefweb.int/files/resources/201905013_urban_compendium.pdf, last accessed 3/6/19.

Vellinga, M. (2013). The noble vernacular. *The Journal of Architecture*, *18*(4), 570–590.

Watson, V. (2013). African urban fantasies: Dreams or nightmares? *Environment and Urbanization*, 26(1), 215–231, doi:10.1177/0956247813513705

Willis, K. (2019). Whose right to the smart city? In Kitchen, R., Cardullo, P., Di Feliciantonio, C. (eds) *The Right to the Smart City*. Bingley: Emerald Publishing, pp. 27–41.

Wilson, E. (1995). The rhetoric of urban space. *New Left Review*, 1/209, Jan–Feb.

9

Governmentality and urban control

Rob Kitchin, Claudio Coletta and Gavin McArdle

Introduction

Since the 1950s digital technologies have been used by governments in the global North for the purposes of managing populations and delivering services. Over time, with the development of more sophisticated hardware and software, and the creation of the internet, the use of digital technologies for the purposes of governance has expanded in scope and depth. This has included the use of computation to monitor, manage and govern urban infrastructures and systems (Mitchell, 1995; Graham and Marvin, 1996). From the late 2000s, networked digital technologies that algorithmically produce, manage, analyse and act on streams of big data to augment and mediate the operation and governance of urban systems and life were branded as smart city technologies (Townsend, 2013). Such technologies include: city operating systems, integrated control rooms, coordinated emergency management systems, intelligent transport systems, smart energy grids, smart lighting and parking, sensor networks, building management systems, social and locative media, and city apps.

A number of scholars have argued that a key transformative effect created through the adoption of smart city technologies is the reconfiguring of urban governmentality and the practices of governance (Kitchin and Dodge, 2011; Braun, 2014; Gabrys, 2014; Klauser *et al.*, 2014; Vanolo, 2014; Davies, 2015; Sadowski and Pasquale, 2015; Luque-Ayala and Marvin, 2016; Krivy, 2018). The general conclusion is that algorithmic forms of governance are producing a shift from disciplinary forms of governmentality towards social control. For Foucault (1991), governmentality is the logics, rationalities and techniques that render societies governable and enable government and other agencies to enact governance. Every society is thus organised and managed through a system of government and governance underpinned by a mode of governmentality. The nature of governmentality mutates over time and periodically its form can shift fundamentally in character, for example, in the shift from a feudal society to modern society, wherein more systematised means for managing and regulating individuals through centralised and institutionalised control were introduced.

Through a series of essays, Foucault (1977, 1978, 1991) argued that modern governmentality – through its interlocking apparatus of institutions, administration, law, technologies, social norms and spatial logics – exercises a form of disciplinary power designed to corral and punish transgressors and instil particular habits, dispositions, expectations and self-disciplining. A key aspect of disciplinary governmentality is that people know that they are subject to monitoring and enrolment in calculative regimes. This has entailed the rollout of procedures and technologies for the systematic, wide-scale generation and assessment of data about them and their actions. In general, monitoring has been periodic, somewhat haphazard, and enacted by people working for institutions.

The implementation of algorithmic forms of governance greatly intensifies the extent and frequency of monitoring and shifts the governmental logic from surveillance and discipline to capture and control (Deleuze, 1992; Agre, 1994) through the use of systems that are distributed, ubiquitous, generate and utilise big data, and increasingly automated, automatic and autonomous in nature and work in real-time (Dodge and Kitchin, 2007; Kitchin, 2017). Here, people become subject to constant modulation through software-mediated systems, such as a transport network controlled by an intelligent transport system or a fuel consumption gauge in a car that continually displays miles per litre, in which their behaviour is directed explicitly or implicitly nudged rather than being (self)disciplined. In these examples, driving is modulated by the traffic light sequencing and the act of driving itself becomes a site of administration (Dodge and Kitchin, 2007; Braun, 2014). In other words, governmentality is no longer solely about subjectification (moulding subjects and restricting action) but about control (modulating affects, desires and opinions, and inducing action within prescribed comportments) (Braun, 2014; Krivý, 2018).

Rather than power being spatially confined and periodic, 'exercised across a network of heterogeneous institutional enclosures – each one possessing its own self-enclosed monitoring system that envelops the targeted population in a homogeneous disciplinary effect', systems of control are distributed, interlinked, overlapping and continuous, enabling institutional power to creep across technologies and pervade the social landscape (Martinez, 2011: 205). For example, as Davies (2015) notes with respect to Hudson Yards (a smart city development in New York that is being saturated with sensors and embedded computation), residents and workers will be continually monitored and modulated across the entire complex by an amalgam of interlinked systems. The result will be a quantified community with numerous overlapping calculative regimes designed to produce a certain type of social and moral arrangement, rather than people being regulated into conformity within certain institutional enclosures (such as schools and work places) (Martinez, 2011; Davies, 2015; Shepherd, this volume).

Many smart city technologies enact social control because they are cybernetic systems that function through dense and simultaneous feedback that modulates the performance of an infrastructure and those captured within it (Braun, 2014; Davies, 2015; Krivý, 2018). From this perspective, the city becomes a system of systems, as initially argued by cybernetians 50 years ago (Forrester, 1969) and reanimated more recently in smart city discourses (Townsend, 2013). Krivý (2018) contends that contemporary smart city systems are forms of second-order cybernetics that utilise positive feedback in a continuous process of self-organisation. That is, they recognise: the open, non-linear, emergent and complex properties of cities and expect unintended consequences and side-effects; and that people act as 'sensors' that feedback and shape the unfolding management rather than simply being acted upon. Within these systems '[t]he cumulative character of data streaming effectuates positive feedback loops whereby certain behaviours are amplified while others are hindered', and social change occurs through the 'accumulation of multifarious but infinitesimal behavioural adjustments' (Krivý, 2018: 23). As noted by Sadowski and Pasquale (2015), this shift to control has also been accompanied by a shift from a social contract between the state and citizens, to corporate contract wherein city services are delivered through public–private partnerships or private entities only.

The tactics and techniques of governmentality are highly varied, utilising a range of technologies, each of which can be configured and deployed in different ways. More fundamentally, the nature of governmentality can be diverse, with several related and overlapping forms of governmentality enacted and promoted by different entities at work at the same time (Ong, 2006). For example, just as disciplinary power never fully replaced sovereign power, control might supplement rather than becoming dominant to discipline (Davies, 2015; Sadowski and Pasquale, 2015). In other words, just as there are varieties of capitalism (Peck and Theodore, 2007) and varieties of neoliberalism (Larner, 2003; Brenner *et al.*, 2010) – shaped by national and local political economies, political ideology, state policies, institutional cultures, market practices, legal frameworks, public sentiment, etc. – there are varieties of governmentalities (Ong, 2006). Indeed, Ong (2006) argues that contemporary governmentalities have mutable logics which are abstract, mobile, dynamic, entangled and contingent, being translated and operationalised in diverse, context-dependent ways. From this perspective, forms of power and control invested in and enacted by smart city technologies are mutable, even within classes of technologies, driven by differing value systems and dependent on local and national institutional politics and policies and practices of deployment.

The challenge then is to map the forms and practices of governmentality with respect to the smart city – what Vanolo (2014) terms 'smartmentality' – detailing the mutable ways in which the logics of power and control are formulated and enacted. We provide an initial exploration of such a position through a case study examination of two smart city technologies: urban control rooms and city dashboards. In the first section, we document, in general terms, the use of these technologies in urban governance. We then extend this analysis and critique by considering the logics of power and control embedded within and exerted by each and their (re)production of certain modes of governmentality. We do this through an examination of the Dublin Traffic Management and Incident Centre (TMIC) and its use of SCATS (Sydney Coordinated Adaptive Traffic System) to control the flow of traffic in the city, and the Dublin Dashboard, a public, analytical dashboard that displays a wide variety of urban data. Our analysis is based upon ethnographic research conducted in TMIC in 2015/16 (see also Coletta and Kitchin, 2016) and building and operating the Dublin Dashboard (see Kitchin *et al.*, 2016).

Smart cities and urban governmentality

Urban control rooms

Accompanying the embedding of computation into the fabric of cities has been the rollout of urban control rooms of varying kinds (e.g., security, transport, utilities) capable of generating, processing, analysing and acting on real-time data. Control rooms utilising Supervisory Control and Data Acquisition (SCADA) systems can be traced back to the mid-twentieth century, but have multiplied with the growth of smart urbanism (Luque-Ayala and Marvin, 2016). Early control rooms had a limited focus, usually to monitor and intervene in real-time into the performance of a closed system, for example the operation of an electricity grid. Since the 1980s, the remit of control rooms has expanded to include more open, second-order cybernetic systems in which there are external actors who have their own ability to make decisions and create feedback loops. For example, CCTV centres for monitoring public spaces, wherein control was enacted in part through self-disciplining; or traffic control rooms which mediate the production of space and time, synchronising and optimising the space-time rhythms of vehicular and pedestrian movement and minimising disruption; or emergency management control rooms wherein control is exerted through coordination and direction of resources and personnel (Norris and Armstrong, 1999; Coletta and Kitchin, 2016; Luque-Ayala and Marvin, 2016). In general, control rooms work in the background, out of sight of public view, thus black-boxing the logic and operations of maintaining and regulating urban systems (Luque-Ayala and Marvin, 2016).

Since the late 2000s control rooms have been changing in two respects. First, with the advent of software-mediated systems, control rooms are shifting from human-in-the-loop configurations (where operators make critical decisions on system performance) to more automated forms that enact automated management; that is, they utilise computation to monitor and regulate systems in automated, automatic and autonomous ways, wherein decision-making is ceded to algorithms (Dodge and Kitchin, 2007). Second, a new breed of integrative city control room is being deployed, wherein several systems and their data are corralled into a single centre, with the walls between data and system silos collapsed to enable a more holistic and integrated view of city services and infrastructures.

The example par excellence of an integrated urban control room is the Centro de Operações Prefeitura do Rio in Rio de Janeiro, Brazil (COR). COR is a data-driven city operations centre in which software and human operators continuously monitor the city and it also acts as a coordinated emergency management centre. COR pulls together into a single location real-time data streams from 32 agencies and 12 private concessions (e.g., bus and electricity companies), including traffic and public transport, municipal and utility services, emergency and security services, weather feeds, information generated by employees and the public via social media, as well as administrative and statistical data (Luque-Ayala and Marvin, 2016). Each agency located in COR is autonomous and continues to maintain its own control room, operative systems and response protocols, with the COR providing a site of coordination and horizontal integration (Luque-Ayala and Marvin, 2016). Luque-Ayala and Marvin (2016) contend that this new type of integrated control room produces a new specific form of governmentality by altering the logic of control in four ways. First, as noted the COR draws together several domains and data flows, thus providing a coordinated meta-infrastructure that extends the logic of the control room to the totality of the city. Second, it collapses together control of the everyday (continual maintenance) and the emergency (discontinuous response to specific events), effectively managing the city as a site of perpetual crises. Third, it inverts the usual 'black box' character of control rooms by making its work visible to the public through daily media reports, its website and enabling the public to visit the centre. Critically, the centre is not positioned as a locus of surveillance, policing, discipline and law-enforcement (indeed, it is generally not used for these activities), but as a means to maintain infrastructure performance, minimise disruption to everyday life and share information. Fourth, it enrols the public as a 'citizen sensors' (they supply information to the centre through social/locative media). In the latter case, the public 'engage in the labour of being watched' (Monahan and Mokos, 2013), and are active participants in a system that is beyond their control and modulates their behaviour.

What this discussion highlights is that, while the logics of control articulated by control rooms share similarities, the means by which control is exercised within and mobilised through them is mutable across domain, systems and location, and evolves with new governmental arrangements and technological configurations. Importantly, how power is exercised through urban control rooms varies in three ways. First, there are different practices of control being exerted: intervention, self-disciplining, mediation, coordination, direction, optimisation and co-option. Each is designed to produce particular regulatory outcomes. For example, intervention seeks to directly control and shape an outcome; self-disciplining seeks individuals to self-modulate behaviour in a desired way; optimisation aims to produce an optimal, efficient system; and so on. Second, how systems are configured and operated varies across sites depending on management practices and governance context. Third, the extent of automated management varies with respect to the role of human operators in mediating their work. Some systems, or selected aspects of them, are configured to be human-in-the-loop (algorithms identify issues and suggest solutions but key decisions have to be made by the human operator), some human-on-the-loop (the system is automated but under the oversight of a human operator who can over-ride or take-over the system) and some human-off-the-loop (fully automated) (Coletta and Kitchin, 2016).

City dashboards

City dashboards use visual analytics – dynamic and/or interactive graphics (e.g., gauges, traffic lights, metres, arrows, bar charts, graphs), maps, 3D models and augmented landscapes – to display information about the trends, performance, structure and patterns of cities. Selected data about cities are displayed on a screen using data visualisations, which in many cases are interactive (e.g., selecting, filtering and querying data; zooming in/out, panning and overlaying; changing type of visualisation or simultaneously visualising data in a number of ways). Most data within city dashboards are sampled data generated on a set schedule (e.g., monthly, annually). Increasingly, big data are being incorporated; that is, data that are produced in real-time by the Internet of Things (IoT), automatically produced by sensors and actuators, but also by people through their participation in crowdsourcing and use of locative and social media. In some dashboards data are 'consolidated and arranged on a single screen so the information can be monitored at a glance' (Few, 2006: 34). Here, a city dashboard operates like a car dashboard or plane cockpit display providing critical information in a single view (Gray et al., 2013). Analytical dashboards are more extensive in scope and are hierarchically organised to enable a plethora of interrelated dashboards to be navigated and summary-to-detail exploration within a single system (Dubriwny and Rivards, 2004). Both types of dashboard are used in urban control rooms, but they are also increasingly being displayed in mayor's offices, public buildings, and made accessible to the general public via dedicated websites along with the associated data (see Figure 9.1). In the latter case, citizens are able to use the data to conduct their own analyses and build city apps.

The power and utility of urban dashboards is twofold. First, they act as cognitive tools that improve the user's 'span of control' over a large repository of voluminous, varied and quickly transitioning data (Brath and Peters, 2004). As such, they enable a user to explore the characteristics and structure of datasets and interpret trends without the need for specialist analytics skills (the systems are point and click and require no knowledge of how to produce such graphics). Second, they purport to show in detail and often in real-time the state of play of cities. Urban dashboards seemingly enable users to know the city as it actually is through objective, trustworthy, factual data that can be statistically analysed and visualised to reveal patterns and trends and to assess how it is performing vis-à-vis other places (Kitchin et al., 2015). They supply a rational, neutral, comprehensive and commonsensical media for monitoring and evaluating the effectiveness of urban services and policy, and to learn and manage through measurement. In so doing, dashboards facilitate the notion that it is possible to 'picture the totality of the urban domain', to translate the messiness and complexities of cities into rational, detailed, systematic, ordered forms of knowledge (Mattern, 2014). In other words, they provide a powerful realist epistemology for monitoring and understanding cities, underpinned by an instrumental rationality in which 'hard facts' trump other kinds of knowledge and provide the basis for formulating solutions to urban issues (Kitchin et al., 2015;

Figure 9.1 An at-a-glance (London Data Dashboard) and analytical city dashboard (Dublin Dashboard)

Source: http://citydashboard.org/london/ and www.dublindashboard

Mattern, 2014, 2015). As such, they seemingly provide a neutral and value-free media through which to make sense of, govern and plan a city. And they expand the capacity to govern by extending forms of power/knowledge.

City dashboards are becoming increasingly important mechanisms for evaluating and guiding the work of city administrations and regimes of urban governance, though how they are implemented differs in ethos and form between cities. In general, initiatives fall into two broad camps, which together reveal the inherent tension within schemes between seeking to facilitate accountability, transparency and democracy, and enact forms of discipline, regulation and control (Hezri and Dovers, 2006; de Waal, 2014). In some municipalities, city dashboard initiatives form the bedrock for performance management systems that are used to guide operational practices with respect to specified targets; to provide evidence of the success or failure of schemes, policies, units and personnel; and to guide new strategies, policy and budgeting (Craglia *et al.*, 2004; Behn, 2014; Kitchin *et al.*, 2015). Since 1999, Baltimore has used a system called CitiStat to implement a balanced scorecard approach to actively monitor the performance of city departments and guide the development of new policies and programmes and then assess the success of their implementation (Behn, 2014). Every week city managers meet to review performance and set new targets for the city as a whole and for each department, and discipline under-achievement (Gullino, 2009). Dozens of other US cities have deployed similar systems. Such an approach is supported by an instrumental rationality that believes that continual monitoring will positively influence the performance, quality and productivity of city staff and services by reshaping behaviours and disciplining and rewarding actions with respect to targets (Hezri and Dovers, 2006).

In other cases, municipalities use city indicator projects and associated dashboards in a more contextual way to provide robust city intelligence, which complements a variety of other information such as staff input, citizen/community feedback, consultancy reports and expert opinion, to help inform policy-making and implementation. In these cases, cities are understood to consist of multiple, complex, interdependent systems that influence each other in often unpredictable ways. Moreover, governance is seen as being complex and multi-level in nature requiring consensus building and cooperation across actors and scales, with the performance of systems and staff not easily reducible to performance metrics and targets. In other words, the city is not a machine that can be fine-tuned and managed through a set of simple data levers (Innes and Booher, 2000). Dashboard information, however, is seen to provide valuable contextual insight that facilitates coordination, integration and interaction across municipality departments and stakeholders by detailing trusted and authoritative datasets for the city and reducing uncertainty and insecurity in decision-making (Van Assche et al., 2010). In other words, dashboards and their data act as a normative and rational bridge between knowledge and policy (Hezri and Dovers, 2006). A long-standing example of a contextual city indicator approach is that employed within Flanders, Belgium, where since the late 1990s a number of cities have employed a common City Monitor for Sustainable Urban Development, consisting of nearly 200 indicators, to provide contextual evidence for policy-making (Van Assche et al., 2010).

Urban power and control

The Dublin Traffic Management and Incident Centre

TMIC provides a single, integrated, 24/7 control room to house the core traffic management systems for monitoring and controlling the road transportation network and traffic flow in the Greater Dublin Area, including dealing with major events and incidents (Figure 9.2). To monitor and regulate the traffic flow the centre uses a network of 380 CCTV cameras, 800 sensors (inductive loops), a small number of Traffic Cams (traffic sensing cameras) used when inductive loops are faulty or the road surface is not suitable for them, a mobile network of approximately 1,000 bus transponders (controllers can also directly contact drivers if needed), phone calls and messages by the public to radio stations and operators, and social media posts. These networks of the Internet of Things and citizen sensors produce a continuous flow of real-time data which are used to dynamically manage the road system.

The core means by which the data are parsed and used to control traffic flow is via the adaptive traffic management system, SCATS. SCATS is an automated and adaptive system whose primary role is to manage the dynamic timing of signal cycles and phases at junctions for vehicles, cycles and pedestrians in order to ensure the optimal flow, minimise congestion and accidents, and manage incidents. The system is adaptive in the sense that it automatically calibrates the cycles and phases dependent on a set of programmed rules and the flow, speed and density of traffic for each lane of traffic in previous cycles and phases. For example, the number of cars and the gaps between them as detected by the inductive loops denotes if a phase was too short or long, with the timing of the next phase recalculated automatically by the system. Public buses benefit from prioritisation, so as they approach a junction the phasing will alter to accommodate their passage. By pressing a pedestrian crossing button at junction, people produce a temporary break in the phasing, closing down the main phase in order to run the pedestrian phase. Cycles are set to last a minimum of 40 seconds to a maximum 130 seconds, but in practice they rarely exceed 80 seconds or go below 60 seconds. This calculation is based on the pragmatic evaluation that the waiting time for a pedestrian crossing above 80 seconds would be too long. Given that alterations in cycles

Figure 9.2 A view from a controller's desk in the Dublin traffic control room
Source: Chapter author(s)

and phases flux, altering traffic flows across the system, changes in one location can sometimes produce congestion elsewhere, and the system seeks to minimise such disruption by balancing competing demands across junctions.

By monitoring patterns over time, the TMIC staff can configure the setting of SCATS to take into account whether it is a weekday or weekend, as well as seasonal/daily rhythms and when schools are closed. In addition, operators can intervene and override the present or original SCATS settings. In this sense, SCATS is a human-on-the-loop system, wherein automation is used monitor and regulate traffic flow but operators oversee and can manually override its work. To oversee SCATs, a controller is presented with a dashboard-like interface (Figure 9.3). The right hand part of the screen displays a junction, the various traffic lanes and their phases, and the left hand part the length of time for each phase. Interventions are circumscribed by the initial configuration of the system by Intelligent Transportation System staff, who in turn refer to the Traffic Signs Manual by the National Roads Authority that sets rules on the minimum and maximum times for phases. If, for example, operators try to go below the minimum safety times for green or red time on a different phase, SCATS will automatically override the modification attempt with the original configuration.

As well as directly altering the phasing of junctions and the rhythms of traffic flow, much of the data utilised in the traffic control room is shared with the public via a number of channels, enabling people to see and interpret the data themselves and self-regulate their interactions with the traffic system and to manage time-based decisions for journey planning.

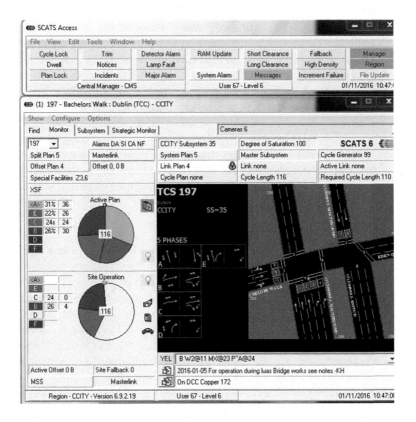

Figure 9.3   The SCATS interface
Source: TMIC screenshot by chapter author(s)

For example, real-time information about the expected arrival time of buses and trams are shared via smartphone apps, websites and on-street dynamic signs. Details of congestion and traffic accidents are shared via radio bulletins (one of the desks in the control room is reserved for an AA Roadwatch operator who communicates traffic news to radio stations throughout the day, and three desks host Dublin City FM's live broadcast of traffic news and music between 7 and 10 am and 4 and 7 pm, Monday to Friday (www.dublincityfm.ie/). These broadcasts inform and pacify travellers and enable them to seek alternative routes. Data from the systems are openly shared via Dublinked (the city's open data portal) and displayed in the Dublin Dashboard.

SCATS is a second-order cybernetic system that enacts a human-on-the-loop form of automated management to dynamically modulate the movement of people while also enrolling them as citizen sensors. At present, the Dublin TMIC is not used for routine surveillance or policing. While the centre pulls together data from a number of different systems, including from citizens, and one desk is reserved for Gardai (police) use, the centre does not generate or store indexical data (e.g., it does not employ automatic number plate recognition cameras), nor does it record video footage. Nor is it responsible for speed, red light or bus lane cameras. That said, the Gardai do have their own access to the camera network, which they use for policing. Moreover, the centre can be used for managing major events and emergencies. As Monahan (2007) details, traffic control rooms are particularly susceptible to control creep, with the purpose of managing traffic flow being extended to include routine surveillance, policing and security work, and the data generated shared with other state agencies. With this control creep the governmentality enacted by traffic control rooms shifts. In addition, he notes that while traffic control rooms are portrayed as socially and political neutral and impartial in how they manage traffic, they nonetheless support certain values and socially sort the users of city space, 'valorizing certain mobilities over others, while normalizing unequal experiences of space' (Monahan, 2007: 373). As such, intelligent transport systems sustain 'ongoing neoliberal development patterns by emphasizing "pipes" over places, maximizing the flow of privately owned vehicles through those pipes' (385), and privileging the support for certain mobilities over others (private over public transportation, driving over walking or bicycling).

The Dublin Dashboard

The Dublin Dashboard (www.dublindashboard.ie, Figure 9.1) is an extensive, open, analytical dashboard launched in September 2014. It provides citizens, planners, policy-makers and companies with an extensive set of data and interactive data visualisations about Dublin City, including real-time information, indicator trends, inter and intra-urban benchmarking, interactive maps, location-based services, a means to directly report issues to city authorities, and links to city apps. The data used in the dashboard are open and available for anyone else to build their own apps. Like the London dashboard in Figure 9.1, the Dublin Dashboard was initiated as a university research project, with the aim of exploring the praxes and politics of developing such a dashboard (Kitchin et al., 2016). However, most urban dashboards are initiated and produced internally or by a third-party supplier (either as a bespoke product or using a template solution such as Socrata) (Kitchin et al., 2015).

Shortly after initiation, the project started to work with Dublin City Council, with city officials supplying data and providing feedback on its development. Although the dashboard is presented as a stable, authoritative and technical assemblage of networked infrastructure, hardware, operating systems, assorted software and data, achieved through neutral, objective processes of scientific conception, engineering and coding, it is also thoroughly social and political. Indeed, it is a complex socio-technical assemblage of actors (e.g., university researchers, city officials, other stakeholders) and actants (e.g., data, software, servers, standards) that work materially and discursively within a set of social and economic constraints, existing technologies and systems, and power geometries to assemble, produce and maintain the system (Kitchin et al., 2016). During development, ideas and choices concerning the aims, principals and technical approach were debated, refined, rescinded, reinstated and revisioned. For example, there were debates concerning what indicators to include and how they should be presented, with the city officials conscious of the potential media and political messaging such data might produce. As such, whilst the narrative spun by companies, and often also by city management, suggests that the transition to a smart city is a smooth path of rollout and integration, the reality is a set of iterative processes of debate and compromise.

Over time, the dashboard has continued to evolve and mutate in terms of the data included, the tools and modules produced, and design interface. In its second phase of development, post-2016 it has been completed revamped to be entirely open source (open code and open data) and cater for different levels of data literacy. What this contingency and relationality means is that power and control are never fixed in either the creation or ongoing operation of city dashboards (Kitchin et al., 2016). That said, initial design does provide a certain degree of path dependency in how the dashboard is organised and its functionality, and how it develops over time. As such, of particular importance to the logic of power and control in the development and use of the Dublin Dashboard were its aims and principles.

Initially, the aim of the site was to provide a contextual, rather than a performance-management, dashboard for the city. The underlying principles related to its development were: there would be no closed modules, with all of the visualisations on the site accessible to everyone; all of the data used on the site would be open in nature, enabling others to build their own apps; as much data as possible, regardless of source or type, would be made available through the site; where possible it would use open source tools; existing resources and apps would be used if they did a good job to remove duplication of effort; the site would be easy to use, with users requiring no mapping or graphing skills; and the site would be interactive allowing users to explore the data.

The Dublin Dashboard then sought to enact an approach that aligned with the principles of open government and the open data movement aimed at producing transparency, participation, empowerment, accountability and evidence-informed decision- and policy-making. However, it did not embrace targets or performance-driven metrics designed to implement a form of city managerialism-by-data. This was one area of negotiation between the research team and the local authority. However, in the absence of already established targets, the lack of mechanisms to guide and react to performance vis-à-vis such targets, and the political nature of the project arbitrarily imposing targets, it became a moot point. The logic of the system was then neither discipline nor control-orientated in a direct way, but rather sought to provide evidence for citizens and city workers with respect to key aspects of everyday life. Nonetheless, the ability to make sense of, mobilise and act on the data presented through the dashboard varies across individuals and organisations, and the data are still used to assess the performance of the city administration and to pressure for reforms and change through political and media campaigns. Moreover, because the data tools are used to shape policy formulation and to justify and underpin modes of governmentality, they inevitably shape and reproduce how people are governed. For all city dashboards, although the updating of the data and visualisation tools can be automated, any control actions are usually human-in-the-loop in nature. That is, the translation of information into knowledge and action, and forms of governance and governmentality, are performed by people rather than algorithms.

Conclusion

There is little doubt that smart city technologies are changing urban governmentality and governance. At a broad level, the embedding of computation into the fabric of cities and its use to manage city services and infrastructures are shifting governmentality from disciplinary forms to those of social control. Here, rather than governmentality concentrating on moulding subjects and restricting action within spatial enclosures, it seeks to modulate affects and channel action across space. However, as our discussion of urban control rooms and city dashboards reveals, this transformation in governmentality is uneven and diversely constituted. Indeed, there is much variety in the configuration and deployment of socio-technical assemblages – even within particular technological domains such city dashboards and control rooms – and in their logics of control, tactics and operational techniques. For example, systems vary with the extent to which they implement forms of automated management (with humans in-, on-, off-the-loop) and in how they seek to enact governmentality: through modes of surveillance and discipline or capture and control; through systems that are 'black-boxed' or transparent; and through regulatory techniques such as coercion, co-option, self-disciplining, punish, modulation, intervention, mediation, coordination, direction and optimisation.

The logics of control articulated by smart city technologies are diverse and mutable across domain, system, location and context. In a city with a range of smart city systems, several related and overlapping forms of governmentality can be enacted and promoted by different entities and be at work at the same time. Both the Dublin TMIC and the Dublin Dashboard pull together and integrate a range of urban data. Both seek to enrol citizen sensors to crowdsource additional data, and both openly share some of their underlying data with the public. The Dublin TMIC seeks to produce a continuous modulation in the flow of traffic using SCATS, a second-order cybernetic system of control, and has a human-on-the-loop configuration. It aims to optimise the performance of the road network, minimising congestion and maximising flow and speed of movement. In contrast, the Dublin Dashboard is rooted in an open data/government ethos and set of principles designed to promote openness, transparency and contextually based, evidence-informed policy-making, and it is very much a human-in-the-loop configuration with regards to action. This diversity in ethos, purpose and logics of control multiplies across the range of smart city technologies deployed in a city.

At any one time then, there are varieties of governmentalities at work in the neoliberal smart city (Ong, 2006). However, heeding caution from Brenner et al. (2010) about overstating the diversity and mutability of governmentality and divorcing its forms from their wider context, it is important, we believe, to consider a number of related questions. These include: How do smart city technologies and initiatives and their associated governmentality and logics of control fit together? Do they work in concert or in opposition to one another? How do they dovetail with other assemblages and practices of governance? Do they work within, reproduce and evolve the wider political economy and regulatory context?

What is required then is for the governmentalities of specific smart city assemblages to be unpacked and to chart how they work in collaborative concert (see Vanolo, 2014; March and Ribera-Fumaz, 2016; Datta, 2018; Wiig, 2018). This needs to be complemented with a mapping in detail of the wider overlapping governmentalities of the smart city, how initiatives interlink and work together to legitimise and (re)produce technocratic forms of governance, and how the practices and governmentalities of smart city endeavours coalesce with and extend those framed and enacted through other means (Shelton et al., 2015; Karvonen et al., 2018; Cardullo et al., 2019). Such a mapping, we believe, needs critical attention if we are to understand the logics of control of smart cities, how they work to produce particular formulations of the neoliberal city, and how we might envisage and create a different kind of smart city.

Acknowledgements

The research for this chapter was provided by a European Research Council Advanced Investigator Award, 'The Programmable City' (ERC-2012-AdG-323636) and a Science Foundation Ireland grant, 'Building City Dashboards' (15/IA/3090).

References

Agre, P. (1994) Surveillance and capture: Two models of privacy. *Information Society* 10(2): 101–127.
Behn, R.D. (2014) *The PerformanceStat Potential: A Leadership Strategy for Producing.* New York: Brookings Institution Press/Ash Center.
Brath, R. and Peters, M. (2004) *Dashboard Design: Why Design Is Important.* DM Direct, October.
Braun, B.P. (2014) A new urban dispositif? Governing life in an age of climate change. *Environment and Planning D: Society and Space* 32: 49–64.

Brenner, N., Peck, J. and Theodore, N. (2010) Variegated neoliberalization: Geographies, modalities, pathways. *Global Networks* 10(2): 182–222.

Cardullo, P., Di Feliciantonio, C. and Kitchin, R. (eds) (2019) *The Right to the Smart City*. Bingley: Emerald.

Coletta, C. and Kitchin, R. (2016) Algorhythmic governance: Regulating the 'heartbeat' of a city using the Internet of Things. *Big Data and Society* 4: 1–16.

Craglia, M., Leontidou, L., Nuvolati, G. and Schweikart, J. (2004) Towards the development of quality of life indicators in the 'digital' city. *Environment and Planning B* 31(1): 51–64.

Datta, A. (2018) Postcolonial urban futures: Imagining, governing and reclaiming India's urban age. *Environment and Planning D: Society and Space*. Online first. doi: 10.1177/0263775818800721

Davies, W. (2015) The chronic social: Relations of control within and without neoliberalism. *New Formations* 84(84–85): 40–57.

de Waal, M. (2014) *The City as Interface: How New Media Are Changing the City*. Amsterdam: Nai010 Publishers.

Deleuze, G. (1992) Postscript on the societies of control. *October* 59 3–7.

Dodge, M. and Kitchin, R. (2007) The automatic management of drivers and driving spaces. *Geoforum* 38 (2): 264–275.

Dubriwny, D. and Rivards, K. (2004) *Are You Drowning in BI Reports? Using Analytical Dashboards to Cut Through the Clutter*. DM Review, April 2004, www.advizorsolutions.com/press/Cut%20Through%20The%20Clutter.pdf (last accessed 4 June 2014).

Few, S. (2006) *Information Dashboard Design: The Effective Visual Communication of Data*. Sebastopol: O'Reilly.

Forrester, J.W. (1969) *Urban Dynamics*. Cambridge: MIT Press.

Foucault, M. (1977) *Discipline and Punish*. London: Allen Lane.

Foucault, M. (1978) *The History of Sexuality*, Volume One. New York: Random House.

Foucault, M. (1991). Governmentality, In Burchell, G., Gordon, C. and Miller, P. (eds) *The Foucault Effect: Studies in Governmentality*. Chicago: University of Chicago Press, pp. 87–104.

Gabrys, J. (2014) Programming environments: Environmentality and citizen sensing in the smart city. *Environment and Planning D* 32(1): 30–48.

Graham, S. and Marvin, S. (1996) *Telecommunications and the City: Electronic Spaces, Urban Places*. London: Routledge.

Gray, S., Milton, R. and Hudson-Smith, A. (2013) *Visualising Real-time Data with an Interactive iPad Video Wall*. Talisman. www.geotalisman.org/files/2013/05/MethodsNewsSpring2013_DRAFT_Part7.pdf (last accessed 29 August 2016).

Gullino, S. (2009) Urban regeneration and democratization of information access: CitiStat experience in Baltimore. *Journal of Environmental Management* 90: 2012–2019.

Hezri, A.A. and Dovers, S.R. (2006) Sustainability indicators, policy and governance: Issues for ecological economics. *Ecological Economics* 60: 86–99.

Innes, J. and Booher, D.E. (2000) Indicators for sustainable communities: A strategy building on complexity theory and distributed intelligence. *Planning Theory & Practice* 1(2): 173–186.

Karvonen, A., Cugurullo, F. and Caprotti, F. (eds) (2018) *Inside Smart Cities: Place, Politics and Urban Innovation*. London: Routledge.

Kitchin, R. (2017) The realtimeness of smart cities. *Tecnoscienza* 8(2): 19–42.

Kitchin, R. and Dodge, M. (2011) *Code/Space: Software and Everyday Life*. Cambridge, MA: MIT Press.

Kitchin, R., Lauriault, T. and McArdle, G. (2015) Knowing and governing cities through urban indicators, city benchmarking and real-time dashboards. *Regional Studies, Regional Science* 2: 1–28.

Kitchin, R., Maalsen, S. and McArdle, G. (2016) The praxis and politics of building urban dashboards. *Geoforum* 77: 93–101.

Klauser, F., Paasche, T. and Söderström, O. (2014) Michel Foucault and the smart city: Power dynamics inherent in contemporary governing through code. *Environment and Planning D: Society and Space* 32 (5): 869–885.

Krivý, M. (2018) Towards a critique of cybernetic urbanism: The smart city and the society of control. *Planning Theory* 17(1): 8–30.

Larner, W. (2003) Neoliberalism? *Environment and Planning D: Society and Space* 21: 509–512.

Luque-Ayala, A. and Marvin, S. (2016) The maintenance of urban circulation: An operational logic of infrastructural control. *Environment and Planning D: Society and Space* 34(2): 191–208.

March, H. and Ribera-Fumaz, R. (2016) Smart contradictions: The politics of making Barcelona a self-sufficient city. *European Urban and Regional Studies* 23(4): 816–830.

Martinez, D.E. (2011) Beyond disciplinary enclosures: Management control in the society of control. *Critical Perspectives on Accounting* 22(2): 200–211.

Mattern, S. (2014) Interfacing urban intelligence. *Places Journal.* https://placesjournal.org/article/interfacing-urban-intelligence/ (last accessed 13 Jan 2017).

Mattern, S. (2015) Mission Control: A history of the urban dashboard, *Places Journal* https://placesjournal.org/article/mission-control-a-history-of-the-urban-dashboard/ (last accessed 13 Jan 2017).

Mitchell, W. J. (1995) *City of Bits: Space, Place and the Infobahn.* Cambridge: MIT Press.

Monahan, T. (2007) 'War rooms' of the street: Surveillance practices in transportation control centers. *The Communication Review* 10(4): 367–389.

Monahan, T. and Mokos, J.T. (2013) Crowdsourcing urban surveillance: The development of homeland security markets for environmental sensor networks. *Geoforum* 49: 279–288.

Norris, C. and Armstrong, G. (1999) CCTV and the social structuring of surveillance. *Crime Prevention Studies* 10(1): 157–178.

Ong, A. (2006) *Neoliberalism as Exception: Mutations of Citizenship and Sovereignty.* Durham, NC: Duke University Press.

Peck, J. and Theodore, N. (2007) Variegated capitalism. *Progress in Human Geography* 31(6): 731–772.

Sadowski, J. and Pasquale, F. (2015) The spectrum of control: A social theory of the smart city. *First Monday* 20(7). http://journals.uic.edu/ojs/index.php/fm/article/view/5903 (accessed 6 Jan 2017).

Shelton, T., Zook, M. and Wiig, A. (2015) The actually existing smart city. *Cambridge Journal of Regions. Economy and Society* 8(1): 13–25.

Townsend, A. (2013) *Smart Cities: Big Data, Civic Hackers, and the Quest for a New Utopia.* New York: WW. Norton & Co.

Van Assche, J., Block, T. and Reynaert, H. (2010) Can community indicators live up to their expectations? The case of the Flemish City Monitor for Livable and Sustainable Urban Development. *Applied Research Quality Life* 5: 341–352.

Vanolo, A. (2014) Smartmentality: The smart city as disciplinary strategy. *Urban Studies* 51(5): 883–898.

Wiig, A. (2018) Secure the city, revitalize the zone: Smart urbanization in Camden, New Jersey. *Environment and Planning C: Politics and Space* 36(3): 403–422.

10

How smart is smart city Lagos?

Taibat Lawanson and Olamide Udoma-Ejorh

Introduction

The 'smart city' concept has gained prominence in global urban discourse in recent times. Though coined in the 1990s with a focus on the role information and communication technology (ICT) can play in modern infrastructure within cities, it is largely being promoted as the key to Africa's third revolution (Huet, 2016). Smart cities are envisioned as the response to the challenges of rapid urbanization across the continent and an opportunity for African cities to leapfrog into the mid-21st century (Delloitee, 2016).

Even though a universal definition of the concept remains elusive, the notion has been substituted with several analogies: digital city, creative city, intelligent city, knowledge city, sustainable city, virtual city, learning city, hybrid city and information city (Pérez-Torregrosa, Martín and Ibáñez-Cubillas, 2017), many of which predate the use of the term 'smart' in describing urban processes (Batty, 1995; Willis and Aurigi, 2017). However, a city can be defined as 'smart' when investments in human and social capital, and traditional (transport) and modern (ICT) communication infrastructure fuel sustainable economic development and a high quality of life, with a wise management of natural resources (Deloitte, 2014).

Various cities all over the world have been designated as 'smart cities' for a variety of reasons (Caragliu, Del and Nijkamp, 2011). In Nigeria, the concept of 'smart city' is subject to various interpretations ranging from utopian visions of modernist infrastructure to the infusion of technology into public policy and practice. Nowhere is this more apparent than in Lagos – Nigeria's commercial capital which is currently on a mission to transit 'from a megacity to a smart city' (Dina, 2017). The Lagos state government has instituted and/or supported various programmes and projects to achieve this transition.

A review of literature on the smart city concept revealed the dominance of studies with geographical emphasis on cities of the Global North (Graham and Marvin, 2001; Kitchin, 2015), Europe (Caragliu, Del and Nijkamp, 2011) and South East Asia (Moe, 2011) theorizing about the emergent smart city as a socio-technical process. Cities investigated include Dubai (Khan *et al.*, 2017), Singapore (Foo and Pan, 2016) and Barcelona (Bakici *et al.*, 2012). Research on African smart cities is less prevalent and includes the work of Watson (2015) as well as focal studies on South African cities (Du Plessis and Marnewick, 2017),

Nairobi (Sen *et al.*, 2016) and Kigali (Doherty, 2013). Slavova and Okwechime (2016) who attempted a synthesis of the smart city concept and the African Union's Agenda 2063 recommended faster technology adoption across the continent. Olokesusi and Aiyegbajeje's (2017) study on e-democracy as a pathway for inclusive development discovered that the poor electricity infrastructure constitutes an impediment to the success of e-democracy in Lagos, while Soyinka *et al.*'s (2017) study of smart infrastructure in Lagos concluded that the development of smart power solutions is critical to the development of a smart city.

However, the studies have not considered the linkages between the global narrative on the smart city concept and its framing vis-à-vis urban development in Lagos. What is the smart city agenda for Lagos? Who benefits? Specifically, this chapter will interrogate the deployment of smart city principles in urban infrastructure, economic development and governance efforts and the effects of these changes on everyday experiences of residents of the city. It will begin by correlating global narratives and local institutional interpretations of the smart city, and then examine three case studies of how local residents are responding to the emerging smart city paradigm.

African smart cities and the pressures of globalization

The 'Africa Rising' narrative is gaining currency around the world, with smart urbanization being touted as a veritable means of achieving this (Huet, 2016). In fact, according to Deloitte (2016), Africa is the next big market for technology, and African cities can through the successful adoption of the ideology and technology of the smart city agenda become globally competitive. The smart city concept and its relevance to the African urban context will be dissected along its three major framings: technology, people and systems (see: Alabi, 2018; Mejer and Bolivear, 2016).

Through the lens of technology, a smart city is an intelligent and interconnected city which has the competence to collect and link real-life data through the use of meters, personal devices, sensors and appliances (Kitchin, 2015) and Vanolo (2014). This position is promoted by international technology corporations such as IBM, Huawei, Siemens AG and Cisco Systems, who are aggressively seeking to expand their product uptake to the emerging African market (MoUD, 2014). However, Greenfield. (2013) has argued that solely viewing and designing smart cities from technological lens shuns the real knowledge on how cities function and the 'people' factor in a city.

The 'people' factor as emphasized in the creative city concept claims that for a city to be smart, education, knowledge and learning all have central roles to play (Thuzar, 2011). Thus the smart city is defined as a city with advanced education, better educated people, skilled and trained labour force (Comunian, 2010; Florida, 2002), and that the more educated and skilled people are in a city, the smarter the citizens and the city itself. This perspective has been adopted by many international consulting firms such as McKinsey & Company (2013) and Deloitte (2014) who posit that Africa is growing economically and thus is set to urbanize rapidly with an emerging middle class. Hence, marketing strategies of international property development corporations (such as Rendeavour, the promoters of Alaro and Tatu smart cities) target Africa's creative class and urban elite (Datta, 2015a; Rappoport, 2015; Wiig, 2016). A criticism of this approach has been on the emphasis on providing elitist luxury housing that does not respond to the intense demand for low-cost housing and infrastructure, hence engendering exclusionary development, gentrification and urban inequality (Ravindran, 2015; Lawanson., 2017). In fact, Florida (2017) articulated this position, effectively rebuking his earlier (Florida, 2002) one about the potentials of the creative class.

The third category of smart city visioning is the systems perspective, which synthesizes the smart city as a series of actions which cities undertake towards innovation in management, technology and policy (Gil-Garcia, 2012; Gil-Garcia et al., 2015; Anthopoulos, 2017). Through this approach, the smart city ecosystem comprises of eight defining components: economy, people, services, governance, infrastructure, mobility, environment and living (Giffinger et al., 2007) that establish a cyber-physical integration of urban life. Other definitions within the systems approach combine smart services with smart government, while smart infrastructure is combined with smart environment. This framing of smart urbanization brings together two problematic neoliberal urban visions – first that the use of ICT will drive economic growth and urban prosperity; second that the use of ICT can make urban governance more efficient, manageable, transparent and hence equitable (Kitchin, 2015).

This systems view of the smart city notion aligns with the growing world-class and global city rankings (Yadav and Patel, 2015), in tandem with the neo-liberal underpinnings of city categorizations (Robinson, 2002). In fact, the network theory of smart cities (Komninos, 2006) emphasizes that the success and prosperity of a city are highly dependent on its virtual position within global networks, in which urban areas are linked via several globalized flows including foreign direct investments, information, knowledge, ideas and people (Sassen, 2011; Söderström et al., 2014; Wall and Stavropoulos, 2016). IESE 2018 Cities in Motion project (Centre for Globalization and Strategy, 2019), which is sponsored by the University of Navarra's Center for Globalization and Strategy, and IESE Business School's Department of Strategy analyzed the level of development of 165 cities, across nine dimensions considered key to being a smart, sustainable city. In the 2018 IESE Smart City index, all African cities ranked poorly from Tunis (134), Cape Town (143), Casablanca (148), Rabat (155), Johannesburg (156), Cairo (157), Nairobi (161) to Lagos (164). Among the variables included are the number of terrorist attacks and the number of Apple stores in each city. A similar scenario played out in the AT Kearney Global Cities Index in which Lagos placed 133 on the global city outlook index among 135 cities ranked. Some indicators include human capital (foreign born population, international schools), business activity (Fortune 500 companies, global service firms), information exchange (broadband subscribers, online presence), cultural experience (museums, international sporting events) and political engagement (embassies, political conferences and international organizations).

While some of these smart city indicators are important, and even necessary to track development progress, the reality is that there is a clear disjuncture between the perception and reality of what a smart city should be between proponents in the Global North and South. For example, according to Nakiguli (2015), in the Global North perceptions of smart cities include the availability of digital infrastructure, robotic assistance, human–robot discussions and systems working with systems; while in many cities of the Global South, more fundamental challenges of access to basic services, infrastructure and governance determine how smart and/or sustainable a city is. Yet scholarship on smart cities is dominated by a 'one-size-fits-all' critique where broader theoretical arguments are seen to stand for and reveal the discursive and material realities of actually existing smart city developments (Kitchin, 2015). For example, studies have shown that many of these projects are fundamentally flawed and others are often incomplete (Cugurullo, 2013; Wiig, 2015). Furthermore, the fact that many of these smart city initiatives in Africa are backed by investment and technology corporations in the Global North corroborates the positions of scholars like Datta (2015b) and Watson (2014), who argue that smart cities represents a techno-utopian fantasy driven by corporate interests in Western contexts (Pollio, 2016).

There is therefore a need to explore differential expressions across the Global North and South, thus the next section will consider how the institutional interpretation of the smart city concept in urban development projects and programmes in Lagos, Nigeria. They will be discussed within the three major dimensions – technology, people and systems. It will go further to examine through case studies how residents of the city are responding to the emerging paradigm.

Smart city Lagos? Institutional interpretations

There is no specific government policy on smart city in Nigeria, either at national or sub-national levels. The first formal acknowledgement was at the 2016 Intercessional Panel of the United Nations Commission on Science and Technology for Development where Nigeria's contribution to the panel as stated by Tukuma (2016) was:

> Smart cities no doubt are money spinning avenues for government at all levels, the opportunity must be explored to address the problems of slums in our cities. Presently, Nigeria is making efforts at developing its own smart cities in some states of the Federation including the Federal Capital Territory (FCT), Abuja through coordinated and concerted opportunities by the government in partnership with the Private Sector. Prominent amongst the cities being developed is the Eko Atlantic City and that of Lekki which is jointly being handled by the Lagos State Government, Commercial Banks and Private Investors. In the FCT, one of such smart city being constructed is the Centenary City currently being built by the Federal Capital Development Authority in collaboration with the Private Sector.

This framing of the federal government's narrative on smart cities aligns with the commercial undercurrents of the *creative city* context. Both Eko Atlantic and Centenary cities are promoted by foreign investments, with no provision for lower middle-class and low-income earners in the layout, thus contradicting the '*smart cities provide the opportunity for addressing the problem of slums in our cities*' notion.

The Nigerian minister of Science and technology in aligning with the technology framing of smart city concept stated at the Nigeria 2017 Smart City Summit that

> More people now have access to smart phones. Experts said that smart phones users in Africa will increase from 75 million in the first quarter of 2016 to 512 million 2018. Smart Cities would promote open data is one of the advantages of smart city. Smart City helps to manage government resources. It also helps to get more revenue for the government. In order to encourage healthy competition among Cities and States, the Ministry in collaboration with its Partners, will soon launch '*The Smart City Challenge*'. The idea is to encourage Local Governments, Cities, States Governments, and the built industry to take revolutionary steps to go to the level of digital technology.[1]

Interestingly, the minister also announced a partnership between the federal government and Huawei (a major smartphone technology conglomerate) during the event. The fact that the Smart Cities Summit was sponsored by Huawei lends credence to the submissions of Datta (2015b) and Watson (2014) on the intense international interests influencing emerging techno-utopian fantasies of African smart cities

Likewise, the Lagos State Development Plan (LSDP) (2012–2025) (Lagos State Government, 2013), while not explicitly mentioning 'smart city', prioritizes technology, global competitiveness and economic growth in its projects and programmes, across various ministries and public agencies. This aligns strongly with the 'systems approach' to smart cities. The Lagos State Ministry of Science and Technology (2017) is the coordinating agency for implementing smart city initiatives across the state, identified the priority areas for technology deployment: public security, emergency responses, waste management, e-governance, smart energy and integration of the public service, as well as revenue generation. An assessment of how much progress Lagos has made in the context of the defining components of the smart city ecosystem,[2] global indicators of smart city development[3] and the targets of the LSDP appears in Table 10.1.

An overview of the institutional manifestation of the various dimensions of the smart city in Lagos has shown the state government has instituted programmes and projects across them all. There has been significant mobile phone penetration across the city, and the extent of the population of the city with tertiary education has been due to the contribution of over ten different higher education institutions owned by the Lagos state government. The e-government platforms still lean heavily on freely available social media platforms, while the CODE Lagos and Lagos State Employment Trust Fund (LSETF) programmes have contributed significantly to the entrepreneurship and innovation thrusts of the economic dimension of smart city. Extensive deployment of smart city technology has been introduced in the transport and mobility sectors, though with low performance due to lack of integration with existing emergency management and security systems. Furthermore, many of the personnel responsible for its implementation are yet to be trained. Regarding environment and infrastructure, green energy initiatives have been largely through the installation of solar power systems in local schools and health centres funded by international aid, though poor maintenance has stifled impact. Smart urban planning has also resulted in many model city and regional development plans being developed, however there is yet to be considerable impact on the environment as implementation of the provisions of the plans is not prioritized.

Smart city Lagos? Residents responses

Residents of Lagos continue to interpret and appropriate the concept of smartness in a variety of ways to suit their needs as they navigate the city. Some of these are discussed below under the broad contextualization of smart infrastructure, smart economy and smart mobility in contradistinction to smart government and services. The thread of smart people runs through the three sub-topics.

Smart infrastructure: the rise of new cities

A seeming solution to the housing challenge in Lagos has been the proliferation of smart city urban development projects across Lagos. One example is the Imperial International Business City (IIBC), a commercial luxury real estate development on the Lekki Peninsula. It is tagged as the first 'eco-friendly, smart business city in Africa', with the anticipation of becoming a popular tourist attraction (Channeldrill Resources Limited, 2017). It is heavily reliant on foreign consultants such as Cisco, Mott MacDonald, Royal Haskoning DHV and Cordros. It is a partnership between the Elegushi Royal Family (the land owners) and Channeldrill Resources, an infrastructure company set up in 2014 to deliver

Table 10.1 Smart city assessment of Lagos

	Component of smart city	Global indicators of smart city development	Target of Lagos state government	Remarks
1	People	• 21st century education • Inclusive society • Embraced creativity	• % of population with tertiary degree • Introduction of ICT in school programmes • Internet penetration	• The state-owned Lagos State University founded in 1983 has about 35,000 students in three campuses. There are about ten other state-owned higher educational institutions including polytechnics and colleges of education • 85% have smartphones; 76% have internet access; 66% use computers (Idowu, 2018) • 79% of internet connections in Nigeria were from a mobile phone; 25% of all mobile phones sold in Nigeria were in Lagos, mostly low-cost smartphones (Jumia, 2018) • The most active apps are WhatsApp and Facebook • 12% of the population use mobile banking apps
2	Government and services	• Increased efficiency in public services such as e-government, open data and transparency • Dependent on customer data and intelligent processes	• E-government platform for government–citizen interaction • Deploying 10,000 high definition cameras to support traffic management and emergency management services	• The state government maintains an active Facebook page and many agencies interact with the public via Twitter • The state government and many agencies have websites, though information is often outdated

No	Dimension			
		• Services are connected to products through technology • Connecting different urban spaces and actors to create inclusive and sustainable cities		• The state publishes official statements and news via its websites • Close circuit cameras have been deployed across the state for real-time traffic and security surveillance; integration with emergency services – 50% completed. Integration with policing network is not operational yet • There are many smart boards providing timely traffic information • Some toll gates have unmanned points with e-card enabled access
3	Economy	• Entrepreneurship and innovation • Smart financing • Local and global connectedness • E-commerce • Employment in technology services • Circular economy	• Setting up of the Lagos State Employment Trust Fund for micro, small and medium enterprises • Setting up of CODE Lagos project to train 1 million young people to code by 2030	• As at 2018, 10,000 beneficiaries of loans from Lagos State Employment Trust Fund • As at 2017, 60,000 people have been trained in CODE Lagos centres • Many public schools do not have ICT laboratories or power to support the CODE Lagos • Tech-hubs are growing but none are supported by state funding • Government provides free WiFi services in some public parks
4	Mobility	• Mixed-modal access • Intelligence traffic management • Clean and non-motorized options	• Auto Inspector – online system with vehicle data • E-platform for verification of vehicle status • Central Billing System	• The state government runs Bus Rapid Transit and skeletal ferry services • Auto Inspector, e-platform and the Central Billing System have been installed but hardly function due to

(Continued)

Table 10.1 (Cont.)

	Component of smart city	Global indicators of smart city development	Target of Lagos state government	Remarks
		• Integrated ICT	• Vehicle inspection service centres • Lagos Bus Rapid Transit app • Traffic cameras installed • Traffic management system • Growth of e-hailing and commuting apps	challenges with decentralization of the technology and lack of adequately trained personnel • 10 out of 57 vehicle inspection centres are operating • Bus Rapid Transit app is not widely used • E-hailing apps such as Uber and Taxify dominate the private taxi services
5	Environment and Infrastructure	• Green buildings • Green energy • Green urban planning	• Establish independent power projects across the state using a standardized and sustainable framework for power • Use best practice for the design, execution, maintaining and improving public lighting projects in Lagos state • Integrate independent power projects with public lighting schemes using innovative approaches • Energy smart meters and smart grids	• There are two LEED (Leadership in Energy and Environmental Design) rated buildings in Lagos • Recycling is championed in the private sector with many small and medium enterprises emerging • Energy from the grid is largely erratic and petrol- and diesel-powered generators are the main alternative to grid supply • Renewable energy – solar energy and energy banking (inverters) are slowly penetrating

No.		Objectives	Indicators	Status
6	Living	• Enhanced quality of life through healthcare and safety • Quality of housing • Culturally vibrant and happy/ social cohesion	• % penetration of primary health care • % penetration of basic education • % of slum like condition • % gini coefficient • 60% skilled formal employment by 2020 • youth unemployment • Public service provision for healthcare, safety, housing and education	• UK government/Department for International Development supported the Lagos Solar Power Project to provide solar electrification in schools, hospitals and local communities. Many are currently non-functional due to poor maintenance • Model city and sub-regional master plans have been commissioned though level of compliance and public uptake is still quite low • Less than 30% have access to primary healthcare • Most citizens access basic services such as water, sanitation and health services through the informal economy • In 2017, informal employment accounted for 65% of employed citizens • Gini coefficient of 0.64 is reflective of high inequality

aspirational luxury infrastructure. According to the managing director of Channeldrill Resources, Femi Akioye:

> The Head of Elegushi Royal Family wanted a legacy project that will redefine the real estate sector. Channedrill Resources on its part wants to be the face of the next century when it comes to real estate in Africa and beyond.[4]

The marketing brochure sells the idea as:

> The IIBC is a city for forward-thinkers, by forward-thinkers. It is a city designed for those no longer satisfied with the status quo but yearn for the next-level. It is designed for those who do not just want to experience history being made but want to make history. This is a city for makers, creators, designers, thinkers, leaders, and visionaries. This is a city for the greats.

It claims to be self-sustaining, climate proof, smart, secure and eco-friendly. In fact, IIBC boasts in its marketing plan of world-class infrastructure and smart technology, including a water treatment plant, independent electricity (gas generated), fibre optics cabling, cloud-enabled communication network, smart city/house infrastructure, a smart shopping mall, cloud-enabled 24 hour spy security connected to a central security centre and a world-class hospital within a dedicated healthcare zone (Vanguard, 2017).

Apart from the shiny packaging of being a technology-enhanced island, the island is being sold as a solution for the housing crisis in Lagos, albeit catering to the high-income market (SEO Properties Limited, 2017). It is designed as an 'exclusive private island'[5], currently selling at N125,000 (US$350) per square metre[6]. In reality however, IIBC will only serve to exacerbate the already evident socio-spatial segregation being experienced in Lagos. According to Akioye (cited in Ojiako, 2018: 31):

> We were also quite passionate about bringing luxury living to Lagosians by creating a serene environment that connects work, living and recreational environments of Lagos in a wholesome mix'. The word affordable is relative, but I can tell you that hard working Nigerians and international firms will be able to make IIBC their abode.

While the land that IIBC sits on presently used to be a Otodo-Gbame community of 36,000 low-income inhabitants who were forcefully removed through a series of violent eviction exercises between 2015 and 2017 (Emeka, 2017), the project is simply unaffordable to most Nigerians. The price of land at IIBC ranges from N64,350,000 to N429,000,000 (US$177,356 to $1,182,370). There is an incremental payment plan with an initial deposit of N10,725,000 ($29,541) for the smallest plot size (SOE Properties Limited, 2017). Furthermore, the design ideas are borrowed from the Middle East and Europe and therefore not a reflection of local realities and culture. In fact, all the design and development consultants are European, with a track record of similar projects executed in the Middle East. The Dutch engineering firm on IIBC Project – Royal Haskoning DHV – has delivered similar projects in Thua Thien Hue, Jordan, Qatar, Myanmar and Indonesia.[7]

IIBC and other emerging new city developments across Lagos do not adequately contribute to resolving the housing crisis in Lagos. Lagos has a population of 23 million people and a housing deficit of about 3 million units, mainly in the low-income sector (UNHABITAT, 2017; PISON, 2016). Local community-based smart solutions being practiced are more impactful in improving housing and infrastructure outcomes. Cooperative societies have been able to provide contextualized housing solutions for many residents, while residents associations have also pooled resources to provide neighbourhood security services, solar-powered street lights and drainage infrastructure (Lawanson, 2017, 2018). This is in addition to many households installing energy storage (inverter) systems homes as a response to endemic power shortages.

Smart economy: from markets to malls

The Lagos State Ministry of Physical Planning and Urban Development, under the 2005 Lagos State Regional Planning Laws, is mandated to plan, facilitate and organize a safe, green, dynamic, economically and culturally vibrant and sustainable city (Filani, 2012). The major milestones support optimal land use and market redevelopment. In the last decade market regeneration and development have become an important part of urban renewal and traditional markets have become collateral damage. It is necessary to note that many of the market men and women consist of Lagos' working poor. The working poor, earning less than $1.9 daily make up 79.4 percent of the population, while middle-income (earning $2–10 daily) and high-income (earning above $10 daily) earners are 18.7 percent and 1.8 percent of all Lagos residents respectively (AfDb, 2017). According to the Lagos State Waste Management Authority (LAWMA) there are 425 markets in Lagos state and Komolafe (2016) estimates that 5 million people, of which 70 percent are women and their children, primarily engaged in informal sector economic activities.

However, under the pretext of urban renewal, market redevelopments have occurred to establish what have been named malls or 'ultra-modern' markets, often through public–private partnerships (PPPs). The Lagos state government has actively supported the construction of malls, with former Governor Babatunde Fashola quoted as saying in 2011, 'The number of new small businesses that are opening or expanding their outlets are the measures of real growth in any economy. Everywhere you see a mall opening; it is a positive sign for that economy'.[8] What is left unsaid is that these market redevelopments in fact hurt existing small businesses, and that major international retail chains are found in these places. For example, some affected markets such as Oyingbo and Tejuosho, which used to serve the entire West African sub-region with items ranging from foodstuffs, local crafts and fabrics to household materials (Lawanson, 2014) are now a shadow of themselves.

Built in the early 1970s, Tejuosho market was gutted by fires in 2004 and 2007. It was redeveloped during the tenure of Governor Fashola in 2014 as a 'world-standard ultra-modern market' and renamed as the Main Tejuosho Shopping Complex. What used to be a local market with fewer than 800 shops is now a five-storey complex of over 4,000 shops, underground parking, elevators, a mini-power plant and waste disposal and water systems. The design of the market is also such that the shops back the street, rather than the conventional market design where shops face the street in order to attract customers. By 2018, less than 40 percent of the shops were inhabited, due largely to exorbitant rates (Alabi, 2018). New occupants in the market are largely formal sector operators such as banks, offices and the international retail chain, Spar. Many of the former occupants who are not able to afford shops in the market have relocated to cheaper stores in adjoining streets. The smaller vendors who used to rent stalls in the former market now operate in the thriving informal market in the open space in front of the market, which attracts daily rental fees for a 1 square metre floor space (Figure 10.1). Others have left the market area entirely, resorting to leveraging the internet for performing their commercial transactions.

Figure 10.1 Left and right: informal market in the compound of Tejuosho market
Source: blackandyellowurbanism.com

The rise of the online market is a clear indication of a citizen-led emerging smart economy, in which small businesses are responding to the challenges of accessible and affordable space for commercial activity. Online businesses require minimal infrastructure: regular internet access, a social medial handle and an active manager to drive content and activity on the handle. For those selling tangible items, a storage space (usually home based) and a delivery service (usually local commercial motorcycle) are also required. While some vendors utilize their social media platforms (WhatsApp, Instagram and Facebook), digital platforms such as Jumia, Konga, Iso and Deal-Dey are alternative marketplaces that enable sellers and buyers to interact (Figure 10.2). In fact, Jumia[9] reports an estimated number of 700,000 visitors weekly, with 57 percent accessing the platform via mobile web and 18 percent via mobile app. Customers prefer the digital platform because of convenience and affordability.

According to Jumia, 25 percent of all mobile phones sold in Nigeria were in Lagos, mostly low-cost smartphones, with the most active apps being WhatsApp and Facebook. The 2018 report by Terragon Group (Egenuka, 2019) estimates about 22 million active social media users in Nigeria, which is a huge and immediate client base for the numerous entrepreneurs who utilize social media as a digital marketplace (Figure 10.2). Even though there are not much data on the pattern of these businesses and their contribution to the economy, there is growing reliance on them across all sectors. For example, while specialized professional service pitches are on LinkedIn and Twitter, product sales thrive more on Facebook, Instagram (Figure 10.3) and WhatsApp (Figure 10.4).

Ultra-modern malls are not fulfilling their intended purpose of improving the economy and/or supporting small businesses. In fact, many erstwhile occupants of local markets are now reduced to street trading (mobile and stagnant), while new malls springing up around the city are mostly utilized for recreational purposes. Cursory observations across various malls in Lagos reveal that only the cinemas, entertainment centres and food courts generate significant patronage. Conversely, the mobile marketplace is thriving, with many small businesses preferring the convenience and affordability it provides.

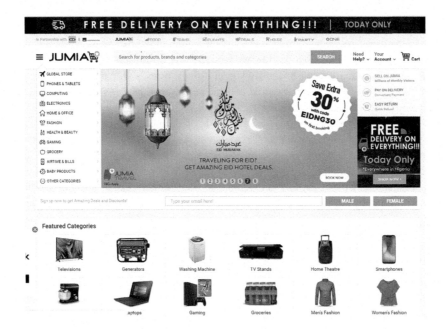

Figure 10.2   Product sales on digital market place, Jumia
Source: Screenshot by chapter author(s)

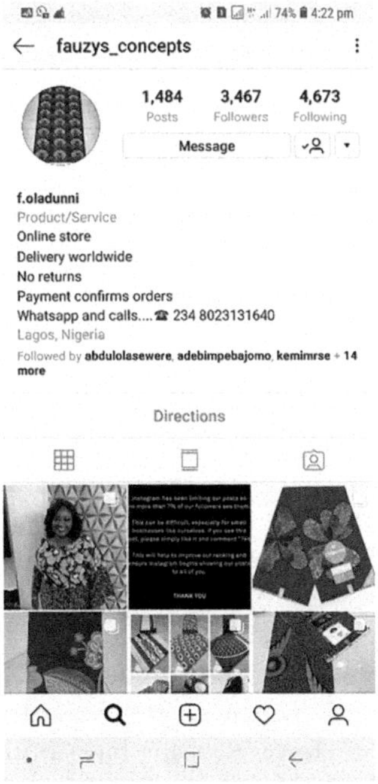

Figure 10.3   Product sales on Instagram
Source: Screenshot by chapter author(s)

Figure 10.4   Product sales on WhatsApp
Source: Screenshot by chapter author(s)

In order to move towards a smart economy, it is important that market redevelopments are done in a manner that enhances the welfare and opportunities accruable to small businesses. A functional market with a growing customer base is the primary objective and therefore the needs and perspectives of the users of the market must be incorporated into the redevelopment process. The vision of modernity has led to a government push of 'ultra-modern' malls which are failing to meet the needs of the communities they are situated in and instead push people to innovate and find new ways to sell. Whether online or offline, traders contribute to the formal economy (through taxes and fines), though they are still described as informal and are yet to be captured in the governments vision of a 'smart city'. A vision of Lagos needs to capture and enhance what already exists and not eradicate to build from scratch because as can be seen from the Tejuosho example, it is not likely to succeed.

Smart mobility: integrated transport and traffic management

The transport system in Lagos is such that available infrastructure is grossly inadequate. Road continues to be the primary mode of transport, with 90 percent of total passengers and goods moved via the road network (Lawanson, 2018). The state-owned Bus Rapid Transit (BRT) scheme is only able to move 1.5 percent of traffic along Lagos roads, while informal mini buses (known as danfos) and private cars account for 45 percent and 11 percent, respectively. The continuous increase of private vehicles and the reliance on informal public transport have led to extreme traffic congestion in the city. Despite the clear need to diversify and the Lagos state government's integrated and multi-modal transport aspirations, the use of rail and water transport modes is minimal.

The Lagos State Development Plan 2012–2025 aims to 'create a safe, reliable and efficient integrated and multimodal transportation system, that included rail, bus and ferry, for sustainable socio-economic development of Lagos State' (Oshodi and Salau, 2016, p. 77). The Lagos Strategic Transport Master Plan (STMP) was developed by the Lagos Metropolitan Area Transport Authority (LAMATA) as a long-term plan focused on integration of the mass rapid transit system, developing the waterways and rail networks, installing a modern intelligent transport system and introducing walking and cycling facilities (Oshodi and Salau, 2016, pp. 83–84). The modern intelligent transport system proposed in the STMP project includes an electronic integrated fare and ticketing system and a unified information system for all modes of transport that uses and manages real-time events efficiently. As of 2019, the Blue Line which is the first of six rail lines and one monorail planned for the city is yet to be completed, seven years after the initial take off date of 2012, thus stalling the integrated multi-modal transport integration.

The BRT system is highlighted in the Lagos State Development Plan 2012–2025 and stretches across 35km, carrying an estimated 120,000 passengers per day. It is the first on the African continent and is projected as a smart transport solution for Africa, and has been introduced into the mass transit systems of cities such as Cape Town, Addis Ababa and Dar es Salaam.[10] However, due to operational challenges, the system suffered several setbacks. The BRT is estimated to carry 150,000 passengers daily. However, many buses of the initial fleet have been grounded due to poor maintenance. The ongoing Lagos bus reform initiative seeks to correct some of these errors by the construction of 13 bus terminals, 100 bus stops and 5,000 new commercial buses to replace the informal 'danfo' buses. These new buses,

800 of which were launched in May 2019, will be maintained in dedicated maintenance yards that are currently being installed.

According to Lagos state Governor Akinwunmi Ambode, the danfo bus drivers will be absorbed into the new bus system.[11] However, there is no official documentation as to how this will be done. The major option culled from some press releases of the government is for danfo drivers through their union – the Nigerian Union of Road Transport Workers – to apply for a franchise and purchase buses (minimum 50 buses) in order to participate in the new bus system.[12] As such, it has increased the livelihood vulnerabilities of danfo drivers who are faced with an uncertain future and possible transition from small business owner to employee status. Furthermore, the environmental consequences of discarding or repurposing 145,000 danfo buses are yet to be considered.

The Lagos BRT mobile phone application was launched in 2017 as a journey planner offering real time, up to the minute information on arrival and departure times of various buses. However, as of March 2018, on the APP Store it is not rated because there are 'not enough ratings or reviews to display a summary', therefore suggesting that not many people use it. Even though Lagos has a large population of over 20 million, the BRT mobile phone application has a low download rate (about 1,000 as of October 2018),[13] probably because many BRT passengers are among the many Nigerians who are have no access to the internet. In 2018, Nigeria had 92.3 million internet users and an internet penetration rate of 47.1 percent of the population.[14] Moreover, the cost of smart phones and data is still out of the financial reach of many, according to Jumia. While Infinix is the highest selling budget smart phone, the cheapest brand costs $95 (N32,900),[15] way higher than the national minimum wage of $51 (N18,000).[16]

Some other smart initiatives have also been introduced into the transport system, though they are largely private sector (international investor) driven. The Lagos state government, through the Motor Vehicle Computerized Inspection Service (LACVIS), introduced the 'Auto Inspector' in 2011 to provide computerized vehicle diagnostics, as well as real-time vehicle information and driver identification for police checks and traffic offender apprehension. The technology, though available, is hardly utilized by LACVIS, primarily because the training of operators as well as fitting of supporting devices in police vehicles and integration with other agencies is yet to be done. As such, the smart solution is not being decentralized and is not being used effectively or efficiently. Similar to the Auto Inspector initiative, in 2017 a technology billing system was created using Automatic Number Plate Recognition (ANPR), Radio Frequency Identification (RFID) and Central Billing System (CBS) technologies. The e-platform was to allow the Vehicle Inspection Service to perform on-the-spot verification of vehicle status. The aim was for defaulters to have bills automatically raised by the CBS and delivered to their registered address. However, neither the billing system nor the auto-check technology is yet fully operationalized. Furthermore, as of February 2019, only 10 of the 57 proposed computerized vehicle inspection service centres have been built.[17]

There have also been some citizen-led interventions in providing smart transport solutions for the everyday challenges faced by Lagos residents. These include Lara.ng, Shuttlers and Max.ng. Lara.ng is a crowdsourced online journey planner that helps Lagosians plan their journeys using both informal and formal transport. Data are collected from the informal commercial buses, tricycle and motorcyles (danfo, keke maruwa and okadas) as well as the formal (BRT) modes of transport. The advantage it has over the BRT app is the fact that it adopts a holistic view to the solution it provides for travelling around the city, acknowledging the limitations of the BRT as a sole mode of travel for the average Lagosian. Another private company that is attempting to ease commuting in Lagos is Shuttlers. This is a company that provides mass transit services for individuals and companies between the central business areas of Ikoyi, Victoria and Lagos Islands and residential areas, including the peri-urban areas of Ikorodu and Ibeju Lekki via a mobile phone application.

Smart people: social media and governance

Worldwide, social media is increasingly being used as a way to bring decision-makers closer to their constituents and allow for a more transparent relationship. If used efficiently, social media can drive citizen engagement with limited resources and budgets. It allows for real-time data collection and instant feedback. This could lead to improved services within cities and communities. Social media can also be used as a tool to transform public perception, maximize awareness of agency goals and gain the trust of citizens by becoming more authentic and transparent. During elections or when implementing a project, social media can also help test messaging and ideas.

There is yet to be a structured e-governance framework in Nigeria, though Lagos state government and some of its agencies leverage the beneficial roles of social media primarily for information dissemination. The Lagos State Emergency Management (LASEMA) and Traffic Management (LASTMA) agencies both have active Twitter handles that interact with citizens, providing timely information and responding to distress calls. Thus, they are able to mobilize citizen support to effectively perform their duties. By contrast, the Lagos state government's official social media handles barely respond to posts from citizens, practicing uni-directional communication and highlighting the fact that technology does not necessarily enhance governance as being promoted by the systems view of smart city. Victoria Okoye stated that 'If governments are not interested in listening to citizens in the first place, internet and open data won't change this'.[18]

Social media has also been used in Nigeria as a tool for citizen mobilization. Though the threat of fake news persists (Hitchen et al., 2019), examples such as the Occupy Nigeria protest of 2012 in which over 20,000 Nigerians were mobilized through social media to stand against petrol price increases and the 2018 ENDSARS campaign which resulted in police service reforms are indicative. More recently, social media outrage regarding the Lagos waste management sector played a major role in escalating the matter to one with political implications (Lawanson, 2018).

Conclusion

The chapter so far has revealed that the notion of smartness goes beyond the incorporation of technology into urban systems and processes. In the case of Lagos, there are glaring differences between the formal institutional (government, technology companies, urban developers) conceptualizations of what a smart Lagos should be, and how citizens are appropriating technology for solving everyday challenges. While the institutional approach is top-down and guided oftentimes by aspirational modernist aspirations that seek to position Lagos as a world-class city, that of citizens is bottom-up and often geared towards survival and adaptation to the everyday challenges of Lagos life. An amalgam of both will put Lagos on the path to becoming a truly smart city, which has been defined as a city with sustainable, people-centred governance and administrative systems that are supplemented by technology (Nsibidi Institute, 2016). In reinforcing the truly smart Lagos is the need to promote both hi-tech (institutional, international) and low tech (citizen led, local) contextualized solutions which are co-created in response to the urgent challenges of a complex city.

In order to achieve this, there is an urgent need for improvements in the public governance and administrative systems. The introduction of technology does not translate to smart urbanism when the human element capacity to operate the technology is lacking. The cases of the Auto Inspector and the CBS are smart solutions which are yet be operationalized due to bureaucratic challenges and capacity gaps. The BRT system is also operating sub-par due to poor maintenance of vehicles and a failure to recognize the contribution of informal systems in the current transportation value chain, while the struggles of the Tejuosho market redevelopment can be attributed to a failure to take into consideration the lived experiences of traders in the redesign of the market. Furthermore, true civic engagement can only be achieved when the uni-directional approach to interaction with the public is lifted in favour of the bi-directional approach being practiced by LASEMA and LASTMA. Evidence points to increased public confidence in both agencies due to such interactions.

While citizen-led approaches to smart urbanism are not without challenges, they have to a large extent responded positively to the challenges of urban life, including those resulting from the modernist aspirations of urban smartness. The lara.org app is more useful to the average commuter than the BRT app due to its recognition of the various ways everyday citizens navigate the city.

Therefore, in conclusion, the quest for smart city Lagos can only be achieved when both human and material efforts are deployed to meet everyday challenges of Lagosians and technology is utilized as a means to a more sustainable society, and not an end in itself.

Notes

1 http://www.nta.ng/news/infrastructure/20180412-smart-cities-in-nigeria-an-achievable-initiative-fg/; www.itnewsafrica.com/2017/08/nigerian-government-and-huawei-partner-on-smart-cities-initiative/.
2 The components of the smart city ecosystem are economy, people, services, governance, infrastructure, mobility, environment and living (Giffinger et al., 2007).
3 Indicators of global smart city development in this context are extracted from McKinsey Global Institute 2018 Smart Cities Report: Digital Solutions for a more liveable future, IESE 2018 Smart City Index and AT Kearney 2018 Global Cities Index.
4 www.thisdaylive.com/index.php/2018/08/10/imperial-international-business-city-is-a-legacy-project-akioye/.
5 The phrase exclusive private island is used twice and the word exclusive is used six times in The Investors' Brochure (SOE Properties Limited, 2017).
6 www.nigeriapropertycentre.com/for-sale/land/mixed-use-land/lagos/lekki/lekki-phase-1/185441-imperial-international-business-city-lekki-1.
7 www.royalhaskoningdhv.com/en-gb/projects.
8 www.tundefashola.com/archives/news/2011/12/14/20111214N01.html.
9 www.jumia.com.ng/mobile-report/.
10 www.ired.org/modules/infodoc/files/english/lagos_sets_standard_for_urban_transport_in_africa.pdf.

11 https://lagosstate.gov.ng/blog/2017/03/12/danfo-well-kick-start-bus-reform-initiative-with-n30bnambode.
12 https://lagosstate.gov.ng/blog/2017/03/03/lagosand-the-danfo-quagmire.
13 https://play.google.com/store/apps/details?id=de.ivu.realtime.app.lamata&hl=en.
14 www.statista.com/statistics/183849/internet-users-nigeria/.
15 www.jumia.com.ng/infinix/?gclid=Cj0KCQjwz6PnBRCPARIsANOtCw3kuZSn9gelzbmhFFQioTCDViDHpbTu0zL437eqOCh5ezK6DGn_JQoaAp6EEALw_wcB.
16 https://tradingeconomics.com/nigeria/minimum-wages. The federal government announced a new minimum wage of N30,000 which is yet to be operationalized.
17 www.lacvis.com.ng/about-us/.
18 https://themetropole.blog/2019/04/29/decision-making-and-future-thinking-in-lagos/.

References

AfDb (2017). Nigeria: Economic Outlook. Available at afdb.org
Alabi. S. (2018). *Strategic Planning and Implementation of the Tejuosho Market Upgrade*. Research report submitted to Department of Urban and Regional Planning, University of Lagos.
Anthopoulos, L.G. (2017). *Understanding Smart Cities: A Tool For Smart Government or an Industrial Trick?*. Springer: Cham, Switzerland, 2017.
Bakici, T., Almirall, E., & Wareham, J. (2012). A Smart City Initiative: The case of Barcelona. *Journal of the Knowledge Economy*, 2(1), 1–14.
Batty, M. (1995). The computable city. *International Planning Studies*, 2, 155–173.
Caragliu, A., Del Bo, C., & Nijkamp, P. (2011). Smart cities in Europe. *Journal of Urban Technology*, 18 (2), 65–82.
Centre for Globalization and Strategy (2019) IESE Cities in Motion Index – 2018 Smart city index. Available at media.iese.edu/research
Channeldrill Resources Limited. 2017. Imperial International Business City, Pre-Sale Teaser [BOOKLET].
Comunian, R. (2010). Rethinking the Creative City. *Urban Studies*, 48(6), 1157–1179. 10.1177/0042098010370626
Cugurullo, F. (2013). How to Build a Sandcastle: An Analysis of the Genesis and Development of Masdar City. *Journal of Urban Technology*, 20(1), 23–37.
Datta, A. (2015a). New urban utopias of postcolonial India: Entrepreneurial urbanization in Dholera smart city, Gujarat. *Dialogues in Human Geography*, 5, 3–22.
Datta, A. (2015b). *The Smart Entrepreneurial City: Dholera and a 100 other utopias in India. Smart Urbanism: Utopian Vision or False dawn?*. C. McFarlane, S. Marvin and A. Luque-Ayala London: Routledge.
Delloitee 2016. Africa Outlook 2016 Summary report. Available at: www2.deloitte.com/.
Deloitte. 2014. Africa is ready to leapfrog the competition Through Smart Cities Technology. [ONLINE] Available at: www2.deloitte.com/.
Dina, A. 2017. Ambode: Above The Euphoria Of Megacity. [ONLINE] Available at: https://lagosstate.gov.ng/blog/2017/08/04/ambode-above-the-euphoria-of-megacity/
Doherty, K (2013), "Kigali – remaking the city", Cityscapes Vol 3, pages 30–31, available at www.cityscapesdigital.net/.
Du Plessis, H., & Marnewick, A. L. (2017). A roadmap for smart city services to address challenges faced by small businesses in South Africa. *South African Journal of Economic and Management Sciences*, 20(1), 1–18.
Egenuka, N. (2019) Firm Empowers SMES on social media investments returns. *The Guardian (Nigeria)*, 22 February 2019.
Emeka, M. (2017). So You Think You Know Lagos III: How Demolition And Eviction By Lasg Is Tearing Families Apart. *Disrupting Education*, [Online] Available at: https://ynaija.com/so-you-think-you-know-lagos-iii-how-demolition-and-eviction-by-lasg-is-tearing-families-apart-disrupting-education/ [Accessed 10 MARCH 2018].
Federal Government of Nigeria. (2016). Contribution of the federal republic of Nigeria to the cstd 2015-16 priority theme on 'smart cities and infrastructure. *Intersessional panel of the united nations commission on science and technology for development (cstd) Budapest, Hungary*, 11-13, January 2016.
Filani, M. O. (2012). *The Changing Face of Lagos From Vision to Reform and Transformation*. Cities Alliance: Brussels.

Florida, R. (2017). *The New Urban Crisis: How our Cities Are Increasing Inequality, Deepening Segregation, and Failing the Middle Class-And What We Can do About It*. New York: Basic Books. AQ2.

Florida, R. L. (2002). *The Rise of the Creative Class: And How It's Transforming Work, Leisure, Community and Everyday Life*. New York: Basic books.

Foo, SL, Pan, G. (2016). Singapore's vision of a smart nation: Thinking big. *Starting Small and Scaling Fast. Asian Management Insights*, 3, 77–82.

Giffinger, R., Fertner, C., Kramar, H., Kalasek, R., Pichler-Milanovic, N., & Meijers, E. (2007). *Smart cities. Ranking of European Medium-Sized Cities*. Vienna UT: Centre of Regional Science.

Gil-Garcia, J. R., Pardo, T. A., & Nam, T. (2015). What makes a city smart? Identifying core components and proposing an integrative and comprehensive conceptualization. *Information Polity*, 20(1), 61–87. http://dx.doi.org/10.3233/IP-150354

Gil-Garcia, M. P. (2012). Towards a smart state? Inter-agency collaboration, information, integration, and beyond. *Information Polity*, 17(3), 269–280. http://dx.doi.org/10.3233/IP-2012-000287

Graham, S. and Marvin. S. (2001). *Splintering Urbanism: Networked Infra- Structures, Technological Mobilities, And the Urban Condition*. London: Routledge. 2014.

Greenfield.A. (2013). *Against the Smart City*. London: Verso.

Hitchen, J., Hassan, I., Fisher, J. and Cheeseman, N. (2019) WhatsApp and Nigeria's elections: Mobilizing the people, protecting the vote. Research Report. University of Birmingham Centre for Democracy and Development, Abuja https://www.cddwestafrica.org/whatsapp-nigeria-2019-elections/

Huet, J. (2016). *Smart Cities: The Key To Africa's Third Revolution*. BearingPoint Institute. www.bearing point.com/en/our-success/thought-leadership/smart-cities-the-key-to-africas-third-revolution/

Idowu. Z. (2018). *Assessing Smart City Agenda in Lagos State*. Final year project report Department of Urban and Regional Planning, Univeristy of Lagos.

Jumia. (2018). Mobile Report Nigeria 2018, [ONLINE] Available at: https://www.jumia.com.ng/mobile-report-2018/.

Nsibidi Institute (2016) Open city Lagos – The City for All. https://ng.boell.org/2015/04/02/city-all

Kearney, A. T. (2019) *2018 Global Cities Report*. www.atkearney.com/2018-global-cities-report

Khan, M.S., Woo, M., Nam, K. and Chathoth, P.K. (2017). Smart Tourism: A Case of Dubai. *Sustainability*, 9, 2279.

Kitchin, R. (2015). Making sense of smart cities: Addressing present shortcomings. *Cambridge Journal of Regions, Economy and Society*, 8, 131–136.

Komninos, N. (2006). *The Architecture of Intelligent Cities*. Intelligent Environments, Institution of Engineering and Technology, 13–20. www.urenio.org/wp-content/uploads/2008/11/2006-The-Architecture-of-Intel-Cities-IE06.pdf

Komolafe, G. (2016). Mbenga The Poor Also Must Live!" Market Demolition, Gentrification and the Quest for Survival in Lagos State, [ONLINE] Available at: www.wiego.org

Lagos State Government (2013). Lagos state development plan (2012–2025). Ministry for Economic Planning and Budget. Available at mepb.lagosstate.gov.ng

Lagos State Government. (2017). *Lagos State Smart City Project Overview*. Flier produced by Lagos State Ministry of Science and Technology.

Lagos State Government (2017). Lasg Completes Two Computerised Vehicle Inspection Centres. [ONLINE] Available at: https://lagosstate.gov.ng/blog/2017/08/09/lasg-completes-two-computerised-vehicle-inspection-centres/. [Accessed 9 March 2018].

Lawanson, L. (2018) Transit Futures and infrastructure: Case study of Lagos Paper presented at Urban Age Conference, LSE-Cities November 2018. Available at www.youtube.com/watch?v=IkOd7pcEhJY

Lawanson, T. (2014). Illegal Urban Entrepreneurship?. The Case of Street Vendors in Lagos, Nigeria, (2014). *Journal of Architecture and Environment*, 13(1), 33–48. April 2014.

Lawanson, T. (2017). *Exploring Alternative Urbanisms for Lagos*. Nigeria: Urban Talk Public Lecture, Technical University Berlin. Novemeber 2017.

McKinsey & Company, How to make a city great, September 2013.

Meijer, A., & Bolívar, M.P.R. (2016). Governing the smart city: A review of the literature on smart urban governance. *International Review of Administrative Sciences*, 82(2), 392–408. https://doi.org/10.1177/0020852314564308.

Moe, T. (2011) Urbanisation in South East Asia: Developing smart cities for the future? Regional Outlook. pp. 96–100. Doi 10.1355/9789814311694-022.

MoUD. (2014). *Draft Concept Note on Smart City Scheme.* updated September 2014 Delhi: Ministry of Urban Development.

Nakiguli. H. (2015). Smart Sustainable Cities Concept in Developing Nations. presented at ITU Regional Standardization Forum For Africa. Dakar. *Senegal,* 24-25, March 2015. www.itu.int/en/ITU-T/Work shops-and-Seminars/bsg/042015/Documents/Presentations/S6P2-Helen-Nakiguli.pptx

Ojiako, C. (2018) Imperial International Business City is a Legacy Project – Akioye. *In This Day,* 10 August, 31.

Olokesusi. F and Aiyegbajeje. F. 2017. e-democracy for smart city Lagos. In Kumar.V. ed (2017) *E-democracy for Smart Cities: Advances in 21st Century Human Settlements.* Singapore: Springer Nature, 51–70.

Oshodi, L. & Salau, T. 2016. Urban Mobility & Transportation. Heinrich Boll Foundation & Fabulous Urban. ed. *Urban Planning Processes.* Abuja: Heinrich Boell Stiftung Nigeria. DOI 978-978-966-316-3.

Pérez-Torregrosa, A.B., Díaz-Martín, C., & Ibáñez-Cubillas, P. (2017). The use of video annotation tools in teacher training. *Procedia – Social and Behavioral Sciences, 237*(1), https://doi.org/10.1016/j.sbspro.2017.02.090.

PISON (2016) The State of Lagos Housing Market Report, 2016. http://pisonhousing.com/wp-con tent/uploads/2016/05/The-State-of-Lagos-Housing-Market-Report-TEASER-N75000-PER-COPY.pdf.

Pollio, A. (2016). Technologies of austerity urbanism: The 'smart city' agenda in Italy (2011–2013). *Urban Geography,* 37, 514–534.

Ravindran, T. (2015). Beyond the Pure and the Authentic: Indigenous modernity in Andean Bolivia. *AlterNative: An International Journal of Indigenous Peoples, 11*(4), 321–333. https://doi.org/10.1177/117718011501100401.

Rapoport, E. (2015). Globalising sustainable urbanism: The role of international masterplanners. *Area,* 47, 110–115.

Robinson, J. (2002). Global Cities and World Cities: A view from off the map. *International Journal of Urban and Regional Research,* 26(3), 531–554.

Sassen, S. Talking Back to Your Intelligent City; McKinsey Publishing: New York, NY, USA, 2011.

Sen, J., Mboup, G., Minervino, G., Paldino, S., and Lecoque, G. (2016). Smart Economy. *Smart Cities Bulletin,* 11, 1.

Slavova, M.; Okwechime, E. (2016). African smart cities strategies for agenda 2063. *Afr. J. Manag.,* 2, 210–229.

Söderström, O., et al. (2014). Smart cities as corporate storytelling. *City,* 18(3), 307–320.

SOE Properties Limited. 2017. The Investors Brochure. [ONLINE] Availabe at: http://soeproperties.com/wp-content/uploads/2017/11/IIBC-Investment-Broc-Final.pdf

Soyinka, Oluwole & Wai Michael Siu, Kin & Lawanson, Taibat & Adeniji, Femi. (2017). Assessing smart infrastructure for sustainable urban development in the Lagos metropolis. *Journal of Urban Management,* 5, 10.1016/j.jum.2017.01.001

Thuzar, M. 2011. Urbanization In Southeast Asia: Developing Smart Cities For The Future? Regional Outlook: Southeast Asia; Singapore (2011/2012): 96–100,183.

Tukuma, A. (2016) Smart Cities and Infrastructure: A development pathway for Nigeria. Contribution of the Federal Republic of Nigeria to the CSTD 2015-16 Priority theme on 'Smart Cities and Infrastructure'. Presented at the Intersessional Panel of the United Nations Commission on Science and Technology for Development.

UNHABITAT (2017) City Prosperity index: Lagos, Nigeria.

Vanguard. 2017. Work commences on Imperial international BUSINESS CITY, AS INVESTORS GET ASSURANCE. {ONLINE} AVAILABLE AT: www.vanguardngr.com/2017/09/work-com mences-imperial-international-business-city-investors-get-assurance/ [ACCESSED 8 March 2018].

Vanolo, A. (2014). Smartmentality: The smart city as disciplinary strategy. *Urban Studies,* 51(5), 883–898.

Wall, R. S. & Stavropoulos, S. (2016). *Smart Cities within World City Networks. Applied Economics Letters.* DOI: 10.1080/13504851.2015.1117038

Watson, V. (2014). African urban fantasies: Dreams or nightmares?. *Environment and Urbanization,* 26, 215–231.

Watson.V. (2015). The allure of smart city rhetoric: India and Africa. *Dialogues in Human Geogrphapy,* 5 (1) 36–39. DOI: 10.1177/2043

Wiig, A. (2015). IBM's smart city as techno-utopian policy mobility. *City,* 19(2–3), 258–273.5(1: 36–39).

Wiig, A. (2016). The empty rhetoric of the smart city: From digital inclusion to economic promotion in Philadelphia. *Urban Geography*, 37(4), 535–553.

Willis, K. S. and A. Aurigi. (2017). *Digital and Smart Cities*. London: Routledge.

Yadav, Praveen & Patel, Sejal. (2015). Sustainable city, Livable city, Global city or Smart City: What value addition should smart city bring to these paradigms in context of global south? Paper presented at Conference: APSA 2015, At Malaysia Technological University, Johar Baru.

11

Smart citizens in Amsterdam
An alternative to the smart city

Judith Veenkamp, Frank Kresin and Max Kortlander

Introduction

The smart city approach has been top down in its implementation; it is a surveillance model in which the data produced by citizens are neither owned nor used by them, but rather by large tech firms and, in some cases, by public administrations.

Waag, an Amsterdam-based non-profit organization, takes a 'smart citizens' approach as a direct challenge to the 'smart city'. The smart citizens approach embraces the potential for technology to improve the city in terms of quality of life, environment, democratic transparency, and administrative effectiveness. Taking this approach, the process of developing technology for the city must be *open* and *citizen led*. In addition to all data and software being open source, the development process must be radically co-creative: citizens are involved from start to finish; prioritizing their needs, ideating solutions, and actually building, prototyping, and implementing those solutions alongside public administrators and experts. Public spaces and the data created within them are considered as 'commons'. Thus, the data resulting from a smart citizens approach are shared resources managed by communities with the aim of assuring their sustainability, inclusivity, or other shared benefits.

This chapter discusses the development and execution of a smart citizens approach in Amsterdam. We first outline Amsterdam's historical background as an open, tolerant, and progressive city, and how this political character provided a fertile foundation for the Digitale Stad. This was an effort in the early 1990s to leverage the infant internet to open up Amsterdam's public databases and allow for direct communication between citizens and lawmakers. We then discuss how the same ecosystem of artists, hackers, and activists that created the Digitale Stad helped to pave the way culturally, politically, and technologically to continue exploring various ways to leverage the internet to foster a more open, fair, and inclusive society. Finally, we outline our vision for the future of the commons, and show how this is underpinned by Waag's 'technology and society' focus towards public research – advocating and building practical solutions to advance citizen participation, data ownership, privacy, and democracy.

The Amsterdam approach

Amsterdam has a rich history that led to the free-spirited use of technology as it is championed by Waag. The city originated in the late 12th century and became world famous and opulently rich during the 17th century in what is called the Dutch Golden Age, mainly resulting from its trading innovations, among which the first stock market exchange and an extensive spice and slave trade emerged. It became home to a large number of immigrants with various cultures and religions, who found the city a welcoming place regardless of their backgrounds. For example, when a religion was formally banned, its members could continue to prosper in minimal disguise—which started the still fashionable Dutch paradox called '*gedogen*', or tolerance, as is practiced for example in its well-known laissez-faire politics governing the recreational use of soft-drugs. As the city flourished, it became well known for its painters, philosophers and musicians. During the 19th and early 20th centuries, the city expanded rapidly. As it recovered from the blows of the Second World War from the 1960s onwards, Amsterdam gave birth to several avant-garde movements that practiced mild and humorous forms of civil disobedience, attacking the dominant top-down hierarchical culture, including a 'white bikes plan' (De Wildt 2015) as an early (and short-lived) form of free bike-sharing. In the 1970s and 1980s, a vibrant squatters' movement developed as Amsterdam experienced a severe housing crisis. The squatters regularly teamed up with artists that practiced various critical art practices, transmitted through a myriad of underground television and radio channels. Into this potent mix of free-thinkers, artists, and activists, the invention of the internet fell into fertile grounds: the well-established culture of progressive pragmatism in Amsterdam. The 'Amsterdam Approach', developed out of this rich cultural and political history, refers to how the city—residents and government alike—are willing to experiment with alternative and sometimes radical approaches to address emerging societal issues (e.g. see D'Antonio *et al.* 2018). In the next section we will outline how the internet and later the open data movement developed in Amsterdam, and in particular the emergence of the Digitale Stad; an early platform that pioneered new approaches to Amsterdam's relationship with technology.

Digitale Stad/digital city

In 1993, the use of the internet, which had been reserved for government and education, was opened by the US government to companies and individuals. At the same time, the Mosaic web browser that popularized the World Wide Web and the internet became available. Many people realized that the internet had an enormous potential for public services and democracy, which resulted in a wave of new initiatives around that time. In Amsterdam, hackers and programmers joined the mix of free-thinkers, artists, and activists and started to explore the medium as a way to stimulate encounters, debate, and expression. In 1994, this culminated in the so-called Digital City: a virtual city made out of bits, as a discursive counterpart to the atom-based physical city (Stikker 2013; Willis and Aurigi 2018, pp.13–14) (Figures 11.1 and 11.2). The platform allowed people to access the internet through a public interface. Users could have their own homepages, send and receive mail, and have access to large amounts of data that had previously been inaccessible in the city archives. The Digital City exploded, as if the Amsterdam population had silently waited for it to happen, into a large network of like-minded pioneers from all walks of life: the new medium of online communication was born never to be silenced again.

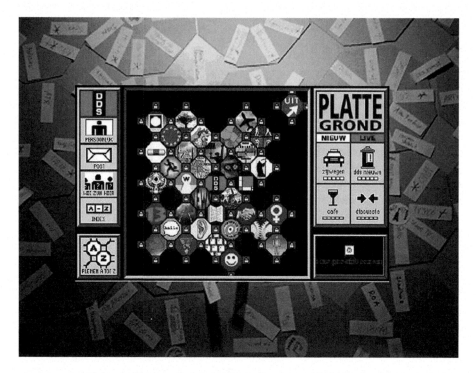

Figure 11.1 The Digitale Stad DDS

Source: Waag (BY-NC-SA)

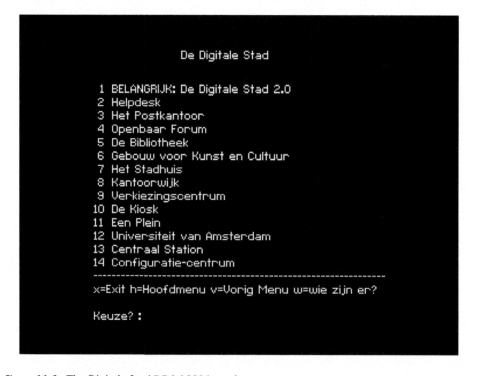

Figure 11.2 The Digitale Stad DDS 1993/remake

Source: Waag (BY-NC-SA)

One of the Digital City's founders was Marleen Stikker, who also established the Waag foundation later that year (it was originally called the Society for New and Old Media). Waag has its roots in the Digital City, and started as an independent media lab, and its first projects aimed to make technology more accessible. For example, it advocated for more publicly available bandwidth, created a public reading space where people could access the internet, and worked with disadvantaged groups to find out, together, how technology could be an enabling force for them. From those early days, Marleen recalls (cited in Stikker and Kresin 2010) how the Digitale Stad opened up new connections between citizens and city government:

> Back in 1993 I founded the Data Liberation Front, that later resulted in the first free access community on the Internet, called the Digital City. I visited newspapers, cultural organisations, libraries, political parties and the city council, educating them about the wonders of the Internet and meanwhile collecting floppy disks full of documents. Although most people had no clue about the Internet, they were all intrigued by the idea to build a digital city together. It was a real citizen-driven initiative, designed and facilitated by artists, hackers and activists. A powerful combination, as history shows again and again.
>
> When a civil servant of the City of Amsterdam provided me with the disks of the complete governmental information system, both he and I could hardly understand what would be the consequences of his action. Official documents used to be confined to the city network. From the moment we put them on the servers of the Digital City on the 15th of January 1994, the documents where open for anyone to access. Within a few weeks it led to situations in which citizens were better prepared on topics than the city council members and city officials were themselves. It opened a completely new playing field for the checks and balances between citizens and politicians. It was a small step for a hacker, but a giant leap for democracy.

These early initiatives provided the initial impetus for Amsterdam's current relationship with technology in two major ways. Firstly, they opened and introduced the internet to the public, with an interface that people could use and with resources that people found valuable and wanted to use. Secondly, these efforts demonstrated the benefits of technological openness to the local city government. As the Digital City required collaboration with the city government, it opened the door to future collaborations between the local city government and this community of artists, hackers, and activists in an emerging symbiotic relationship.

Open data

In the first decade of the 21st century, the open data movement gained ground and altered the relation of society and government with regards to information. The open data movement, striving to make data available and free to use, share or mash-up by anyone, attracted attention both globally and in Amsterdam. It put pressure on governments to become more transparent and work from the principle of 'open, unless' in considering their datasets, which expanded the power of citizens to make use of digital public resources. The concept underpinning it was that governments should not wait until they know what the data will be used for, since the whole point of open data was to open a field of possibilities for others. Adopting this approach in Amsterdam, slowly but surely, more and more data were opened up to the public. Initiatives like 'Apps for Amsterdam',[1] launched in 2011 by a collaboration between Waag and the City of Amsterdam, helped to create spaces for the development of solutions, software, and public interfaces by other parties in society. 'Apps for Amsterdam' specifically called upon developers to submit applications focused on the use of real estate, tourism and culture, democracy, mobility, safety, and energy. These themes reflected growing areas of interest for local citizens, and helped open the door for citizens to start addressing issues collaboratively along with the local city government.

Smart citizens

As citizens became more involved with addressing pressing local issues into the 2010s, urgency grew amongst both citizens and public administrations to undertake deeper collaborative action together to solve these challenges. One of the most complex challenges urban areas face is air quality and a healthy living environment. The UN estimates that in 2050 68% of the people will live in cities (United Nations 2018). At the same time, research shows the indisputable evidence on the harmful effects of air pollution on people's health and wellbeing. For example, data show that in the Netherlands 12,000 people die prematurely every year, and people are being hospitalized everyday due to air pollution.[2] In Amsterdam, one day of simply breathing the air has the equivalent lung damage of smoking 6.4 cigarettes[3] (Kruyswijk 2016).

Raising awareness on air quality and inviting citizens to be partners with official institutes and governments is much needed. In Amsterdam, official institutes like the National Institute for Public Health and Environment have traditionally been responsible for official measurements of air quality. Official measurements are used for political decisions and policy design, such as the creation of an environmental zone, limiting traffic in city centres, putting in place speed limits on highways, and siting industrial and economic activities in certain areas. However, the official government air quality measuring network is limited in scale, with few official measurement stations, while the air quality can show great variety from street to street and from day to day. Therefore, the idea of a dense, decentralized citizen sensing network is a promising method for both citizens and public institutions to collect air quality data.

In the mid-2000s, sensor and networking hardware started to become cheaper and smaller, and physical computing became more accessible. Mapping services like Google maps became available, allowing for cheap and quick integration of real-world data into an essentially free, relatively accurate map. From around 2007 onwards, Waag became inspired by several citizen sensing initiatives where motivated citizens used low-cost technology to measure aspects of the environment such as noise around Amsterdam's Schiphol airport,[4] air pollution in Paris,[5] and nuclear radiation in Japan.[6]

These developments in citizen sensing coincided with the global emergence of so-called 'smart city' programmes that were increasingly pushed on city governments by multinational technology developers like IBM, CISCO, and Huawei. These programmes promoted the use of networked technologies, with the promise that cities would become smarter and more efficient (Söderström et al. 2014). However in practice they reinforced a singular top-down, control focused, privatized, and black-box system of control (Kitchin 2014). But a potent countermovement to the smart city was unleashed by initiatives linking people from the do-it-yourself, bottom-up technology groups (communities of practice) to ordinary citizens who care about the place they live and work (communities of interest). These efforts took the city and the monopolized data acquisition back into civic hands, as is detailed in the Manifesto for Smart Citizens (Kresin 2013):

> We, citizens of all cities, take the fate of the places we live in into our own hands. We care about the buildings and the parks, the shops, the schools, the roads and the trees. But above all, we care about the quality of the life we live in our cities. Quality that arises from the casual interactions, uncalled for encounters, the craze and the booze and the loves we lost and found. We know that our lives are inter-connected, and what we do here will impact the outcomes over there. While we can never predict the eventual effect of our actions, we take full responsibility to make this world a better place.
>
> Therefore, we refuse to be consumers, client and informants only, and reclaim agency towards the processes, algorithms and systems that shape our world. We need to know how decisions are made, we need to have the information that is at hand; we need to have direct access to the people in power, and be involved in the crafting of laws and procedures that we grapple with every day.
>
> Fortunately, we hold all the means in our hands. We have appropriated the tools to connect at the touch of a button, organise ourselves, make our voices heard. We know how to measure ourselves and our environment, to visualise and analyse the data, to come to conclusions and take action. We have continuous access to the best of learning in the world, to powerful phones and laptops and software, and to home-grown labs that help us make the things that others won't. Furthermore we were inspired by such diverse examples as the 1% club, Avaaz, Kickstarter, Couch-surfing, Change by Us and many, many more.
>
> We are ready. But, as yet, our government is not. It was shaped in the 18th Century, but increasingly struggles with 21st Century problems it cannot solve. It lost touch with its citizens and is less and less equipped to provide the services and security it pledged to offer. While it tries to build 'Smart Cities' that reinforce or strengthen the status quo that was responsible for the problems in the first place—it loses sight of the most valu-able resource it can tap into: the Smart Citizen.

Smart Citizens:

- Take responsibility for the place they live, work and love in;
- Value access over ownership, contribution over power;
- Ask forgiveness, not permission;
- Know where they can get the tools, knowledge and support they need;
- Value empathy, dialogue and trust;
- Appropriate technology, rather than accept it as is;
- Help the people that struggle with smart stuff;
- Ask questions, then more questions, before they come up with answers;
- Actively take part in design efforts to come up with better solutions;
- Work agile, prototype early, test quickly and know when to start over;
- Will not stop in the face of huge barriers;
- Unremittingly share their knowledge and their learning, because this is where true value comes from.

All over the world, smart citizens take action. We self-organise, form cooperations, share resources and take back full responsibility for the care of our children and elderly. We pop up restaurants, harvest renewable energy, maintain urban gardens, build temporary structures and nurture compassion and trust. We kick-start the products and services we care about, repair and upcycle, or learn how to manufacture things ourselves. We have even coined new currencies in response to events that recently shook our comfortable world, but were never solved by the powers that be.

Until now, we have mostly worked next to governments, sometimes against them, but hardly ever with them. As a result, many of the initiatives so far have been one-offs, inspiring but not game changing. We have put lots of energy into small-scale interventions that briefly flared and then returned to business as usual. Just imagine what will happen if our energy, passion and knowledge are teamed up by governments that know how to implement and scale up. Governments that take full responsibility for participating in the open dialogue that is needed to radically rethink the systems that were built decades ago.

To get ourselves ready for the 21st Century, we have to redefine what 'government' actually means. We ARE our government. Without us, there is nobody there. As it takes a village to raise a child, it takes people to craft a society. We know it can be done; it was done before. And with the help of new technologies it is easier than ever. So we actively set out to build truly smart cities, with smart citizens at their helms, and together become the change that we want to see.

(Kresin 2013)

Waag joined this countermovement when it established its Smart Citizens Lab.[7]

Smart Citizens Lab

The Smart Citizens Lab explores tools and applications that help make sense of the world around us. Waag works with citizens, scientists, and designers to tackle environmental issues ranging from air and water quality to noise pollution.

In recent years, improved access to open hardware tools and makerspaces, as well as the creation of online data sharing platforms, has made possible the design of low-cost, open-source sensors that citizens can use to measure the environmental health of their neighbourhoods and take action: from starting a campaign, to co-creating solutions and influence policy.

Waag believes the public interest should be at the heart of innovation, and therefore society is the ideal research community. This is what Waag refers to as public research. Starting with 'curiosity driven research' where ideation and formulation requirements are central, Waag moves to 'context research'—building prototypes and piloting them, with the aim to eventually move to sustainable innovation solutions that meet needs in society with 'transformation focused research' (Figure 11.3).

This process for public research provides a blueprint for the evolvement of the work on air quality within the Smart Citizens Lab. Starting from curiosity, small experiments took place with the distribution of a hundred 'Smart Citizen Kits'. The small and cheap sensors delivered low-quality data results, however the participants were highly engaged and enthusiastic. More and more partners came on board, including the City of Amsterdam, the SenseMakers community,

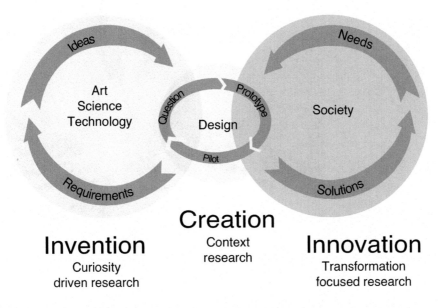

Creation

Invention Context research Innovation

Curiosity driven research Transformation focused research

Figure 11.3 Waag's approach to transformation-focused research

the National Institute for Public Health and the Environment, and the Amsterdam Institute of Advanced Metropolitan Solutions. May 2017 marked the official start of the Smart Citizens Lab.

After the 'invention' phase that led to the establishment of the Smart Citizen Lab, the partners and citizens joined together to work further on the topic of air quality and low-cost sensors (Figure 11.4). Sixty citizens, scientists, and technicians dived into the complex issues of air quality and sensors, with the needs and worries of citizens as the guiding force. Armed with new knowledge they sought help from experts and searched for the best way to collect relevant data. They formulated research questions, selected and tested sensors, or custom made them when necessary. Citizen-led sensing and data collection still experienced challenges, from technological problems to a lack of skills on data analysis, but the lab proved it could actively involve citizens and enable them to be better informed.

One of the key projects in the development of the Smart Citizens Lab presented itself in 2015 with Waag's involvement in the Making Sense project. This project was designed to explore how open-source software, open-source hardware, digital maker practices, and open-source design could be used effectively by local communities to appropriate their own sensing tools to make sense of their environments and address pressing environmental problems. Building upon the knowledge and experiences of the lab, Waag initiated a local Dutch pilot as part of this wider project, Urban AirQ, which brought together a consortium with European partners and managed to research and iterate citizen sensing on air quality.[8] In this project the eight-step method for citizen sensing (Woods *et al.* 2018) was developed that included the following steps (Figure 11.5): start with *scoping* of project, then undertake *community building* and *planning* of sensing activities, followed by the *sensing* and data collection, then raise *awareness* on the data

Figure 11.4 A citizen sensor is assembled in Amsterdam

Source: Waag (BY-NC-SA)

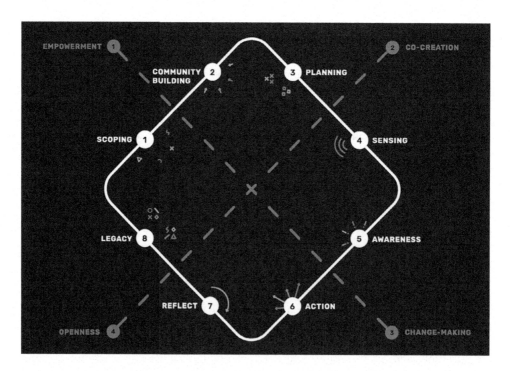

Figure 11.5 The 'making sense' framework in the 'Citizen Sensing: A Toolkit' developed as part of the Making Sense project

Source: Coulson 2018 p. 820

collected and formulate next steps (*action*), and finally *reflect* on the citizen sensing and make sure that the knowledge built during the citizen sensing is documented and can be shared with others through a *legacy*.

The Making Sense pilot delivered answers to concrete questions posed by citizens—like 'What is the air quality on my street?' and 'Should I open my front or back window to air out my house?'—and had a local impact on policy-makers and raised awareness in the neighbourhood. Making Sense had two wider impacts: firstly, formal institutions observed the power of smart citizens and it provided an catalyst for open data initiatives; and secondly it inspired the development of a knowledge base on sensor technology. Using the lessons learned in the design phase, the Smart Citizens Lab aims to have lasting impact on air pollution through the creation of a citizen sensing network on a much larger scale and with a longer lifespan than in previous projects. In this way Waag aims to facilitate societal change.

For example, the Hollandse Luchten (Dutch Skies) project builds upon earlier experiments and pilots to deploy 200 sensors in three pilot locations in the Province of North Holland (Figure 11.6). One of these pilots is located in the industrial IJmond region, where Tata Steel, the second largest steel factory of Europe, is located. Citizens have been concerned with air quality and their health for decades, but due to recent graphite 'rains' in the autumn of 2018, the concerns sky-rocketed. Previous projects were often experimental and focused on designing and prototyping sensor technology and community engagement on data collection. Now, with the government and the steel factory as a partner to the project,

Figure 11.6 Sensor demo during the launch of Hollandse Luchten

Source: Waag (BY-NC-SA) Photographer: Ilyas bin Sarib

Waag aims to focus on a new type of impact: the path is clear to actually use the data collected by citizens to improve design of policy.

Smart city: a responsible digital city

As the 'smart city' is now starting to be more fully implemented, concerns are increasingly raised with the technology provided by private corporations, platforms, and consultancies (Hollands 2015; Wylie 2017). The potential risk lies in the fact that private corporate values of the companies developing and deploying the technology will replace public and inclusive values that should underpin city life. Smart city projects need to acknowledge that technology is not a neutral (Haraway 1991) set of instruments that merely solves problems but instead contains implicit social expectations and policy directions of its owners and developers. Organizations like the World Economic Forum and the Dutch Social Economic Board (SER) have both identified the necessity to shift from an instrumental perspective on technologies to an approach in which technological potential is explicitly designed and articulated with the public values a city wishes to embody and achieve (Caragliu *et al.* 2011).

The discussions on public values and principles within Amsterdam were solidified in 2017 after several organizations and citizens had become critical of the deployment of technology and data collection in city life. This led to the Tada initiative[9] and the 'Tada-data disclosed' manifesto of which Waag was an initiator. The manifesto elaborated on six principles that should be applied in the digital city:

(1) Inclusive
(2) Control (by residents)
(3) Tailored to the people
(4) Legitimate and monitored
(5) Open and transparent
(6) From everyone—for everyone

With the municipality of Amsterdam committed to Tada, there is consensus on the principles that should be the starting point for technological development and implementation. The next step is to translate Tada into the daily practice of local government. The Tada trajectory demonstrates how collaboration between government, citizens, and local organizations can help to make concepts like open data, data privacy, and data ownership concrete within a city.

The bright future of the commons

Where data are being produced and used in both 'smart city' and 'smart citizen' models, each model requires some kind of framework for data governance. In a smart city model, this governance framework is often a closed form of ownership by the companies and governing bodies who own the technical infrastructure for gathering and working with data (Kitchin 2014). Waag takes the view that this type of ownership, which excludes those citizens who produce the data, is not appropriate for a smart citizens model.

The opportunities for smart citizenship and citizens sensing are accompanied by the revival of the idea of the commons. Waag defines commons as 'shared resources managed by communities with an aim of assuring their sustainability and inclusivity'. They foster bottom-up initiative and community self-determination, while keeping a close watch on the needs of the wider public. Thinking about the city as a commons, where air quality, water quality, and general liveability are the shared resources, sets the stage for a governance model that aligns with the concept of smart citizens as an equal partner in designing the way forward.

Not only can the governance of physical space follow the logic of the commons, but the same logic can be applied to management of data, especially when these data are collected by citizens with their own sensors. Data commons is a citizen-centric approach to data governance. Here, data are self-controlled and available for broader communal use, with appropriate outcomes for privacy protection and value distribution.

Conclusion

Who owns the city? This is the central question in the debate between smart cities and smart citizens. It can be taken as a given that technology will continue to play an increasing role in our cities. Frameworks that do not address issues of ownership and access in smart cities pose a fundamental threat to those who champion democracy, fairness, and civil rights. As cities become more digitized, those who own and control that digitization will increasing own and control the city itself.

The amount of data about our cities, and about us as individuals, will increase; there is little we can do to change this development. The precise manner in which this will take place—how citizens will be made aware of and have access to the data that are produced about and by them, and to what extent those data will be used to address societal challenges that citizens deem relevant—is still taking shape. This is the area in which the smart citizen approach taken by Waag and the City of Amsterdam aims to make an impact, by keeping control of the city and its data in the hands of citizens.

Notes

1 https://waag.org/nl/project/apps-amsterdam
2 www.longfonds.nl/buitenlucht-en-je-longen/ongezonde-lucht
3 www.parool.nl/nieuws/amsterdammer-rookt-ruim-6-sigaretten-per-dag-mee-door-luchtvervuiling~bf5755aab/
4 www.sensornet.nl/sensornet/geluidsnet
5 http://fing.org/?Green-Eyes-Montre-verte-CityPulse&lang=fr
6 https://blog.safecast.org/
7 https://waag.org/en/project/amsterdam-smart-citizens-lab
8 http://making-sense.eu/campaigns/urbanairq/
9 https://tada.city/en/about/

References

Caragliu, A., Del Bo, C., & Nijkamp, P. (2011). Smart cities in Europe. *Journal of Urban Technology*, 18(2), pp. 65–82.
D'Antonio, S., Patti, D., & Polyak, L. (2018). Digital cities: Amsterdam's ecosystem of cooperation. URBACT. https://urbact.eu/digital-cities-amsterdam-ecosystem-cooperation
De Wildt, A. (2015). White Bicycle Plan. Amsterdam Museum. https://hart.amsterdam/nl/page/49069/witte-fietsenplan. Accessed 20 May 2019.
Haraway, D. (1991). A Cyborg Manifesto: Science, Technology, and Socialist-Feminism in the Late Twentieth Century. In Haraway, D., *Simians, Cyborgs and Women: The Reinvention of Nature*. New York, Free Association Books. pp. 149–181.

Hollands, R. G. (2015). Critical interventions into the corporate smart city. Cambridge Journal of Regions. *Economy and Society*, 8(1), pp. 61–77.

Kitchin, R. (2014). The real-time city? Big data and smart urbanism. *GeoJournal*, 79(1), pp. 1–14.

Kresin, F. (2013) A Manifesto for Smart Citizens. De Waag https://waag.org/nl/article/manifesto-smart-citizens. Accessed 1 May 2019.

Kruyswijk, Marc (2016). Amsterdammer rookt ruim '6 sigaretten' per dag mee door luchtvervuiling. Het Parool, 17 August 2016. www.parool.nl/nieuws/amsterdammer-rookt-ruim-6-sigaretten-per-dag-mee-door-luchtvervuiling~bf5755aab/Accessed 4 June, 2019.

Söderström, O., Paasche, T., & Klauser, F. (2014). Smart cities as corporate storytelling. *City*, 18(3), pp. 307–320.

Stikker, M. (2013). Public Domain 4.0. In Hemment, D., Thompson, B., de Vicente, J.L., Cooper, R. (eds). *Digital Public Spaces*. Manchester, Future Everything, pp. 32–33.

Stikker, M. & Kresin, F. (2010). Open Data - Open Government: Designing Principles for the Public Internet. Waag Society Publication, Netzpolitischer Kongress, 12–13. Nov 2010. Deutscher Bundestag, Berlin. https://waag.org/sites/waag/files/Publicaties/Open_Data.pdf

United Nations (2018). 68% of the world population projected to live in urban areas by 2050, says UN. www.un.org/development/desa/en/news/population/2018-revision-of-world-urbanization-prospects.html. Accessed 1 May 2019.

Willis, Katharine S. & Aurigi, Alessandro. (2018). *Digital and Smart Cities*. London: Routledge.

Woods, M. (2018). Citizen Sensing: A Toolkit. In Woods, M., Balestrini, M., Bejtullahu, S., Bocconi, S., Boerwinkel, G., Boonstra, M., Boschman, D.-S., Camprodon, G., Coulson, S., Diez, T., Fazey, I., Hemment, D., van den Horn, C., Ilazi, T., Jansen-Dings, I., Kresin, F., McQuillan, D., Nascimento, S., Pareschi, E., Polvora, A., Salaj, R., Scott, M. & Seiz, G. *Citizen Sensing: A Toolkit*. Making Sense. https://doi.org/10.20933/100001112.

Wylie, B. (2017). Civic tech: On google, sidewalk labs, and smart cities. The Torontoist. https://torontoist.com/2017/10/civic-tech-google-sidewalk-labs-smart-cities/. Accessed 3 June 2019.

12

Governing technology-based urbanism

Technocratic governance or progressive planning?

Chiara Garau, Giulia Desogus and Paola Zamperlin

Introduction

Literature on 'smart urbanism' has broadly underlined how the utopian idea of a smart city paved the way for an apolitical understanding of good governance, in which cities are organized, developed and governed for a better future. Verrest and Pfeffer (2018) outline the emerging academic topic of 'smart urbanism' and underline how its policies drive contemporary cities. They also add how, under the label of smart urbanism, a series of 'technocratic solutions to urgent urban problems' (Verrest and Pfeffer, 2018, p. 2) are adopted, without taking into consideration different political responses 'to political conflicts that reflect discourses on what urban problems are, what appropriate solutions are and what urban development is desired' (ibid.). Wiig (2015) notes how smart urbanism has been integrated with a technologically driven governance model, where cities use as a strategy to 'sell' themselves in a globalized word. In this context, the city is seen as a promising location for multinational corporations and foreign investment, instead of recognizing the intrinsic and extrinsic benefits for the residents (Wiig, 2015, p. 260). McFarlane and Söderström (2017, p. 2) establish that smart urbanism is not pure rhetoric and has real impacts on 'both in the urban policies of national governments and municipalities and in the grass-roots initiatives and social movements that disturb, resist or create their versions of smart urbanism'.

Many cities aspire to solve everyday urban problems, with a combination of the integration of information and communication technology (ICT) and an acknowledgment of the characteristics that make it unique, such as its geographical position, history and culture. Different cities are therefore developing strategies that may derive inspiration from other contexts but are as unique and specific as the city itself, even if often, the literature identifies digital and technology-driven focus approaches as a universal solution in different cities (Verrest and Pfeffer, 2018). To achieve these goals, the literature indicates that the sub-dimensions of governance (Garau *et al.*, 2015; Giffinger *et al.*, 2007; Rodrigues and Franco,

2019) and ICT (Caragliu *et al.*, 2011; Ferro *et al.*, 2013; Sepasgozar *et al.*, 2019) are a fundamental pillar in the smart cities paradigm. However, as underlined by Akter *et al.* (2019), efficient and good governance requires not only the latest technologies (multi-source big data, real-time processing for complex data, sensors and so on), transparency of processes, networks, security, communication policy regulations and strategic planning in order to improve the efficiency of cities, but also a 'long-term perspective on what is needed for sustainable development and how to achieve the goals of such development' (Akter *et al.*, 2019, p. 37).

These new dynamics inside cities make the close relationship between ICTs and the smart dimension perceptible, and in so doing can reveal different strategies and procedures in the governance process (Nel *et al.*, 2018). Nevertheless, the benefits and externalities provided by ICT are visible, as well as the challenges and opportunities of initiatives in technology-driven smart cities, in terms of synergies between all public and private actors (Angelidou, 2014; Valencia *et al.*, 2019); network integration (Internet of Data, Internet of Things (IoT), Internet of Services and Internet of People); flexibility and open attitudes in governance networks of all actors involved (Sol *et al.*, 2018); inter- and intra-city transfer and share of knowledge, and easier access to information (Rodrigues and Franco, 2019).

The mission to be a smart city has been seen as technocratic, due to a focus on technological solutions and business interests that promote the empty rhetoric of 'citizen-centered approaches' and 'user-generated data' (Cardullo and Kitchin, 2019; Greenfield, 2013). These criticisms advance the view that smart city initiatives promote forms of algorithmic governance that control and regulate citizens, and are driven by choices guided by market-led solutions and individual autonomy. The justification for these smart initiatives is made by a simplification in management practices (for city users) and a civic paternalism (for smart cities marketeers who want to do the best for citizens) promulgated by the political class. These put the city, as a common good with its civil, social and political rights, in second place (Cardullo and Kitchin, 2019).

Taking this approach, the chapter aims to offer an overview of how urban governance is changed within the paradigm of smart cities, by providing theoretical conceptualizations of difference governance models and linking these to an analysis of a series of smart city projects. It seeks to understand smart governance by addressing a series of questions. Firstly, we focus on the theoretical transition from smart cities 1.0 to smart cities 2.0 by answering to the question; 'Which aspects of the smart city influence smart governance?' Secondly, we will translate this theoretical discourse to an analysis of local-level initiatives in a selection of Italian cities by answering to the question 'What is smart governance in the context of Italian cities?' Finally, the chapter discusses risks, challenges and future research by responding to the question 'Do smart cities represent technocratic governance or progressive planning?'

From smart cities 1.0 to smart cities 2.0: what about the governance?

Critical reviews of literature on the first generation of smart cities, which has been termed 'smart cities 1.0' (Trencher, 2019), focus on the importance of the technological aspects of a smart city for urban innovation in order to solve problems associated with rapid urbanization. In smart cities 1.0, the literature identifies smart governance as not only closely associated with the use of ICTs, but also as one of the key pillars of a smart city with the components of smart economy, smart mobility, smart environment, smart people and smart living (Azzari *et al.*, 2018; Caragliu *et al.*, 2011; Giffinger *et al.*, 2007; Mistretta and Garau, 2013). Other elements of smart governance in this first generation include extensive use of technologies, improvement in intra-governmental coordination (Willke, 2007), increased participation in decision-making (Giffinger *et al.*, 2007), renewal of organizational structures, the widespread use of open big data (Clarke and Margetts, 2014), and a city focused on single-issue technical agendas such as transport and energy (Trencher, 2019).

However, the governance framework of smart cities and its structural correlations with ICTs and ICT policy are still underdeveloped (Cardullo and Kitchin, 2019; Nam and Pardo, 2011; Pereira *et al.*, 2018). In this first phase of the smart cities paradigm (smart cities 1.0), Hollands outlines how cities sought to define themselves as smart cities, whilst often lacking a holistic understanding of the governance reorganization required in a smart context (Hollands, 2008).

The second-generation of smart cities, so-called 'smart cities 2.0', has led to further development of smart cities, and is characterized by a focus on smart governance frameworks with a people-centric and decentralized approach. These initiatives move beyond the techno-economic objectives of smart cities 1.0, in order to use technologies in an efficient and effective manner to address social problems, serve citizen needs and measure and enhance the effectiveness of urban governance and policy-making (Šiugždinienė *et al.*, 2017; Broccardo *et al.*, 2019; Trencher, 2019). This second-generation approach has a broader perspective in the way in which smart governance influences and, simultaneously, is influenced by the tools, people, principles and capacities appropriate to the urban context. This constantly evolving understanding of what constitutes smart governance challenges the widely accepted definition of the term, and reflects a range of critical issues in defining the meaning of a smart city.

To understand the typologies of smart city governance, the governance frameworks and the degree of government and societal transformation in smart cities, we undertook an analysis of the literature on the governance of smart cities through well-known databases (Google Scholar; Science Direct; ISI Web of Science; IEEE Xplore; Scopus; SpringerLink) from 2007 until to 2019. This research initially produced 652 results, which we subsequently refined to 18 research papers that we categorized as most relevant to the topic of smart governance. As a result, we identified a number of different distinct conceptualizations of smart city governance in the so-called smart cities 2.0 as follows:

1. Traditional government: this conceptualization of smart city governance sees existing governance as a form of advocacy of the smartness of a city and does not require transformations or changes (Meijer and Bolívar, 2016).
2. Informing urban governance (Acuto *et al.*, 2019): this conceptualization is based on governing the city through the power of technology rather than the restructuring of organizations: big data, sensors, IoT, Internet of Everything (IoE) for monitoring, controlling and managing urban developments, recourses, urban infrastructures, risks and people. All of these appear in city dashboards, sensor networks or centralized control rooms. In this case, urban governance is part of a data-driven urbanism and the expected changes are exclusively in decision-making processes.
3. Electronic governance for smart public administration (Bolívar, 2019; Edelenbos *et al.*, 2018): this conceptualization focuses on the capacity for change of the public administrators. In fact, governing through this way can be potentially modifiable because the

administrators are aware that citizens cannot accept these new experiments and for this reason the strategies can be addressed in new creative forms of smart urbanism.

4. Collaborative smart governance (Pereira *et al.*, 2017): this conceptualization reaffirms the central role of citizens and focuses on smart interactions between various stakeholders in the different contexts of smart city initiatives. Specifically, Bartenberger and Grubmuller-Regent (2014) analyzed the more restrictive concept of collaborative governance to distinguish the concept of 'smart city governance' from the broader concept of participatory democracy. Pereira *et al.* (2017) analyzed the same concept as a key factor with the central role of citizens and with ICT, for supporting smart cities initiatives. Verrest and Pfeffer (2018) underline the importance of ICT strongly associated with non-technical and bottom-up initiatives. Such initiatives, also labelled as collaborative smart governance in a smart urbanism, 'mobilize technology as enabler in the knowledge production process recognizing varieties of knowledge or operate without ICT at all, highlighting creativity, social learning or alternative ways to achieve' (Verrest and Pfeffer, 2018, p. 6).

Although the four approaches above are all found within the smart cities 2.0 generation, we argue that the second is the evolution of the first and so on, and consequently the 'collaborative smart governance' is the most developed conceptualization (Figure 12.1).

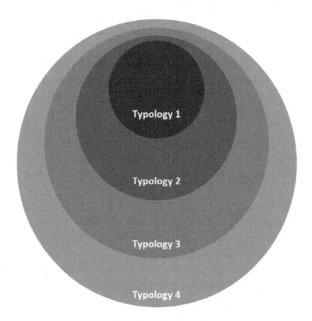

Figure 12.1 Conceptualizations of smart city governance in the smart cities 2.0: traditional governance (Typology 1); informing urban governance (Typology 2); electronic governance for smart public administrators (Typology 3); collaborative smart governance; (Typology 4)

Source: Chapter author(s)

However, there is still a mismatch between urban governance structures (with its regulatory processes) and civil society governance initiatives (where a city's inhabitants have the ability to equip themselves with shared tools and methods for its overall management). In the next section we will apply this conceptualization framework to an analysis of a series of smart city governance initiatives. It focuses on the Italian smart city and, through a comparative analysis of smart city projects, seeks to identify which of the four smart governance types the initiatives belong to.

Research methodology on comparative analysis of Italian smart cities initiatives

Why Italian smart city projects?

The decision to analyze Italian cities in this chapter is based on evidence from the European Parliament (2014) that identifies Italy as one of the leading European countries in smart governance and with the largest number of smart cities involved in implementing smart governance projects (Vázquez and Vicente, 2019, p. 164). In addition, the National Association of Italian Municipalities (ANCI)—with its Smart City National Observatory founded in 2012—highlights how smart strategies and actions succeed in spreading not only in the largest Italian cities, but also in the smaller ones (Testa, 2016).

According to the latest data from the Italian National Institute of Statistics (ISTAT, 2018), Italy is a country where only two cities have more than 1 million inhabitants (Rome 2,872,800 and Milan 1,366,180), and there are 5,497 municipalities with under 5,000 inhabitants (which represents 69.45% of the total number of 7,915 Italian municipalities). In Italy, the city governments of the smallest municipalities are partially decentralized because the areas of influence of a single urban center tend to extend beyond the administrative limits of a single municipality (Garau *et al.*, 2015; Verrest and Pfeffer, 2018). Therefore, the governance of smaller municipalities typically extends into a wider geographical area and includes several smaller municipalities and their territories, where decentralization prevails. Typically, larger Italian cities, as well as metropolitan cities, are organized with a multi-level governance structure, with a range of powers and competences devolved from the central government to local administrative bodies (Zamperlin and Garau, 2017).

This phenomenon has consolidated the formation of widespread urban polarities and—despite it being difficult to see and define the boundaries between neighboring cities and between built and rural spaces—it has not led to the definition of national smart systemic initiatives. This is despite the fact that a shared theoretical idea of smartness has been acknowledged in the Italian context, since the paradigm of smart cities appeared. In addition, the Italian model of decentralized territorial and administrative organization has led to the development of a bottom-up approach in the smart urbanism paradigm. This is in part due to funding mechanisms supporting Italian smart cities, since each city has tried to plan and execute its own smart program, through applying for European Union funding, in order to circumvent central government funding. In the next section we discuss this further.

Methodological approach and application to Italian smart city projects

As mentioned above, Italian cities have typically not adopted a unified national program that defines smart initiatives priorities, and consequently the goals that have guided funding in the various cities varied from city to city. The Italian Urban Agenda's platform, which is organized and promoted by the National Association of Italian Municipalities (Associazione Nazionale dei Comuni italiani, ANCI[1]) and by the Institute for Finance and Local Economy (L'Istituto per la Finanza e l'Economia Locale, IFEL[2]), gives an overview of the distribution of the different municipalities involved. It categorizes projects by the total funding invested and correlates projects in relation to the different pillars or sectors of smart cities programmes (living, energy, environment, people, planning, economy, mobility, government). Table 12.1 shows the current implementation phase for a total of 157 Italian municipalities, 1,314 projects and total invested funding of €3,809,120,361.00, subdivided in different sectors. Table 12.1 indicates that the largest proportion of funding (26.7%) was allocated to the planning sector, however, this involves the smallest number of total municipalities (29.3%) and funded projects (8.0%). Conversely, the governance sector, with a minimum investment (3.1%) involves about 32% of the total of municipalities, for almost 13% of projects.

After identifying the smart city initiatives in Italy, we focused on an analysis of governance projects, by considering the projects under the 'government' sector (170 projects) and under the sub-sector of 'planning' (22 projects). These were selected because they represent the most relevant sectors to be included globally as examples of smart governance in Italian cities. The result produced a detailed analysis of 180 projects,[3] summarized in Table 12.2. This analysis was useful to obtain a first schematization of the complete framework of the smart cities initiatives under smart governance sectors present in Italy and also to identify for each municipality involved the number of projects, the recipients and the type of innovation brought by the project (Table 12.2).

Subsequently, we focused on the cities with the highest national ranking in Italy (Smart City Index, 2018), particularly those belonging to the first band (national ranking between 1 and 39) in the 'positioning of cities in the rankings by strata and fields' (Smart City Index, 2018, p. 13). The results obtained for each selected city enabled an analysis based on the identification of:

Table 12.1 Analysis of funding for Italian smart city initiatives divided by the smart cities sectors: living, energy, environment, people, planning, economy, mobility, government

Sector	Total funding: €3,809,120,361		Total municipalities involved: 157		Total number of projects: 1314	
	Funding €	% funding	No. municipalities	% municipalities	No. projects	% projects
Living	309,584,287	8.1%	51	32.5%	170	12.9%
Energy	644,341,700	16.9%	58	36.9%	143	10.9%
Environment	286,828,706	7.5%	68	43.3%	191	14.5%
People	161,973,421	4.3%	54	34.4%	183	13.9%
Planning	1,017,653,753	26.7%	46	29.3%	105	8.0%
Economy	464,101,497	12.2%	51	32.5%	116	8.8%
Mobility	805,927,115	21.2%	56	35.7%	236	18.0%
Government	118,709,882	3.1%	50	31.8%	170	12.9%

Source: http://agendaurbana.it/

Table 12.2 Smart cities initiatives in the smart governance sector in Italian cities

Municipalities	No. projects	Projects recipients	Type of innovation
Campagna*	4	Administrations, Companies, Municipal structure, Citizens, City user, Third sector	Technological, of service, organizational/business
Cinisello Balsamo	4	Administrations, Companies, Municipal structure, Citizens, City user	Procedural, technological, of service
Brescia	4	Administrations, Companies, Municipal structure, Citizens, City user, Third sector, Other	Procedural, technological, of service, organizational/business, of technologies
Desio	1	Citizens, City user	Procedural, technological, of technologies, organizational/business, of service
Ferrara*	5	Administrations, Companies, Municipal structure, Citizens, City user, Third sector, Other	Procedural, technological, of technologies, of service, organizational/business
Cagliari*	5	Administrations, Companies, Municipal structure, Citizens, City user, Third sector	Technological, of technologies, of service, procedural
Roma*	5	Administrations, Companies, Municipal structure, Citizens, City user, Third sector	Procedural, technological, of technologies, of service
Tavagnacco	1	Companies, Municipal structure, Citizens, City user	Procedural
Bergamo*	3	Administrations, Companies, Municipal structure, Citizens, City user, Third sector	Technological, of service, organizational / business
Torino	10	Administrations, Companies, Municipal structure, Citizens, City user, Third sector, Other	Procedural, of service, of technologies
Milano	11	Municipal structure, Citizens, City user	Technological, of service
Venezia*	5	Municipal structure, Citizens, City user	Technological, of service, of technologies, organizational/business
Trento	4	Administrations, Companies, Citizens, City user	Procedural, of service, technological
Tavagnacco	4	Companies, Municipal structure, Citizens, City user	Technological, of service
Siena	1	#	#
San Pietro a Maida	1	Citizens	Technological, of service
San Giovanni in Persiceto	1	Citizens	Technological
San Giovanni in Persiceto	1	Municipal structure, Citizens	Technological

(Continued)

Table 12.2 (Cont.)

Municipalities	No. projects	Projects recipients	Type of innovation
Rosignano Marittimo	5	Municipal structure, Citizens	Technological, of service, procedural
Rieti	6	Municipal structure, Citizens, City user	Technological, of service, procedural
Ravenna*	8	Municipal structure, Citizens	Technological, of service, procedural.
Pordenone	9	Companies, Municipal structure, Citizens	Of service
Pavia*	5	Municipal structure, Citizens	Technological, of service
Palermo	6	Companies, Municipal structure, Citizens	Technological, of service, procedural
Padova	2	Municipal structure	Of service
Oriolo Romano	1	Municipal structure, Citizens	Of service, procedural
Modena	3	Companies, Citizens	Technological, of service.
Martignacco	1	#	#
Lumezzane	2	#	#
Livorno	5	Companies, Municipal structure, Citizens	Technological, of service
Lecce	10	Administrations, Companies, Municipal structure, Citizens, City user, Third sector	Of technologies, organizational / business, technological, of service
Latina	2	Administrations, Municipal structure, Citizens	Technological, of service
La Spezia*	6	Administrations, Companies, Municipal structure, Citizens, City user	Technological, of service, procedural
L'Aquila	1	Municipal structure, Citizens	Technological, of service
Imola	4	Citizens, Other, #	Of technologies, #
Genova*	2	Municipal structure, Citizens, City user	Technological, of service
Formia	1	Companies, Citizens	Technological, of service
Fiumicino	1	Citizens	Technological, of service, procedural
Firenze*	4	Companies, Municipal structure, Citizens, #	Of service, procedural, organizational/ business, #
Fabriano	1	Municipal structure, Citizens	Of service
Solarolo	1	Municipal structure, Citizens	Technological, of service
Riolo Terme	1	Municipal structure, Citizens	Technological, of service
Faenza	2	Companies, Municipal structure, Citizens	Procedural, technological, of service

(Continued)

Table 12.2 (Cont.)

Municipalities	No. projects	Projects recipients	Type of innovation
Castel Bolognese	4	Municipal structure, Citizens	Technological, of service
Casola Valsenio*	3	Companies, Municipal structure, Citizens	Technological, of service
Collesalvetti	3	#	#
Crosia	1	Citizens, City user	Technological, of service
Bari	2	Municipal structure, Citizens	Procedural, of service
Cava de' Tirreni	1	Companies, Citizens	Technological, of service
Baronissi	3	Companies, Citizens	Procedural, technological, of service
Reggio nell'Emilia*	1	Municipal structure, Citizens,	Procedural, organizational/ business
Ragusa*	1	Municipal structure	Of service
Formia*	1	Municipal structure, Citizens, Third sector	Of service
Brisighella*	1	Municipal structure	Technological, of service
Total	180		

Source: www.agendaurbana.it/
* planning projects, # data not found

(1) The size of city (small, medium or metropolitan city)
(2) The number of projects to which a letter of recognition is assigned (A, B, C, D etc.)
(3) The municipalities and the type of innovation (Table 12.2)

Through these parameters and after a careful reading of all the projects, the type of governance conceptualizations of the smart cities 2.0 was assigned. Table 12.3 identified the correlation between the Italian Urban Agenda projects indicated in Table 12.2 and the four governance conceptualizations of the smart cities 2.0 from Figure 12.1:

• Traditional government (Typology 1)
• Informing urban governance (Typology 2)
• Electronic governance for smart public administrators (Typology 3)
• Collaborative smart governance (Typology 4)

The outcome of this analysis led to the analysis of 104 projects for 20 municipalities. The recipients of smart-type initiatives belong to seven types:

• Administrations (17 projects out of 103)
• Citizens (79 projects)
• Companies (26 projects)
• Municipal Structure (62 projects)
• City User (26 projects)

Table 12.3 Correlations between (1) smart cities initiatives under smart governance sectors in Italian cities, (2) the national ranking between 1 and 39 in the Smart City Index 2018, and (3) the four typologies of governance identified by authors

Municipalities	City type	No. projects	Projects recipients	Type of innovation	Typologies
Brescia	MEDIUM-SIZED CITIES	4	A,B,C,D	Administrations (prog A,C,D) Citizens (prog A,B,D) Companies (prog A,C) Municipal structure (prog A,B,C,D) Other (prog C) City user (prog C) Third sector (prog C,D)	Of service (prog A,B,C,D) Procedural (prog B,C,D) Technological (prog C,D) Organizational/business (prog C,D) Of technologies (prog C,D)	Tip,1–prog C Tip,2–prog D Tip,3–prog A,B,C Tip,4–prog A
Ferrara	MEDIUM-SIZED CITIES	5	A,B,C,D,E*	Administrations (prog D) Citizens (prog B,C) Municipal structure (prog A,C,E) Other (prog A,C) City user (prog C,D) Third sector (prog B)	Of service (prog A,B,D) Procedural (prog A,B,C,D,E) Technological (prog A,D) Of technologies (prog A) Organizational/business (prog C)	Tip,1–prog E* Tip,2–prog A,B Tip,3–prog C,D Tip,4–prog B
Cagliari	METROPOLITAN CITIES	5	A,B,C,D,E*	Administrations (prog A,D) Citizens (prog A,B,D) Companies (prog A,B,C,D) Municipal structure (prog A,C,D) City user (prog A,D,E) Third sector (prog D)	Of service (prog B,C,D,E,F) Procedural (prog D) Technological (prog A,C,D,F) Of technologies (prog C,F) No data (prog G)	Tip,1–prog A,E* Tip,2–prog B,C Tip,3–prog D Tip,4–prog E*
Roma	METROPOLITAN CITIES	5	A,B,C,D,E*	Administrations (prog A) Citizens (prog A,B,D,E) Companies (prog A,B,D) Municipal structure (prog A,C) City user (prog A,B,D,E)	Of service (prog B,C,D,E) Procedural (prog A) Technological (prog A,B,C,D) Of technologies (prog B,C,D)	Tip,2–prog A,B,E* Tip,3–prog C,D Tip,4–prog B

City	Type	No.	Codes	Stakeholders	Category	Tipology/prog
Bergamo	MEDIUM-SIZED CITIES	3	A,B*,C*	Citizens (prog A,B,C) Administrations (prog B) Municipal structure (prog C) Companies (prog A,B) City user (prog A,B) Third sector (prog B)	Of service (prog A) Technological (prog A,C) Organizational/business (prog B,C)	Tip,2- progA,B* Tip,4- prog C*
Torino	METROPOLITAN CITIES	10	A,B,C,D,E,F,G,H, I,L	Administrations (prog B,C,D,F) Citizens (prog A,B,D,E,F,G,H,I,L) Companies (prog B,C,D,E,F) Municipal structure (prog B,C,D,G, H,I) City user (prog B,D,E,F,G,H,I,L) Third sector (prog H)	Of service (prog A,B,D,E,F,G,I,L) Procedural (prog A,C,G,H,I,L) Of technologies (prog C,H)	Tip,1- prog E,I Tip,2-prog B, D,G, L Tip,3-prog A,C,H Tip,4- prog F
Milano	METROPOLITAN CITIES	11	A,B,C D,E,F G,H,I, L,M	Citizen (prog A,B,D,E,F,G,H,I,L,M) Municipal structure (prog A,C,F,H,I) City user (prog A,L)	Of service (prog A,B,D,E,F,G,H,I,L, M) Technological (prog A,B,C,G,H,I,L,M)	Tip,1- prog D,F,G, L Tip,2- prog H,I Tip,3- prog A,B,C, M Tip,4- prog E,I
Venezia	METROPOLITAN CITIES	5	A,B,C, D,E*	Citizens (prog A,B,C,D,E) Municipal structure (prog C,E) City user (prog B)	Of service (prog A,B,C,E) Technological (prog A,B,C,E) Of technologies (prog D)	Tip,1- prog B Tip,2- prog E* Tip,3- prog A,B,D Tip,4- prog C
Trento	MEDIUM-SIZED CITIES	4	A,B,C, D	Administrations (prog D) Citizens (prog A,B,C,D) Companies (prog A,B) City user (prog C)	Of service (prog A,B,C,D) Procedural (prog B) Technologies (prog A,C,D)	Tip,1-prog B Tip,3-prog A,C,D
Ravenna	MEDIUM-SIZED CITIES	8	A,,B,C, D,E,F,* G*,H*	Citizens (prog A,B,D,E,F,G,H) Municipal structure (prog A,B,C,D,G,H)	Of service (prog A,B,D,E,F,G,H) Procedural (prog C,G) Technological (prog B)	Tip,1-prog B,H* Tip,2-prog G* Tip,3-prog C,D Tip,4-prog A,E,F*

(Continued)

Table 12.3 (Cont.)

Municipalities	City type	No. projects	Projects recipients	Type of innovation	Typologies
Pordenone	SMALL TOWN	9	A,B,C, D,E,F, G,H,I	Citizens (prog C,D,E,F,G,H,I) Companies (prog A,B) Municipal structure (prog C,D,E,G,H)	Of service (prog C,D,E,F,G,H,I) Procedural (prog A,B,E,G) Technological (prog D,E,F,H,I)	Tip,1- prog D,E,G, H Tip,2- prog C Tip,3- prog A,B,F,I Tip,4- prog I
Pavia	SMALL TOWN	5	A,B,C D*,E*	Citizens (prog A,B,E) Municipal structure (prog C,D)	Of service (prog A,,B,C,D,E) Technological (prog A,B,C,D,E)	Tip,1- prog A,B,E* Tip,3- prog C,D*
Padova	MEDIUM-SIZED CITIES	2	A,B	Municipal structure (prog A,B)	Of service (prog A,B)	Tip,2-prog B Tip,3-prog A
Modena	MEDIUM-SIZED CITIES	3	A,B,C	Citizens (prog A,B,C) Companies (prog A)	Of service (prog A,B,C) Technological (prog A,B,C)	Tip,2- prog A Tip,3- prog B,C
Lecce	MEDIUM-SIZED CITIES	10	A,B,C, D,E,F, G,H,I,L	Administrations (prog A,I) Citizens (prog B,C,D,E,F,G,I,L) Companies (prog I) Municipal structure (prog B,C,E,F,G,H,I,L) City user (prog I) Third sector (prog I)	Of service (prog A,B,C,D,E,F,G,H,L Technological (prog A,B,C,D,E,F,G,L Organizational/business (prog I) Of technologies (prog I)	Tip,1-prog B,D,E,F Tip,2-prog H,L Tip,3-prog A,C Tip,4-prog G,I
La Spezia	MEDIUM-SIZED CITIES	6	A,B,C D,E*,F*	Administrations (prog C,F) Citizens (prog A, B,C,D) Companies (prog C) Municipal structure (prog A,B, C,D,E,F) City user (prog C)	Of service (prog A,B,C,E,F) Procedural (prog C,F) Technological (prog A,B,C,D,E)	Tip,1-prog B,D,F* Tip,2-prog C,E* Tip,3-prog A

Genova	METROPOLITAN CITIES	2	A,B*	Citizens (prog A, B) Municipal structure (prog B) City user (prog A)	Of service (prog A,B) Technological (prog B)	Tip,1-prog A Tip,4-prog B
Firenze	METROPOLITAN CITIES	4	A,B,C, D*	Citizens (prog B) Companies (prog A) Municipal structure (prog B,D) No data (prog C)	Of service (prog A,B) Procedural (prog A) Organizational/business (prog D) No data (prog C)	Tip,1-prog B* Tip,3-prog A,B,C Tip,4-prog C,D*
Bari	METROPOLITAN CITIES	2	A,B	Citizens (prog A) Municipal structure (prog A,B)	Of service (prog A,B) Procedural (prog A)	Tip,1-prog A Tip,3-prog B Tip,4-prog A
Reggio nell'Emilia	MEDIUM-SIZED CITIES	1	A*	Citizens (prog A) Municipal structure (progA) Third sector (prog A)	Procedural (prog A) organizational/business (prog A)	Tip,2-prog A* Tip,4-prog A*
Total		104				

Source: www.agendaurbana.it/
* planning projects, # data not found

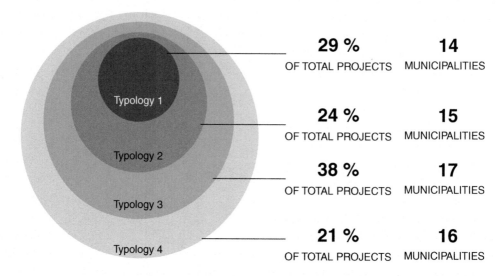

29 % **14**
OF TOTAL PROJECTS   MUNICIPALITIES

24 % **15**
OF TOTAL PROJECTS   MUNICIPALITIES

38 % **17**
OF TOTAL PROJECTS   MUNICIPALITIES

21 % **16**
OF TOTAL PROJECTS   MUNICIPALITIES

Figure 12.2 Subdivision of governance conceptualizations in Italian cities: traditional government (Typology 1); informing urban governance (Typology 2); electronic governance for smart public administration (Typology 3); collaborative smart governance (Typology 4)

Source: Chapter author(s)

- Third Sector (8 projects)
- Other

The types of innovation are classified into 5 types:

- Service (87 projects out of 103)
- Procedural (28 projects)
- Technological (57 projects)
- Technologies (12 projects)
- Organizational/business (5 projects)

In Figure 12.2 we further subdivided the governance conceptualizations of the Italian smart cities 2.0 projects. It is interesting to underline that Typology 3 (Electronic governance for smart public administrators) involves the greatest number of projects and municipalities.

Finally we mapped the same municipalities, taking into consideration the total amount of funding provided for the 104 projects that fell under the sectors 'government' and 'governance in planning'. The results do not constitute a complete picture of the Italian context, since the collection and cataloging of design interventions on smart cities throughout the country is constantly in progress and therefore not exhaustive, but they do allow for some interesting observations to be drawn.

Figure 12.3 clearly shows that cities of central and northern Italy have a greater per capita investment (in euros) in smart governance actions, while the southern regions tend to invest less in these issues, with the exception of Cagliari and Bari.

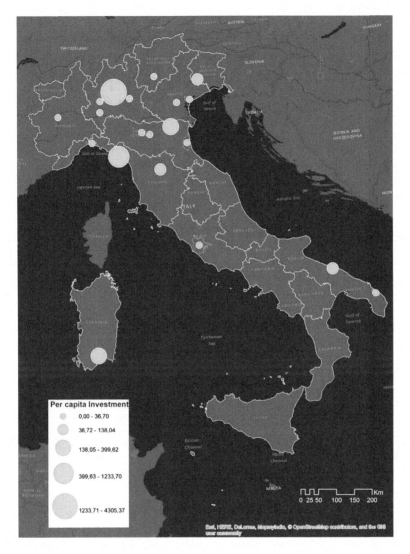

Figure 12.3 Per capita investment calculated in relation to the projects financed on the themes of 'government' and 'planning'

Source: IFEL-ANCI, Agendaurbana.it

Conclusion

In this chapter we presented a framework for comprehensive critical policy analysis of smart governance, which we identified as an overlooked topic in the emerging academic field of smart urbanism. In summary, smart urbanism, which we have defined as the transformation of smart city initiatives from smart cities 1.0 to smart cities 2.0, is characterized by a focus on smart governance. To demonstrate this, we studied the implementation through the governance of smart cities 2.0 and developed the definition of four conceptualizations applicable in society.

In order to investigate this further we undertook a comparative review of Italian smart city projects through a process of analysis that sought to understand which types of governance applied in different cities that have been useful for the transformation into smart cities 2.0. Despite the Italian smart cities initiatives being characterized by a bottom-up approach—in which each city tried to plan and execute its own smart program—the outcomes of the analysis established a changing governance model of smart cities. The analysis found that the typology of 'collaborative smart governance', which we consider the best conceptualization in relation to smart cities governance initiatives, currently has the lowest percentage of projects but involves a significant number of municipalities (16 out of 20 in the study). This implies that governance based on the technology-based approach is transforming into a progressive people-centric governance, in which the technology appears to be instrumental. Future research plans to extend the methodology to a comprehensive study of smart governance initiatives in Italy in order to define how these initiatives have been advanced by the smart cities 2.0 generation.

Acknowledgments

This study was supported by the MIUR (Ministry of Education, Universities and Research [Italy]) through a project entitled *Governing the smart city: a governance-centred approach to smart urbanism – GHOST* (Project code: RBSI14FDPF; CUP Code: F22I15000070008), financed with the SIR (Scientific Independence of Young Researchers) program. We authorize the MIUR to reproduce and distribute reprints for governmental purposes, notwithstanding any copyright notations thereon. Any opinions, findings and conclusions or recommendations expressed in this material are those of the authors and do not necessarily reflect the views of the MIUR. This chapter is the result of the joint work of the authors. 'Research methodology on comparative analysis of Italian smart cities initiatives' was written jointly by the authors. Chiara Garau wrote the 'From smart cities 1.0 to smart cities 2.0: what about the governance?', Giulia Desogus wrote the 'Introduction' and Paola Zamperlin wrote the 'Conclusion'. Chiara Garau revised the whole paper and checked for its comprehensive consistency.

Notes

1 http://agendaurbana.it/.
2 http://fondazioneifel.it/.
3 The final number was 180 because repetitive projects were excluded from the analysis.

References

Acuto, M., Steenmans, K., Iwaszuk, E., & Ortega-Garza, L. (2019). Informing urban governance? Boundary-spanning organisations and the ecosystem of urban data. *Area, 51*(1), 94–103.

Akter, S., Molla, M., Islam, S. R., Kabir, R., & Alam, F. (2019). Unequivocal ICT in Enhancing the Essence of Democracy and Good Governance. *Journal of Modern Accounting and Auditing, 15*(1), 34–39.

Angelidou, M. (2014). Smart city policies: A spatial approach. *Cities, 41*, S3–S11.

Azzari, M., Garau, C., Nesi, P., Paolucci, M., & Zamperlin, P. (2018). Smart City Governance Strategies to Better Move Towards a Smart Urbanism. In Gervast, O. et al. (eds) *International Conference on Computational Science and Its Applications* (pp. 639–653). Cham: Springer.

Bartenberger, M., & Grubmuller-Regent, V. (2014). The enabling effects of open government data on collaborative governance in smart city contexts. *eJournal of eDemocracy and Open Government, 6*, 36–48.

Bolívar, M. P. R. (2019). Public Value, Governance Models and Co-Creation in Smart Cities. In Bolivar, R. (ed.) *Setting Foundations for the Creation of Public Value in Smart Cities*, Cham: Springer. pp. 271–280.

Broccardo, L., Culasso, F., & Mauro, S. G. (2019). Smart city governance: Exploring the institutional work of multiple actors towards collaboration. *International Journal of Public Sector Management*, *32*(4), 1–22.

Caragliu, A., Del Bo, C., & Nijkamp, P. (2011). Smart cities in Europe. *Journal of urban technology*, *18*(2), 65–82.

Cardullo, P., & Kitchin, R. (2019). Being a 'citizen'in the smart city: Up and down the scaffold of smart citizen participation in Dublin, Ireland. *GeoJournal*, *84*(1), 1–13.

Clarke, A., & Margetts, H. (2014). Governments and citizens getting to know each other? Open, closed, and big data in public management reform. *Policy & Internet*, *6*(4), 393–417.

Edelenbos, J., Hirzalla, F., van Zoonen, L., van Dalen, J., Bouma, G., Slob, A., & Woestenburg, A. (2018). Governing the complexity of smart data cities: Setting a research agenda. In Bolivar, R. (ed.) *Smart Technologies for Smart Governments*, Cham: Springer. pp. 35–54.

Ferro, E., Caroleo, B., Leo, M., Osella, M., & Pautasso, E. (2013). The role of ICT in smart cities governance. In *Proceedings of 13th international conference for E-democracy and open government. Donau-Universität Krems* (pp. 133–145).

Garau, C., Balletto, G., & Mundula, L. (2015). A critical reflection on smart governance in Italy: Definition and challenges for a sustainable urban regeneration. In Bisello, A., Vettorat,o D., Stephens, R., & Elisei, P. (eds) *International conference on Smart and Sustainable Planning for Cities and Regions* (pp. 235–250). Cham: Springer.

Giffinger, R., Fertner, C., Kramar, H., Kalasek, R., Pichler-Milanovic, N., and Meijers, E., 2007, October. Smart cities – ranking of European medium-sized cities. Centre of Regional Science, Vienna. Available from: http://smart-cities.eu [Accessed April 18, 2019].

Greenfield, A. (2013). Against the smart city (1.3 edition). Berlin: Do projects.

Hollands, R.G. (2008). Will the real smart city please stand up?. *City*, *12*(3), 303–320,.

ISTAT (2018) Italian National Institute of Statistics. Data available at the following link: https://www.istat.it/it/

McFarlane, C., & Söderström, O. (2017). On alternative smart cities: From a technology-intensive to a knowledge-intensive smart urbanism. *City*, *21*(3–4), 312–328.

Meijer, A., & Bolívar, M. P. R. (2016). Governing the smart city: A review of the literature on smart urban governance. *International Review of Administrative Sciences*, *82*(2), 392–408.

Mistretta, P., & Garau, C. (2013). Città e sfide. Conflitti e Utopie. Strategie di impresa e Politiche del territorio. Successi e criticità dei modelli di governance, Cagliari: CUEC.

Nam, T. & Pardo, T.A., (2011). Smart city as urban innovation: Focusing on management, policy and context. In Estevez E., & Janssen M. (eds) *Proceedings of the 5th International Conference on Theory and Practice of Electronic Governance*, New York. pp. 185–194.

Nel, D., Du Plessis, C., & Landman, K. (2018). Planning for dynamic cities: Introducing a framework to understand urban change from a complex adaptive systems approach. *International Planning Studies*, *23*(3), 250–263.

EU Parliament, 2014. Mapping Smart Cities in the EU. Available from: https://researchitaly.it/uploads/8408/IPOL-ITRE_ET.pdf?v=08ef5aa [Accessed April 18, 2019].

Pereira G. V., Cunha, M. A., Lampoltshammer, T. J., Parycek, P., & Testa, M. G. (2017). Increasing collaboration and participation in smart city governance: A cross-case analysis of smart city initiatives. *Information Technology for Development*, *23*(3), 526–553.

Pereira, G. V., Parycek, P., Falco, E., & Kleinhans, R. (2018). Smart governance in the context of smart cities: A literature review. *Information Polity*, *23*(2), rreprint 1–20.

Rodrigues, M., & Franco, M. (2019). Measuring cities' performance: Proposal of a composite index for the intelligence dimension. *Measurement*, *139*, 112–121.

Sepasgozar, S. M., Hawken, S., Sargolzaei, S., & Foroozanfa, M. (2019). Implementing citizen centric technology in developing smart cities: A model for predicting the acceptance of urban technologies. *Technological Forecasting and Social Change*, *142*, 105–116.

Šiugždinienė, J., Gaulė, E., & Rauleckas, R. (2017). In search of smart public governance: The case of Lithuania. *International Review of Administrative Sciences*, *85*(3), 1–20.

Smart City Index. 2018 POLIS 4.0 Available at the following link: https://ey.com/Publication/vwLUAssets/Smart_City_Index_2018/$FILE/EY_SmartCityIndex_2018.pdf [Accessed May 08, 2019].

Sol, J., van der Wal, M. M., Beers, P. J., & Wals, A. E. (2018). Reframing the future: The role of reflexivity in governance networks in sustainability transitions. *Environmental Education Research*, *24*(9), 1383–1405.

Testa, P. (2016). Italian Smart Cities from the ANCI's National Observatory standpoint. *TECHNE-Journal of Technology for Architecture and Environment*, *11*, 40–44.

Trencher G. (2019), Towards the smart city 2.0: Empirical evidence of using smartness as a tool for tackling social challenges. *Technological Forecasting and Social Change*, *142*, 117–128.

Valencia, S. C. et al. (2019). Adapting the Sustainable Development Goals and the New Urban Agenda to the city level: Initial reflections from a comparative research project. *International Journal of Urban Sustainable Development*, *11*, 1–20.

Vázquez, A. N., & Vicente, M. R. (2019). Exploring the Determinants of e-Participation in Smart Cities. In Bolívar R., & Muñoz L.A. (eds) *E-Participation in Smart Cities: Technologies and Models of Governance for Citizen Engagement*, Cham: Springer. pp. 157–178.

Verrest, H., & Pfeffer, K. (2018). Elaborating the urbanism in smart urbanism: Distilling relevant dimensions for a comprehensive analysis of Smart City approaches. *Information, Communication & Society*, *22*, 1–15.

Wiig, A. (2015). IBM's smart city as techno-utopian policy mobility. *City*, *19*(2–3), 258–273.

Willke, H. (2007). Smart governance: Governing the global knowledge society. New York, NY: Campus Verlag.

Zamperlin, P., & Garau, C. (2017). 'Smart region': analisi e rappresentazione della smartness delle città metropolitane. *Bollettino dell'Associazione Italiana di Cartografia*, *161*, 59–71.

Part II
Smart city development

Section 1
Creative, smart or sustainable?

13

Will the real smart city please stand up?

Intelligent, progressive or entrepreneurial?

Robert G. Hollands

Editor's introduction

Would the real ( ) please stand up?

Idiom: *Would the true, genuine, or authentic person please identify themselves?* Used rhetorically to suggest that the person in question has up until now been disingenuous or failed to live up to their own potential or ideals.

(Idioms, 2015)

Geographer Paul Chatterton, in his article 'Will the real creative city please stand up?' (2000), uses the above idiom to reflect on what was behind the new language of creativity being used to drive and inform urban policy and development. Chatterton sets out the motivation for the article with the following reflection on the emergence of 'creative city' labelling of urban development projects: 'something has been troubling me. It is that, apparently, the cities we live in have become creative' (Chatterton, 2000, p. 390). Drawing on Landry's creative city work, Chatterton describes how cities such as Barcelona, Cologne, Bologna and even Huddersfield had adopted the tag 'creative city' (Florida, 2005a; Landry, 2000b), and then unpacks whether they are in fact delivering on the claim. Eight years later, Hollands (2008) asks the same rhetorical question of smart cities in his, now seminal, article; 'Will the real smart city please stand up?' and sets out to explore whether smart cities live up to their claimed potential. The smart city agenda is grounded in a very different lineage to creative cities: the 'Wired Cities' (Dutton, Blumler and Kraemer, 1987), 'Intelligent Cities' (Komninos, 2008) and 'Digital Cities' (Willis and Aurigi, 2017) of the later twentieth century. But, Holland's article identifies the fact that the smart city agenda is part of this long-standing practice of 'urban labelling' that hides a number of contradictions and assumptions. The article explores to what extent smart city labelling can be understood as a new variation of the 'entrepreneurial city', and counters this with speculation on what would be needed to make smart cities more inclusive and progressive.

In 2007, one year before Holland's article was published, IBM launched its 'Smarter Cities Challenge' (Dirks and Keeling, 2009). The Challenge, a label self-adopted by IBM, ran a high-profile campaign using numerous slick online promotional materials, white papers and policy reports[1] to set out a vision of delivering transformative change. Despite its claims of societal benefits, Wiig highlights how the IBM smarter cities challenge acted as a mask for entrepreneurial governance strategies (2015). Whilst setting out a promotional strategy for urban transformation may not in itself be problematic, what is necessary to consider beyond the rhetoric of such a vision is whether the delivery of the vision is actually achievable and more importantly, who benefits. According to a study of Barcelona smart city, the smart city label has the potential to become an 'empty, hollow signifier, as has already happened with other concepts such as the sustainable city or the resilient city (Vale, 2013), built in the image of capital and of the political elites' (March and Ribera-Fumaz, 2014, p. 826). In a long lineage of promotional city concepts, from creative city, eco-city, sustainable city and resilient city, the smart city label may simply be an empty placeholder that has little base in the reality of realizing positive urban development. Hollands' article deftly explores how some of the underlying assumptions result in a rather normative and celebratory evaluation of the label and sought to assess whether the promise of the smart city was actually being realized in the growing number of global cities that claimed smartness.

'Intelligent, progressive or entrepreneurial?' is the subtitle of Hollands' article and this categorizes how the smart city label is used in three general ways: firstly to describe the use of networked infrastructures to improve economic and political efficiency and to enable social, cultural and urban development; secondly for business-led urban development; and thirdly to progress both social and environmental sustainability agendas. Underlying this use of labels for urban development projects is the fact that the technological smart city agenda is adopted as a mask to hide a more limited political agenda of high-tech urban entrepreneurialism (Karvonen et al., 2019; Kitchin et al., 2019).

Central to this critique is the now widely acknowledged understanding that the 'smart city' label, that has been adopted by a range of actors, from IT companies, to city marketing departments through to urban policy makers, is ambiguous at best and at worst highly problematic (Cardullo et al., 2019; Datta, 2015; Gaffney and Robertson, 2018; Greenfield, 2013; Luque-Ayala and Marvin, 2015; March and Ribera-Fumaz, 2014; McFarlane and Söderström, 2017; Shelton, Zook and Wiig, 2015; Vanolo, 2013). In the context of urban development strategies, smart city agendas have been shown to mask many of the underlying assumptions and contradictions hidden within this process, and as Hollands points out:

> the characterization of these changes through the use of the term smart cities can create certain assumptions about this transformation, as well as play down some of the underlying urban issues and problems inherent in the labelling process itself (Begg, 2002).
>
> (Hollands, 2008, p. 304)

In fact, the 'real' smart city has been shown to not only mask urban problems and issues, but to deepen social divisions and contribute to patterns of exclusion. Hollands draws on Harvey's notion of the global 'spatial fix' (1989) to demonstrate that capital investment in urban infrastructure often results in diversion of public resources to help attract global capital with the effect of increasing social polarization. The public–private partnership model of many smart city initiatives results in the privatization of public infrastructure and services (from waste collection to information technology (IT)/WiFi provision and transport services etc.), where the financial benefit of both the running and maintenance often flows to global IT companies and is not underpinned by any socially progressive urban agenda. This leaves the role of 'smart' in the city as an optimizing rather than transformative one.

The smart cities rhetoric presents a challenge of testing the promise against the reality in the 'actually existing' smart city (Shelton *et al.*, 2015). Hollands concludes by making the case for a progressive smart city that is based on three aspects. The first is a recognition of human capital and a shift from smart cities to smart communities, where people's existing knowledge and skills are put centrally in smart city developments. The second is the use of IT to transform urban governance into a much more participatory and bottom-up process. Hollands speculates as to how this might be used to 'enhance democratic debates about the kind of city it wants to be and what kind of city people want to live in' (Hollands, 2008, p. 315). Finally, Hollands makes a plea for smart cities to make central an agenda that addresses issues of power and inequality by embracing diversity and viewing the marginalized as a social and cultural resource. This depiction of what might constitute a progressive smart city outlines the potential of the label to transcend its current model of urban entrepreneurialism and instead become a more inclusive, diverse and 'real' model of urban development. Put simply, smart cities need to speak the truth.

References

Cardullo, P., Feliciantonio, C. D. & Kitchin, R. (eds.) (2019) *The Right to the Smart City*. Bingley: Emerald Publishing.

Chatterton, P. (2000) 'Will the real creative city please stand up', *City* 4(3), pp. 390–397.

Datta, A. (2015) "The Smart Entrepreneurial City: Dholera and a 100 other utopias in India", in McFarlane, C., Marvin, S. and Luque-Ayala, A. (eds.) *Smart Urbanism: Utopian Vision or False Dawn?*, pp. 52–70. London: Routledge.

Dirks, S. & Keeling, M. (2009) *A Vision of Smarter Cities: How Cities Can Lead the Way Into a Prosperous and Sustainable Future*. New York: IBM Institute for Business Value. Available.

Dutton, W., Blumler, J. & Kraemer, K. (eds.) (1987) *Wired Cities: Shaping the Future of Communications*. New York: G.K. Hall.

Florida, R. (2005) *Cities and the Creative Class*. Abingdon: Routledge.

Gaffney, C. & Robertson, C. (2018) 'Smarter than Smart: Rio de Janeiro's Flawed Emergence as a Smart City', *Journal of Urban Technology* 25(3), pp. 47–64.

Greenfield, A. (2013) *Against the Smart City*. New York: Do Projects.

Harvey, D. (1989) 'From managerialism to entrepreneurialism: The transformation in urban governance in late capitalism', *Geografiska Annale* 71B(1), pp. 3–17.

Hollands, R. G. (2008) 'Will the real smart city please stand up?: Intelligent, progressive or entrepreneurial?', *City* 12(3), pp. 303–320.

Idioms, F. D.O. (2015) *Farlex Dictionary of Idioms.* https://idioms.thefreedictionary.com/will +the+real+someone+please+stand+up: Farlex.

Karvonen, A., Cugurullo, F. & Caprotti, F. (eds.) (2019) *Inside Smart Cities: Place, Politics and Urban Innovation.* London: Routledge.

Kitchin, R., Coletta, C., Evans, L. & Heaphy, L. (2019) 'Creating smart cities', in Kitchin, R. Coletta, C. Evans, L. and Heaphy, L. (eds.) *Creating smart cities* Abingdon: Routledge. pp. 1–18.

Komninos, N. (2008) *Intelligent Cities and Globalisation of Innovation Networks.* London and New York: Routledge.

Landry, C. (2000) *The Creative City: A Toolkit for Urban Innovators.* Abingdon: Earthscan.

Luque-Ayala, A. & Marvin, S. (2015) 'Developing a critical understanding of smart urbanism?', *Urban Studies* 52(12), pp. 105–2116.

March, H. & Ribera-Fumaz, R. (2014) 'Smart contradictions: The politics of making Barcelona a Self-sufficient city', *European Urban and Regional Studies* 23(4), pp. 816–830.

McFarlane, C. & Söderström, O. (2017) 'On alternative smart cities: From a technology-intensive to a knowledge-intensive smart urbanism', *City* 21(3-4), pp. 312–328.

Shelton, T., Zook, M. & Wiig, A. (2015) ''The 'actually existing smart city' '', *Cambridge Journal of Regions, Economy and Society* 8(1), pp. pp. 13–25.

Townsend, A. (2013) *Smart Cities: Big Data, Civic Hackers, and the Quest for a New Utopia.* New York: W. W. Norton & Company.

Vanolo, A. (2013) 'Smartmentality: The Smart City as Disciplinary Strategy', *Urban Studies* 51, pp. 883–898.

Wiig, A. (2015) 'IBM's smart city as techno-utopian policy mobility', *City* 19(2–3), pp. 258–273.

Willis, K. S. & Aurigi, A. (2017) *Digital and Smart Cities.* London: Routledge.

Introduction

Debates about the future of urban development in many Western countries have been increasingly influenced by discussions of smart cities (American Urban Land Institute, 2007; Coe *et al.*, 2000; Eger, 1997; New Zealand Smart Growth Network, 2000; Thorns, 2002), and there have been numerous examples of cities designated as smart in recent years. In the USA, information and communication technologies (ICTs) are seen as major factors in shaping and ensuring the success of San Diego as a 'City of the Future', while in Canada, Industry Canada injected CA$60 million into its nationwide 'Smart Communities' initiative, including Ottawa's 'Smart Capital' project involving enhancing business, local government and community use of internet resources. In the UK, Southampton claims to be the country's first smart city by virtue of the development of its multi-application smartcard, while in Southeast Asia, Singapore's IT2000 plan was designed to create an 'intelligent island', with information technology (IT) transforming work, life and play (Wei Choo, 1997). Numerous other examples abound from across the globe—from Bangalore, India's own Silicon Valley (Graham, 2002), Brisbane, Australia's 'sustainable' brand of smart urbanism, to a whole host of cities pursuing culturally based initiatives emphasizing the arts (Eger, 2003a), digital media, and culturally creative industries more generally (Florida, 2005a). These handful of examples are far from atypical. The 1997 World Forum on Smart Cities suggested that around 50,000 cities and towns around the world would develop smart initiatives over the next decade.

While it is obvious that IT and creative industries can and indeed have transformed many urban areas economically, socially and spatially (see Florida, 2002; Graham and Marvin, 2001, 1996), it might equally be argued that the characterization of these changes through the use of the term smart cities can create certain assumptions about this transformation, as well as play down some of the underlying urban issues and problems inherent in the labelling process itself (Begg, 2002). Part of the problem concerns the manner in which and variety of ways the term 'smart' is employed. For example, while the adjective smart clearly implies some kind of positive urban-based technological innovation and change via ICTs, analogous to the wired (Dutton, 1987), digital (Ishido, 2002), telecommunications (Graham and Marvin, 1996), informational (Castells, 1996) or intelligent city (Komninos, 2002), it has also been utilized (not unproblematically) in relation to 'egovernance' (Eurocities, 2007; Van der Meer and Van Winden, 2003), communities and social learning (Coe *et al.*, 2000), and in addressing issues of urban growth and social and environmental sustainability (Smart Growth Network, 2007; Polese and Stren, 2000; Satterthwaite, 1999). Further terminological confusion arises around the link between IT, knowledge, and the culturally creative industries (arts, media, culture), in discussions about the knowledge economy (Carrillo, 2006; Wolfe and Holbrook, 2002), and debates about creative cities[2] (see Eger, 2003a; Florida, 2002; Hall, 2000; Landry, 2000a). Finally, it might be argued that the problematic mapping of the smart label onto a series of other seemingly progressive debates and concepts concerning the technological and creative city, creates not only definitional problems, but also hints at some of the more normative and ideological dimensions of the concept/label. Not surprisingly, there are few analyses of smart city discourse from the point of view of more critical urban perspectives, such as ideas surrounding the 'entrepreneurial city' (Harvey, 1989), the growing domination of neo-liberal urban activities and spaces (Peck and Tickell, 2002), not to mention the existing literature on urban place marketing (Begg, 2002; Short *et al.*, 2000).

Due to its definitional impreciseness, numerous unspoken assumptions and a rather self-congratulatory tendency (what city does not want to be smart or intelligent?), the main aim of this chapter is to provide a preliminary critical polemic against some of the more rhetorical aspects of cities labelled as smart.[3] In performing this task, there are a number of important qualifications and caveats to make. First, the purpose of this chapter is not to provide a clearer and more empirically verifiable definition of what a smart city actually is, but rather to explore how some of its underlying assumptions can result in a rather normative and celebratory evaluation of the label. Second, whilst alluding to some of the wider commentary and critique surrounding IT, cities and spatiality (see Graham and Marvin, 2001; Webster, 2002; Castells, 1996 for instance), rather than directly review or assess this wider literature, the chapter specifically focuses its critique on cities which have been 'designated/labelled' as smart. As such, its purpose is not to empirically define or prove that smart cities do or do not exist, nor is it to assess to what degree such cities are successful or not at being smart. Such a task would require a measurable comparative[4] and/or case study method. Rather, the focus here is on the 'labelling process' itself adopted by a range of smart cities, with a view to problematizing aspects of this so-called 'new' urban form, as well as question some of the underlying assumptions/contradictions hidden within this process. To aid this critique, the chapter also explores to what extent such labelled smart cities can be seen as a 'high-tech' variation of urban entrepreneurialism (see Jessop, 1997), introduces a social justice element into the debate (Harvey, 2000), and hints at some general principles which might characterize a more progressive and inclusive smart city (Chatterton, 2000).

The first two sections of the chapter critically interrogate some definitions and elements of smart cities by briefly exploring their roots in wider debates, as well as teasing out some constituent elements through reference to numerous examples of cities which publicly market themselves as smart. A third section further develops a polemical critique of these self-promotional examples by stressing their underlying pro-business and neo-liberal bias, including questioning their various assumptions about transformations in urban governance and rhetoric of community participation, as well as raising hidden questions about social justice and sustainability. The main argument advanced is that smart urban labelling plays down some of the negative effects the development of new technological and networked infrastructures are having on cities (see Graham and Marvin, 2001), whilst overlooking alternative critical analyses of urban development associated with the entrepreneurial city (Harvey, 2000) and the growing domination of new-liberal urban space (Peck and Tickell, 2002).

Smart cities: Difficulties of definition

In today's modern urban context, we appear to be constantly bombarded with a wide range of new city discourses like smart, intelligent, innovative, wired, digital, creative, and cultural, which often link together technological informational transformations with economic, political, and socio-cultural change. One of the difficulties is separating out the terms themselves, which often appear to borrow on one another's assumptions, or in some cases, get conflated together. A second problem with such urban labelling is separating out the hype and use of such terms for place marketing purposes (Begg, 2002; Harvey, 2000; Short et al., 2000) as opposed to referring to actual infrastructural change or evidence of workable and effective IT policies. In essence, the disjuncture between image and reality here may be the real difference between a city actually being intelligent, and it simply lauding a smart label. A third problem with many of these terms is that they often imply, by their very nature, a positive and rather uncritical stance towards urban development. Which city, by definition, does not want to be smart, creative, and cultural?

Many of these points appear to apply to the smart city discourse. For example, Komninos (2002, p. 1) in his attempt to delineate the intelligent city, (perhaps the concept most closely related to the smart city), cites four possible meanings. The first, concerns the application of a wide range of electronic and digital applications to communities and cities, which effectively work to conflate the term with ideas about the cyber, digital, wired, informational or knowledge-based city. A second meaning is the use of information technology to transform life and work within a region in significant and fundamental ways (somewhat akin to the smart communities idea in the literature—i.e. see Coe *et al.*, 2000; Roy, 2001). A third meaning of intelligent or smart is as embedded information and communication technologies in the city, and a fourth as spatial territories that bring ICTs and people together to enhance innovation, learning, knowledge, and problem solving (the last of these being related somewhat to the smart growth agenda—see below). Overall then, Komninos (2006, p. 13) sees intelligent (smart) cities as 'territories with high capacity for learning and innovation, which is built-in the creativity of their population, their institutions of knowledge creation, and their digital infrastructure for communication and knowledge management'.

While this definition of intelligent (smart) cities initially appears to be a useful way of categorizing and indeed combining different aspects of the term, it also hints at some of the problems cited earlier. First, there is a clear problem conflating smart cities with a range of terms like cyber, digital, wired, knowledge cities etc., when in fact these various ideas themselves have somewhat different meanings. For example, wired cities (Dutton, 1987) refer literally to the laying down of cable and connectivity (not in itself necessarily smart), digital cities often infer virtual reconstructions of cities (i.e. like the virtual Digital City of Amsterdam; also see Ishido, 2002 on 'digital Kyoto'), and knowledge-based cities frequently focus on the relation of universities and academic knowledge and their links to the business world (Deem, 2001; Slaughter and Rhoades, 2004), relations which don't only depend on ICT infrastructure (although they often do, see Carrillo, 2006). The use of the terms innovation and creativity in the above definition also hints at the relationship between IT, knowledge, and media/cultural industries, problematically invoking at least some of the discourses of the creative city (see Peck, 2005). Second, while all of these terms imply that IT has a significant impact on cities, which it clearly does, they also emphasize quite different aspects of this relationship. For example, some aspects may be more technologically driven (i.e. cables and wires) and determinist (i.e. embedded systems of technology), others refer to types of information and human networks (i.e. academic knowledge, business innovation, etc.), while still others emphasize more human capital approaches to do with skills, education, competencies, and creativity.

Similarly, the 'smart growth' agenda[5] has been described as a rather wide-ranging approach which can be typified as those urban regions seeking to utilize innovative ITCs, architectural planning and design, creative and cultural industries, and concepts of social and environmental sustainability, in order to address various economic, spatial, social, and ecological problems facing many cities today (see Thorns, 2002). Different aspects of this agenda have focused in on innovative forms of 'e' or 'virtual' governance and citizen participation (Eger, 2003b; Eurocities, 2007; Van der Meer and Van Winden, 2003), smart communities and social learning approaches (Coe *et al.*, 2000; Paquet, 2001; Roy, 2001), and social and environmental sustainability in urban regions (Inoguchi *et al.*, 1999; Polese and Stren, 2000; Satterthwaite, 1999). Yet, even within more progressive sounding models of smart communities and smart growth there are inherent hidden assumptions and ideological contradictions. For instance, the notion of IT transforming life and work within a region, found within the smart communities literature (Coe *et al.*, 2000; as well as in Komninos, 2002, second definition of intelligent cities above), not only begs the question 'how and in what way is it being transformed?' but it also automatically assumes that there is some kind of community 'consensus' and involvement in the transition, and that such a change is inherently positive. Similarly, what if some smart initiatives which started out as publicly funded and with social inclusion as a goal, become overtaken by private sector concerns whose goal becomes purely profit-making? What happens to 'balance' with the smart growth agenda, for instance, when community interests are superseded by developer's interests, or the requirements of capital accumulation do not easily square with environmental and social sustainability?

In attempting to pin down what is smart about the smart city, one finds that not only does it involve quite a diverse range of things—IT, business innovation, governance, communities and sustainability—it can also be suggested that the label itself often makes certain assumptions about the relationship between these things (i.e. regarding consensus and balance discussed earlier for instance). The point here is not to try and offer a better definition, or argue that all smart cities are essentially the same. Nor is it to prove or disprove how smart they are according to some empirical criteria. Instead, the emphasis of the next section is to critically focus on numerous examples of places using the label smart (designated or self-designated), in order to practically untangle some of the elements involved in making them up, and critically explore what the relationship between these elements is, or what they are assumed to be.[6] This polemical exercise and analysis is deemed necessary to counter some of the taken-for-granted and self-congratulatory rhetoric of the smart-label bandwagon.

'Unwrapping' the smart city label

One of the key elements which stands out in the smart (intelligent) city literature is the utilization of networked infrastructures to improve economic and political efficiency and enable social, cultural, and urban development (Komninos, 2006; Eger, 1997). While this involves the use of a wide range of infrastructures including transport, business services, housing, and a range of public and private services (including leisure and lifestyle services), it is ICTs in particular that undergird all of these networks and which lie at the core of the smart city idea (see Graham and Marvin, 2001; Komninos, 2002). As Graham (2002, p. 34) argues, ICTs—including mobile and land line phones, satellite TVs, computer networks, electronic commerce, and internet services—are one of the main economic driving forces in cities and urban regions, producing numerous social and spatial effects. Smart cities, by definition, appear to be 'wired cities', although this cannot be the sole defining criterion, as it will later be argued. The Canadian city of Ottawa with 65 per cent of the population connected to the internet (not to mention its clustering of numerous software firms) is one example, while Blacksburg, USA, a university town of 38,000 which has a 100 per cent hook-up rate is another case in point. Andrew Michael Cohill, of Virginia Tech University and director of the Blacksburg Electronic Village project has argued telecommunications 'is the highway system of the twenty-first century' (cited in Evans, 2002), and many towns and cities across North America, Europe, and in the developing world are increasingly wedded to the idea that they have to be connected in order to be competitive in the new global economy (Graham and Marvin, 2001).

While there are numerous well-known examples of cities and regions developing through this route, including Singapore (Wei Choo, 1997), Silicon Valley and more recently San Francisco's 'Multimedia Gulch' in the US, and Bangalore (Asia's own Silicon Valley) (see Graham, 2002), the interesting thing is the degree to which many 'ordinary' cities have taken up the mantra that IT equals urban regeneration. For example, Newcastle Upon Tyne's economic strategy reflected in the document, 'Competitive Newcastle' (whose by-line is 'a dynamic entrepreneurial city at the heart of a knowledge based regional economy') has prioritized digital technology and the creative industries as one of its eight main business clusters. The idea of becoming an 'E-City' is also mentioned on the city council website, involving investing in broadband infrastructure, smart cards, e-commerce, and portal-based electronic service delivery, as is a joint £10 million partnership project between Newcastle City Council and Digitalbrain Plc to turn Newcastle into Europe's first

'Digital City' (Newcastle City Council, 2006). Another interesting North American example here is Halifax, on Canada's traditionally deprived East coast. In a speech titled 'Smart Growth for a Smart City: A New Economic Vision for Halifax', Brian Crowley, the president of the Atlantic Institute for Market Studies, states that location is no longer the key to economic success because 'the three most important things now affecting the future prosperity and development of human communities are technology, technology, and technology' (quoted in Siemiatycki, 2002). A final example here is San Diego. Because of its highly educated workforce and mix of high-tech industry and recreational assets, a marketing consortium of high-tech industries has dubbed San Diego 'Technology's Perfect Climate' (City of San Diego, 2007).

A second element characterizing many self-designated smart cities is their underlying emphasis on business-led urban development. There is a general world-wide recognition (and indeed acceptance) of the domination of neo-liberal urban spaces (Brenner and Theodore, 2002), a subtle shift in urban governance in most Western cities from managerial to entrepreneurial forms (Harvey, 1989; Quilley, 2000), and cities being shaped increasingly by big business and/or corporations (Gottdiener, 2001; Klein, 2000; Monbiot, 2000). This is no less true for self-designated smart cities. As the Edmonton, Canada, Smart City webpage (City of Edmonton, 2006) states, a smart city is characterized by 'a vibrant economy where businesses want to locate and expand'. It is interesting to note that six out of the ten features mentioned on their web pages mention or imply 'business-led' or 'business-friendly' criteria. And under the category of smart business and industry, the Edmonton webpage focuses perhaps predictably on technology sectors, including things like IT and bio-tech industries, as well as highlighting the advantages of having clusters of high-tech companies together and possessing an advanced telecommunications infrastructure. Another example comes from the economic development section of the city of San Diego's website, whose logo is 'San Diego—The Perfect Climate for Business' (City of San Diego, 2007). Even the progressive sounding smartgrowth.org website admits: 'Only private capital markets can supply the large amounts of money needed to meet the growing demand for smart growth developments' (Smart Growth Network, 2007). Often this element is couched in terms of talking about business as a whole, including small and medium-sized enterprises (SMEs) and through the language of businesses 'cooperation' and 'consultation' with local government ('public–private partnerships') and communities, rather than representing this relationship as one of potentially conflicting interests and contradictions (Harvey, 2000).

There is also a developing link between business-driven urban development, technology, and the changing role and function of urban governance (Harvey, 1989) in the smart city. While the UK city of Southampton's 'SmartCities' project (part financed by the European Commission) focuses around a smartcard system of accessing local government services such as libraries and leisure services (Southampton City Council, 2006), and hence makes reference to issues of social inclusion, it also intends to create 'a unified interface between city, authorities, commercial organizations and citizens'. Furthermore, the contention that the 'integration of commercial applications, such as loyalty card schemes, will further develop commercial relations between citizens and private organizations' (Kirkland, nd) hints at a rather different type of market-led smart agenda. In Edmonton, high-tech business-led growth and development are seen to require local government support in terms of providing a 'strong pro-business environment', and 'reasonable taxes and low cost to live and to do business' (City of Edmonton, 2006), as well as a providing a highly skilled and educated workforce, and creating partnerships between education, business, and government. San Diego ('Technology's Perfect Climate') boasts one of the most competitive sales tax rates in California (7.75 per cent) and its business tax rate is lower than any of the 20 largest US cities (City of San Diego, 2007). All examples here very much echo Harvey's (1989) discussion of the role of local government as civic boosters and aiding urban entrepreneurial-ism, through providing public–private partnerships and knowledge transfer through higher education institutions (Wolfe and Holbrook, 2002). It also ties in with the more peripheral literature on growth coalitions (Logan and Molotch, 1987), urban regimes (Elkin, 1987; Stone, 1993), and urban place marketing (Short *et al.*, 2000).

Of course there exist other models of e-governance that are more directed towards intra-city cooperation, while others lean towards social learning, inclusion, and community development. For example, with respect to e-governance, the European Digital Cities (EDC) programme, which started in 1996, was designed to share information and good local government practice amongst European cities through common internet portal sites (Komninos, 2002). Eurocities now has working groups on e-citizenship and e-governance, and is clearly committed 'to ensur-ing that everyone can have access to ICTs and participate in the Knowledge Society' (Eurocities, 2007). The latter emphasis on social learning/community development is best represented per-haps in the idea of smart community initiatives in Canada (Coe *et al.*, 2000). Komninos (2002, p. 188) describes smart communities as where business, government, and residents use new tech-nology to transform life and work in their region, while Roy (2001, p. 7) defines it as:

> a holistic approach to helping entire communities go online to connect to local govern-ments, schools, businesses, citizens and health and social services in order to create specific services to address local objectives and to help advance collective skills and capacities.

The key question such interesting initiatives raise is how to effectively balance the needs of the community with both those of local government and the needs of business, particularly corporations (Monbiot, 2000).

While two of the main aspects of designated smart cities are the use of new technologies and a strong pro-business/entrepreneurial state ethos, a related concern is with particular high-tech and creative industries such as digital media, the arts, and the cultural industries more gen-erally (see Eger, 2003a; Florida, 2005; Hall, 2000; Scott, 2000). In Europe, the work of Landry and Bianchini (1995) has emphasized the issues for the creative city of the future will focus upon its 'soft infrastructure', including such things as knowledge networks, voluntary organiza-tions, safe crime-free environments, and a lively after-dark entertainment economy. Similarly, in the USA, Richard Florida's creativity schema is popularly represented by the three 't's' of economic development—tolerance, technology, and talent—and his concern with catering for the creative classes lifestyles and needs (Florida, 2002). Although Florida includes a technology measure here, and discusses various fractions of the creative classes working in IT, science, and the digital media, he broadens the notion of creativity to the cultural industries more generally. He also emphasizes the importance of other characteristics of the population, including diver-sity, tolerance, and even 'bohemia' (defined as the concentrations of writers, designers, musi-cians, and artists in a city—see Florida, 2002). In essence, the bulk of writers in the creative city discourse emphasize the social and human dimensions of the city (see Landry, 2000b), as much, if not more than, the technological emphasis at the core of the smart city. There is also generally more of a focus here on how alternative cultures can help fuel urban growth (although see Peck, 2005 for a trenchant critique of Florida's work here), rather than relying on new technology or corporate businesses (although some emphasize how the arts/culture environment can also contribute to the 'new economy' of cities, see Eger, 2003a, pp. 14–15).

This more 'humanist' emphasis ties in with other related discourses of smart communities, including the importance of social leaning, education, and social capital for developing the smart city (Eger, 2003b). For example, the City of Brisbane, has adopted a ten-year Smart

City vision aimed at addressing and promoting the following: information access, lifelong learning, the digital divide, social inclusion, and economic development (Siemiatycki, 2002). Coe et al. (2000, p. 13) also generally admit while the emphasis of smart cities is very much on economic growth, they 'are not possible outside of the development of smart communities—communities that have learned how to learn, adapt and innovate'. Similarly, the role of social capital, defined as the construction of social relations and networks of trust and reciprocity (see Carley *et al.*, 2001), is considered necessary in order to engage all stakeholders to participate and engage with a smart city. Connection rates are only a limited measure of success. It is also recognized that technology has to be utilizable and understandable by the communities that it is supposed to serve (Evans, 2002), and that ordinary people and communities need to have the skills necessary to utilize ICTs.

Finally, present within some smart city agendas is a concern with both social and environmental sustainability. Social sustainability implies social cohesion and sense of belonging (Carley *et al.*, 2001), while environmental sustainability refers to the ecological and 'green' implications of urban growth and development (Gleeson and Low, 2000; Inoguchi *et al.*, 1999). With respect to the first type, it is recognized by some that the smart city has to be an inclusive not just technological city (Helgason, 2002). As Coe et al. (2000, p. 21), argue, 'local community partnerships—not wires—are the fibres that bind smart communities'. With respect to the second type of sustainability, it is equally recognized that while cities may be drivers of economic growth, they are also great consumers of resources and creators of environmental waste (Low *et al.*, 2000; Satterthwaite, 1999). For example, it is estimated that urban areas consume around 75 per cent of the worlds resources (80 per cent of fossil fuels) and produce most of its waste (Baird, 1999). All told then, self-designated smart cities project different emphases and can mean different things to different people. However, it might also be suggested that not all the elements mentioned here have equal weighting in the labelling process. The next section provides a critique of the interplay between these various aspects by looking deeper into some the self-designated smart cities already discussed.

Critiquing self-designated smart cities

In order to further assess the labelled smart city, it is important to step back and look more critically at some of its main assumptions, and query the positive spin given to its main elements. For example, in unproblematically adopting some of the assumptions from the IT model of urban development (Eger, 1997), some smart cities might be critiqued as being technologically determined. In a word, undue influence can be attributed solely to urban technological advancements in explaining what happens in cities and how they are currently being shaped. While there is no denying the impact of ICTs on the urban form (Graham and Marvin, 1996), and of course this process may be viewed critically (i.e. see Graham and Marvin, 2001; Webster, 2002), there can be a more conservative application here that implies that somehow IT itself will deliver the smart city a priori—a kind of technological 'Field of Dreams' scenario (see Eger, 1997; Dutton, 1987 for instance).

However, some recognize that smart cities have to be more than just broadband networks. As Chris Wilson of the University of Ottawa Centre on Governance has argued 'Being connected is no guarantee of being smart' (quoted in Evans, 2002). Similarly, Paquet (2001) suggests that although technology is an enabler, it is not necessarily the most critical factor in defining the smart city. One of the best examples of the mismatch between developing technologies and low take up comes from Graham's (2002) discussion of the South American city of Lima. Despite increasing rates of telecommunication diffusion, in 1990 less than half of all households in the city had a phone and only 7 per cent had access to the internet, with the poorest 50 times less likely to have the internet (Graham, 2002, p. 43). In other words, having the technology does not always lead to its take-up, nor are take-up rates always equitable. Technological determinism with respect to ICTs, through advertising and magazine articles, suggest, argues Graham (2002, p. 35), some 'value-free technological panacea offering instant, limitless access to some entirely separate and disembodied on-line world'. A less charitable analysis might suggest that it offers up yet another urban form dominated not by industrial capital this time but by technological and knowledge capital. The main idea here is that the technological smart city becomes a smokescreen for ushering in the business-dominated informational city.

For example, while local governments from around the globe all stress they are concerned with how residents and communities utilize the new technologies, their 'bottom line' economic imperative appears to be to attract capital, particularly knowledge and informational capital to their city. For example, despite the fact that much of Ottawa's economy is derived from government sources, even it acknowledges that 'individual companies drive a city's prosperity' (City of Ottawa, 2006). In San Diego's General Plan for the city they state: 'Economic prosperity is a key component of quality of life. The structure of the City of San Diego's economy influences the City's physical development and determines the City's capacity to fund essential services' (City of San Diego, 2007). And yet, while much of rhetoric about business and capital in the smart city is linked to small-scale IT companies and providing local employment opportunities, the fact of the matter is that huge chunks of this industry are controlled and dominated by multi-national firms which are highly mobile (Shiller, 1999).

The history of Singapore's IT revolution is a good example of the ideological shifts smart cities can undergo. It has been suggested that such a revolution unfolded in three phases (Wei Choo, 1997). First, a public sector-funded IT initiative from 1981–85 to computerize government ministries, improve public service,s and produce a good stock of computer experts. Second, a shift from the public to private sector through the National Technology Plan (1985–90) designed to 'develop a strong export-oriented IT industry and to improve business productivity through IT' (Wei Choo, 1997, p. 48). And finally a third phase begun in 1991 entitled the IT2000 masterplan in which the city/state was to be transformed into an 'intelligent island', where IT permeates every aspect of the society—home, work, and play. The stated goals of the masterplan are to enhance national competitiveness and to improve the quality of life of citizens (Wei Choo, 1997, p. 49). What is interesting about this example is first the financial shift from the public to the private sector and second, a more ideological shift towards merging business competitiveness with social well-being.

The 'ideological turn' expressed here has an effect on the development of the urban form, as cites can be seen increasingly to serve global mobile IT businesses as opposed to looking after stationary ordinary citizens (Amin et al., 2000). As Graham and Marvin (2001) put it, the diffusion of information technology across cities is actually having an effect which can only be described as 'splintering urbanism'—a fragmentation and polarization of whole urban regions, both economically and socially. While the effects are numerous, Graham (2002) provides a host of examples such as the targeting of particular IT services to 'high-end' wealthy customers and the creation of fortified high-tech enclaves in places like Sao Paulo, Kuala Lumpur, Bangalore, and Singapore, as well as the development of gentrified urban neighbourhoods to house smart workers, such as in San Francisco.

This latter point leads to a further critique of smart cities along similar lines to that of the creative city (i.e. see Peck, 2005). While the creative city envisioned by Florida (2005) consists of trying to recruit and retain the 'creative classes' generally (see Florida, 2002), the idea of the smart city is to presumably attract and cater for smart workers. One of the inevitable by-products of either urban form, by definition, is social polarization (Harvey, 2000). For instance, despite being a relatively rich country, aided partly through its advanced technological infrastructure, Singapore's poverty level is estimated to be in the region of 25–30 per cent of the population. Perhaps even more telling is that during the height of its IT boom, the city/country became even more polarized. In 1990 the richest 10 per cent of households earned 15.6 times more than the poorest 10 per cent, but by 2000, the gap widened further with the richest earning 36 times more than the poorest (Singapore Democratic Party, nd). Similarly, poverty rates in San Diego, despite it having relatively high labour force participation rates and low levels of unemployment over the past decade, have actually risen during their so-called high-tech boom, suggesting that rhetoric about the digital revolution reaching everyone is wildly optimistic. For example, child poverty rates (under 18 years of age) in the city actually increased from 1990 to 2002 from 15.6 per cent to 17.5 per cent (City of San Diego, 2007).

The smart/creative city can become not only more economically polarized, but also socially, culturally, and spatially divided by the growing contrast between incoming knowledge and creative workers, and the unskilled and IT illiterate sections of the local poorer population (Peck, 2005; Smith, 1996). Urban gentrification in this regard refers not just to housing and neighbourhoods as it once did (see Butler, 1997), but increasingly to consumption, lifestyle, and leisure in the city (see Chatteron and Hollands, 2002). Chatterton and Hollands (2003), for example, have studied the gentrification and social polarization of UK nightlife, tracing it back to changes in the urban economy, including the impact of IT and service employment on cities. For instance, the transformation of the UK city of Leeds from a manufacturing city to a service-based urban form, has resulted in the creation of a range of up-market bars and nightclubs, which work to exclude whole sections of the local population (Hollands and Chatteron, 2004). The impact of the gentrified smart/creative city then goes far beyond creating inequalities of work, housing, and neighbourhood, and extends to areas such as inequitable city space (Byrne, 1999) and entertainment provision (Chatterton and Hollands, 2003).

Despite representations in smart city discourses about the importance of local communities and social learning, an overall emphasis on business-driven technology and gentrification could be interpreted to imply that this urban form is relatively unconcerned with class inequality (i.e. particularly the uncreative classes, see Peck, 2005), inclusion (Byrne, 1999), and social justice (Harvey, 1973). Even the more humanist rationale of the smart/creative city is predicated on attracting educated people by providing a creative infrastructure of work, community, and leisure (Florida, 2002). Edmonton's smart city approach here is to offer 'an exceptional arts and entertainment scene' (City of Edmonton, 2006), presumably mostly for the middle classes. Eger (2003a, p. 14), quoting the National Governor's Association in the US, states that arts programmes contribute 'to a region's "innovation habitat", thus improving quality of life— *making it more attractive to the highly desirable knowledge-based employees*' (my emphasis). The issue here is how does this provision relate to the 'less' smart/creative sections of the local population? What can the smart city offer them? And what impact does catering for knowledge-based employees have on arts provisions for the less well off? So, while smart cities may fly the banner of creativity, diversity, tolerance, and culture, the balance appears to be tipped towards appealing to knowledge and creative workers, rather than using IT and arts to promote social inclusion (Sibley, 1995; Solnit and Schwartzenberg, 2000).

Part of the response to this dilemma lies in some of the discussion surrounding smart communities in North America (Coe *et al.*, 2000) or various inclusion measures through ICTs in the USA (Phipps, 2000) and the UK (Talbot and Newman, 1998). While many of these measures appear progressive and there are numerous examples of 'successful' participatory IT projects, looked at more critically, many of these programmes could be viewed as neo-liberal attempts to incorporate the local community into the entrepreneurial city (Harvey, 1989). Notions of smart communities and the importance of social learning/social capital, in this view, seem less progressive and more ideological. Education within capitalism has always been necessary to reproduce the workforce. In this instance, it has simply been reoriented towards the new information economy, primarily through training local people to serve the needs of the new creative and informational classes (Peck, 2005). The irony is that many such social learning and training programmes, often funded by national and local government money, may actually work to subsidize the training needs and requirements of multi-national companies which cities hope to lure to town (Harvey, 2000). The emphasis of smart cities, looked at in this light, shifts from discourses about inclusion and human capital to more of a 'culture of contentment' idea (Galbraith, 1993), with unskilled local labour servicing the leisure and lifestyle needs of the new incoming knowledge and creative workers.

Finally, what can one make of those self-designated smart cities which emphasize environmental sustainability as their smart feature? The key question here is to what extent are economic growth and environmental sustainability compatible (Gleeson and Low, 2000), and is the information city automatically all that eco-friendly? As Graham (2002, p. 34) perceptively points out, despite notions that ICT work can potentially conquer space through increased 'homeworking', this practice is relatively rare, hence information workers still have to get to the office. Therefore, at least two of the outcomes created by urban ITC clusters—transport and car parking space—are not particularly environmentally friendly (Newman and Kenworthy, 1999). Additionally, the information technology revolution is perhaps not as clean as it initially appears. Researchers at the United Nations university in Tokyo, for example, have estimated that the production of a new computer demands ten times its weight in fossil fuels and chemicals, as opposed to two times for the production of an automobile, and in the future the world could face a computer 'waste mountain' as people constantly upgrade their technology (Sample, 2004, p. 2).

Brisbane, Australia, is a useful example of some the contradictions between smart cities being committed to economic growth and the environment, simultaneously. For example, the city has utilized the smart label in conjunction with notions of the 'sustainable city' with regard to a unique water recycling programme (Local Government Focus, 2004). Yet as the city website makes explicit: 'Brisbane is a great place to do business. It has low taxes and charges, excellent infrastructure, great support networks and a forward-thinking local administration to support you in your business venture' (Brisbane City Council, 2005). The key question is what happens when there are not enough resources to cater for both of these things? Or what happens when the focus on environmental sustainability itself begins to be seen as a new branch of capitalistic opportunity? For example, Smart-Cities.net is a web-portal site which currently promotes urban sustainable development by providing a platform for information exchange and interaction between Asian Cities and European environmental solution providers (Smart Cities.net, 2002). While its focus on urban environmental challenges is laudable, its website might also be 'read' as a future stepping stone for ecological business opportunities. The question is, can cities accord the same priority to all aspects of the smart city agenda, or do some elements automatically take precedence over others (i.e. business needs over environmental ones, see Gleeson and Low, 2000; Inoguchi et al., 1999)?

Underneath the rather self-congratulatory surface of self-designated smart cities are some unspoken assumptions and continuing urban problems. Issues concerning the splintering effects of the informational city, the limits of urban entrepreneurialism, problems created by the creative classes for local communities, including deepening social inequality and urban gentrification, not to mention the conflict between environment sustainability and economic growth, loom in the background behind the smart city label. The conclusion picks up on some of these issues and discusses how the smart city discourse might be moved in more progressive directions.

Conclusion: towards more 'progressive' smart cities?

This chapter began with a critical interrogation of the concept of smart (intelligent) cities, and through an analysis of a range of (self-)designated examples has subjected the idea to a polemical critique. Many cities from around the globe have been keen to adopt the smart city mantle and emphasize its more acceptable face for self-promotional purposes. In addition to assuming there is an automatically positive impact of IT on the urban form, the smart city label can also be said to assume a rather harmonious high-tech future. However, it might be argued that beneath the emphases on human capital, social learning, and the creation of smart communities, lay a more limited political agenda of 'high-tech urban entrepreneurialism'. Analyses of some designated smart cities here reveal examples of prioritizing informational business interests and hiding growing social polarization (Harvey, 2000), features more reminiscent of the 'entrepreneurial city' (Jessop, 1997), and 'neo-liberal' urbanism more generally (Peck and Tickell, 2002). Of course, this assertion requires further study and in-depth analyses of specific urban cases. All cities differ somewhat in their history, economic and political makeup, and cultural legacy. They are also influenced by national boundaries and indigenous government policies and laws.[7] However, the apparent ascendance of the entrepreneurial city (Harvey, 1989; Quilley, 2000) and its high-tech variant (i.e. the smart city) belies a set of underlying shortcomings and contradictions.

First, is the urban problematic revealed by Harvey's theoretical notion of the global 'spatial fix'. As Harvey (1989) argues, capitalist investment in urban infrastructure, while necessary, is no guarantee to further capital accumulation. And while such investment may temporarily act to boost an area's profile and create employment, it can also mean a diversion of public (welfare) resources to help lure in mobile global capital thereby creating social polarization. Furthermore, the 'spatial fix' inevitably means that mobile capital can often 'write its own deals' to come to town, only to move on when it receives a better deal elsewhere. This is no less true for the smart city than it was for the industrial, manufacturing city. Investment in ICTs, human capital, social learning, and smart communities, while seemingly laudable aims for any city or urban region wanting to regenerate, also holds no guarantees. Public–private partnerships and investment in these areas may in fact backfire, as information technology capital may flow elsewhere depending on what advantages are available to aid further capital accumulation. Perhaps one of the best illustrations of this process concerns the city of Ottawa and its boom bust cycle of high-tech industries. While the Canadian government has poured some CA\$6.4 billion into the Technology Partnerships Canada programme (effectively loans to multi-national companies), it is expected that only about a third of that money will have been repaid by 2020, which is in effect a public subsidy (Aubry, 2002). At the same time, it is now felt by some that the city is losing control of its high-tech industry, reverting to its former role as a technology research and development site servicing multi-nationals based elsewhere (Bagnell, 2003; Hill, 2002).

Additionally, as the previous analysis shows, self-designated smart cities face the interminable difficulty of how to deal with the issue of widening inequality and social polarization, a problem brought on partly by its own 'success', so to speak. Rather than raising standards of living for all urban dwellers, IT has been shown to deepen social divisions in cities (Graham, 2002). The attraction of educated, mobile, middle-class professionals and IT workers (part of the 'creative classes', Florida, 2002), can result in the production of highly gentrified neighbourhoods and leisure/entertainment provision, thereby excluding traditional communities and poorer residents. Furthermore, it is often understated that smart cities requires a sizable secondary workforce needed to service the entertainment and leisure needs of professionals and information workers (Peck, 2005), thereby contributing to entrenched labour market inequalities. So, while much of the smart city discourse gives emphases to the creation of smart communities and the raising of everyone's access to urban IT, education, and governance, ironically it can actually contribute to the two-speed or 'dual city'. The dominance of the entrepreneurial version of smart cities does not of course preclude the existence of different smart urban forms or examples, or the future development more progressive models. The remainder of the conclusion briefly explores what aspects a more progressive smart city might strive for.

First and foremost, progressive smart cities must seriously start with people and the human capital side of the equation, rather than blindly believing that IT itself can automatically transform and improve cities. To some extend this is already recognized (see Eger, 2003b). As Paquet (2001, p. 29) has argued regarding the creation of smart communities, 'The critical factor in any successful community has to be its people and how they interact'. The important aspect of information technology is not its capacity to automatically create smart communities, but its adaptability to be utilized socially in ways that empower and educate people, and get them involved in a political debate about their own lives and the urban environment that they inhabit. As Raymond Williams (1983) always reminded us, while technology (of any kind) is never neutral, it has the potential and capacity to be used socially and politically for quite different purposes. In this vein, perhaps some of the best instances of where ICTs have been utilized most progressively would be the development of community telecentres, particularly those ones that attempt to link up IT to socially marginalized groups (see Graham, 2002, p. 50). While there are numerous worldwide examples to draw upon (see Phipps, 2000; Talbot and Newman, 1998), perhaps one of the most revealing cases concerns Rathgeber's (2002) study of community telecentres in Africa to help women, in particular, to enhance their job prospects and opportunities. The most telling aspect of this research was that because initial attempts to set up such telecentres were 'technological' (about hardware/software) and business-led, rather than social and people-led, they were largely ineffective and inaccessible, and hence failed. Rathgeber's (2002) study showed that such centres were seen rather as a social resource by the target group of African women to help run their daily lives, rather than as a technological/economic resource. This specific example demonstrates the pressing need to start with people's existing knowledge and skills, not with technology per se.

Second, the progressive smart city needs to create a real shift in the balance of power between the use of IT by business, government, communities, and ordinary people who live in cities (Amin et al., 2000), as well as seek to balance economic growth with sustainability. As Coe et al. (2000, p. 13) argue, while the emphasis on smart cities is very much about economic growth, and competitiveness in the global knowledge economy, smart communities can also 'provide an opportunity for enhancing citizen participation in and influence over local decision making'. In a word, the 'real' smart city might use IT to enhance democratic debates about the kind of city it wants to be and what kind of city people want to live in a type of virtual 'public culture', to redefine a term from Sharon Zukin (1995). Zukin basically defines public culture as where all possible interests and priorities of a whole range of citizens are placed on the agenda and discussed and debated. While IT might make the conditions for developing a 'virtual public culture' possible, there must be the political will to make this happen and the digital divide must be addressed.

Such shifts, would involve the progressive smart city addressing issues of power and inequality in the city (Harvey, 2000), as well as begin to seriously respect diversity and build a democratic urban pluralism (Sandercock, 1998). Part of the difficulty here undoubtedly concerns how one understands and comes to term with the variety of inequalities that exist in cities (Fincher and Jacobs, 1998; Harvey, 1989; Keith and Pile, 1993). One thing that is patently clear, however, is the degree to which cities have become more unequal through IT (Graham, 2002), the processes of globalization (Harvey, 2000), changes in urban labour markets (Peck, 2005), and increased gentrification (Smith, 1996). While the smart entrepreneurial city 'successfully 'caters for the rich, mobile, creative businessman, through the creation of corporate informational portals and services, not to mention through luxury hotels, restaurant, bars, and global business transport links, by definition, it also simultaneously ignores the welfare needs of its poorer residents (Byrne, 1999; Graham, 2002). And while these economic hierarchies are not in dispute, urban feminists and multi-cultural theorists argue that they differentially impact on gendered and ethnic populations. For example, while the entrepreneurial smart city might cater for the small number of professional and creative females working in the IT sector, the majority of working women are left to service the largely male business city—waiting on, cleaning, and servicing its dominant male make-up (Jarvis, 2005). Furthermore, according to theorists like Sandercock (2003), our urban ethnic minorities and migrants are simultaneously feared, ignored or exploited, rather than viewed as a social and cultural resource. Finally, the talents of many young people in cities are wasted under the rubric of a social problem discourse, rather than seen through the lens of cultural creativity (see Chatterton and Hollands, 2003), while many alternative political groups such as environmentalists, squatters, third sector groups and cooperatives, and/or urban political movements such as 'reclaim the streets' and 'critical mass' go un-noticed or are seen as public nuisances (Chatterton, 2000).

In essence the smart progressive city needs and requires the input and contribution of these various groups of people, and cannot simply be labelled as smart by adopting a sophisticated IT infrastructure or through creating self-promotional websites. Cities are more than just wires and cables, smart offices, trendy bars, and luxury hotels, and the vast number of people who live in cities deserve more than just these things. Because the smart city label can work to ideologically mask the nature of some of the underlying changes in cities, it may be a partial impediment toward progressive urban change. Real smart cities will actually have to take much greater risks with technology, devolve power, tackle inequalities, and redefine what they mean by smart itself, if they want to retain such a lofty title.

Notes

1 According to Townsend, IBM is estimated to have invested hundreds of million dollars in the smarter cities campaign (Townsend, 2013).
2 While the discourse of smart cities has certain parallels with that of the creative city, and hence is open to similar criticism (see Peck, 2005), it is distinguished by its particular focus on information and communication technologies as the driving force in urban transformation (Eger, 1997), rather than creativity in a more general sense (see Florida, 2002). However, as I shall go on to argue, there are selective borrowings in some of the smart city discourses regarding the role IT increasingly plays in the arts, culture, and media (see Eger, 2003a).
3 As such, I would liken the aim of this chapter to Peck's (2005) critique of Florida's (2005, 2002) work on the creative city, albeit it is critiquing a somewhat different literature. In other words, the point of both articles is not to prove or disprove the existence of the creative or smart city, but rather to critically explore some of the assumptions and rhetoric behind these labels as well as examine some examples of cases where the term is applied.

4 There are methods developed which claim to help measure smartness/intelligence and innovation—see Intelligent Community Forum (2007) which lists the five main elements of intelligent communities and the OECD and Eurostat (2005) Oslo Manual designed to provide guidelines for measuring innovation. As this is not the purpose of this chapter, I do not really make any further reference to these measurement criteria.

5 While there is clearly some overlap here between the use of the term smart in relation to smart cities and the smart growth agenda (particularly as they both relate to ICTs and how these can transform work and life in a region), the two terms should not be completely conflated. The smart growth agenda is a somewhat more wide-ranging urban approach, with a strong emphasis on policy prescriptions and problem-solving. There are also specific national variations, such as the smart growth agenda emanating out of the USA (American Urban Land Institute, 2007), which has developed in response to specific urban problems such as sprawl, inner-city decline, and a lack of community in suburban areas (see Smart Growth Network, 2007). Despite these differences, I would content that smart cities and the smart growth agenda tend to share some similarities when it comes to emphasizing the underlying importance of IT and business-led initiatives when solving urban problems. In this chapter I reserve the term smart cities to refer to those urban regions which publicly label themselves as smart, whilst focusing in on the labelling process they adopt.

6 In this regard I examine a range of cities from around the world that have been designated (either through award or competition) or have self-designated themselves as smart cities. In examining this labelling process I look particularly at city websites as this is one of their main promotional vehicles and hence reveals what kinds of things are emphasized and which things are hidden from view.

7 It is generally recognized that North American cities in particular have always been more shaped by pro-business influences, so it is hardly surprising to see smart city discourses here more nakedly influenced by 'neo-liberalism' (for example in the case of San Diego and Edmonton). At the same time, it is clear that the Smart Capital initiative in Ottawa did partly achieve a balance of IT initiatives that cut across business, government, and community interests (Ottawa Centre for Research and Innovation, 2007), hence the need for more specific case studies. European cities, by contrast, have historically, at least, been more welfare-oriented in their urban policy-making and generally been more concerned with social inclusion, although as Harvey (1989) argues, they too have embraced urban entrepreneurialism in the last couple of decades (see also Quilley, 2000), and many are competing with one another via various creative indexes (see Florida and Tinagli, 2004). Meanwhile, political transformations in Eastern Europe have meant rapid change in cities as they have made the rather rapid transition from a socialist to an entrepreneurial urban form (see Sykora, 1999).

References

American Urban Land Institute (2007) Material from their website, www.uli.org/AM/Template.cfm?Section=Home&CONTENTID=92882&TEMPLATE=/CM/ContentDisplay.cfm (accessed 10 September 2007).

Amin, A., Massey, D. and Thrift, N. (2000) *Cities for the Many Not the Few*. Bristol: Policy Press.

Aubry, J. (2002) 'Billions in tech loans will remain unpaid', Ottawa Citizen, 21 October, pp. 1–2.

Bagnell, J. (2003) 'Entrust in the news', National Post, 22 August, www.entrust.com/news/reprints/off_the_map.htm, accessed 10 September 2007.

Baird, V. (1999) 'Green cities', *New Internationalist* 313, www.newint.org/issue313/keynote.htm (accessed 10 September 2007).

Begg, I. (ed.) (2002) *Urban Competitiveness: Policies for Dynamic Cities*. Cambridge: Polity Press.

Brenner, N. and Theodore, N. (eds.) (2002) *Spaces of Neo-Liberalism*. Oxford: Blackwell.

Brisbane City Council (2005) See www.ourbrisbane.com/business/doingbusiness/(accessed 10 August 2005).

Butler, T. (1997) *Gentrification and the Middle Classes*. Aldershot: Ashgate.

Byrne, D. (1999) *Social Exclusion*. Buckingham: Open University Press.

Carley, M., Jenkins, P. and Small, H. (2001) *Urban Development and Civil Society: The Role of Communities in Sustainable Cities*. London: Earthscan.

Carrillo F.J. (2006) *Knowledge Cities: Approaches, Experiences and Perspectives*. New York: Elsevier Butterworth Heinemann.

Castells, M. (1996) *Rise of the Network Society: The Information Age*. Cambridge: Blackwell.

Chatteron, P. and Hollands, R. (2002) 'Theorising urban playscapes: Producing', *Regulating and Consuming Youthful Nightlife City Spaces'*, *Urban Studies* 39(1), pp. 95–116.

Chatterton, P. and Hollands, R. (2003) *Urban Nightscapes: Youth Cultures, Pleasure Spaces and Corporate Power*. London: Routledge.

City of Edmonton (2006) See www.smartcity.edmonton.ab.ca/smart1.html (accessed 20 September 2005).

City of Ottawa (2006) See www.ottawa.ca/2020/es/1_0_en.shtml (accessed on 20 September 2006).

City of San Diego (2007) Economic development, www.sandiego.gov/environmental-services/sustainable/pdf/survey_answers.pdf (both accessed 10 September 2007).

Coe, A. and Paquet, G. and Roy, J. (2000) 'E-goverance and smart communities: A social learning challenge', Working Paper 53, Faculty of Administration, University of Ottawa, October.

Deem, R. (2001) 'Globalisation', *New Managerialism, Academic Capitalism and Entrepreneurialism in Universities: Is the Local Dimension Still Important?'*, *Comparative Education* 37(1), pp. 7–20.

Dutton, W.H. (1987) *Wired Cities: Shaping the Future of Communications*. London: Macmillan.

Eger, J. (1997) *Cyberspace and cyberplace: Building the smart communities of tomorrow*. San Diego: San Diego Union-Tribune, Insight.

Eger, J. (2003a) 'The creative community', White paper on cities and the future, San Diego State University, San Diego, www.smartcommunities.org/creative-1.htm (accessed on 12 August 2005).

Eger, J. (2003b) 'Smart communities: Becoming smart is not so much about developing technology as about engaging the body politic to reinvent governance in the digital age', *Urban Land* 60(1), pp. 50–55.

Elkin, S. L. (1987) *City and Regime in the American Republic*. Chicago: University of Chicago Press.

Eurocities (2007) Knowledge society, www.eurocities.org/main.php (accessed 9 February 2007).

Evans, S. (2002) 'Smart cities more than broadband networks', *Ottawa Business Journal*, 25 September.

Fincher, R. and Jacobs, J. (eds.) (1998) *Cities of Difference*. London: The Guilford Press.

Florida, R. (2002) *The Rise of the Creative Class: and How it's Transforming Work, Leisure, Community and Everyday Life*. New York: Basic Books.

Florida, R. (2005a) *Cities and the Creative Class*. New York: Harper Business.

Florida, R. (2005b) *Cities and the Creative Class*. Abingdon: Routledge.

Florida, R. and Tinagli, I. (2004) *Europe and the Creative Age*. Pittsburgh and London: Carnegie Mellon Software Industry centre and Demos.

Local Government Focus (2004) See www.locgov-focus.aus.net/2001/june/bris13.htm (accessed on 10 February 2004).

Galbraith, J.K. (1993) *The Culture of Contentment*. London: Penguin.

Gleeson, B. and Low, N. (2000) 'Cities as consumers of worlds environment', In N. Low, B. Gleeson, I. Elander and R. Lidskog (eds.) *Consuming Cities: The Urban Environment in the Global Economy after the Rio Declaration*, pp. 1–29. London: Routledge.

Gottdiener, M. (2001) *The Theming of America*. Boulder, CO: Westview Press.

Graham, S. (2002) 'Bridging urban digital divides: urban polarisation and information and communication technologies (s)', *Urban Studies* 39(1), pp. 33–56.

Graham, S. and Marvin, S. (1996) *Telecommunications and the City: Electronic Spaces, Urban Places*. London: Routledge.

Graham, S. and Marvin, S. (2001) *Splintering Urbanism: Networked Infrastructures, Technological Mobilities and the Urban Condition*. London: Routledge.

Hall, P. (2000) 'Creative cities and economic development', *Urban Studies* 37(4), pp. 633–649.

Harvey, D. (1973) *Social Justice and the City*. Baltimore: John Hopkins University Press.

Harvey, D. (2000) *Spaces of Hope*. Edinburgh: Edinburgh University Press.

Helgason, W. (2002) 'Inclusion through a digital lens'. Paper presented at the conference Thinking Smart Cities, Carleton University, Ottawa, Canada, 15 November.

Hill, B. (2002) 'Fastest-50 list includes only three Ottawa firms', *Ottawa Citizen* 26(September), pp. D1.

Hollands, R. and Chatteron, P. (2004) 'The London of the north?: Youth cultures, urban change and nightlife in Leeds', in R. Unsworth and J. Stillwell (eds.) *TwentyFirst Century Leeds: Geographies of a Regional City*. Leeds: Leeds University Press.

Inoguchi, T., Newman, E. and Paoletto, G. (1999) *Cities and the Environment: New Approaches for Ecosocieties*. New York: UN University Press.

Intelligent Community Forum. (2007) See www.intelligentcommunity.org accessed on10 September 2007.

Ishido, T. (2002) 'Digital city Kyoto', *Communications of the Acm* 45(7), pp. 78–81.

Jarvis, H. (2005) *Work/Life City Limits: Comparative Household Perspectives*. New York: Palgrave.

Jessop, B. (1997) 'The entrepreneurial city: Re-imagining localities, redesigning economic governance or restructuring capital', In N. Jewson and S. McGregor (eds.) *Transforming Cities*, pp. 28–41. London: Routledge.

Keith, M. and S. Pile. (eds.) (1993) *Place and the Politics of Identity*. London: Routledge.

Kirkland, D. (nd) Smart cities—a smarter approach, www.publicservice.co.uk/pdf/detr/winter2000/p24. pdf (accessed on 10 September 2007).

Klein, N. (2000) *No Logo*. London: Flamingo.

Komninos, N. (2002) Intelligent Cities: Innovation, Knowledge Systems and Digital Spaces. London: Spon Press.

Komninos, N. (2006). The architecture of intelligent cities integrating human, collective, and artificial intelligence to enhance knowledge and innovation. In *2nd International Conference on Intelligent Environments, Institution of Engineering and Technology* (pp. 13–20).

Landry, C. (2000a) The Creative City: A Toolkit for Urban Innovation. London: Earthscan.

Landry, C. (2000b) *The Creative City: A Toolkit for Urban Innovators*. Abingdon: Earthscan.

Landry, C and Bianchini, F. (1995) *The Creative City*. London: Demos.

Logan, J. and Molotch, H. (1987) *Urban Fortunes: The Political Economy of Place*. Berkeley: University of California Press.

Low, N., Gleeson, B., Elander, I. and Lidskog, R. (eds.) (2000) *Consuming Cities: The Urban Environment in the Global Economy after the Rio Declaration*. London: Routledge.

Monbiot, G. (2000) *The Captive State*. London: Macmillan.

Smart Growth Network. (2007) 'Smart growth online', http www.smartgrowth.org accessed10 September 2007.

New Zealand Smart Growth Network. (2000) *Smart Growth: Intelligent Development in a New Century*. Rotorua: New Zealand Smart Growth Network.

Newcastle City Council (2006) See www.newcastle.gov.uk/compnewc.nsf/a/home (accessed 10 June 2006).

Newman, P. and Kenworthy, J. (1999) *Sustainability and Cities: Overcoming Automobile Dependency*. Washington DC: Island Press.

OECD and Eurostat (2005) Oslo Manual: guidelines for collecting and interpreting innovative data. 3rd edition, a joint publication of OECD and Eurostat, http://epp.eurostat.cec.eu.int/cache/ITY_PUB LIC/OSLO/EN/OSLO-EN.PDF (accessed 10 September 2007).

Ottawa Centre for Research and Innovation (2007) Smart capital projects, www.ocri.ca/smartcapital/ sc_subprojects.asp (accessed10 September 2007).

Paquet, G. (2001) 'Smart communities', *LAC Carling Government's Review* 3(5), pp. 28–30.

Peck, J. (2005) 'Struggling with the creative class', *International Journal of Urban and Regional Research* 29(4), pp. 740–770.

Peck, J. and Tickell, A. (2002) 'Neo-liberalising space', *Antipode* 34(3), pp. 380–404.

Phipps, L. (2000) 'New communication technologies—a conduit for social inclusion', *Information, Communication & Society* 3(1), pp. 39–68.

Polese, M. and Stren, R. (2000) *The Social Sustainability of Cities: Diversity and the Management of Change*. Toronto: University of Toronto Press.

Quilley, S. (2000) 'Manchester first: From municipal socialism to the entrepreneurial city', *International Journal of Urban and Regional Research* 24(3), pp. 601–615.

Rathgeber, E. (2002) 'Gender and telecentres: What have we learned?', *World Bank Group*, www.world bank.org/gender/digitaldivide/Eva%20Rathgever.ppt (accessed on 12 June 2004).

Roy, J. (2001) 'Rethinking communities: Aligning technology & governance', *LAC Carling Government's Review*, Special Edition, 6–11 June.

Sample, I. (2004) 'PCs: The latest waste mountain', *The Guardian* 8, March pp. 2.

Sandercock, L. (1998) *Towards Cosmopolis: Planning for Multicultural Cities*. Chichester: John Wiley.

Sandercock, L. (2003) *Cosmopolis II: Mongrel Cities of the 21st Century*. 2nd London: Continuum.

Satterthwaite, D. (ed.) (1999) *The Earthscan Reader in Sustainable Cities*. London: Earthscan.

Scott, A. (2000) *The Cultural Economy of Cities: Essays on the Geography of Image-producing Industries*. London: Sage.

Shiller, D. (1999) *Digital Capitalism: Networking the Global Market System*. Massachusetts: MIT Press.

Short, J.R., Breitbach, C., Buckman, C.S. and Essex, J. (2000) 'From world cities to gateway cities: Extending the boundaries of globalization theory', City 4(3), pp. 317–340.

Sibley, D. (1995) *Geographies of Exclusion*. London: Routledge.

Siemiatycki, M. (2002) 'Smart cities, whats next?' *Paper presented at the conference Thinking Smart Cities*, Carleton University, Ottawa, Canada, 15 November.

Singapore Democratic Party (nd) See www.singaporedemocrat.org/poverty.html (accessed on 10 September 2007).

Slaughter, S. and Rhoades, G. (2004) *Academic Capitalism and the New Economy: Markets, State and Higher Education*. Baltimore: Johns Hopkins University Press.

Smart Cities.net (2002) See www.smart-cities.net (accessed on 10 September 2007).

Smith, N. (1996) *The New Urban Frontier: Gentrification and the Revanchist City*. London: Routledge.

Solnit, R. and Schwartzenberg, S. (2000) *Hollow City: The Siege of San Francisco and the Crisis of Urban America*. London: Verso.

Southampton City Council (2006) Southampton On-line, www.smartcities.co.uk/InterestGroup/ (accessed on20 September 2006).

Stone, C. N. (1993) 'Urban regimes and the capacity to govern: A political economy approach', *Journal of Urban Affairs* 15, pp.1–28.

Sykora, L. (1999) 'Processes of socio-spatial differentiation in post-communist Prague', *Housing Studies* 14 (5), pp. 679–701.

Talbot, C. and Newman, D. (1998) 'Beyond access and awareness—evaluating electronic community networks', The British Library Board: British Library Research and Innovation Centre Report 149/ Queens University Belfast On-line, www.qub.ac.uk/mgt/cicn/beyond/(accessed 10 June 2005).

Thorns, D. (2002) *The Transformation of Cities: Urban Theory and Urban Life*. Basingstoke: Palgrave.

Vale, L. J. (2013). 'The politics of resilient cities: Whose resilience and whose city?' *Building Research & Information* 42(2), pp. 191–201.

Van der Meer, A. and Van Winden, W. (2003) 'E-governance in cities: A comparison of urban policies', *Regional Studies* 37(4), pp. 407–419.

Webster, F. (2002) *Theories of the Information Society*. London: Routledge.

Wei Choo, C. (1997) 'IT2000: Singapore's vision of an intelligent island', In P. Droege (ed.) *Intelligent Environments: Spatial Aspects of the Information Revolution*, pp. 48–65. Amsterdam: Elsevier Science.

Williams, R. (1983) *Towards 2000*. London: Chatto and Windus.

Wolfe, D. and Holbrook, J. (eds.) (2002) *Knowledge, Clusters and Regional Innovation: Economic Development in Canada*. Kingston: Queen's School of Policy Studies and McGill-Queen's University Press.

Zukin, S. (1995) *The Culture of Cities*. Oxford: Blackwell.

14

Smart to green

Smart eco-cities in the green economy

Federico Caprotti

Introduction

In the late 2010s, smart urbanism has become a key guiding concept for sustainable urban development. City and national governments refer to the smart city as a solution to a panoply of problems, including but not limited to the challenges of deindustrialisation, stimulating urban economies, technological and economic growth, and governance experiments in an increasingly urbanising society. In some national settings, such as in rapidly developing countries, smart urbanism is branded as a solution for keeping gross domestic product (GDP) growth rates high, as seen in the following extract from a 2012 *China Daily* article: 'At a time when China is experiencing slower economic growth, building smart cities will be a huge driving force for not only the information industry but also related industries' (China Daily, 2012, np). At the same time, in national contexts facing economic crisis, smart cities are presented as ways of facing up to the economic challenges of stagnation or even contraction. In a landscape of post-2008 austerity measures, for example, the smart city became seen as 'a political device that was used to frame a number of new national and local policies as "technical" solutions to low budgets and economic stagnation' (Pollio, 2016, 514). Nonetheless, the common thread linking smart urban strategies and roadmaps is that of an interest in specific (albeit sometimes contrasting) visions of future urbanism. As the US National Science and Technology Council (NSTC) states: 'Motivated by a vision of ubiquitous, smart infrastructure, systems, and services, many cities and communities view advances in networking and information technology as a way to increase efficiency, reduce costs, and improve quality of life for their residents (NSTC, 2017: 5). What can be seen in this statement is a clear interest in smart urbanism because of its promises in terms of efficiency, speed, analytical power and governance and market opportunities. What is often missing in both policy and scholarly accounts of the smart city, however, is a discussion of how plans for smart urbanism contribute to, and intersect with, one of the key driving forces behind national and international agendas: environmental sustainability in the context of sustainable development.

Although there is, currently, a strong and evident focus on high-tech, digital urbanism, green agendas have not disappeared from the urban scene. They been replaced by the buzz around ubiquitous computing, the Internet of Things (IoT) and digital economies. Eco-urbanism, in its multiple guises, is a continuing and enduring area of focus at a variety of scales (Caprotti, 2015; Joss, 2015), in large part because of the real, material and deepening challenges associated with anthropogenic climate change and its impact on current and future cities. Indeed, in the late 2010s eco-urbanism arguably has a more material effect on cities worldwide than smart urbanism. This effect can be felt through international policymaking, national policies and directives, environmental regulation, low-carbon policies and incentives, and citizen and grassroots interest in greener urban environments. Indeed, the 2016 United Nations Paris Agreement on climate change identified cities as important actors in shaping sustainable futures (Roberts, 2016), and the New Urban Agenda, heralded at the 2016 UN-Habitat III conference in Quito, Ecuador, included cities as a focus for sustainable development (Caprotti *et al.*, 2017).

A key challenge facing urban policymakers and scholars today is how to reconcile 'smart' and 'eco' urban agendas. In order to do so, it is important to recognise that these agendas have specific (if diverse) roots, and that there are areas of distinct overlap and shared intent in their logics. The chapter begins by tracing the roots of eco-urbanism and then smart urbanism, focusing on their contextualisation within a deeply modern project to *reshape* the city, and citizens, often according to market logics. The green economy is then introduced as a concept that potentially holds together both smart and eco-urbanism, through what can be called the 'smart eco-city' (Caprotti *et al.*, 2016). The chapter then concludes by focusing on the ways in which more socially sensitive approaches to smart-eco futures could help to challenge hegemonic technocratic views of the future city.

Cutting the Gordian knot between nature and the city: tracing the genealogies of the eco-urbanism

The genealogy of eco-urbanism can, in many ways, be directly traced to the development of environmental consciousness. The roots of eco-urbanism are, therefore, multiple. They are not reconcilable to a single point of origin, although it is sometimes tempting to read urban environmental history as issuing from one tradition, whether it be scholarly, or more socio-historically grounded in a particular place or period. Nonetheless, and briefly, contemporary trends in eco-urbanism draw strongly on the 1970s environmental movement, as well as on broader 20th century attempts to establish a science of cities and their ecology and metabolism (Melosi, 1993). These attempts range from those more influenced by natural and physical sciences, such as studies of cities' urban metabolism that attempted to materially and statistically account for material, energy and other urban inflows and outflows; to studies of urban metabolism from a critical and radical standpoint that focus on how the city, as a metabolic entity, both produces and reproduces socio-environmental inequalities and injustices at a variety of scales (Swyngedouw, 2009). Overall, what characterises these attempts to develop an understanding of eco-urbanism is an interest in the relationship between nature and the city.

While the latter half of the 20th century saw key interventions in the development of eco-urbanism from both a policy and scholarly standpoint, eco-urbanism is also rooted much more deeply in the broader context of modernity (Kaika, 2005). This can be seen rather strikingly in the oft-used example of Ebenezer Howard's Garden City concept, an attempt to reconcile the positive aspects of countryside living with those of urban life, while reducing or eliminating the negatives associated with either (Ward, 2016). Howard's concept was a response to the inequalities and socio-environmental degradation and excesses of the Industrial Revolution and its aftermath: an attempt to shape new cities in the image of a kinder, gentler and more humane urban industrial capitalism that would also be more sensitive to the natural environment of, and around, the city.

Garden Cities are an early example of the variety of ways in which the relationship between nature and the city has been tackled in modernity. Resolving and untying the Gordian knot of nature–city relations was, in many ways, a driving part of the urban project of modernity. The 'knot', in this context, is the problem of how to reconcile the growth of industrial economies, the attendant emergence and expansion of cities throughout the industrialised and industrialising world, and the effect this has had on natural ecosystems, landscapes and the environmental landscape. Attempts to unravel the difficult link between cities and their 'negative' effects on the natural world have happened in a multiplicity of ways, often experimental, in different settings. For example, the New Towns built in the Soviet Union (Bolotova, 2012), or in fascist Italy (Caprotti, 2007), were in large part attempts to reshape the city and re-orient urban structure so as to fit more closely into metabolic trajectories decided by the state. This was carried out through the tools and techniques of urban planning, urban design, architecture, engineering and other areas of urban technique through which specific political-ideological views came to be expressed and materialised. The emphasis on (re)shaping the city in the image of a new, greener society can also be seen in more recent projects. This is the case, for example, with China's eco-city construction programme (Jong et al., 2016), which numbers over 100 cities with plans for eco-city construction.

The eco-city concept is, like the broader trend of eco-urbanism within which it is situated, genealogically rooted in deeply modern developments occurring over decades. While all these strands cannot be summarised here (but see Roseland, 1997), the eco-city concept has its roots in Richard Register's 1975 Urban Ecology think tank and in his *Eco-City Berkeley* (Register, 1987), in which the broad outlines of an eco-city as an urban area that was sensitive to its ecological environment were codified (Roseland, 1997). The eco-city concept became further enshrined and broadened, as well as accepted by broader policy audiences, in the 1990s with the start of the EcoCity World Summits, held every two years. This partially enabled the eco-city to become a key guiding concept for sustainable urban development, lending itself especially to projects for newly build cities in geographical contexts from China, to the Gulf, to North America. Some of the key, canonical eco-city examples completed or under construction by the end of the 2010s include the Sino-Singapore Tianjin Eco-City in China (Caprotti et al., 2015; Chang et al., 2016), Treasure Island in California (Joss, 2011), Eko Atlantic in Nigeria (Watson, 2014) and Masdar eco-city in Abu Dhabi (Chakravarty, 2017; Cugurullo, 2016). More broadly, eco-city principles influenced national sustainable urban development strategies, from China's eco-city programme mentioned above, to the UK's late 2000s eco-town projects, to France's nationwide Écoquartier strategy.

Tracing the genealogies of smart urbanism

When compared with eco-cities, it can appear at first glance as if the smart city concept is completely new and novel. Indeed, the notion of the smart city has only become popular and mainstreamed from the 2000s onwards (Hollands, 2008), and the term 'smart' is now synonymous with notions of innovative, new and ground-breaking urban trajectories and projects. Smart urbanism refers to a panoply of initiatives, technologies and approaches, from national strategic policies aimed at stimulating the digital economy, to e-governance, urban environmental, security, transport and social sensor systems, driverless vehicles, the sharing and 'gig' economy, and the integration of socio-economic activities through smartphone technologies. Corporations, from IBM to the increasingly well-established technology leaders of Asia (such as Samsung, Alibaba, Tencent, Huawei and Baidu) have taken centre stage in much of this rapid development of interest in smart urbanism. Indeed, it has been argued that smart cities 'describe cities that, on the one hand, are increasingly composed of and monitored by pervasive and ubiquitous computing and, on the other, whose economy and governance is being driven by innovation, creativity and entrepreneurship, enacted by smart people' (Kitchin, 2013, 1).

Notwithstanding the at times frothy focus on ever-increasing production of new technologies, products and applications, the smart city, like the eco-city, has roots that stretch back decades, or even longer – certainly far before the digital era. Indeed, the focus on data, efficiency and processing power can be seen as stemming from the focus on increasing production and economic efficiency that is a characteristic of the industrial era. Taylorism, the production line and the use of scientific methods to control industrial production (much of it aimed at urban markets) can in turn be seen as contributing to smart urbanism's focus on efficiency, speed, pervasiveness and control. Indeed, the transparent and highly ordered, scientifically organised urban society described by Zamyatin in his 1924 novel *We*, in many ways prefigures the city of smart glass and Big Data of the 21st century (Zamyatin, 2013). In turn, the wired technologies of electricity grids and telephone networks, as well as radio communications, heralded an increasingly networked society that was always at least partially wireless.

Another set of historical roots to the smart city can be seen around the design and importance of the smart city control rooms that are celebrated as futuristic today (such as that of the Rio Operations Centre, or the Glasgow Operations Centre). These are, in turn, similar to the factory control rooms introduced in industrial plants, or the power plant control rooms present in nuclear power stations from their inception. Just as a nuclear plant is an attempt to control and harness the atom through its control room, the smart city control room can be likened to an attempt to gain control of the often uncontrollable: social life in the city. Likewise, the architecture and design of these smart urban spaces often materialises previous imaginative visions: NASA space programme control rooms from the 1970s, for example, resemble today's smart city control rooms (Picon, 2015). At the same time, the specific screen-based, macro view of the city enabled through smart city technologies, sensors and Big Data can be seen as symbolic references to earlier visual imaginations of future society, from videogames, to science fiction. And yet, it could be argued that control rooms and videological approaches to the smart city are part and parcel of a liquid modernity (Bauman, 2013), functioning as islands of illusory technological stability in the shifting seas of techno-social change. While these spaces and views of the smart city seem to function to render the dynamic smart city visible and controllable, they are also rendered almost instantly obsolete through the mere fact of their solid construction.

Overall, then, smart urbanism has become a leading, guiding concept in ways of thinking about and potentially directing urban change. In terms of the circulation of discourses around sustainable urbanism, the smart city concept has overtaken eco-urban ideas (such as the eco-city) in terms of usage and spread (de Jong *et al.*, 2015). The emergence to discursive and policy prominence of smart urbanism can be seen as part and parcel of the rise of the networked society (Castells, 2013). With this development comes a potential risk of overlooking environmental sustainability, if only because mainstream visions and strategies for smart cities seem to exclude, or at least sideline, alternatives that may bring environmental (and social) sustainability more to the fore of planning, policymaking and corporate smart product marketing. If smart cities are a way of exploring the 'symbiotic relationship between cities and information technology' (Townsend, 2013, 4), environmental and social concerns are striking through their absence.

Nonetheless, smart urbanism is, in the late 2010s, discussed and debated widely at conferences, conventions and practitioner fora. Smart cities garner interest at the level of the grassroots, of communities, neighbourhoods and municipalities, as well as being expressed in national programmes aiming to promote future forms of urban development. An example of this is the 2012–13 Future Cities Demonstrator competition, run in the UK by the UK government's Technology Strategy Board. The competition awarded over £24 million in funding to cities (namely Glasgow, Bristol, London and Peterborough) that developed plans for becoming smart city demonstrators, and can be seen as an example of the broad appeal of smart urban ideals in the second decade of the 21st century (Caprotti and Cowley, 2019).

Smart and eco-urbanism in the green economy

The chapter's focus on smart and eco-urbanism as two dominant trends in urban development would be incomplete without considering one of the key themes in sustainable development over the last few decades: that of the *green economy*. The ideal of moving the global economy towards greener, more ecologically sensitive outcomes has become widespread. Indeed it is enshrined in the 2012 Rio+20 UN Conference on Sustainable Development as a guiding principle for contemporary sustainable development (Bernstein, 2013). Prior to this, the green economy was already a concept commonly used and deployed at a variety of scales. Indeed, after the 2008 financial crisis, the green economy was the focus of 'green recovery plans' in the USA, Germany, France, South Korea, South Africa, Mexico and elsewhere. In South Korea, for example, 79% of the financial stimulus made available by the state in the aftermath of the crisis was part of a 'green stimulus' package (UNEP, 2009).

There has been widespread critique of mainstream ideals around the green economy as based on slightly changed 'business as usual' scenarios (Schulz and Bailey, 2014). Scholars have highlighted how dominant green economy discourses produced by national and international policy actors tend to replicate and not challenge mainly neoliberal discourses that enshrine economic growth (and its socio-environmental costs) as an unassailable given (Wanner, 2015). An example of these types of discourses can be found in the United Nations Environment Programme's (UNEP) broad definition of what is understood by reorienting the economy towards 'green' futures:

> Greening the economy refers to the process of reconfiguring businesses and infrastructure to deliver better returns on natural, human and economic capital investments, while at the same time reducing greenhouse gas emissions, extracting and using less natural resources, creating less waste and reducing social disparities.
>
> *(UNEP, 2009, 1)*

What is highlighted in the above quote is a focus on the green economy as a way to both lessen the environmental impact of economic-industrial activity, and also as a way of taking ownership of future economic opportunities around green technology by focusing on the potential for economic growth in these sectors and industries (Georgeson *et al.*, 2014). It is not simply the financial crisis of 2008 that has become a justificatory logic for the green economy. A range of other issues, including Peak Oil, the rise of emerging markets, increasing environmental regulation and rising rates of urbanisation, are all presented as examples of challenges and developments that can be ameliorated through the green economy.

The city, as a sub-national actor around which national economies are largely organised, has become identified as a key site for experimentation with visions for a green economy future. This is the case, for example, with Masdar, an eco-city in Abu Dhabi that was built from scratch from 2008. Its guiding aim is to reduce carbon, waste and energy use while at the same time functioning as a node within the Masdar Free Zone (MFZ). The MFZ functions as a Special Economic Zone (SEZ) aiming to attract corporations active in the renewable energy and sustainable technologies sectors. Masdar can be seen as an example of a new urban project that uses the urban sphere as a focus point for eco-urban and green economy initiatives. In the case of Masdar, the urban future is allegedly guaranteed through the (green) technologically contingent development made possible through transition towards green economic futures. Nonetheless, it is important to note the multiple critiques made of eco-cities such as Masdar in terms of their broader social performance (Cugurullo, 2013). This is paralleled by the existence of other high-profile cases – such as Dongtan, China (Chang and Sheppard, 2013) – of eco-city projects which promised green urban futures, but which did not initially succeed, for a variety of reasons.

Enter the smart eco-city

The chapter has highlighted how eco-urbanism and smart urbanism have both been rooted in deeply modern attempts to reshape the city. Smart city ideals have risen to prominence since the 2000s, and are now a dominant way of talking about urban futures. At the same time, eco-urban ideals are characterised by their permanence: this is because while smart urbanism has emerged, there has not been a concurrent decrease in interest in low-carbon or sustainable urbanism. Rather, what has occurred is an expansion of the discursive envelope around sustainable urbanism so as to now include 'smart' urban trajectories.

It is this chapter's contention that the green economy functions as a catalyst through which eco-urban and smart urban trajectories coalesce into sustainable urbanism. Indeed, various contemporary national and city-scale urban projects attempt to approach sustainable urban and economic development from a standpoint that attempts to hold together both environmental amelioration and economic growth. This is the case, for example, in Bristol, UK (Burton *et al.*, 2018). The city was awarded the European Green Capital award in 2015, and several urban projects have attempted to experiment with low-carbon urbanism. This has included experimental initiatives, such as Bristol's 3e Houses project, that aimed to help council house tenants in a sample of 100 houses reduce energy usage through the roll-out of Toshiba smart tablets used for tracking energy usage. At the same time, Bristol has invested in large-scale smart urban projects that have focused on Big Data and creating an urban IoT network and open data. Thus, Bristol is an example of a city that focuses both on eco- and smart urban themes.

It seems clear that in a context of rising interest in smart urbanism, the eco-urban component of sustainable urban development has not ceased to exist, but rather has persisted and in some cases merged with smart city initiatives. It is here that we can note the emergence of what has been called the 'smart eco-city' (Caprotti *et al.*, 2016) as a way of achieving environmentally amenable objectives in part through and within smart urban agendas. Furthermore, and as seen in the Bristol example cited above, the smart eco-city can be placed within a broader context of concern with harnessing Big Data, the IoT, digital lifestyles and various infrastructures to connect the urban sphere to green economy visions, strategies and pathways. This means, in turn, that the smart eco-city is in many ways an attempt to encapsulate eco-urban priorities and themes within a broader remit of integration of environmental aims with new (digital) forms of governance and economic organisation.

The blending of eco-urban and smart digital priorities into the smart eco-city may appear as though the latter is a confused and vague construct, difficult to grasp conceptually, and empirically slippery. This, however, does not seem to be the case with regards to several smart eco-city initiatives underway at the time of writing. Indeed, several of these initiatives are in fact highly defined through recourse to distinct physical, geographical and discursive boundedness. For example, the Euratlantique project to redevelop a 738-hectare area of central Bordeaux is clearly defined through negotiated and clearly marked boundaries: lines on a map that limit the spatial extent of Euratlantique as an experimental project. The project itself is anchored by a redeveloped station for the new Paris–Bordeaux rail line, opened in 2017. The project has environmental aims (such as around green building standards and energy use), as well as a clear focus on promoting digital enterprises and economic growth. Another example of a smart eco-city is the Sino-Singapore Tianjin Eco-City (SSTEC). It differs from many other smart eco-city cases in that its key defining identity, as seen through policy and corporate documentation, lies squarely in the 'eco' bracket. Nonetheless, the city's developments plans have grown to include and incorporate smart characteristics, such as a focus on smart grids, and the development of a smart city master plan in 2017. Economically, SSTEC attempts to attract industries and firms active in the green economy as well as the smart economy, as seen through the attraction of digital creative industries. Thus, SSTEC is an example of a smart eco-city developed with an eco-urban identity but now clearly incorporating smart elements in its development and design. At the same time, both SSTEC and Bordeaux's Ecoatlantique are examples of projects that attempt to connect to green economic agendas by developing different economic growth trajectories in the city.

Between mainstream and alternative: urban social sustainability

In summary, eco-urban and smart agendas have coalesced into urban constructs that can be termed smart eco-cities in that they combine elements of green planning and digital innovation with a focus on greening the municipal and wider economy. And yet, criticisms of these types of approaches to sustainable urban development are widespread. Briefly, critiques point to the fact that today's smart eco-cities, and mainstream notions of the green economy, represent 'business as usual' scenarios. Thus, the economic and political bases for the generation of the urban and environmental 'problems' that the smart eco-city is meant to solve are not tackled at root. The smart eco-city, then, has been presented as a way of dealing with the symptoms of unbridled economic growth. At the same time, many smart and eco-urban projects have been critically analysed in light of their clearly commercial or profit-driven characteristics (Luque-Ayala and Marvin, 2015). Scholars have highlighted how notions of a smart and eco-urban future are more often than not produced by coalitions of powerful actors (corporations, governments, consultants and others) (Hollands, 2015; McNeill, 2015; Ren, 2017; Wiig, 2015), often with little or no reference to local or grassroots contexts. Similar critiques are made of eco-city and eco-urban projects as examples of green capitalism and a desire to find financial returns in new (eco-)markets (Rapoport, 2014). In urban planning terms, the disjuncture between digitally augmented cities and physical urban space has also been highlighted (Aurigi, 2013).

Critiques of 'business as usual' scenarios in the smart eco-city are largely centred on the ecologically modernising character of smart and eco-urban initiatives, which rely on logics based in the market, regulation and technology to deliver desired outcomes. In turn, outcomes are often conceptualised as economic, technical, technological and governance-based – they are rarely defined in terms of citizens or citizenship (Joss et al., 2017). An ecologically modernising smart eco-city (based on the ideals of economic growth and gradual technological improvements delivering more environmentally amenable outcomes), therefore, can be seen as playing a part in reshaping and rethinking the city for the future, but in specific ways constrained by the logics of profit, the bottom line and the drivers behind the agency of powerful corporate and policy actors. A key critique of smart eco-cities rooted in visions of transitions towards the (ill-defined) green economy, then, is that these approaches and new projects do little to deal with the core reasons behind the development of unequally distributed socio-environmental externalities in the first place. In so doing, they risk deepening and replicating these problems, while generating new ones.

The core issue remains the rootedness of smart, eco-urban and green economy approaches in deeply modern visions of the city as a product of distinct binaries (such as that between nature and the city). The institution and reinforcement of these binaries effectively functions to fetishise specific aspects of the city, while overlooking or eliding others. Thus, eco-urbanism can be described as fetishising the environment, while smart urbanism fetishises technology and technique (Ellul, 1973). Cities' social dimensions – arguably harder to conceptually grasp than digital infrastructure networks and flows of $CO2$ and energy – remain absent in this dualistic perspective.

What is to be done? It is clearly desirable to move beyond facile dualisms and binary oppositions in conceptualising and operationalising sustainable urban development. And yet what is striking, as noted by Hemani and Das (2016) in their discussion of urban sustainability in India, is the lack of focus on the *human* element in projects aiming to bring about smart and eco-urban futures. A focus on urban social sustainability as a reference point from which to contextualise smart urbanism and eco-urbanism may be a fruitful way of both thinking about, and shaping, cities of the future. This is because holding the *human city* at the core of, and as a starting point for, urban sustainable development may help to reframe both technology and environmental goals. This reframing has the potential of becoming progressive and inclusive when urban development is carried out with human welfare in mind. It is clear that the details of how this can work in practice needs research, theorising and operational examples, but the current literature on urban social sustainability (Dempsey *et al.*, 2011) provides useful entry points into attempts to turn smart eco-urbanism into a more socially sustainable vision of future urban development. While the smart eco-cities of the contemporary era are expressions of elite power and agency, one wonders what the smart eco-cities of the future could look and feel like if the starting point was planning for the most vulnerable citizens and for enabling human development.

References

Aurigi, Alessandro. 2013. 'Reflections towards an Agenda for Urban-Designing the Digital City'. *Urban Design International* 18 (2): 131–144. doi:10.1057/udi.2012.32

Bauman, Zygmunt. 2013. *Liquid Modernity*. New York: John Wiley & Sons.

Bernstein, Steven. 2013. 'Rio+20: Sustainable Development in A Time of Multilateral Decline'. *Global Environmental Politics* 13 (4): 12–21. doi:https://doi.org/10.1162/GLEP_e_00195

Bolotova, Alla. 2012. 'Loving and Conquering Nature: Shifting Perceptions of the Environment in the Industrialised Russian North'. *Europe-Asia Studies* 64 (4): 645–671. doi:10.1080/09668136.2012.673248

Burton, Kerry, Andrew Karvonen, and Federico Caprotti 2018. 'Smart goes green: Digitalising environmental agendas in Bristol and Manchester'. In Karvonen, Andy, Federico Cugurullo and Federico Caprotti (eds) *Inside Smart Cities: Place, Politics and Urban Innovation*. London, Routledge: 117–132.

Caprotti, Federico. 2007. 'Destructive Creation: Fascist Urban Planning, Architecture and New Towns in the Pontine Marshes'. *Journal of Historical Geography* 33 (3): 651–679. doi:10.1016/j.jhg.2006.08.002

——— 2015. *Eco-Cities and the Transition to Low Carbon Economies*. London: Palgrave Macmillan.

Caprotti, Federico, and Robert Cowley 2019. 'Varieties of Smart Urbanism in the Uk: Discursive Logics, The State, and Local Urban Context'. *Transactions of the Institute of British Geographers*. doi:10.1111/tran.12284

Caprotti, Federico, Robert Cowley, Ayona Datta, Vanesa Castán Broto, Eleanor Gao, Lucien Georgeson, Clare Herrick, Nancy Odendaal, and Simon Joss 2017. 'The New Urban Agenda: Key Opportunities and Challenges for Policy and Practice'. *Urban Research & Practice* 10 (3): 367–378. doi:10.1080/17535069.2016.1275618

Caprotti, Federico, Robert Cowley, Andrew Flynn, Simon Joss, and Li Yu. 2016. 'Smart-Eco Cities in the UK: Trends and City Profiles 2016'. Exeter: University of Exeter (SMART-ECO Project). www.smart-eco-cities.org/wp-content/uploads/2016/08/Smart-Eco-Cities-in-the-UK-2016.pdf

Caprotti, Federico, Cecilia Springer, and Nichola Harmer 2015. '"Eco" For Whom? Envisioning Eco-Urbanism in the Sino-Singapore Tianjin Eco-City, China'. *International Journal of Urban and Regional Research* 39 (3): 495–517. doi:10.1111/1468-2427.12233

Castells, Manuel. 2013. *Communication Power*. Oxford: Oxford University Press.

Chakravarty, Surajit. 2017. 'Buying (into) Sustainability: Technocratic Environmentalism in Abu Dhabi'. In Federico Caprotti and Li Yu (eds), *Sustainable Cities in Asia*. Abingdon, Routledge: 135–145.

Chang, I-Chun Catherine, Helga Leitner, and Eric Sheppard 2016. 'A Green Leap Forward? Eco-State Restructuring and the Tianjin–Binhai Eco-City Model'. *Regional Studies* 50 (6): 929–943. doi:10.1080/00343404.2015.1108519

Chang, I-Chun Catherine, and Eric Sheppard 2013. 'China's Eco-Cities as Variegated[1] Urban Sustainability: Dongtan Eco-City and Chongming Eco-Island'. *Journal of Urban Technology* 20 (1): 57–75. doi: https://doi.org/10.1080/10630732.2012.735104

China Daily, 2012. 'Smart City' initiatives to boost economy. *China Daily*, 24 July 2012. http://www.chinadaily.com.cn/business/2012-07/24/content_15935686.htm

Cugurullo, Federico. 2013. 'How to Build a Sandcastle: An Analysis of the Genesis and Development of Masdar City'. *Journal of Urban Technology* 20 (1): 23–37. doi:https://doi.org/10.1080/10630732.2012.735105

Cugurullo, Federico. 2016. 'Urban Eco-Modernisation and the Policy Context of New Eco-City Projects: Where Masdar City Fails and Why'. *Urban Studies* 53 (11): 2417–2433. doi:10.1177/0042098015588727

Dempsey, Nicola, Glen Bramley, Sinéad Power, and Caroline Brown 2011. 'The Social Dimension of Sustainable Development: Defining Urban Social Sustainability'. *Sustainable Development* 19 (5): 289–300. doi:10.1002/sd.417

Ellul, Jacques. 1973. *The Technological Society*. Extensive Underlining edition. New York: Random House USA Inc.

Georgeson, Lucien, Federico Caprotti, and Ian Bailey 2014. '"It's All a Question of Business": Investment Identities, Networks and Decision-Making in the Cleantech Economy'. *Geografiska Annaler: Series B, Human Geography* 96 (3): 217–229. doi:10.1111/geob.12047

Hemani, Shruti, and Amarendra Kumar Das 2016. 'Humanising Urban Development in India: Call for a More Comprehensive Approach to Social Sustainability in the Urban Policy and Design Context'. *International Journal of Urban Sustainable Development* 8 (2): 144–173. doi:10.1080/19463138.2015.1074580

Hollands, Robert G. 2008. 'Will the Real Smart City Please Stand Up?' *City* 12 (3): 303–320. doi:10.1080/13604810802479126

——— 2015. 'Critical Interventions into the Corporate Smart City'. *Cambridge Journal of Regions, Economy and Society* 8 (1): 61–77. doi:10.1093/cjres/rsu011

Jong, Martin de, Simon Joss, Daan Schraven, Changjie Zhan, and Margot Weijnen 2015. 'Sustainable–smart–resilient–low Carbon–eco–knowledge Cities; Making Sense of a Multitude of Concepts Promoting Sustainable Urbanization'. *Journal of Cleaner Production* 109 (December): 25–38. doi:10.1016/j.jclepro.2015.02.004

Jong, Martin de, Chang Yu, Simon Joss, Ronald Wennersten, Li Yu, Xiaoling Zhang, and Xin Ma 2016. 'Eco City Development in China: Addressing the Policy Implementation Challenge'. *Journal of Cleaner Production*, Special Volume: Transitions to Sustainable Consumption and Production in Cities, 134 (October): 31–41. doi:10.1016/j.jclepro.2016.03.083

Joss, Simon. 2011. 'Eco-City Governance: A Case Study of Treasure Island and Sonoma Mountain Village'. *Journal of Environmental Policy & Planning* 13 (4): 331–348. doi:10.1080/1523908X.2011.611288

——— 2015. *Sustainable Cities: Governing for Urban Innovation*. Palgrave Macmillan. http://westminsterresearch.wmin.ac.uk/15027/.

Joss, Simon, Matthew Cook, and Youri Dayot 2017. 'Smart Cities: Towards a New Citizenship Regime? A Discourse Analysis of the British Smart City Standard'. *Journal of Urban Technology*, 24 (4): 29–49.

Kaika, Maria. 2005. *City of Flows: Modernity, Nature, and the City*. London: Routledge.

Kitchin, Rob. 2013. 'The Real-Time City? Big Data and Smart Urbanism'. *GeoJournal* 79 (1): 1–14. doi:10.1007/s10708-013-9516-8

Luque-Ayala, Andrés, and Simon Marvin 2015. 'Developing a Critical Understanding of Smart Urbanism?'. *Urban Studies* 52 (12): 2105–2116. doi:10.1177/0042098015577319

McNeill, Donald. 2015. 'Global Firms and Smart Technologies: IBM and the Reduction of Cities'. *Transactions of the Institute of British Geographers* 40 (4): 562–574. doi:10.1111/tran.12098

Melosi, Martin V. 1993. 'The Place of the City in Environmental History'. *Environmental History Review* 17 (1): 1–23. doi:10.2307/3984888

NSTC, 2017. *Smart Cities and Communities Federal Strategic Plan: Exploring Innovation Together*. Draft for public comment. https://www.nitrd.gov/drafts/SCC_StrategicPlan_Draft.pdf

Picon, Antoine. 2015. *Smart Cities: A Spatialised Intelligence*. Chichester: John Wiley & Sons.

Pollio, Andrea. 2016. 'Technologies of Austerity Urbanism: The "Smart City" Agenda in Italy (2011–2013)'. *Urban Geography* 37 (4): 514–534. doi:10.1080/02723638.2015.1118991

Rapoport, Elizabeth. 2014. 'Utopian Visions and Real Estate Dreams: The Eco-City Past, Present and Future'. *Geography Compass* 8 (2): 137–149. doi:10.1111/gec3.12113

Register, Richard. 1987. *Ecocity Berkeley: Building Cities for a Healthy Future*. Berkeley, CA: North Atlantic Books.

Ren, Xuefei. 2017. 'Green as Urban Spectacle in China'. In Federico Caprotti and Li Yu (eds), *Sustainable Cities in Asia*. Abingdon: Routledge: 77–85.

Roberts, Debra. 2016. 'The New Climate Calculus: 1.5°C = Paris Agreement, Cities, Local Government, Science and Champions (PLSC2)'. *Urbanisation* 1 (2): 71–78. doi:10.1177/2455747116672474

Roseland, Mark. 1997. 'Dimensions of the Eco-City'. *Cities*, Sustainable Urban Development 14 (4): 197–202. doi:10.1016/S0264-2751(97)00003-6

Schulz, Christian, and Ian Bailey 2014. 'The Green Economy and Post-Growth Regimes: Opportunities and Challenges for Economic Geography'. *Geografiska Annaler: Series B, Human Geography* 96 (3): 277–291. doi:10.1111/geob.12051

Swyngedouw, Erik. 2009. 'The Antinomies of the Postpolitical City: In Search of a Democratic Politics of Environmental Production'. *International Journal of Urban and Regional Research* 33 (3): 601–620. doi:10.1111/j.1468-2427.2009.00859.x

Townsend, Anthony M. 2013. *Smart Cities: Big Data, Civic Hackers, and the Quest for a New Utopia*. New York: W. W. Norton & Company.

UNEP 2009. Global Green New Deal: An Update for the G20 Pittsburgh Summit. Nairobi: UNEP.

Wanner, Thomas. 2015. 'The New "Passive Revolution" of the Green Economy and Growth Discourse: Maintaining the "Sustainable Development" of Neoliberal Capitalism'. *New Political Economy* 20 (1): 21–41. doi:10.1080/13563467.2013.866081

Ward, Stephen. 2016. *The Peaceful Path: Building Garden Cities and New Towns*. Hatfield: University of Hertfordshire Press.

Watson, Vanessa. 2014. 'African Urban Fantasies: Dreams or Nightmares?'. *Environment and Urbanization* 26 (1): 215–231. doi:10.1177/0956247813513705

Wiig, Alan. 2015. 'IBM's Smart City as Techno-Utopian Policy Mobility'. *City* 19 (2–3): 258–273. doi:10.1080/13604813.2015.1016275

Zamyatin, Yevgeny. 2013. *We*. New York: Momentum.

15

Towards ethical legibility

An inclusive view of waste technologies

Dietmar Offenhuber

Introduction

The collection and management of municipal solid waste (MSW) is perhaps the most elementary urban service and is generally considered a responsibility of local government. At the same time, waste management is a global system: a networked, heterogeneous assemblage of various processes, facilities, regulatory systems, social practices, and economic relationships. Its overall shape remains largely opaque since many of its processes are informal or remain otherwise undocumented. Furthermore, the system presents itself differently when viewed at different scales. At the local scale, the waste system appears fragmented into a multitude of different models. Municipal waste collection may be performed by a public works department, through private contractors, cooperatives, community organizations, or through autonomous scavengers. Each actor is interwoven with the larger economy and the global supply chain. The livelihood of a waste picker in São Paulo partially depends on the spot market price of cardboard, which is determined by the global trade of goods and the demand for packaging material. The U.S. recycling system is currently in disarray because China, until recently the main buyer of scrap materials, has closed its borders to imports of recovered plastics and comingled recyclables (Resource Recycling, 2018).

The waste system involves a wide range of social, economic, cultural, environmental, and political complexities and controversies. Technological approaches to managing waste therefore need to be discussed in their wider societal and historical context. The concept of legibility, as used in this chapter, can be understood as the combination of transparency and interpretability. Rendering legible the diverse processes comprising a complex system is a central concern for managing technical systems. The smart city and its technologies of legibility such as GPS trackers,[1] data exchange standards, and various approaches to machine interpretation, however, are often criticized for painting a partial picture that emphasizes efficiency while, for example, hiding the environmental burdens of underprivileged communities. An ethical approach to legibility requires to consider the various ethical, social, and political implications of what is shown and what remains hidden in the waste system. Following this premise, this chapter provides an overview of waste technologies with regard to the kinds of problems they address. Each section will provide a brief context and will address critiques of the respective approach.

Conceptual models of waste systems

Planners and engineers have developed a number of conceptual models for waste systems. The field of *urban metabolism* conceptualizes the waste system as part a larger system of material flows and energy, comparable to the metabolic cycle of an organism (Kennedy et al., 2007; Wolman, 1965). The perspective considers all material and energetic requirements of a city as part of a closed system, following the material and energetic transformation upstream from raw material extraction downstream to disposal. Ideally, the residues are captured and transformed into secondary raw materials and folded back into the system rather than disposed of. The concept of *industrial ecology* builds upon the material flow approach. It aims to take advantage of the fact that the waste of one industrial process can be a valuable raw material for another. Therefore, waste can be prevented by connecting industrial processes across firms and sectors so that a maximum of material and energy is preserved (Ayres and Ayres, 1996). A similar approach has been popularized by McDonough and Braungart under the concept of *cradle to cradle* (2002). Other conceptual models focus on the logistics of waste. Waste management involves a considerable amount of transportation and is shaped by the forces of economics. As Rick Porter reminds us, "all waste is traded" (Porter, 2002). Nevertheless, the waste system has long been neglected by the theory and practice of supply chain logistics—partly due to the lack of economic incentives to optimize waste transportation and partly due to weak regulations and enforcement. The field of *reverse logistics* aims to rectify this situation and integrate the collection of waste materials with the processes of the supply chain (de Brito and Dekker, 2004). This requires closing the information gap between producers and end-of-life treatment, which is especially wide in the case of consumer products.

As tools for policy decisions, these conceptual models depend on a significant amount of data. These data, however, are rarely available in sufficient quality and coverage (Tchobanoglous and Kreith, 2002). When comparing different waste systems across the globe, data quality and availability can in fact be used as proxies for the quality of the system itself (Wilson *et al.*, 2012). An especially problematic domain is waste transportation at both the local and international levels. While material streams such as nuclear and certain hazardous wastes are tightly monitored in industrial countries, other kinds of waste are entirely absent from official statistics. These blind spots raise the concern that waste may travel the route of convenience and least resistance, which tends to lead to the neighborhoods of disadvantaged communities.

Technologies of waste systems: a brief historical context

In 1912, sanitation engineer Rudolf Hering was worried by the effects of urbanization which led to ever-increasing amounts of garbage and decreasing sites for disposal. To solve the urban crisis, he declared, "we must have more special data and statistics before we can indicate the best methods for the disposal of a particular town's refuse" (1912, 909). The preceding decades saw profound technological changes in the form of new waste and sanitation systems, driven by a public health crisis in industrial cities. From early on, waste systems have been a site for technological ambitions that transformed the shape of the city. In this regard, the ambitions of today's smart city programs have much in common with those of the 19th and early 20th century sanitation reformers. The similarities include the strong preference for system-wide interventions over piecemeal improvements. In the 19th century, this preference had its roots in the flawed environmental determinism of the miasma theory, which helped to establish comprehensive planning as a modern value (Tarr, 1996). Both periods also show an overarching concern for efficiency. In the early 20th century, with the public health crisis averted, the design of large-scale networked sanitation systems has become primarily seen as an engineering problem (Melosi, 2004). Third, both periods were influenced, if not motivated, by economic goals and were, as Hering's proposal demonstrates, strongly data-driven. The early recycling system established by New York sanitation commissioner "colonel" Harding, complete with an early version of a material recovery center, was driven by the goal to cover the costs of the sanitation system (Melosi, 2008). At the same time, however, it also removed the source of income for informal scrap dealers (Strasser, 2000). Unfortunately, then as today, social implications are often overlooked in technological system building.

Contemporary waste management approaches are deeply rooted in the historical context. The following sections provide a discussion of emerging methods, approaches, and technologies. They can broadly be grouped into three problem areas, focusing on making waste systems:

(1) More efficient and environmentally sustainable
(2) More accountable and transparent
(3) More participatory, just, and ethical

These are of course not exclusive categories—quite the contrary: good waste policy, management, and infrastructure design requires all of them sufficiently addressed.

Efficiency and environmental sustainability

Environmental sustainability and waste management efficiency are generally considered to be closely linked, but their relationship also involves paradoxical aspects—a highly efficient waste system may make waste more invisible and therefore lead to more consumption and waste production. It is also worth noting that most environmental and public health issues arise from decisions during the design, production, and transportation of products rather than their disposal. Of the famous three R's in the waste hierarchy (Reduce, Reuse, Recycle), only the last depends on technological efficiency. Nevertheless, considering the size and extent of the global waste system, even moderate improvements of its operational efficiency and the stringent enforcement of violations promise substantial environmental benefits. Both management and enforcement require information about the state of the waste system: which materials are in the waste stream, where were they collected, by whom, and how far were they transported. Digital sensor technologies can potentially help generating information about material provenance and waste composition, which are currently difficult to come by. Those discussed in this section focus on the use of sensor and data technologies in the area of collection and material recovery.

Improving collection

Waste collection and transportation takes up a significant portion of municipal waste budgets, contributes to traffic and noise pollution and produces non-negligible carbon emissions. As both waste materials and recycling methods become more complex, collection becomes an increasing logistic challenge.

Several technologies aim to improve waste collection and minimize its environmental impacts. One of the older systems is the *Automated Vacuum Collection* (AVAC) technology, consisting of sealed underground vacuum tubes that move waste from residential buildings and public receptacles to a central collection place. Pioneered by Swedish company Envac in

1960s, this technology is used in areas where truck collection is difficult, for example, due to narrow streets. It has been implemented in certain neighborhoods of Montreal, New York City, Barcelona, and Toronto.

A second approach aims to optimize collection by reducing the vehicles miles. This can involve switching from a static to a dynamic collection schedule servicing only bins that need to be emptied. Smart trash cans, often used by cities for waste receptacles in public space, use sensors to detect when the bin is full and sends an alert to a central platform. For example, the *Big Belly* system comprises solar-operated waste bins that include sensors and a waste compactor.[2] (Figure 15.1) Other systems, such as *Smartbin* collect detailed information about volume or humidity of material in the bin.[3] The industrial company Compology offers location-aware cameras in large waste containers that allow estimating fullness, composition, and movement of the container.[4] Such systems allow, for example, the adjustment of collection schedules based on actual need, and uses smaller collection vehicles that service bins on demand using dynamic routing calculated on the basis of sensor data. Humidity and material composition allow further to direct the contents of a container into the appropriate material stream.

A third model aims to minimize waste generation and through *Pay As You Throw* (PAYT) schemes that charge consumers dynamic service fees based on the amount of waste

Figure 15.1 Big Belly smart waste bin in Philadelphia
Source: Wikimedia

generated, therefore encouraging waste reduction (Folz and Giles, 2002). This can be accomplished by enforcing the use of certified waste bags sold at the premium that includes the collection fee.[5] A second approach to waste metering involves the tagging waste bins with smart labels and weighing their contents during collection. A mix of both approaches is currently used in countries such as Taiwan and South Korea.

The three approaches to minimizing the impact of waste collection—automated vacuum collection, dynamic collection models, and PAYT schemes—differ in their scope, infrastructure requirements, and integration with social practices. AVAC systems are comfortable to use, but are infrastructure intense and often limited to a few neighborhoods, potentially introducing spatial inequalities in public service provision. They also don't promote waste reduction unless combined with PAYT schemes. Dynamic collection models may make waste collection leaner, but also have social side effects as cities often embrace them as a means to reduce their workforce for waste collection. PAYT schemes offer incentives for waste reduction, but their data collection is sometimes experienced as intrusive, and they are only socially just if not only consumers, but also producers and retailers contribute to the cost of waste collection.

Improving material and energy recovery

After collection, materials are fed into distinct material streams such as MSW, comingled recyclables, or compost for further processing. Recovered recyclable materials need to be sorted and cleaned. This happens usually in a material recovery facility (MRF, Figure 15.2). An important criterion for both value and recyclability of almost any material is its purity based on the thoroughness of sorting. The popular single-stream curbside collection model that comingles paper,

Figure 15.2 Material recovery facility in New England
Source: Chapter author(s)

cardboard, plastic, metal, and glass in a single container is problematic in this regard due to cross-contamination. Cardboard, for example, becomes unusable for recycling if it gets in contact with grease, liquids, and food residues. The contamination of mixed recyclables played a major role in China's ban of scrap material imports (Recycling Today, 2013).

Over the past years, the technologies used in MRFs for separating materials have seen large improvements. Separation uses mechanical properties such as weight, size, density, or magnetism, but increasingly also computer vision (CV), machine-learning techniques, and robotics to recognize objects and sort them by color, shape, or other characteristics. Despite the technological advances of MRFs, the facilities still face very elementary vulnerabilities such as plastic bags, which can get entangled in the machinery and need to be manually picked out by workers from the material stream. The nature of these jobs, especially the working conditions in so-called "dirty MRFs" that separate recyclable materials from solid waste, has raised social and environmental justice concerns (Pellow, 2004). Sifting through solid waste, workers are exposed to unsanitary and dangerous working environments, including medical scraps and infectious substances.

Waste sensor data and the implications for privacy

From PAYT models to the effective recovery of recyclables, most of the discussed models and technologies rely on previously unavailable data about the provenance of materials and the composition of the waste stream. This datafication of the waste system has inspired many projects, ranging from the futuristic to the whimsical. A smart waste bin recognizes waste and automatically orders new supplies when an empty package is thrown away.[6] Another waste bin product promises to automatically sort trash using computer vision and artificial intelligence.[7] Yet another version of a smart waste bin shares the amount of generated waste on Facebook as a form of self-policing (Comber et al., 2013). Several groups work on proposals and methods for automatic litter recognition and robotic trash collection (Chiang, 2015; Fuchikawa et al., 2005; Rad et al., 2017). Similar concepts have been proposed for debris in rivers and oceans (Valada et al., 2014; Wang et al., 2014).

Considering the range of data collection efforts, it is important to remember that all data about waste are inherently linked to human behavior and consumption and therefore reveal more than their immediate purpose, optimizing waste collection. Data sets from smart waste bins such as the Big Belly system are reliable proxies for human activity in public space, indicating the presence of pedestrians and their consumption behavior over time (Figure 15.3). The contents of waste bins and landfills provide a surprisingly accurate record of human behavior, as the archaeologist Bill Rathje demonstrated in his landfill excavations, or the gonzo journalist A.J. Weberman in his investigations into the garbage of celebrities (Rathje and Murphy, 2001; Weberman, 1980). Beyond waste-specific data, sensor-driven public waste bins offer themselves also as a convenient infrastructure for other kinds of sensors that record the presence of pedestrians and potentially identify individuals[8]—often concealed in WiFi access points or advertisement screens. Such an *internet of bins* raises privacy concerns, since those sensors can be used to identify individuals and track their paths through the city. A pilot project with public waste bins with integrated information displays in London led to controversy after it was discovered that the bins track the network addresses of smartphones for purposes of targeted advertising (Miller, 2013). Alarming in this context is that many providers of smart waste bins increasingly understand themselves as data companies. In many cases, data collection in public space can serve the public interest, and the additional information about the city that can be gleaned from waste data can increase this benefit. In the

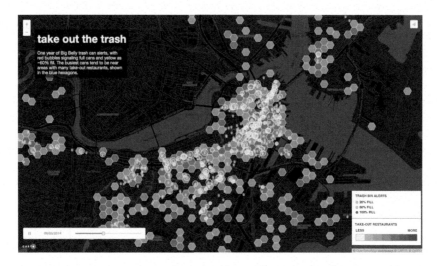

Figure 15.3 Volume data from smart waste bins, correlated with the density of take-out restaurants

Source: Project by David Lee, https://hubhacks2.devpost.com/submissions/35312-take-out-the-trash

case of data harvesting for private interests such as advertising, such a benefit may be more doubtful. Establishing ethical legibility requires considering the implications of collected data and their dissemination methods.

Transparency and accountability

A second site of technological intervention includes efforts to make waste systems more transparent and accountable in order to enforce environmental laws, prevent corruption, and investigate environmental crimes. This section discusses monitoring infrastructures and emerging waste tracking methods employed by public authorities, private actors, and members of the civic society.

Comprehensive data infrastructures for waste accountability emerged during the 1970s and 1980s in the wake of new *public right to know* legislations. The US Resource Conservation and Recovery Act (RCRA) and its amendments mandated a central database for monitoring the release of toxins by facilities and companies that process, store, or transport hazardous waste materials. The law, however, has two important gaps. First, municipal solid waste is exempt from monitoring requirements and second, international exports of waste is not covered since the U.S. has not ratified the Basel Convention controlling of transboundary movements of hazardous wastes. The European Union has more comprehensive reporting and tracking requirements for many types of waste, but also in this case, comprehensive monitoring and enforcement of violations is a weak spot.

Polluter pays policies, supply chain tracking, and the internet of materials

For certain monitoring tasks, supply-chain data can be utilized. The current paradigm of recycling is not uncontroversial, since it places the cost and burden of collection and valorization on consumers and municipalities, while manufacturers are exempt from this responsibility (MacBride, 2012). Extended producer responsibility (EPR) policies are based on the *polluter pays* principle and require manufacturers and retailers to cover the cost of recycling in order to nudge them towards more sustainable product and packaging designs. Over the past decade, EPR schemes have gained traction and have been implemented in different forms. A well-known implementation is the German *Green Dot* program, which involves a separate reverse logistics stream for packaging materials, operated and paid for by retailers and manufacturers.

EPR schemes require the identification of individual waste items in order to connect them to their producers. To turn supply chains into such closed-loop systems requires sharing information across the system, and more specifically, using the extensive data resources of the supply chain to inform end-of-life treatment. In supply-chain logistics, tracking objects typically involves sensor infrastructures built around radio-frequency identification (RFID) sensing infrastructures (Saar and Thomas, 2002). Since those databases contain proprietary information, attempts to establish open RFID data standards to enact EPR schemes have so far been met with reluctance from manufacturers. Beyond utilizing product labels, RFID tags, and existing databases, the *internet of materials* represents a more extensive concept that involves pervasive tracking of material flows across actors and system boundaries in order to record histories and trajectories of every material.[9]

This vision can only be accomplished using distributed sensor networks on vehicles and facilities, using the arsenal of the *Internet of Things* (IoT) or even *smart dust* (Warneke *et al.*, 2001). The use of blockchain technology, increasingly used in the logistics sector, allows keeping track of objects and transactions across multiple systems without the need for a centralized database (Badzar, 2016; Saberi *et al.*, 2018). Since the de-centralized data storage on blockchains is energy and computation intense and leaves a significant environmental footprint, it remains to be seen whether its use for waste tracking offers any environmental advantage or exceeds the benefits of recycling. The approach might, however, play a role in tracking the movement of hazardous waste.

Bottom-up waste tracking methods

Active location sensing allows recording the path of objects without any of the infrastructures discussed in the previous paragraphs. It is the tool of choice for investigative journalists and activists who do not have access to internal information about waste systems (Greenpeace International, 2008). Active location sensing may involve radio beacons or GPS receivers (Lee and Thomas, 2004). To collect the data, the global cell-phone system can be used, as in a project by the MIT Senseable City Lab that attached 2800 GPS trackers to individual items garbage (Offenhuber *et al.*, 2014). Such waste tracking methods have obvious limitations—the cost and labor necessary for instrumenting a statistically significant number of waste items can be prohibitively expensive. If, instead of individual items, larger units such as bales of recyclable materials or containers are tracked, a large part of the journey is omitted. Sensing and transmission of location data are also a technical challenge, not the least part of which is the sensors' battery life.

Nevertheless, for smaller deployments a number of use cases are plausible. Municipalities have explored waste tracking as a way to analyze their system and evaluate policies, such as the preferred model for collecting e-waste. Companies may use it to monitor their waste management subcontractors, for whose conduct they are responsible under strict liability laws. Activists and investigators such as the Basel Action Network (BAN) use active location sensing to collect evidence of illegal waste exports (Lee *et al.*, 2018). Waste tracking is especially powerful when the results can be matched with other available data sources such as municipal contracts, administrative statistics, or databases of facilities that participate in the transportation, handling, storage, or disposal of waste.

Cities can use waste tracking to evaluate and compare different collection models, especially for materials that are difficult to collect in a centralized way, such as electronic and hazardous waste. Waste tracking offers cities a way to monitor compliance of contractors and estimate the global impact of the municipal waste system. When staged as a public experiment, waste tracking can also be a powerful tool for public engagement. Similar to how citizen feedback systems offer a glimpse into the world of infrastructure maintenance, waste tracking could offer citizens a lens to experience the waste system in its global scope. Besides cities, also investigative journalists and environmental justice advocates increasingly explore waste tracking as a covert method of uncovering environmental injustices and violations that otherwise might not be investigated.

Social accountability and community-centered approaches

The work of BAN and other environmental groups is an example of social accountability (Malena *et al.*, 2004), a strategy used when public accountability instruments are experienced as ineffective or insufficient. Citizen reporting apps such as FixMyStreet[10] were initially conceptualized as crowdsourced social accountability tools which citizens could use to send complaints to their local representatives (King and Brown, 2007). In the following years, cities embraced these projects as a way to collect feedback that could be used to improve public service delivery and integrated smartphone apps into their already existing constituent relationship management (CRM) platforms. A significant part of these reports typically includes sanitation and waste-related issues. As citizen perception of infrastructure issues normally diverges from that of public work departments, citizen feedback apps are a way to mediate between different perspectives and negotiate infrastructure governance (Offenhuber, 2014).

Participation and social issues

Citizen feedback apps are promoted by cities as a way to make the process of public services such as waste collection more participatory (Figure 15.4). First, they allow cities to convey the complexities of infrastructure maintenance, and second, allow them to voice concerns or submit suggestions for improvement. Despite the rhetoric, however, the level of participation

Figure 15.4 Interfaces of different citizen feedback apps available in 2013
Source: Screenshot by chapter author(s)

and control that those platforms allow is limited to collecting information (Offenhuber, 2014). As mechanisms for complaints, they tend to foreground aesthetic issues that have limited environmental impact, such as litter on street. Furthermore, service requests often express private interests and individual grievances rather than issues relevant to the wider public. As a result, citizen feedback systems tend to be less effective when it comes to address issues of justice and equity, since neighborhoods with lower service quality often complain less than neighborhoods with an already high level of service (Martinez, Pfeffer, and van Dijk, 2011; Pak, Chua, and Vande Moere, 2017).

Summary

Existing data infrastructures for monitoring waste systems originate from *Right to Know* provisions of environmental laws and policies of the 1970s and 1980s. While still a central source of information, especially in the U.S. these data infrastructures have significant loopholes and gaps when it comes to detecting and enforcing environmental crimes or informing comprehensive waste reduction policies. The slow but steady emergence of EPR schemes and the increasing complexity of reverse logistics systems illustrates the need for more detailed data sources—from the governance of waste systems to the enforcement of environmental crimes. IoT technologies, widely used in the logistics sector, have also taken hold in waste management. A challenge remains concerning the proprietary status of most of these data sources, which are therefore not accessible for public agencies and other actors in the waste system. Waste tracking approaches, used by cities, companies, or private initiatives, allow probing waste systems across system boundaries. Crowdsourced citizen reports, collected through citizen feedback apps, have become another important data source used by cities to collect information about their infrastructure and to engage citizens in issues of infrastructure maintenance.

Environmental justice

This last section addresses the role of data in environmental justice and conflicts around issues of hazardous waste. The influential 1987 report *Toxic Wastes and Race in the United States* has statistically shown that toxic waste facilities tend to be sited in close proximity to disadvantaged black communities (United Church of Christ, 1987). The report, which became a founding document for the environmental justice (EJ) movement, was based on federal data about hazardous waste sites that were only released shortly before. Since the 1960s, non-governmental organizations (NGOs) and community activists have used a range of strategies to investigate and address a wide range of hazardous waste and pollution issues (Brown *et al.*, 1997). In recent years, the traditional citizen science toolbox has been extended with sophisticated participatory sensing approaches for monitoring the impacts of toxic waste pollution including DIY sensor kits or remote sensing using balloon and kite mapping (Bryson *et al.*, 2013; Wylie *et al.*, 2017).

Coproduction of services and inclusion of informal recycling

Not only poor communities exposed to pollution stemming from hazardous waste are a concern, but also informal participants in the waste system. Many cities in both the developed and the developing world have a thriving informal recycling sector. For example, in countries like Brazil the informal sector captures the overwhelming majority of recycled material such as aluminium or cardboard (Medina, 2010).

Modernization efforts of the waste system often disadvantage informal recyclers and waste pickers by making valuable material inaccessible; corresponding waste policies often criminalize their activity. However, modernization based on industrial-scale material collection and recycling has often failed in cities of the developing world—due to insufficient budgets, narrow streets unsuitable for large trucks, or poor accountability and enforcement of the contractors.

In 2002, the city of Cairo attempted an aggressive modernization of its MSW system through contracts with multinational companies. The system displaced the traditional door-to-door collection system run by the Zabaleen community. After a few years, however, complaints about poor service in the new system started to emerge and recycling rates dropped sharply (Fahmi and Sutton, 2013). To avoid such failures and recognize the value informal actors provide, many cities experiment with inclusive recycling systems, which recognize and respect informal actors and support their work and wellbeing (Scheinberg, Savain, and Countries, 2015). In the context of inclusive recycling and the cooperative movement, waste picker organization and unionization has been an important catalyst. Cooperatives are experimenting with technology and data-driven methods to support not only the management of their organization but also the valorization of materials, since in EPR schemes, e-waste and other materials yield higher prices if their origin is well-documented (Offenhuber, 2017, 115). Beyond the organization of the waste system, inclusive recycling gives workers access to healthcare, a minimal level of social security, and most importantly, dignity.

Crowdsourcing and critiques of pseudo-participation

Inclusive recycling and citizen sensing emphasize participatory modes for producing public services, and are therefore part of a larger world-wide trend of departing from centralized models of infrastructure service provision. All waste systems depend on user participation, but in traditional models this means mostly compliance with the intentions of the system builders and policy makers. To raise recycling rates, cities engage in awareness campaigns, educational efforts, and increasingly try to persuade users through means of social media, gamification, and nudging (Deterding et al., 2011; Thaler and Sunstein, 2008). While participation is almost universally seen as a positive value, some of the claims deserve closer scrutiny. Not only do many apps limit the voice of the user to being a source information, they also promote the individualization of responsibility, can distract from systemic issues, and placate citizens who "want to do the right thing" (MacBride, 2012; Maniates, 2001). Participation can also be a burden that is often unequally distributed, and serves often as a convenient narrative for cutting public services.

Conclusion: towards ethical legibility

To make sense of the various possibilities to render the waste system legible through data, only a few of which are described above, it is helpful to differentiate between three levels of transparency. The first level is narrowly limited to issues that are subject to environmental legislation and involves data that are generated by agencies, cities, and companies in fulfilment of these laws. Over the past 40 years, these data have proven to be tremendously important for environmental movements and social justice initiatives, yet are fairly limited in their scope. The second, more expansive level could be described as operational transparency (Buell et al., 2014). It provides a broader view of the processes and places involved in the management of waste and allows for a wider range of investigations. Operational transparency includes information generated by citizen feedback systems and, more broadly, repositories collected under open data policies that make data collected by administrations open by default. Such repositories offer a glimpse into the processes and implications of infrastructure maintenance and governance, but are at the same time the result of an opportunistic rather than targeted approach to data collection—using data that happen to be readily available. At the third level, ethical legibility would involve reflection beyond the opportunistic approach about which issues should be made transparent, and conversely, areas where transparency can introduce an unjust burden, for example by prioritizing a particular aspect while neglecting another. In this sense, an intense focus of individual waste generation would raise questions when the impacts of manufacturing do not receive equal scrutiny. Ethical legibility requires attention to the implications for communities affected by waste management processes and related issues of environmental pollution, acknowledging that public health and welfare issues are often difficult to establish unambiguously. It would also have to consider the conditions and assumptions under which data are collected. While the waste system suffers from a lack of universal data standards, standardization efforts should include communities to determine how they and their concerns are represented in these standards. Emerging debates around data sovereignty involve questions how indigenous groups can have more voice regarding the definition of collection protocols, storage, and ownership of data pertaining to them (Kukutai and Taylor, 2016). The need for more stringent environmental regulations on national and international levels comes with the recognition that data about the waste system cannot be exclusively defined top down, but should also recognize the efforts of grassroots science groups who develop their own methods of environmental monitoring, expanding the right to know to a right to investigate.

For many *smart city* system builders, the inhospitable environments of the waste system towards digital technologies are a seductive technical challenge and, more importantly, a welcome opportunity to demonstrate their commitment to environmental sustainability and social welfare. Their arguments benefit from the perceived opacity of the waste system and the fact that waste touches almost all aspects of everyday life. For these and other reasons, most smart city solutions spotlight novel approaches to waste management, which may involve novel tracking and collection systems, sensor networks, robots, or vacuum tubes. At the same time, the smart city has a reputation problem, often criticized for an overly technocratic focus on efficiency, surveillance, and control (Luque-Ayala and Marvin, 2016). While the approaches discussed in this chapter go beyond the typical smart city applications, the technologies associated with the smart city paradigm can in fact be found in the first section focusing on sustainability and efficiency. Granted, efficiency is certainly not undesirable for a waste system, if the choice is to either open a new landfill or make material recovery more efficient. Consistency in data standards and monitoring infrastructures allow addressing waste issues not only at the local, but also the global scale. The logic of efficiency, however, is less applicable to civic discourse and the governance of public services (Gordon and Walter, 2016). Considering this contentious social history of waste management, it has been repeatedly shown that exposure to toxins often affects underprivileged communities with little voice in the governance of the system. It is a serious shortcoming that smart city projects have so far largely neglected justice aspects. In the same way that smart city projects emphasize a technical legibility, they would benefit from promoting an ethical legibility to facilitate democratic and equitable infrastructure governance.

Notes

1 Short for Global Positioning System, a satellite-based technology for sensing geographic location, see www.gps.gov or their Russian, European, and Chinese counterparts GLONASS, Galileo, and BeiDou-2.
2 See http://bigbelly.com.

3 See www.smartbin.com.
4 See http://compology.com/.
5 See http://wastezero.com/.
6 See www.genican.com.
7 See the Bin-E product page http://bine.world.
8 Bluetooth and WiFi sniffers that record the unique hardware addresses of smartphones are the most popular method.
9 See http://endswapper.com.
10 www.fixmystreet.com.

References

Ayres, Robert U., and Leslie Ayres. 1996. *Industrial Ecology: Towards Closing the Materials Cycle*. Cheltenham: Edward Elgar.

Badzar, Amina. 2016. "Blockchain for Securing Sustainable Transport Contracts and Supply Chain Transparency - An Explorative Study of Blockchain Technology in Logistics." lup.lub.lu.se. http://lup.lub.lu.se/student-papers/record/8880383.

Brito, Marisa P. de, and Rommert Dekker. 2004. "A Framework for Reverse Logistics." In R. Dekker, M. Fleischmann, K. Inderfurth, L.N. Van Wassenhove (eds) *Reverse Logistics*, 3–27. Berlin, Heidelberg: Springer.

Brown, Phil, Edwin J. Mikkelsen, and Jonathan Harr. 1997. *No Safe Place: Toxic Waste, Leukemia, and Community Action*. Berkeley, CA: University of California Press.

Bryson, Mitch, Matthew Johnson-Roberson, Richard J. Murphy, and Daniel Bongiorno. 2013. "Kite Aerial Photography for Low-Cost, Ultra-High Spatial Resolution Multi-Spectral Mapping of Intertidal Landscapes." *PloS One* 8 (9): e73550.

Buell, Ryan W., Ethan Porter, and Michael I. Norton. 2014. "Surfacing the Submerged State: Operational Transparency Increases Trust in and Engagement with Government." *SSRN Scholarly Paper ID 2349801*. Rochester, NY: Social Science Research Network. https://papers.ssrn.com/abstract=2349801.

Chiang, C. H. 2015. "Vision-Based Coverage Navigation for Robot Trash Collection Task." In *2015 International Conference on Advanced Robotics and Intelligent Systems (ARIS)*, 1–6.

Comber, Rob, Anja Thieme, Ashur Rafiev, Nick Taylor, Nicole Krämer, and Patrick Olivier. 2013. "BinCam: Designing for Engagement with Facebook for Behavior Change". In Linda Little, Elizabeth Sillence, and Adam Joinson (eds), *Human-Computer Interaction – INTERACT 2013*, 181–194. Lecture Notes in Computer Science. Berlin, Heidelberg: Springer.

Deterding, Sebastian, Dan Dixon, Rilla Khaled, and Lennart Nacke. 2011. "From Game Design Elements to Gamefulness: Defining Gamification." In *Proceedings of the 15th International Academic MindTrek Conference: Envisioning Future Media Environments*, 9–15. MindTrek '11. New York, NY, USA: ACM.

Fahmi, Wael, and Keith Sutton. 2013. "Cairo's Contested Waste: The Zabaleen's Local Practices and Privatisation Policies". In María José Zapata Campos and C. Michael Hall (eds), *Organising Waste in the City*, 159–180. Bristol: Policy Press.

Folz, David H., and Jacqueline N. Giles. 2002. "Municipal Experience with 'Pay-as-You-Throw' Policies: Findings from a National Survey." *State & Local Government Review* 34 (2):105–115.

Fuchikawa, Yasuhiro, Takeshi Nishida, Shuichi Kurogi, Takashi Kondo, Fujio Ohkawa, Toshinori Suehiro, Yasuhiro Watanabe, et al. 2005. "Development of a Vision System for an Outdoor Service Robot to Collect Trash on Streets". In *Proceedings of the Elghth IASTED Conference on Computer Graphics and Imaging*, 100–105.

Gordon, Eric, and Stephen Walter. 2016. "Meaningful Inefficiencies: Resisting the Logic of Technological Efficiency in the Design of Civic Systems." In Eric Gordon and Paul Mihailidis (eds), *Civic Media: Technology, Design, Practice*, Cambridge, MA: MIT press, 243–266.

Greenpeace International. 2008. "Following the E-Waste Trail - UK to Nigeria." www.greenpeace.org/international/photosvideos/greenpeace-photo-essays/following-the-e-waste-trail.

Hering, Rudolf. 1912. "The Need for More Accurate Data in Refuse Disposal Work." *American Journal of Public Health* 2 (12): 909–911.

Kennedy, Christopher, John Cuddihy, and Joshua Engel-Yan. 2007. "The Changing Metabolism of Cities." *Journal of Industrial Ecology* 11 (2). MIT Press: 43–59.

King, Stephen F., and Paul Brown. 2007. "Fix My Street or Else: Using the Internet to Voice Local Public Service Concerns." In *Proceedings of the 1st International Conference on Theory and Practice of Electronic Governance*, 72–80. ICEGOV '07. New York, NY, USA: ACM.

Kukutai, Tahu, and John Taylor. 2016. *Indigenous Data Sovereignty: Toward an Agenda*. Vol. 38. Anu Press.

Lee, David, Dietmar Offenhuber, Fábio Duarte, Assaf Biderman, and Carlo Ratti. 2018. "Monitour: Tracking Global Routes of Electronic Waste." *Waste Management* 72 (February): 362–370.

Lee, J. A., and V. M. Thomas. 2004. "GPS and Radio Tracking of End-of-Life Products [recycling and Waste Disposal Applications]." In *Proceedings of the International Symposium on Electronics and the Environment*, 309–312. IEEE Computer Society.

Luque-Ayala, Andrés, and Simon Marvin. 2016. "The Maintenance of Urban Circulation: An Operational Logic of Infrastructural Control." *Environment and Planning D: Society and Space* 34 (2): 191–208.

MacBride, Samantha. 2012. *Recycling Reconsidered: The Present Failure and Future Promise of Environmental Action in the United States*. Cambridge, Mass.: MIT Press.

Malena, Carmen, Reiner Forster, and Janmejay Singh. 2004. "Social Accountability - An Introduction to the Concept and Emerging Practice." 76. *Social Development Papers*. Washington DC: The World Bank.

Maniates, Michael F. 2001. "Individualization: Plant a Tree, Buy a Bike, Save the World?" *Global Environmental Politics* 1 (3): 31–52.

Martinez, Javier, Karin Pfeffer, and Tara van Dijk. 2011. "E-Government Tools, Claimed Potentials/ Unnamed Limitations: The Case of Kalyan–Dombivli." *Environment and Urbanization Asia* 2 (2): 223–234.

McDonough, William, and Michael Braungart. 2002. *Cradle to Cradle: Remaking the Way We Make Things*. New York: North Point Press.

Medina, Martin. 2010. "World's Largest And Most Dynamic Scavenger Movement." *BioCycle*, 51 (10), 32–33.

Melosi, M. V. 2004. Garbage In The Cities: Refuse Reform and the Environment. REV. Pittsburgh Hist Urban Environ. University of Pittsburgh Press.

———. 2008. *The Sanitary City: Environmental Services in Urban America from Colonial Times to the Present*. History of the Urban Environment. Pittsburgh, PA: University of Pittsburgh Press.

Miller, Joe. 2013. "City of London Calls Halt to Smartphone Tracking Bins." *BBC News*.

Offenhuber, Dietmar. 2014. "Infrastructure Legibility - a Comparative Study of Open311 Citizen Feedback Systems". *Cambridge Journal of Regions, Economy and Society*, 8(1), 93–112.

——— 2017. *Waste Is Information: Infrastructure Legibility and Governance*. Cambridge, MA: MIT Press.

Offenhuber, Dietmar, David Lee, Malima I. Wolf, Santi Phithakkitnukoon, Assaf Biderman, and Carlo Ratti. 2014. "Putting Matter in Place." *Journal of the American Planning Association. American Planning Association* 78 (2): 173–196.

Pak, Burak, Alvin Chua, and Andrew Vande Moere. 2017. "FixMyStreet Brussels: Socio-Demographic Inequality in Crowdsourced Civic Participation." *Journal of Urban Technology* 24 (2). Routledge: 65–87.

Pellow, D. N. 2004. *Garbage Wars: The Struggle for Environmental Justice in Chicago*. Cambridge, MA: MIT Press.

Porter, R. C. 2002. *The Economics of Waste*. Washington, DC: Resources for the Future.

Rad, Mohammad Saeed, Andreas von Kaenel, Andre Droux, Francois Tieche, Nabil Ouerhani, Hazim Kemal Ekenel, and Jean-Philippe Thiran. 2017. "A Computer Vision System to Localize and Classify Wastes on the Streets." *arXiv [cs.CV]*. arXiv. http://arxiv.org/abs/1710.11374.

Recycling Today. 2013. "China's 'Green Fence' Continues to Clog Export Markets." *Recycling Today*. March 21, 2013. Accessed: November 23, 2019 http://www.recyclingtoday.com/Article.aspx?article_id=139141.

Resource Recycling. 2018. "From Green Fence to Red Alert: A China Timeline." *Resource Recycling News* (blog). February 13, 2018. https://resource-recycling.com/recycling/2018/02/13/green-fence-red-alert-china-timeline/.

Saar, S., and V. Thomas. 2002. "Toward Trash That Thinks: Product Tags for Environmental Management." *Journal of Industrial Ecology* 6 (2): 133–146.

Saberi, Sara, Mahtab Kouhizadeh, and Joseph Sarkis. 2018. "Blockchain Technology: A Panacea or Pariah for Resources Conservation and Recycling?." *Resources, Conservation and Recycling* 130 (March). Elsevier). : 80–81.

Scheinberg, Anne, Rachel Savain, and Giz Office Tunis Solid Waste Exchange of Information And Expertise Network in Mashreq And Maghreb Countries. 2015. *Valuing Informal Integration: Inclusive Recycling in North Africa and the Middle East*. Deutsche Gesellschaft für Internationale Zusammenarbeit.

Strasser, S. 2000. *Waste and Want: A Social History of Trash*. 1st ed. A Holt Paperback: Henry Holt and Company.

Tarr, Joel A. 1996. *The Search for the Ultimate Sink: Urban Pollution in Historical Perspective*. 1st ed. Akron, OH: University of Akron Press.

Tchobanoglous, G., and F. Kreith. 2002. *Handbook of Solid Waste Management*. Handbook. New York N. Y: Mcgraw-hill.

Thaler, Richard H., and Cass R. Sunstein. 2008. *Nudge: Improving Decisions about Health, Wealth and Happiness*. New Haven, CT: Yale University Press.

United Church of Christ. 1987. "Toxic Wastes and Race in the United States." New York: United Church of Christ.

Valada, A., P. Velagapudi, B. Kannan, C. Tomaszewski, G. Kantor, and P. Scerri. 2014. "Development of a Low Cost Multi-Robot Autonomous Marine Surface Platform". In *Field and Service Robotics: Results of the 8th International Conference*, ed. Kazuya Yoshida and Satoshi Tadokoro, 643–658. Springer Tracts in Advanced Robotics. Berlin, Heidelberg: Springer.

Wang, Yu, Rui Tan, Guoliang Xing, Jianxun Wang, Xiaobo Tan, Xiaoming Liu, and Xiangmao Chang. 2014. "Aquatic Debris Monitoring Using Smartphone-Based Robotic Sensors." In *Proceedings of the 13th International Symposium on Information Processing in Sensor Networks*, 13–24. IPSN '14. Piscataway, NJ, USA: IEEE Press.

Warneke, B., M. Last, B. Liebowitz, and K. S. J. Pister. 2001. "Smart Dust: Communicating with a Cubic-Millimeter Computer." *Computer* 34 (1): 44–51.

Wilson, David C., Ljiljana Rodic, Anne Scheinberg, Costas A. Velis, and Graham Alabaster. 2012. "Comparative Analysis of Solid Waste Management in 20 Cities." *Waste Management & Research: The Journal of the International Solid Wastes and Public Cleansing Association, ISWA* 30 (3): 237–254.

Wolman, A. 1965. "The Metabolism of Cities." *Scientific American* 213 (3): 179–190.

Wylie, Sara, Nick Shapiro, and Max Liboiron. 2017. "Making and Doing Politics Through Grassroots Scientific Research on the Energy and Petrochemical Industries." *Engaging Science, Technology, and Society* 3: 393–425.

16

Stand up please, the real Sustainable Smart City

C. William R. Webster and Charles Leleux

Introduction

There are competing discourses in academia and public policy about what constitutes a city that is either Smart (Nam and Pardo, 2011) or sustainable (Haughton and Hunter, 2004). Within these discourses there are assumptions about the positive role played by technology (Sæbø *et al.*, 2008) and its positive impact on the environment (Mitchell and Casalegno, 2009). Whilst strictly speaking a Smart city is not necessarily also a sustainable city, these concepts are increasingly being conflated in the term – the 'Sustainable Smart City' (SSC) (Kramers *et al.*, 2014). In this chapter, we review the conceptual components of an SSC from the perspective of the citizen, and 'unpick' how each of these terms relate to each other in practice. The central argument of the chapter explores the tensions between traditional eGovernment (eGov) approaches, which focus on service delivery at the expense of citizen engagement (Meijer, 2011), and of the sustainability literature, which focuses on environmentalism at the expense of sustainable societies (Robinson, 2004). Sustainability in the SSC context is much more than simply being environmentally friendly, it is about designing organisational structures and processes and institutional norms and values that will exist over time, in a manner that enriches citizens' lives and at the same time prioritises the efficient use of societal resources and the environment. The term 'sustainability' is a contested one and it is evident that the dominant discourse is being shaped by commercial interests, which may be repurposing the concept to suit business needs (Gray, 2010). New digital technologies harnessing information and communication technologies (ICTs), such as apps, social media, the internet and the improved collection of data through sophisticated monitoring systems, all provide opportunities for a new public sector vision of efficiency, often promoted jointly by municipal administrations working with commercial technology providers. This 'vision' also purports to offer new opportunities for streamlined public service provision and a more economic use of scarce resources. The transformative potential of technology also offer long-term sustainable outcomes where it engages citizens actively in participatory activities, and where these activities are linked to the services which they need (Batty *et al.*, 2012). In this respect, new ICTs act as a critical conduit in the co-production of public services and citizen–government engagement mechanisms, in which the citizen is becoming increasingly empowered to participate in SSC governance (Meijer and Bolívar, 2016).

The rise of the Smart city

Although the concept of the Smart city has been in existence since the mid-1990s (Harrison and Donnelly, 2011; Hollands, 2008; Neirotti *et al.*, 2014), its origins and use are subject to different interpretations. It is claimed that the Smart city idea originated from the 'New Urbanism movement in the USA of the 1980s and … the concept of the technology-based intelligent city' (Söderström *et al.*, 2014, p. 310). An alternative view is offered through the 'Connected Sustainable Cities' project (Mitchell and Casalegno, 2009), which aimed to demonstrate how technological innovation can enhance opportunity, equity and cultural creativity in urban areas. This led to city planners designing 'creative cities' where policy interventions revitalised derelict spaces into creative clusters and where the opportunity and conditions were provided for creative industries to flourish (Florida, 2003; Hall, 2000; Pratt, 2010; Scott, 2006). In Europe, the European Commission 'Horizon 2020 Smart Cities and Communities' project (European Commission, 2019), includes Smart Cities 'Lighthouse' projects, which include energy, transport and ICTs as particular areas of focus.

Kitchin (2014) describes a key feature of the Smart city as being the development of a knowledge economy, involving a combination of ICTs and human capital, a point reinforced by Meijer and Thaens (2016), who emphasise the importance of combining social structures and new technologies in order to realise urban innovation. Since the late 2000s, discourse around the term Smart city has been increasingly influenced by large IT companies (Lee and Lee, 2014; Shelton *et al.*, 2015), and IBM adopted the trademark 'smarter cities' in 2011 (Söderström *et al.*, 2014). Subsequently, many multinational IT companies have sought to be involved in the transformation taking place around delivery of public services and renewal of city infrastructure (Hollands, 2015). Smart cities are also synonymous with many other connected terms, such as the intelligent city, the information city, the digital city and the virtual city (Batty, 2013), and these terms are often used interchangeably (Albino *et al.*, 2015). What they share is a positive perspective concerning the transformational potential of digital technology and the benefits to society that can accrue from their use.

Defining the Smart city

Conflicting views on the concept of a Smart city are reflected in the difficulties experienced in attempting to find a suitable definition: 'The label smart city is a fuzzy concept and is used in ways that are not always consistent. There is neither a single template of framing smart city nor a one-size-fits-all definition of smart city' (Nam and Pardo, 2011, p. 283). The application of the term 'Smart' to a city, or to its constituent elements, such as Smart economy, Smart mobility, Smart environment, Smart people, Smart living, Smart governance, etc. (Lee *et al.*, 2013), does not necessarily mean that there will be equanimity of interventions, allocation of resources, or equality of opportunities to all citizens, neighbourhoods and spaces: 'Whatever it means for a city to be "smart", it is also readily apparent that not all spaces of the city will be equally smart, meaning that smart cities will privilege some places, people and activities over others' (Shelton *et al.*, 2015, p. 15). The normative assumptions around the beneficial aspects of a Smart city, including its 'self-congratulatory nature', should be challenged, as should the assumption that ICTs will deliver this transformation (Hollands, 2015, p. 62). New ICTs can include mobile telephone apps, with links to interactive municipal services such as leisure class bookings, entertainment events and real-time citizen information such as traffic or air quality. Additionally, citizens can participate in local affairs more easily, through online citizen discussion forums and surveys. These participatory measures are central to the Smart city vision, and play a critical role in shaping service provision and also in providing new opportunities for greater participation by citizens in public policy, decision making and the co-production of services (Meijer and Bolívar, 2016; Schaffers *et al.*, 2011). The extent to which these initiatives are truly Smart, or simply the rebranding of existing eGov policy and services, is open to question.

The sustainable Smart city

The focus on the sustainability aspect of Smart cities has been gaining traction in recent years, due in part to projected population levels in urban areas (Albino *et al.*, 2015). Increasing population growth places pressure on a city's ability to manage waste, natural resources, air pollution, traffic, infrastructure and governance (Chourabi *et al.*, 2012). In this respect, city administrations have a moral obligation to their citizens, communities and businesses to reflect on, consider and design sustainable futures. The concept of the SSC has grown due to five main developments:

> (1) globalization of environmental problems (climate change and decline in biodiversity); (2) urbanization (rise in the numbers of people living in cities); (3) sustainable development (managing urban environments); (4) ICTs (embedded in everyday life); and, (5) smart cities (the transformative application of ICTs in the public sector).
>
> *(Höjer and Wangel, 2015, pp. 334–337)*

Smart solutions involving the use of new technologies are increasingly being sought in response to these pressures and are contributing to the rise in the currency of the concept of the SSC (Höjer and Wangel, 2015), with an underlying belief that new ICTs will contribute directly to a more sustainable future. The emergence of the SSC also identifies a role for active citizenship as a critical success factor to achieve sustainable outcomes: 'The primary way in which sustainability is to be achieved within smart cities is through more efficient processes and responsive urban citizens participating in computational sensing and monitoring practices' (Gabrys, 2014, p. 32). Sustainability is directly embedded in the concept of the Smart city (Batty *et al.*, 2012; Chourabi *et al.*, 2012; Hancke *et al.*, 2013) and increasingly the term the SSC is being used as a label to capture the new technologically advanced city environment.

At the core of this chapter is an exploration of how SSCs are conceptualised and understood. This includes an analysis of the different components of an SSC and how they interact and have evolved together over time. The three dominant components explored here are (1) sustainability, (2) new ICTs and (3) citizen participation. In public policy terms, and in the rhetoric around Smart cities, these three components appear to be interdependent, despite each being associated with its own logic, public policy and academic literature. In this chapter, we argue that these three perspectives – the sustainability perspective, the technological perspective and the participatory perspective – all promote certain features and characteristics, and that whilst they may be conflated within the term the SSC, they are actually rarely completely separable. The remainder of the chapter is split into six sections. Sections 2, 3 and 4 set out the key components of the sustainability, technological and participation perspectives, including their key focus and institutional logic. Section 5 presents a new SSC conceptualisation that captures the divergence and compatibility of the different perspectives, and highlights the conditions required for a 'true' SSC – namely where the three perspectives discussed here overlap and are fully intertwined. The final section, section 6, offers concluding comments.

The sustainability perspective: delivering environmentalism

The contemporary discourse on sustainability is widely acknowledged to have been influenced by the report of the World Commission on Environment and Development (Brundtland, 1987). The Brundtland report raised awareness amongst policy-makers and practitioners of their societal responsibilities to ensure that the needs of the present day would not compromise the needs of future generations (Baker *et al.*, 2012, p. 3). Against a backdrop of massive urban population growth, sustainable development became a policy priority at local, national and supra-national levels, the last of these involving the World Bank (Kahn, 2014), the United Nations (United Nations, 2015) and the European Union (Lazaroiu and Roscia, 2012). The European Union considers the concept of the Smart city as one that supports environmental sustainability through the use of innovative technologies, for example in the reduction of greenhouse gas emissions (Ahvenniemi *et al.*, 2017). Sustainability and environmentalism have both grown as social phenomena in their own right, and academic approaches developed in the 1990s in the work of Gray (1992), Elkington (1994) and others. Elkington's (1994) seminal work on accounting for sustainability, introduced the term the 'triple bottom line' (TBL), which has been used as a method for introducing greater corporate social responsibility for organisations with regards to accounting for their impact on sustainability (Henriques and Richardson, 2013). The TBL has been referred to as 'economy, social (people, citizenry), and environment' (Kondepudi and Kondepudi, 2015, p. 10) and as a method for full cost accounting for individual and collective social, economic and environmental sustainability (Larsson and Grönlund, 2014). The 'full cost accounting' approach involves the addition of 'technology' as a fourth category to Elkington's model, which is used as a basis for improving the analysis of sustainability in eGov applications (Larsson and Grönlund, 2014, p. 139). Underpinning the sustainability perspective, in both policy and academic discourse, is a link between sustainable practices and their positive impact on the environment.

Of particular interest here is the development and merging of separate phenomena – Smart cities and sustainable cities – into what is now referred to as the SSC. Critical factors in the formation of the SSC concept include the pressures of growing urbanization, population growth, management of finite energy resources and the potential offered by ICTs to offer Smart solutions to these challenges. The role of the private sector has had considerable influence as business approaches have drifted from Smart city discourses into SSC solutions and applications (Deakin, 2013). Kitchin (2015) makes a strong case for the commercial interests of the private sector being the reason why private sector IT companies have changed their corporate language from a managerially focused approach, to an inclusive and citizen-centred one, with more emphasis on social, economic and environmental challenges. Ahvenniemi *et al.* (2017) in their analysis of the differences between sustainable and Smart cities, and their respective performance monitoring systems, point to the limited use of environmental indicators, compared to the more extensive use of economic and social indicators, when analysing Smart city frameworks. They argue for the integration of sustainability and Smart city frameworks and make a case for the use of a more accurate term, 'smart sustainable cities' (instead of Smart cities). As a response to deficiencies in measuring 'smartness', Al-Nasrawi *et al.* (2015) propose a multidimensional model that is context sensitive, whilst Hara *et al.* (2016) propose key performance indicators (KPIs) based on well-being and the quality of life. The United Nations Economic Commission for Europe provide a strategic vision for improving sustainability of communities and offer their own definition of an SSC:

> A smart sustainable city is an innovative city that uses information and communication technologies (ICTs) and other means to improve quality of life, efficiency of urban operation and services, and competitiveness, while ensuring that it meets the needs of present and future generations with respect to economic, social, environmental as well as cultural aspects .
>
> *(United Nations, 2015, p. 3)*

The sustainability perspective plays an important role in shaping Smart cities. It provides a focus on environmental issues and a normative belief structure that organisations and services should be environmentally friendly, and that embedding environmental considerations into societal structures and processes will lead to sustainable societies. This is the institutional logic at the heart of this perspective and which permeates public policy and service delivery in urban environments.

The technological perspective: e-Government and the transformative potential of ICTs

At the heart of the Smart city vision is the notion that new technologies can be utilised to transform the way that services are delivered and citizens live their lives. In this respect, new technology is the engine room of the smart city phenomenon. The rhetoric around technological potential is extremely positivist and very deterministic. This is shaped by the main actors involved, the technological companies wanting to sell their 'kit' and municipalities attempting to sell the concept to make life better, healthier and more sustainable for their citizens. Technology and data are being used as the core of a powerful argument about transforming society for the better (Angelidou, 2015; Nam and Pardo, 2011). Tensions that may exist where new technologies interact with traditional forms and structures of democratic decision-making are underplayed in this vision. Meijer (2017, p. 2), for example, provides a theoretical model of the relationship between technology and urban governance, exploring the complexities around the interface between the 'political community of citizens and urban data infrastructures'.

The transformative potential of ICTs is well recognised and has been credited with facilitating a 'digital revolution' and a new social order often referred to as the 'Information Society' (Castells, 2011). Public policy and services have not been immune to this revolution and have harnessed the characteristics of new ICTs to deliver new and enhanced public services, often referred to as eGov (Taylor and Webster, 1996). Here it is argued, that new information flows, embedded in new ICTs, allow for the delivery of new electronic services that are more timely, personalised, commoditised, more accessible, integrated and more cost efficient (Heeks and Bailur, 2007). At the heart of this vision are information flows and how these flows are changing society, including in the 'Information Polity' (Bellamy and Taylor, 1994). Municipal decision-making has traditionally followed a vertical and arguably bureaucratic hierarchy (Bonsón *et al.*, 2012), and municipalities operating within established decision-making structures have often interacted with historically recognised groups operating within already established neighbourhoods and policy domains. The proliferation of social media and other new digital technologies has opened-up the possibilities for municipal engagement. New 'virtual' and often previously unrelated 'communities' can be created, where there are no predetermined or spatially defined geographical boundaries, or indeed a 'sense of place' which might connect them (Albino *et al.*, 2015). Traditional political and decision-making processes are therefore being supplemented and simultaneously undermined by new information flows facilitated by ICTs.

Whilst there has been the widespread adoption of ICTs to deliver eGov services and electronic citizenship, often utilising the internet, there are still concerns about how to apply universally acceptable criteria for assessing how ICTs will improve public services for all parties involved, namely, policy-makers (elected representatives), service providers (practitioners) and service users (citizens). Successful transition from paper-based systems to ones dependent upon ICTs may not be universally beneficial, particularly for disengaged citizens, the elderly, and those who do not possess basic internet skills. Successful deployment of new ICT systems may be judged using criteria determined by practitioners and IT companies, rather than by the citizens who are consuming and using services. The assessment criteria of the impact of eGov initiatives become a central issue to understanding the rhetoric which surrounds the deployment of new electronically based public services. Kumar *et al.* (2007, p. 73) argue that citizens are the focal point of eGov, and that citizens' user characteristics need to be properly understood, including perceived risks, data security, privacy and their perceived control over the processes involved. Scott *et al.* (2016) emphasise the importance of the value realised by citizens in the engagement process, and point to the impact which new social media technologies can have by changing the ways in which users interact with organisations, and the potential for the co-creation of value between citizens and governments. Meijer and Bekkers (2015) refer to eGov as being an important driver for the modernisation of the public sector, where the use of ICTs can redesign existing information processing and communication practices to achieve better output, especially in relation to electronic service delivery.

The technological perspective is central to the Smart city phenomenon. It focuses on the potential benefits deriving from technological deployment and promotes a belief structure that the optimal use of technology will lead to better societal outcomes, including more sustainable societies. At the core of the technological perspective is the idea that technologies are the vehicle for 'transformation', that they encourage efficiency and the better use of scarce recourses. Here the institutional logic is that technology can be harnessed for the good of society.

The participation perspective: citizen engagement

As Smart city agendas have evolved it has become apparent that sustainable approaches to public policy and services should include some form of citizen engagement. In this perspective citizen participation is important because it ensures that services meet the requirements of local needs, and that it provides legitimacy for the institutions involved in policy and service provision. In this respect, the resilience and sustainability of local communities and institutions determine some form of citizen engagement, including in Smart city environments.

It has long been recognised that involving citizens in decisions about public services and policy can foster legitimacy for services and can ensure that provision is tailored to needs of local communities. Arnstein (1969) introduced a hypothetical model of differing levels of citizen engagement and participation ranging from tokenism to full citizen engagement, where there was a redistribution of power and control over decision-making processes. Yang and Pandey (2011, p. 880) emphasise the importance of effective citizen participation to democratic governance, focusing on the values which can be achieved: 'fostering citizenship values; enhancing accountability; improving trust in government; maintaining legitimacy; achieving better decisions; building consensus'. Michels and De Graaf (2010, p. 488) also emphasise the positivistic democratic effects of citizen participation on the quality of democracy, for example through participatory projects where there is the potential for giving citizens some influence, and 'inclusion, civic skills and virtues, deliberation and democracy'.

There have been long-standing claims that traditional eGov approaches have focused more on service delivery than on meaningful citizen engagement and democratic participatory processes (Bekkers and Homburg, 2005). Furthermore, not only has the focus been on service delivery at the expense of citizen engagement, there is also limited evidence that services have been improved or that cost savings have been on the scale imagined (Bekkers and Homburg, 2005; Norris and Reddick, 2013). Government bodies, public agencies and ICT-providers have not delivered change on the levels anticipated and it has been argued that ICTs have reinforced existing power inequalities and relationships (Bannister, 2010; Chadwick and May, 2003). Part of the reason for these adverse outcomes is due to service providers having values that are at odds with citizens and that it is difficult to find common ground on which to move forward (Bannister, 2010). One of the drivers for the creation of eGov was 'to create a more citizen-focused government' (Reddick and Turner, 2012, p. 1), where the importance was emphasised of citizens' values including trust, leadership, fairness of treatment and competence, leading to the concept of a citizen-centric approach (Nam and Pardo, 2011). Nevertheless, there is a growing perception that with the advent of Smart cities, the extent to which engagement is developing and diversifying is now occurring in more sophisticated ways than previously witnessed. This transformation is attributable, in part, to the role which ICTs are playing in changing public life (Hollands, 2015) and the internet in particular, which promises levels of co-production on an 'unprecedented scale' (Linders, 2012).

Delivery of, and access to, public services and information through electronic methods reflect the growing opportunities for citizens generated by the internet, social media and Smart devices (Clarke, 2013; Ellison and Hardey, 2014; Morgeson et al., 2010). This transformation has been facilitated by the normalisation of electronic communication in society, through social media and the increasing prevalence of online transactions. In the Smart city context there is a plethora of new apps and platforms to report potholes, arrange the collection of garden waste and building materials, or to reserve a seat at a municipality-owned theatre (Höffken and Streich, 2013). Here it is argued that improved levels of trust between citizens and government, and greater transparency, can be achieved through improved levels of citizen use of social media in government activities (Song and Lee, 2016), alongside an increasing expectation from citizens that access to e-services and e-information will be available on a 24/7 basis (Lofstedt, 2012). Whilst there is general agreement about the transformative potential of ICTs in relation to service delivery, it is less clear about the extent to which these technologies provide opportunities to genuinely engage and empower citizens to participate in local affairs: 'If e-government is to be truly transformative of government in terms of citizen engagement and participation, then e-government must be citizen-centred in its development and implementation' (Jaeger and Bertot, 2010, p. 4).

Citizen engagement and participation in the Smart city or SSC may be influenced by the ability or motivation of citizens, depending upon their affluence and education, to articulate their views. Some studies have shown that citizens who live in more affluent areas and are better educated may be more inclined to engage (Hastings et al., 2014). The co-production literature also suggest that citizens are more inclined to engage when they are given the opportunity to address issues which are particular to their own neighbourhood (Alford, 2009). Driskell (2017) offers the view that children and young people are frequently excluded from decision-making when it comes to matters concerning their city or neighbourhood, even in relation to local issues where they are likely to know more than the policy-makers. He argues that children have a legitimate right to be regarded as active and valued partners, and to ignore their views can jeopardise the 'social, economic and environmental quality of communities and neighbourhoods' in the development of policies for poverty reduction, increased equality and social cohesion. Successful outcomes, in terms of citizen engagement and participation, could be defined as both greater levels of participation and improved public services and policy. However, any assessment of what constitutes 'success' may depend upon the ability of municipalities and governments to incentivise citizens to engage and participate. There will have to be meaningful devolution of some aspects of municipal/central government decision-making apparatus to a more decentralised and citizen-centric model, in which there will be benefits and rewards for both parties, in an environment of reciprocity (Webster and Leleux, 2018).

The participation perspective promotes active citizenship and the need for citizens and service users to be involved in policy and decision-making and the co-production of services. It provides a focus on the formation of effective engagement mechanisms and has a normative belief structure that more citizen engagement will lead to better services and policy. In the Smart city, this perspective encourages the use of innovative citizen engagement and co-production mechanisms (Webster and Leleux, 2018). Here the institutional logic is that better and more citizen engagement will lead to more sustainable societies and social structures. However, citizen engagement and participation in the Smart city can be seen as being strongly instrumental and a necessity to making certain systems function, such as those based on big-data. Citizen engagement in this context can be construed as being less proactive and aware, and rather a functional requirement for citizens to be data providers.

Perspectives and institutional logics

Framing the development of SSCs around three perspectives, each with their own institutional logic (Friedland and Alford, 1991), demonstrates the multifaceted nature of the concept and that the origins of the SSC are embedded in pre-existing perspectives with their own activities, organisations, processes, beliefs and norms. Each of the three perspectives identified here – the sustainability, technological and participation perspectives – are absolutely core to the development of SSCs, each has its own logic, promotes certain beliefs and is independent of the others. They each tell us something about how the world around us should be organised and how to deliver sustainability effectively. In many ways, the SSC can be seen as the convergence of these three perspectives in an integrated approach to urban development.

Table 16.1 captures the three perspectives discussed here with a particular focus on their core messages and entrenched beliefs. Presenting the perspectives in this way allows us to untangle the foundations of SSCs and to think about how the different constituent parts fit together.

Table 16.1 Sustainable Smart perspectives

Perspective	The sustainable perspective	The technological perspective	The participation perspective
Emphasis:	• Focus on the climate and environmentalism • Organisations/services should be environmentally friendly • Establish mechanisms and processes to ensure sustainable outcomes • Environmentalism leads to sustainable societies	• Focus on potential of new technology (ICTs) • Promotes the deployment of ICTs • New ICTs will lead to enhanced services • Technology will improve society	• Focus on citizen participation/engagement • Promotes new mechanisms for participation • Greater levels of participation will lead to better public services and policy

The sustainable Smart city: a multifaceted concept

Conceptualising the SSC around a series of dominant perspectives highlights the multifaceted complex nature of the concept, and in particular that SSCs are not just focused on environmental sustainability, they are equally concerned with new technologies and citizen participation. The relationships between each of the perspectives are complicated and are shaped by diverse institutional and local settings. Of course, deconstructing complex social phenomena into a simple heuristic device may aid understanding, but it will not be a complete picture of practice or reality. Under scrutiny, the relationships are neither simple nor easily compartmentalised. Instead, as Smart cities have evolved, the relationships between the perspectives have become more dependent upon each other and are fused together. Figure 16.1 offers a visual representation, in the form of a Venn diagram, of the three constituent perspectives that make up the SSC. Illustrating the SSC in this way highlights the simultaneous independence and interdependence of the three perspectives, and that the SSC is constituted from a number of core elements.

The 'real' SSC is at the heart of the Venn diagram where the institutional logics of the three perspectives converge. This is where the demands of the institutional logics and belief systems are satisfied and coexist. For many Smart cities this convergence is not to be assumed and it may be the case that many Smart city applications meet the needs of two of the perspectives but not the third, illustrated on the Venn diagram by 'dual' overlaps. In this way it is possible to map and assess the characteristics of an individual Smart city application to see if it meets the criteria of a SSC application. In this practical sense the perspectives and any Smart city application have to be rooted and understood in their societal and institutional contexts.

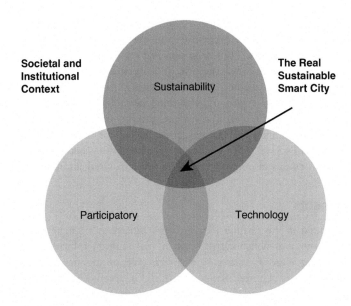

Figure 16.1   The real Sustainable Smart City

Source: Chapter author(s)

Concluding comments

The chapter highlights the core conceptual perspectives supporting the emergence of the notion of a SCC. Central to the analysis presented here is the idea that these perspectives are interdependent, that they are reliant on each other and at the same time and in tension with one another. The SSC offers substantive change in citizen–state relations, it points to new relations being formed around the use of new technologies with an aspiration that local governance and services will be enhanced at the same time as facilitating more sustainable futures. Transition is being assisted by greater citizen awareness of sustainability issues, for example in recycling initiatives and the adoption of home Smart meters. There are sizeable challenges for citizens and policy-makers if widespread citizen engagement in the SSC is to be realised, particularly in connecting with 'hard-to-reach' groups which may be disconnected from mainstream society, and societal values such as sustainability. Academic literature shows that engagement works better when citizens are incentivised to participate and when there is a redistribution of power to communities (Alford and Yates, 2015). New mechanisms of engagement mediated by new ICTs are helping this transformation, for example, hackathons, living labs, maker spaces, gamified public services, 'open data' and crowdsourcing (Webster and Leleux, 2018). However, it must be recognised that these new mechanisms may appeal to 'niche' groups in society which have specific and perhaps transient 'local' interests, and that there will not necessarily be potential for the scalability of these mechanisms to reach wider participatory audiences.

The SSC is a multifaceted and contested concept which at its core has a number of competing intertwined discourses and logics. The move from Smart cities to SSCs signals the importance of integrating sustainability in governance approaches that utilise technology and citizen engagement. Ahvenniemi *et al.* (2017) point to a 'gap' in the use of environmental indicators when assessing Smart city frameworks and make a strong case for the use of the term SSC, instead of Smart cities. Sustainability approaches tend to focus on environmentalism at the expense of sustainable societies, whilst technological solutions have concentrated on functionality and service improvement, at the expense of citizen participation. Smart city aspirations are likely to be constrained if they are driven by technological solutions. If they are merged successfully with citizen engagement mechanisms and sustainable practices, then it is possible to imagine governance and organisational structures that will facilitate sustainable futures. True sustainability and hence the SSC is realised at the nexus of ICT use, citizen engagement and sustainable outcomes. It is an aggregate concept which takes into account the sustainability aspirations of a city and the importance of ICTs to achieve sustainability goals (Höjer and Wangel, 2015). This is a more nuanced conceptualisation of the Smart city, one that emphasises the human and organisational aspects of smart cities, yet at the same time recognises the primacy of technology and sustainability. Whether or not the real SSC exists in practice is an empirical question requiring further investigation.

Acknowledgements

The research presented in this chapter derives from 'SmartGov': 'Smart Governance of Sustainable Cities',[1] a four year collaborative transnational multidisciplinary research project on the value of ICTs for engaging citizens in governance of sustainable cities (2015–2019). Funding Councils in the United Kingdom (ESRC), Netherlands (NWO) and Brazil (FAPESP) have co-funded the research. The three project partners are Utrecht University (Netherlands), University of Stirling (United Kingdom) and Fundação Getulio Vargas, Sao Paulo (Brazil). The ESRC Grant reference number is: ES/N011473/1.

Note

1 SmartGov research project website, URL: http://smartgov-project.com.

References

Ahvenniemi, H., Huovila, A., Pinto-Seppä, I., & Airaksinen, M. (2017). What are the differences between sustainable and Smart cities?. *Cities*, 60, 234–245.

Albino, V., Berardi, U., & Dangelico, R. M. (2015). Smart cities: Definitions, dimensions, performance, and initiatives. *Journal of Urban Technology*, 22 (1), 3–21.

Alford, J. (2009). *Engaging public sector clients: From service-delivery to co-production*. Springer.

Alford, J. & Yates, S. (2015). Co-production of public services in Australia: The roles of government organisations and co-producers. *Australian Journal of Public Administration*, 75 (2), 159–175.

Al-Nasrawi, S., Adams, C., & El-Zaart, A. (2015). A conceptual multidimensional model for assessing smart sustainable cities. *JISTEM-Journal of Information Systems and Technology Management*, 12 (3), 541–558.

Angelidou, M. (2015). Smart cities: A conjuncture of four forces. *Cities*, 47, 95–106.

Arnstein, S. (1969). A ladder of participation. *Journal of the American Institute of Planners*, 35 (4), 216–244.

Baker, S., Kousis, M., Richardson, D., & Young, S. (2012). *The Politics of sustainable development*. London: Routledge.

Bannister, F. (2010). Deep e-government. In Scholl, H. J. (ed). *E-government: Information, technology, and transformation*, Vol. 3, 33–51, Abingdon, Oxon: Routledge.

Batty, M. (2013). Big data, smart cities and city planning. *Dialogues in Human Geography*, 3 (3), 274–279.

Batty, M., Axhausen, K. W., Giannotti, F., Pozdnoukhov, A., Bazzani, A., Wachowicz, M., & Portugali, Y. (2012). Smart cities of the future. *The European Physical Journal Special Topics*, 214 (1), 481–518.

Bekkers, V. J., & Homburg, V. (Eds.) . (2005). *The information ecology of e-government: e-government as institutional and technological innovation in public administration*. Vol. 9 Amsterdam: IOS Press.

Bellamy, C., & Taylor, J. A. (1994). Reinventing government in the information age. *Public Money & Management*, 14 (3), 59–62. 10.1080/09540969409387830.

Bonsón, E., Torres, L., Royo, S., & Flores, F. (2012). Local e-government 2.0: Social media and corporate transparency in municipalities. *Government Information Quarterly*, 29 (2), 123–132.

Brundtland, G. H. (1987). World commission on environment and development (1987): Our common future. *World Commission for Environment and Development*, 17, Doc. 149, WCED/87/6, 1–87.

Castells, M. (2011). *The rise of the network society*. Somerset: John Wiley & Sons.

Chadwick, A., & May, C. (2003). Interaction between states and citizens in the age of the internet: "e-Government" in the United States, Britain, and the European Union. *Governance*, 16 (2), 271–300.

Chourabi, H., Nam, T., Walker, S., Gil-Garcia, J. R., Mellouli, S., Nahon, K., & Scholl, H. J. (2012). Understanding smart cities: An integrative framework. In: *System Science (HICSS), 2012 45th Hawaii International Conference on* (pp. 2289–2297). IEEE.

Clarke, R. Y. (2013). *Smart cities and the internet of everything: The foundation for delivering next-generation citizen services*. Alexandria, VA: Tech. Rep.

Deakin, M. (Ed.). (2013). *Smart cities: Governing, modelling and analysing the transition*. Abingdon, Oxon: Routledge.

Driskell, D. (2017). *Creating Better Cities with Children and Youth: A Manual for Participation*. Abingdon: Earthscan.

Elkington, J. (1994). Towards the sustainable corporation: Win-Win-Win business strategies for sustainable development. *California Management Review*, 36 (2), 90–100.

Ellison, N., & Hardey, M. (2014). Social media and local government: Citizenship, consumption and democracy. *Local Government Studies*, 40 (1), 21–40.

European Commission (2019). Smart cities and communities. *Horizon 2020 programme*. https://ec.europa.eu/inea/en/horizon-2020/smart-cities-communities

Florida, R. (2003). Cities and the creative class. *City & Community*, 2 (1), 3–19.

Friedland, R., & Alford, R. R. (1991). Bringing society back in: Symbols, practices and institutional contradictions. In: Powell, W. W., & Dimaggio, P. J. (eds) *The new institutionalism in organizational analysis*, 232–263, Chicago: University of Chicago Press.

Gabrys, J. (2014). Programming environments: Environmentality and citizen sensing in the smart city. *Environment and Planning D: Society and Space*, 32 (1), 30–48.

Gray, R. (1992). Accounting and environmentalism: An exploration of the challenge of gently accounting for accountability, transparency and sustainability. *Accounting, Organizations and Society*, 17 (5), 399–425.

Gray, R. (2010). Is accounting for sustainability actually accounting for sustainability ... and how would we know? An exploration of narratives of organisations and the planet. *Accounting, Organizations and Society*, 35 (1), 47–62.

Hall, P. (2000). Creative cities and economic development. *Urban studies*, 37 (4), 639–649.

Hancke, G. P., Silva, B. C. & Hancke G., Jr (2013). The role of advanced sensing in smart cities. *Sensors*, 13 (1), 393–425.

Hara, M., Nagao, T., Hannoe, S., & Nakamura, J. (2016). New key performance indicators for a smart sustainable city. *Sustainability*, 8 (3), 206.

Harrison, C., & Donnelly, I. A. (2011). A theory of smart cities. In: *Proceedings of the 55th Annual Meeting of the ISSS-2011, Hull, UK*, 55, 1.

Hastings, A., Bailey, N., Bramley, G., Croudace, R., & Watkins, D. (2014). 'Managing 'the middle classes: Urban managers, public services and the response to middle-class capture. *Local Government Studies*, 40 (2), 203–223.

Haughton, G. and Hunter, C. (2004). *Sustainable cities*. Routledge.

Heeks, R., & Bailur, S. (2007). Analyzing e-government research: Perspectives, philosophies, theories, methods, and practice. *Government Information Quarterly*, 24 (2), 243–265.

Henriques, A., & Richardson, J. (Eds.). (2013). *The triple bottom line: Does it all add up*. Routledge.

Höffken, S., & Streich, B. (2013). Mobile participation: Citizen engagement in urban planning via smartphones. In Carlos Nunes Silva (ed.) *Citizen E-Participation in urban governance: crowdsourcing and collaborative creativity*, 199–225, Portland, OR: Book News Inc.

Höjer, M., & Wangel, J. (2015). Smart sustainable cities: Definition and challenges. In. In Hilty L., & Aebischer B. (eds) *ICT innovations for sustainability: Advances in intelligent systems and computing*, Vol. 310, 333–349, Cham: Springer.

Hollands, R. G. (2008). Will the real smart city please stand up? Intelligent, progressive or entrepreneurial?. *City*, 12 (3), 303–320.

Hollands, R. G. (2015). Critical interventions into the corporate smart city. *Cambridge Journals of Regions Economics and Society*, 8 (1), 61–77.

Jaeger, P. T., & Bertot, J. C. (2010). Designing, implementing, and evaluating user-centered and citizen-centered e-government. *International Journal of Electronic Government Research*, 6 (2), 1–17.

Kahn, M. E. (2014). Sustainable and smart cities. *Policy Research working paper; no. WPS 6878*. Washington, DC: World Bank Group.

Kitchin, R. (2014). The real-time city? Big data and smart urbanism. *GeoJournal*, 79 (1), 1–14.

Kitchin, R. (2015). Making sense of smart cities: Addressing present shortcomings. *Cambridge Journal of Regions, Economy and Society*, 8 (1), 131–136. 10.1093/cjres/rsu027.

Kondepudi, S., & Kondepudi, R. (2015). What constitutes a Smart City?. In A. Vesco and F. Ferrero (Eds) *Handbook of research on social, economic, and environmental sustainability in the development of Smart cities*, 1–25, Hershey: IGI Global.

Kramers, A., Höjer, M., Lövehagen, N., & Wangel, J. (2014). Smart sustainable cities–Exploring ICT solutions for reduced energy use in cities. *Environmental modelling & software*, 56, 52–62.

Kumar, V., Mukerji, B., Butt, I., & Persaud, A. (2007). Factors for successful e-government adoption: A conceptual framework. *The Electronic Journal of e-Government*, 5 (1), 63–76.

Larsson, H., & Grönlund, Å. (2014). Future-oriented egovernance: The sustainability concept in egov research, and ways forward. *Government Information Quarterly*, 31 (1), 137–149.

Lazaroiu, G. C., & Roscia, M. (2012). Definition methodology for the smart cities model. *Energy*, 47 (1), 326–332.

Lee, J., & Lee, H. (2014). Developing and validating a citizen-centric typology for smart city services. *Government Information Quarterly*, 31, S93–S105.

Lee, J. H., Phaal, R., & Lee, S. H. (2013). An integrated service-device-technology roadmap for smart city development. *Technological Forecasting and Social Change*, 80 (2), 286–306.

Linders, D. (2012). From e-government to we-government: Defining a typology for citizen coproduction in the age of social media. *Government Information Quarterly*, 29 (4), 446–454.

Lofstedt, U. (2012). E-government-assessment of current research and some proposals for future directions. *International journal of public information systems*, 1 (1), 39–52.

Meijer, A. (2017). Datapolis: A Public Governance Perspective on "Smart Cities". *Perspectives on Public Management and Governance*, 1-12, gvx017. 10.1093/ppmgov/gvx017.

Meijer, A., & Bekkers, V. (2015). A metatheory of e-government: Creating some order in a fragmented research field. *Government Information Quarterly, 32* (3), 237–245.

Meijer, A. & Bolívar, M. P. R. (2016). Governing the smart city: A review of the literature on smart urban governance. *International Review of Administrative Sciences, 82* (2), 392–408.

Meijer, A., & Thaens, M. (2016). Urban technological innovation: Developing and testing a sociotechnical framework for studying smart city projects. *Urban Affairs Review, 54* (2), 363–387.

Meijer, A. J. (2011). Networked coproduction of public services in virtual communities: From a government-centric to a community approach to public service support. *Public Administration Review,* 71 (4), 598–607.

Michels, A., & De Graaf, L. (2010). Examining citizen participation: Local participatory policy making and democracy. *Local Government Studies,* 36 (4), 477–491. 10.1080/03003930.2010.494101.

Mitchell, W. J., & Casalegno, F. (2009). *Connected sustainable cities.* MIT Mobile Experience Lab Publishing.

Morgeson III, F. V., VanAmburg, D., & Mithas, S. (2010). Misplaced trust? Exploring the structure of the e-government-citizen trust relationship. *Journal of Public Administration Research and Theory,* 21 (2), 257–283.

Nam, T., & Pardo, T. A. (2011). Conceptualizing smart city with dimensions of technology, people, and institutions. In: *Proceedings of the 12th annual international digital government research conference: Digital government innovation in challenging times,* 282–291. ACM.

Neirotti, P., De Marco, A., Cagliano, A. C., Mangano, G., & Scorrano, F. (2014). Current trends in smart city initiatives: Some stylised facts. *Cities,* 38, 25–36.

Norris, D. F., & Reddick, C. G. (2013). Local e-government in the United States: Transformation or incremental change?. *Public Administration Review,* 73 (1), 165–175.

Pratt, A. C. (2010). Creative cities: Tensions within and between social, cultural and economic development: A critical reading of the UK experience. *City, Culture and Society,* 1 (1), 13–20.

Reddick, C. G., & Turner, M. (2012). Channel choice and public service delivery in Canada: Comparing e-government to traditional service delivery. *Government Information Quarterly,* 29 (1), 1–11.

Robinson, J. (2004). Squaring the circle? Some thoughts on the idea of sustainable development. *Ecological economics,* 48 (4), 369–384.

Sæbø, Ø., Rose, J., & Flak, L. S. (2008). The shape of eParticipation: Characterizing an emerging research area. *Government Information Quarterly,* 25 (3), 400–428.

Schaffers, H., Komninos, N., Pallot, M., Trousse, B., Nilsson, M., & Oliveira, A. (2011). Smart cities and the future internet: Towards cooperation frameworks for open innovation. In Domingue, J. (ed.) *The future internet assembly,* 431–446, Berlin Heidelberg: Springer.

Scott, A. J. (2006). Creative cities: Conceptual issues and policy questions. *Journal of urban affairs,* 28 (1), 1–17.

Scott, M., DeLone, W. & Golden, W. (2016). Measuring eGovernment success: A public value approach. *European Journal of Information Systems,* 25 (3), 187–208.

Shelton, T., Zook, M., & Wiig, A. (2015). The 'actually existing smart city'. *Cambridge Journal of Regions, Economy and Society,* 8 (1), 13–25.

Söderström, O., Paasche, T., & Klauser, F. (2014). Smart cities as corporate storytelling. *City,* 18 (3), 307–320.

Song, C., & Lee, J. (2016). Citizens' use of social media in government, perceived transparency, and trust in government. *Public Performance & Management Review,* 39 (2), 430–453.

Taylor, J. A., & Webster, C. W. R. (1996). Universalism: Public services and citizenship in the information age. *Information Infrastructure and Policy,* 5 (3), 217–233.

United Nations. Economic and social council, economic commission for Europe. (2015). *The UNECE-ITU Smart Sustainable Indicators.* Geneva: Economic and Social Council, Economic Commission for Europe. 1–13.

Webster, C. W. R., & Leleux, C. (2018). Smart governance: Opportunities for technologically-mediated citizen co-production. *Information Polity,* 23 (1), 95–110.

Yang, K., & Pandey, S. K. (2011). Further dissecting the black box of citizen participation: When does citizen involvement lead to good outcomes? *Public Administration Review,* 71, 880–892. 10.1111/j.1540-6210.2011.02417.x.

Section 2
Citizen science and co-production

17

Sharing in smart cities

What are we missing out on?

*Christopher T. Boyko, Serena Pollastri,
Claire Coulton, Nick Dunn and Rachel Cooper*

Introduction

The growth in global urbanisation and the rise of information and communications technology (ICT) over the past 20 years or so have had a profound effect on the origin of the smart city (Wang, 2018). We take the definition of the concept from Caragliu et al. who state that:

> We believe a city to be smart when investments in human and social capital and traditional (transport) and modern (ICT) communication infrastructure fuel sustainable economic growth and a high quality of life, with a wise management of natural resources, through participatory governance.
>
> *(2009, p. 50)*

Both urbanisation and ICT have created the stimulus for academics, technology companies, marketing agents and others to ask questions about, and develop solutions to, the large influxes of people to cities, the infrastructural resources that will be needed to accommodate them and further issues around big data, data openness, data security and so forth. One area where smart cities have responded to urbanisation and ICT has been around sharing. Often labelled the sharing economy (see the 'What are sharing cities?' section below), this has been seen by some as one of many 'smart' solutions to the phenomena above, giving people a new perspective on what may be considered an asset, how choice can be exercised and which platforms should be used to share.

Although the sharing economy has many advantages in the smart city, the practice of sharing in cities is not new, nor shows any signs of dying out. In fact, we argue that smart cities need to take account of the 'non-smart' forms of sharing, as they help to provide a richer picture of how people interact in cities and how we might design for increasing urbanisation and ICT development. Doing so also sheds light on those people whom the smart city may exclude, and may contribute to offering them 'rights to the city' (Lefebvre, 1968, 1991) through opportunities to engage people, places, services and things in affordable, face-to-face ways.

This chapter begins by defining sharing and then connects it to the concepts of 'cities' and 'smart' via a description of the sharing economy and case studies of 'sharing cities' from around the world. The authors then discuss what is missing from the current smart cities and sharing cities debates via the introduction of research conducted as part of the Liveable Cities project. We conclude by positing how smart cities can enable sharing practices that are both 'smart' and 'non-smart' through an emphasis on infrastructures.

What is sharing?

If we asked 100 people the question, 'What is sharing?', the majority would be able to give us a good working definition with examples, such as sharing food, like a cake, with somebody. They might also talk about getting a 'fair share', or discuss other terms like borrowing or lending. However, a small, but growing, body of literature has been more critical in its analysis and overview of sharing, suggesting that the concept is more complex than we imagine.

Belk gives one of the most-cited definitions of sharing when he states that it, 'involves the act and process of distributing what is ours to others for their use and/or the act and process of receiving or taking something from others for our use' (2007, p. 126). In this definition, sharing is a social act between two or more people. The act is distributive in that something owned by one person or group is given to, or divided amongst, others. Furthermore, an economic exchange is not necessary for this act to take place; however, McLaren and Agyeman (2015) have argued that while an economic exchange is unnecessary, the act of sharing usually is shaped by cultural, political and social norms and motivations.

Belk's definition covers more obvious examples of sharing which we are all probably well acquainted. Above, we mentioned sharing a cake, but during our lives, and from a very early age, we share any number of things, that can differ in terms of scale, ownership and tangibility. For example, as children, most of us have attended school. In this case, we are all sharing a teacher and a classroom. We do not own the classroom, the school or the teacher, but we are all sharing them. The same goes for a bus journey: we do not own the bus and we are paying to use it, but we are sharing the bus, the journey and the roads with other users. Nicholas (2012) also highlights the idea of sharing something intangible, like communication. For example, somebody may share his or her feelings with you on a subject during your bus journey or in an online setting like Twitter or Facebook.

Another issue raised by Belk's definition centres around the issue of trust and collaboration. If we return to our cake-sharing example, the sharer is presuming that people want to have some cake. These same people are also trusting that the cake tastes good and will not poison them. Thus, trust and collaboration are seen as key elements in sharing (Botsman and Rogers, 2011; Volker and Flap, 2007).

In his discussion of sharing, Belk also coined the terms, *sharing in* and *sharing out* to describe who is doing the sharing and with whom. *Sharing in* refers to people with close, kinship ties, such as a family. For example, a family may all share use of a television and watch a TV show together. *Sharing out* takes place outside these close kinship ties between individuals and groups and the wider community. An example here might be someone sharing their home, or part of their home, through services like Couchsurfing. These terms, though, do not cover the collective ownership of goods. Quilligan (2012) therefore expands on this to include the terms *public goods* and *common goods*. *Public goods* are things we all share collectively, such as services like waste management, or public parks and roads, and that are

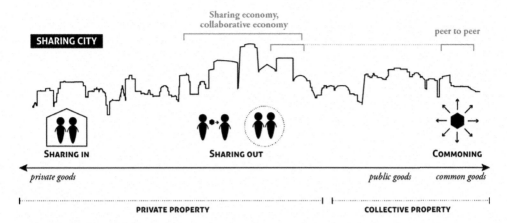

Figure 17.1 Describing sharing

Source: Chapter author(s)

managed by local authorities and governments. *Common goods* are things that we all share, but nobody owns. Here, we imagine the example of the English language. Nobody owns it, but millions of people share in its use. It is collectively managed by negotiation, norms and everyday practices and traditions.

From this discussion, we can see that sharing can take place with material and immaterial things, and that ownership is not always a necessity. Shared things can be individual, like the cake; collective, like the school; or public, like the roads. There also are common things we might share, like the air, the sunshine and the sea. Nobody owns these things, but we all share them. Again, we can view these things being shared as things, services, like the UK's National Health Service, or experiences, such as seeing a play at the theatre with 500 other people (Agyeman et al., 2013). Finally, we must acknowledge that sharing is not always free: some of the things mentioned above, like the bus journey and the theatre, have a cost associated with them. Some of them, though, including the air we breathe, are free. And, sometimes, we collectively contribute towards something we all share by paying taxes (Figure 17.1).

Figure 17.1 brings together the various categories of sharing mentioned so far along a continuum, spanning from private, informal models of sharing, to the communing of collective property, and shows how different models coexist within the city. The figure highlights how the sharing economy represents one part of the spectrum, and does not cover the complexity and richness of all forms of sharing in urban contexts.

With the growing number of definitions and expanding understanding of what constitutes sharing, a number of taxonomies and typologies have been developed (see Agyeman et al., 2013; Bazelon and McHenry, 2014; Botsman and Rogers, 2011; Lamberton and Rose, 2012; Owyang et al., 2014; Schor, 2014). These have provided more nuance to the subject, but they remain descriptive and static. As Boyko et al. (2017) argue, the taxonomies and typologies do not consider broader examples of sharing that might be relevant in cities, such as waste management and energy generation, the governance of cities and how to manage other public resources, such as parks, roads and pavements.

What are sharing cities?

That cities are places of sharing is nothing new. Specific historical circumstances may have brought individual cities into being, but the need of sharing – spaces, resources, culture or personal relations – in the public realm is ultimately one of the core *raison d'être* of any city (Sennett, 2010). Most recently, though, the term 'sharing cities' has come to signify something more specific than the full spectrum of formal and informal sharing of private and collective property described in the previous section.

In the last decade, new models of collaborative economy (Botsman and Rogers, 2011) became increasingly popular and widely adopted, especially by urban dwellers in Europe and North America, and especially by people aged under 45 years old (PricewaterhouseCoopers, 2016). In particular, much has been written – especially in popular media – on sharing economy models, in which companies provide platforms that allow individuals to share idle assets (Botsman, 2013). While the Web plays a key role in contemporary practices of sharing and collaborating (Botsman and Rogers, 2011), most models of sharing have physical interactions at their core. These practices are effectively supported by a combination of platforms in the urban and digital space (Agyeman et al., 2013; McLaren and Agyeman, 2015).

Understanding the relationship between sharing practices and the context in which they take place is key. This became particularly evident in recent years, when actors involved in city governments and sharing economies started confronting each other over regulations that often were unable to deal with these new practices. These confrontations lead to disparate results (McLaren and Agyeman, 2015). In some cities – most notoriously San Francisco – negotiations were dominated by well-established sharing economy companies, lobbying for more supportive policies (Gorenflo, 2012; Hoge, 2014). In other cases, such as Milan, Seoul or Amsterdam, local administrations took proactive actions to rethink their cities as 'sharing cities'. Interestingly, many of the cities that have been branded 'sharing cities' by their local administration are often also referred to, or aspire to be, 'smart cities' (see, for example, Lee and Gong Hancock, 2012). This is because the sharing initiatives that the city governance promotes are largely – although not entirely – dependent on networked, digital platforms.

The city of Seoul is perhaps the most striking example of how *smart* and *sharing* attributes are often married in the governance as well as the academic discourse. One of the cities with the highest rates of broadband internet and smartphone penetration in the world (OECD, 2018), Seoul has been one of the earliest adopters of online city platforms using data and diffused ICT technologies for providing and accessing both public and commercial services. On more than one level, what made it possible for Seoul to become effectively one of the first smart cities was the established, digital network infrastructures that constitute the backbone of services and platforms (Townsend, 2007).

However, the rapid pace of urbanisation also has left Seoul facing important social, environmental and resources problems. To address some of these issues, Seoul Innovation Bureau leveraged on the already-established ICT infrastructure and launched the 'Sharing City, Seoul' initiative in 2012. Funded by the city government, Sharing City Seoul, 'aims to bring the sharing economy to all Seoul citizens by expanding sharing infrastructure, promoting existing sharing enterprises, incubating sharing economy startups, utilizing idle public resources, and providing more access to data and digital works' (Johnson, 2013). A portal, called ShareHub,[1] functions as a central platform for the project from which sharing services and resources can be accessed.

Following the example of Seoul, other cities have been forming public–private partnerships to promote the collaborative and sharing economy. Some of them have joined the

Sharing Cities Alliance (https://sharingcitiesalliance.com), a 'city network operated by an independent foundation that connects cities from all continents, and fosters city-to-city learning, empowering city governments to continuously address the sharing economy'.

Generally speaking, the model that these cities seem to follow is characterised by systems of hybridised initiatives, led by, and providing, services to private citizens as well as public agencies, non-governmental organisations and private companies (Cohen and Muñoz, 2016). Economist and social theorist, Jeremy Rifkin, notes how the potential of these distributed and decentralised networks of diverse actors grows exponentially when supported by a ubiquitous nervous system of integrated Internet of Things devices, and is managed by sensors collecting big data (Rifkin, 2015). Once again, it is in the smart city that the fertile ground for the sharing city seems to be mostly found.

Examples of initiatives in smart cities

There are many examples of ICT-enabled, urban sharing with increasing numbers of Urban Sharing Organisations. A broader programme to enable smart cities by fostering collaboration between cities, industry and communities internationally can be seen in the Sharing Cities 'lighthouse' programme.[2] This is a European Union–funded project that began in 2016 with the aim of introducing replicable, urban digital solutions and collaboration models in European cities, including London, Milan and Lisbon. Solutions include retrofitting buildings, designing shared, electric mobility services and smart lampposts and creating an urban sharing platform that engages with citizens.

In Germany, Austria and Switzerland, foodsharing.de uses open source software to enable local 'food rescuers' to collect excess food from supermarkets and producers before it is consigned to the bin. The food is redistributed in local 'food baskets'. Other foodsharing initiatives, such as olioex.com, the LeftoverSwap app and cropmobster.com, are committed to ending food waste and, in turn, promoting and educating people about sustainable food production.

Smart city initiatives to minimise carbon emissions and improve air quality by encouraging alternative forms of transport have been noted in the proliferation of bike sharing schemes in cities across the world, such as London, Paris and San Francisco. This has not always been successful, as in the case of Manchester, where Mobike withdrew bikes after 'unsustainable losses from theft and vandalism' (Pidd, 2018). Other, 'smart' forms of transport, or micromobility, increasingly are becoming popular in order for travellers to easily finish their *last mile*, where people often get cabs, or the inconvenience forces them to drive. Shareable, electric scooters are now available to use in cities around the world, including Paris, Madrid, Tel Aviv and most US cities. Electric scooters also are undergoing a trial run in the Olympic Park in London, although this is limited due to their being illegal to use on the UK's road network.

In San Francisco, however, Jump (owned by Uber) has recently finalised a permit with the city authorities to enable stationless e-bikes throughout the city with users locating and unlocking the bikes via an app on their smartphone. Data from the bike (e.g., battery status, journey times, routes) are uploaded to the cloud and allow the company to maintain the bicycles and batteries seamlessly.

Beyond the much-discussed model of a 'smart sharing economy', other narratives of sharing cities can be found in the literature. In their book, *Sharing Cities*, McLaren and Agyeman (2015) describe examples of cities focusing on promoting social inclusion by helping communities to thrive (e.g., Medellín, Columbia), or by improving the shared use of urban commons (e.g. Copenhagen, Denmark).

Elsewhere, the term, 'sharing city', is commonly used to define the landscape of grassroots collaborative initiatives that spread in cities around the world. These initiatives – mapped by groups, such as OuiShare[3] and Shareable[4] – often showcase examples of community-generated, sustainable alternatives to mainstream lifestyles. Food groups, time-banks, tool libraries and other examples of collaborative solutions embed elements of sharing in the way everyday problems are solved (Meroni, 2007).

Arguably, the same platforms that have allowed the sharing and collaborative economy to proliferate could be appropriated by informal, grassroot initiatives to scale up and amplify positive examples of innovation (Manzini and Coad, 2015). In their *Hackable Cities Research and Design Manifesto*, Ampatzidou et al. (2015) describe how cities could open up their urban institutions to the public, by allowing for the appropriation of urban infrastructure digital platforms and the empowerment of citizens designing and tinkering with grassroot solutions. While on one hand, this approach may allow for innovative and effective, bottom-up innovation supported by top-down governance (also advocated in Oliveira and Campolargo, 2015), on the other hand, by overlooking the importance of informal interactions, it also runs the risk of '[subsuming] all social relations under the functionalist and commercial "city as a service" logic' (Ampatzidou et al., 2015, p. 13).

Formalised, grassroots initiatives, new examples of the sharing economy, but also informal sharing in close-knit communities and urban commons, and shared spaces and resources, often coexist and affect each other. Collectively, they define the identity of the sharing city. What these example show is that there seems to be much more to the sharing city than the digitally supported sharing economy. For this reason, in this chapter, we ask: by focusing solely on the 'smart+sharing economy' model of sharing cities, what are we missing out on?

Sharing in the city: what is missing?

To get a better sense of how people share in their cities, we conducted two workshops with local residents in Lancaster and Birmingham (UK) in March and September, 2015, respectively. We recruited 40 people in Lancaster and 22 people in Birmingham, inviting them to debate sharing with us and discuss whether sharing could help make cities more sustainable. The workshops were held in local community centres, took about three hours and consisted of two main, designed activities, influenced by co-design (Cruickshank and Evans, 2012), futures (Fry, 1999) and social innovation discourses (Jégou and Manzini, n.d; Manzini, 2010):

- Mapping formal and informal sharing practices: upon arrival at the workshop, participants were asked to write down an example of sharing and pin it to a 'sharing wall', containing categories of sharing. They then sat at tables with four–seven other participants and disclosed their sharing example to the group with additional details (e.g., who is sharing, how it is being shared). Once everyone's sharing example was heard, the participants were asked to make connections between the examples. A member of the research team facilitated discussion at each table and probed for missing, existing and potential connections.
- Envisioning scenarios for the sharing city of the future: after the first part of the workshop, participants were given the task of writing down the worst-case scenario regarding sharing in their city and then communicating it with everyone at the workshop. These scenarios were put inside a box and hidden from sight to remove negativity from the envisioning exercise. Subsequently, participants were asked to design a future city or neighbourhood that fostered sharing. They were given materials to work with, including maps, coloured blocks, tissue paper, pens, string and plastic figures of people. They also were encouraged to consider producing something that was not currently available in their scenario ('create'),

Figure 17.2 Designing the sharing city at the Lancaster workshop, March 2015
Source: Chapter author(s)

adding more of something ('amplify') or removing something they do not currently like ('destroy') (see McLaren and Agyeman, 2015, for discussion of the relevance of using terms like create, amplify and destroy in developing system changes in cities) (Figure 17.2).

Workshop analysis: sharing categories

Based on the information we collected at the workshops, we were able to identify 41 different examples of sharing across 11 sharing categories (Table 17.1). These included:

Of the 41 examples, 33 had some connection with physical assets, such as swapping clothes and sharing fruit trees. Most of these examples were organised and run through informal and communal word of mouth, rather than set up as an economic exchange, like what is encouraged through the 'smart+sharing economy'. Furthermore, only three examples depended on digital platforms to enable sharing activities, again, which is promoted in both the sharing economy and in smart cities discourses.

Workshop analysis: create, amplify, destroy

When we looked at the scenarios that participants envisioned for sharing cities of the future, we saw that new models of social, business and governance needed to be *created* to guide

Table 17.1 Sharing categories, examples of sharing and assets or practices used in sharing from the Lancaster and Birmingham sharing cities workshops

Sharing category	Example	Asset or practice
Experience	Member of Ethical Small Trade Association	Creating poetry while selling tea and hand-made cups in local community
Food	Garden share	Space to grow fruit and vegetables
Governance	Lancaster Arts Partnership 'Lancaster Arts City'	How to set up and sustain arts venues in Lancaster
Ideas	Shout About Brum	An event that brought together large scale and niche arts communities to discuss how to use city spaces to communicate Birmingham's offerings
Mobility	Lancaster Community Car Club	Sharing cars among the community
Resources	Highbury Orchard	Outdoor learning project shares orchard tools with Muddy Puddles Birmingham
Skills	Mosley and Kings Heath Shed	Older people help repair specialist items for the public
Space	Morecambe library	Sharing space with Young People Service
Things	Freegling	Re-using unwanted items
Time	Time banking	Time to give to people who need help with something
Utilities	Distributed, shared energy schemes	Energy to power items in home

primarily non-economic and communal sharing. Issues of transparency, empowerment of people to engage in civic life and trust – both in terms of trusting the person or organisation who could host any sharing networks that would be created, and trust being a value that wanted to be maintained – were discussed.

Both economic and non-economic forms of sharing needed to be *amplified* and made visible. For example, existing places and activities related to sharing should be recognised and valued, like libraries and food growing projects, and more of an effort was required to extend sharing networks into hard-to-reach communities. Also important to the participants was the informal and unstructured nature of sharing in their cities: casual meetings in places that are conducive to sharing, like parks and cafés, should be promoted over more formal sharing networks.

Finally, participants spoke about the need to '*destroy* time', suggesting that a change in mindsets was needed to prioritise the time spent contributing to the wider community versus time spent for paid working (Albinsson et al., 2012). They also mentioned that cars could be destroyed in an effort to transform busy roads into assets that people would enjoy. This included focusing on making places more walkable and cycle-friendly, and creating safer, more sustainable mobility options.

Workshop analysis: looking across the information gathered

We collated all of the information collected from the workshops, and created a visualisation to help us to understand how people, groups and institutions share in the city, and how

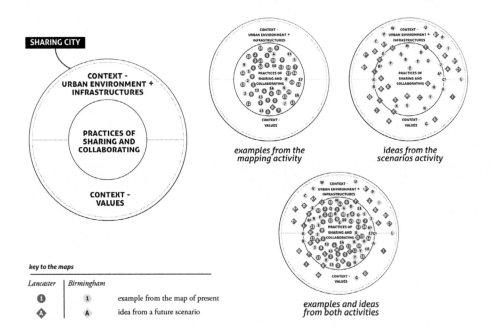

Figure 17.3 Visualised findings from the Lancaster and Birmingham sharing cities workshops
Source: Chapter author(s)

initiatives at different scales relate to each other (Figure 17.3). Across the information, we discovered some particularly interesting findings.

Figure 17.3 shows that envisioning future scenarios helped expand the conversation beyond practices of sharing, to include the context – both as urban environment and intangible values – in which these practices do, or could, take place. In other words, when describing the sharing city, participants did not just talk about what *should* happen, but also about what *should be in place* for different types of sharing to happen. Thinking broadly about sharing cities allowed us to reflect on those aspects that do not directly relate to sharing and collaborating, but that are seen as essential for a 'sharing paradigm' to emerge. The importance of infrastructure, governance and urban design interventions to indirectly promote sharing has been discussed by McLaren and Agyeman (2015), but often is overlooked in studies that are based on surveys, economic analysis and structured interviews.

In addition, we found that sharing occurs both within the realms of the sharing economy and outside it. However, most of the sharing examples could be considered 'non-smart'; that is, they did not use digital platforms, they did not originate with economic gain in mind (McLaren and Agyeman, 2015) and they did not have a critical mass to support a large amount of choices for users to feel satisfied (Botsman and Rogers, 2011). For the most part, these sharing examples were focused more on the *effectiveness* or *quality* of the sharing experience, rather than on efficiency or optimisation (Batty et al., 2012; Zanella et al., 2014), which are hallmarks of smart cities (cf. de Lange and de Waal (2019) hackable cities, which use bottom-up or crowdsourced intiatives as a method to open up urban institutions and infrastructures in the public interest). Key to the quality of many of these sharing experiences was face-to-face, social interaction: formal and informal meetings in urban spaces (e.g., chance meeting at a bus stop) and places (e.g., community centre) acted as physical hubs and bridges to connect people and their sharing practices. Indeed, these less digital 'sharing infrastructures' were seen by participants as integral to sharing and needed to be acknowledged within cities (Boyko et al., 2017).

Related to the above point, some participants mentioned that they preferred face-to-face sharing in their cities because they were not confident with using digital platforms, social media or even email. The opportunity to meet in-person and use spaces and places in their city through sharing meant that they were exercising their 'right to the city' (Lefebvre, 1968, 1991), and were not feeling isolated or disadvantaged (Dillahunt and Malone, 2015) because of their lack of digital presence or experience. Sharing 'offline' and in a 'non-smart' manner meant that participants were actively shaping, designing and making their cities in a more equitable way (McLaren and Agyeman, 2015), something that a 'smarter', more technological approach to city development and governance (Anthopoulos & Vakali, 2012) may be less able to do without disenfranchising some urban inhabitants (de Lang and de Waal, 2019).

Conclusion

This chapter started by defining sharing cities and situating it in the context of 'smart'. We argued that sharing does not necessarily have to be 'smart' to exist in smart cities, and that current discourses on smart cities appear to be missing these 'non-smart' sharing practices. We then presented some original research, done in collaboration with local experts of sharing in Lancaster and Birmingham, to shed light on what might be missing from sharing in smart cities. This work has highlighted that truly 'smart' sharing cities should be much more than cities where digitally enabled sharing practices are encouraged and made possible. The previous sections described the diversity of sharing practices that take place – or might take place – in the city. As a descriptive framework, we proposed a spectrum of sharing practices that spans from informal practices of *sharing in* among family and friends, to formalised examples of *sharing out* (which include traditional services such as libraries, as well as collaborative and sharing economies), to the *commoning* of shared resources in the public realm.

From this perspective, we can ask ourselves, what is missing from 'smart+sharing economy' cities? We believe that more needs to be done to recognise the value that 'non-smart' sharing brings to cities. In particular, the somewhat hidden and disconnected forms of sharing practices that we found in our workshops could be promoted better (Boyko et al., 2017), with infrastructures being designed and maintained that celebrate sharing (i.e., physical hubs and bridges). While the literature on sharing cities has highlighted the importance of networked, digital infrastructures enabling sharing economy initiatives, other types of physical and cultural infrastructures should be included in the conversation. By physical infrastructures, we mean roads, squares, mobility networks and spaces; cultural infrastructures may include education, governance and a value system that rewards sharing practices.

McLaren and Agyeman (2015) propose that designing sharing cities is both about '*understanding* cities as shared spaces, and *acting* to share them fairly' (p. 4). This statement strongly resonates with our experience of researching sharing cities in collaboration with those actors who are most directly involved with sharing *in* the city and who most directly benefit from sharing being part of their 'rights to the city'. As we explore the potential of smart, sharing practices in the urban environment, we must not forget that a lot of sharing and collaborating is already happening in our city. Our question moving forward, then, is: how do we design smart cities that support and encourage different, formalised and informal practices of sharing in the city and of the city?

Based on our research, we believe that cities need to leave space for informality and commoning in the day-to-day interactions between city dwellers. Just like in Ampatzidou et al. (2015) and Oliveira and Campolargo (2015), our workshops casted doubt on the view of the sharing city as a smart platform where efficiency and control are key. We see the focus of the discussion, not so much on how local government could support the development of digitally mediated *solutions* to promote sharing, but on how conditions may be created or preserved to allow informal and mundane sharing – which may have digital components to them – to thrive. For example, current neighbourhood funds might be oriented towards the creation or sustainability of community sharing practices and spaces. Doing so will foster greater inclusivity among the population and demonstrate that both 'smart' and 'non-smart' approaches can be part of a wider suite of tools and methods that allow for plurality and contradiction in the smart city (Dunn and Cureton, 2019).

Notes

1 http://english.sharehub.kr/.
2 www.sharingcit ies.eu/sharingcities/about.
3 www.ouishare.net/.
4 www.shareable.net/.

References

Agyeman, J., McLaren, D., & Schaefer-Borrego, A. (2013). *Briefing. Sharing cities*. London: Friends of the Earth.

Albinsson, P. A., & Yasanthi Perera, B. (2012). Alternative marketplaces in the 21st century: Building community through sharing events. *Journal of Consumer Behavior, 11*, 303–315.

Ampatzidou, C., Bouw, M., van de Klundert, F., de Lange, M., & de Waal, M. (2015). *The hackable city: A reserach manifesto and design toolkit*. Retrieved from www.publishinglab.nl/wp-content/uploads/2016/01/HvA_HackableCities_DEF_spreads-cover.pdf

Anthopoulos, L. G., & Vakali, A. (2012). Urban planning and Smart cities: Interrelations and reciprocities. In Álvarez, F. et al., (ed.), *The future Internet. FIA. 2012. Lecture notes in Computer Science, 7281* (pp. 178–189). Berlin: Springer.

Batty, M., Axhausen, K. W., Giannotti, F., Pozdnoukhov, A., Bazzani, A., Wachowicz, M., Ouzounis, G., & Portugali, Y. (2012). Smart cities of the future. *The European Physical Journal Special Topics, 214*, 481–518.

Bazelon, C., & McHenry, G. (2014). *Spectrum Sharing: Taxonomy and Economics*. Cambridge: The Brattle Group.

Belk, R. (2007). Why not share rather than own? *Annals of the American Academy of Political and Social Science, 611* (1), 126–140.

Botsman, R. (2013). The sharing economy lacks a shared definition. Retrieved 30 April 2015 from www.fastcoexist.com/3022028/the-sharing-economy-lacks-a-shared-definition

Botsman, R., & Rogers, R. (2011). *What's mine is yours: How collaborative consumption is changing the way we live*. London: HarperCollins Business.

Boyko, C. T., Clune, S. J., Cooper, R., Coulton, C. J., Dunn, N. S., Pollastri, S., Leach, J. M., Bouch, C. J., Cavada, M., De Laurentiis, V., Goodfellow-Smith, M., Hale, J., Hunt, D. K. G., Lee, S. E., Locret-Collet, M., Sadler, J. P., Ward, J., Rogers, C. D.F., Popan, C., Psarikidou, K., Urry, J., Blunden, L. S., Bourikas, L., Büchs, M., Falkingham, J., Harper, M., James, P. A. B., Kamanda, M., Sanches, T., Turner, P., Wu, P. Y., Bahaj, A. S., Ortegon, A., Barnes, K.,

Cosgrave, E., Honeybone, P., Joffe, H., Kwami, C., Zeeb, V., Collins, B., & Tyler, N. (2017). How sharing can contribute to more sustainable cities. *Sustainability*, 9 (5), 701.

Caragliu, A., Del Bo, C., & Nijkamp, P. (2009). Smart cities in Europe. *3rd Central European Conference in Regional Science*. Košice, Slovakia.

Cohen, B., & Muñoz, P. (2016). Sharing cities and sustainable consumption and production: Towards an integrated framework. *Journal of Cleaner Production*, *134*, 87–97.

Cruickshank, L., & Evans, M. (2012). Designing creative frameworks: Design thinking as an engine for new facilitation approaches. *International Journal of Arts and Technology*, 5 (1), 73–85.

de Lang, M., & de Waal, M. (2019). *The hackable city: Digital media and collaborative city making in the network society*. Singapore: Springer.

Dillahunt, T. R., & Malone, A. R. (2015). *The promise of the sharing economy among disadvantaged communities*. CHI, April 18–23, Seoul, Republic of Korea.

Dunn, N., & Cureton, P. (2019). Frictionless futures: The vision of smartness and the occlusion of alternatives. In S. Figueiredo, S. Krishnamurthy & T. Schroder (eds.), *Architecture and the smart city* (pp. 17–28). London: Routledge.

Fry, T. (1999). *A new design philosophy: An introduction to defuturing*. Sydney: University of New South Wales Press.

Gorenflo, N. (2012). 'San Francisco announces sharing economy working group'. Retrieved 30 April 2015 from www.shareable.net

Hoge, P. (1 May 2014). 'Critics slam Mayor Lee's phantom "Sharing Economy" working group. Retrieved 30 April 2015 from www.bizjournals.com/sanfrancisco/blog/techflash/2014/05/mayor-lees-sharing-economy-task-force-never-met.html

Jégou, F., & Manzini, E. (n.d.). LOLA brochure. Retrieved 12 November 2013 from www.sustainable-everyday-project.net/blog/library-lola-looking-for-likeky-alternatives/lola_brochure.

Johnson, C. (5 August 2013). Is Seoul the next Great Sharing City? Retrieved 2 February 2018 from https://ourworld.unu.edu/en/is-seoul-the-next-great-sharing-city

Lamberton, C. P. & Rose, R. L. (2012). When is ours better than mine? A framework for understanding and altering participation in commercial sharing systems. *Journal of Marketing*, *76*, 109–125.

Lee, J. -H., & Gong Hancock, M. (2012). *Toward a framework for smart cities: A comparison of Seoul*, San Francisco and Amsterdam. Research paper, Yonsei University and Stanford University. Retrieved 27 November 2019 from www.estudislocals.cat/wp-content/uploads/2016/11/ComparisonSEOUL-SF-AMSTERDAM.pdf.

Lefebvre, H. (1968). *Le droit à la ville*. Paris, FR: Anthropos.

Lefebvre, H. (1991). *The production of space*. D. Nicholson-Smith, Trans. Oxford, UK: Blackwell.

Manzini, E. (2010). Small, local, open, and connected: Design for social innovation and sustainability. *Journal of Design Strategies*, *4*, 8–11.

Manzini, E., & Coad, R. (2015). *Design, When Everybody Designs: An Introduction to Design for Social Innovation*. Cambridge, Massachusetts: MIT Press.

McLaren, D., & Agyeman, J. (2015). *Sharing cities: A case for truly smart and sustainable cities*. Cambridge: MIT Press.

Meroni, A. (ed.), (2007). *Creative communities: People inventing sustainable ways of living*. Milan: Polidesign.

Nicholas, A. J. (2012). Sharing and web 2.0: The emergence of a keyword. *New Media & Society*, *15*, 167–182.

OECD (2018). Korea. Retrieved 30 January 2018 from https://data.oecd.org/korea.htm#profile-innovationandtechnology

Oliveira, Á., & Campolargo, M. (2015). From Smart Cities to Human Smart Cities. *2015 48th Hawaii International Conference on System Sciences*, 2336–2344. https://doi.org/10.1109/HICSS.2015.281

Owyang, J., Samuel, A., & Grenville, A. (2014). Sharing is the new buying. In *How to win in the collaborative economy*. Vancouver: Vision Critical. Retrieved 27 November 2019 from www.slideshare.net/jeremiah_owyang/sharingnewbuying.

Pidd, H. (5 September 2018). 'Mobike pulls out of Manchester citing thefts and vandalism'. Retrieved 24 April 2019 from www.theguardian.com/uk-news/2018/sep/05/theft-and-vandalism-drive-mobike-out-of-manchester

PricewaterhouseCoopers (2016). *Future of the sharing economy in Europe 2016*. Retrieved 31 January 2018 from www.pwc.co.uk/issues/megatrends/collisions/sharingeconomy/future-of-the-sharing-economy-in-europe-2016.html

Quilligan, J. B. (2012). Why distinguish common goods from public goods? Retrieved 21 April 2015 from http://wealthofthecommons.org/essay/why-distinguish-commongoods-public-goods#footnote1_qy0tikc

Rifkin, J. (2015). *The zero marginal cost society*. New York: Palgrave Macmillan.

Schor, J. (2014). Debating the sharing economy. Retrieved 12 November 2015 from www.greattransi tion.org/publication/debating-the-sharing-economy

Sennett, R. (2010). The Public Realm. In G. Bridge & S. Watson (eds.), *The Blackwell City Reader* (2nd ed., pp. 261–270). Chichester: John Wiley & Sons.

Townsend, A. M. (2007). Seoul: Birth of a broadband metropolis. *Environment & Planning B, 34,* 396–413.

Volker, B., & Flap, H. (2007). Sixteen million neighbors: A multilevel study of the role of neighbors in the personal networks of the Dutch. *Urban Affairs Review, 43* (2), 256–284.

Wang, D. (2018). *Not a dashboard, not a sandscastle*. Doctoral thesis, Lancaster University Lancaster: UK.

Zanella, A., Bui, N., Castellani, A., Vangelista, L., & Zorzi, M. (2014). Internet of Things for smart cities. *IEEE Internet of Things Journal, 1* (1), 22–32.

18

Taxonomy of environmental sensing in smart cities

Christian Nold

Introduction

Much of the recent scholarship on smart cities has wrestled with the label of 'smartness' to try and disentangle its discourse and imaginaries (Vanolo, 2013). But what if we leave this to the side and focus on how cities are already being shaped? This is not to disregard the effect of imaginaries but to place a focus on observable urban environmental practices.

The last decade has brought radical transformations in the way environmental sensing of pollutants is taking place. The classic model that has been adopted by governments is to erect expensive, stationary hardware at the sides of urban roads to measure data to identify long-term trends and conform to regulatory standards. Typically, this hardware will cost around GB£10,000 (US$13,000). Yet recently, new kinds of low-cost (£150/$180) and portable 'smart' devices have emerged, built by hobbyists, entrepreneurs and researchers to measure air and noise pollution. These devices are targeted at members of the public who are asked to take part in environmental data gathering by installing them in their homes or carrying them with them every day. The best-known example of this is the Safecast radiation monitoring network (Safecast, 2011). This emerged in response to the Fukushima nuclear disaster where volunteers built hardware and provided vital data for the public, while the government was criticised for lack of sensor coverage and public data access. This incident became an exemplar of bottom-up smart technology and precipitated a global growth in the availability and use of similar sensing devices developed using crowdfunding platforms such as Kickstarter and often designed without input from environmental scientists or experts. A key component of this growth has been an academic and industry discourse valorising these devices as best practice exemplars of the smart cities and the Internet of Things (IoT). Fernandez (2013) for example argues 'the smart city becomes real when people can deal with open technologies to build their own public infrastructure for environmental monitoring' (p. 44). In the same vein the *Fast Company* magazine suggests these sensors are the 'perfect example of how Internet of Things will work in the future' (Captain, 2016, para. 8). The argument is that low-cost sensors make pollution 'visible' for non-expert individuals, allowing them to see their personal exposure and avoid polluted areas as well as change their own behaviour to produce less pollution. On a collective level the devices are seen as producing vast quantities of data that can be aggregated to produce environmental datasets that are not available to governments and institutions, the assumption being that this leads to new knowledge that creates 'information power' (Carton and Ache, 2017). According to Bria, each sensor box 'empowers citizens to improve urban life through capturing and analysing real-time environmental data' (Bria, 2014, p. 2). In return the people are said to become 'Smart citizens' that can renegotiate their relationship with governments and institutions (Townsend et al., 2010; Hemment and Townsend, 2013; Hill, 2013; Kresin, 2013). At the heart of this argument is a positivist knowledge paradigm where technological data are framed as neutral and directly leading to knowledge. In effect, 'more' data are seen as creating 'more' environmental knowledge.

Yet, this argument doesn't reflect on the fact that these low-cost sensing devices might actually be fundamentally different from institutional environmental sensors (Kumar et al., 2015). In fact, the low-quality data generated by these sensors are often not comparable to existing data standards and the sensing process often takes place in ad-hoc settings and contexts. Despite the growing public awareness and importance of these devices, there has been little research on them. Most studies focus on testbed evaluations of sensor hardware (Choi et al., 2009; Mead et al., 2013), yet few examine how sensing deployments take place in practice with communities. Bell and Dourish (2006) suggest that technologists largely frame their devices as future technologies that will function perfectly within a 'proximate future' that is just around the corner. The effect being that technological artefacts are seen as sketches and the community practices they generate are treated as 'irrelevant or at the very least already out-moded' (p. 134). In this way, smart sensors are similarly framed as anticipatory sketches of a future technology (Kinsley, 2012) and the material practices they create in the present are largely not evaluated. As a result, only a few researchers (Pritchard et al., 2018; Zandbergen, 2017) have probed the ways these sensors are creating new kinds of sensing practices. What is needed is an overview of the material practices these smart sensing devices are creating for communities.

Case studies

To engage this gap, this text offers a survey of six case studies of smart environmental sensing projects and devices see Table 18.1.[1]

Table 18.1 shows the breadth of the six case studies, including sensors focused on a range of pollutants/phenomena, funding contexts and targeted participants. Yet, what makes them comparable is their focus on low-cost sensors whose data are uploaded to the internet, their focus on engaging participants and a common reference to notions of 'smartness' and the IoT. The aim of comparing them is to identify the range of sensing practices and develop a taxonomy that might highlight patterns across the devices.

The survey is based on my PhD study (Nold, 2017) that involved multiyear ethnographic observations across Europe of the AirProbe, WideNoise, Air Quality Egg and Smart Citizen Kit. Due to my paid position on the EveryAware team, I was involved in configuring the AirProbe and WideNoise devices and had privileged access to the coordinators of the Air Quality Egg and Smart Citizen Kit. This allowed me to follow the sensing devices across their lifetime (2011–2014) from design, usage with participants and later academic and policy outputs. In addition, the research involved extensive interviews with participants and users of the sensing devices. The Meet je stad! and Pacco Test case studies are based on interviews with the respective project coordinators. Theoretically, the approach of this survey is based

Table 18.1 Overview of the six case studies with the different phenomena being sensed, their funding contexts and participants

Project/device	What is sensed?	Funding context	Participants
AirProbe	Air pollution (carbon monoxide, nitrogen dioxide, gasoline and diesel sensor, ozone, temperature humidity)	Academic	Paid recruits
WideNoise	Sound level, geo location	Academic	Public/recruits
Air Quality Egg	Air pollution (nitrogen dioxide, carbon monoxide, temperature, humidity)	Commercial Kickstarter	Public
Smart Citizen Kit	'Ambient' (nitrogen dioxide, carbon monoxide, temperature, humidity, light & sound level, Wi-Fi)	Commercial Kickstarter	Public
Meet je stad!	Temperature, humidity, geo location	Community	Residents of Amersfoort
Pacco Test	Water quality (pH, dissolved oxygen, electrical conductivity, oxidation reduction potential, temperature)	Community	River stakeholders

on a framework from actor-network theory (Latour, 1987) with attention to the concept of the 'device' as a gathering agent of different agendas (Law and Ruppert, 2013). This social science approach focuses on practices as relationships enacted by humans, technologies and other kinds of nonhuman actors. This empirical approach is particularly useful for highlighting tensions between rhetoric and observed reality.

AirProbe

This device was built as part of 'EveryAware',[2] a European Union-funded academic research project. It combined a hardware sensor box, neural net calibration model, smartphone app and an online gaming platform. Yet, throughout its development there was ambiguity about the goals of the project and what the device should be sensing. Some of the researchers were trying to generate data for air quality modellers as a public health policy instrument, while others wanted to use it as an online platform for running experiments on the public. The final implementation of the sensor system was a mixed reality game where paid participants were asked to carry smart sensors in multiple European cities. Other participants were playing an online game of guessing pollution levels at different city locations. What mattered to the researchers was that they could alter the rules of the game, such as the in-game currency pay-out rate, to see how the users would change their behaviour. Yet, the hardware sensing devices proved unreliable and provided misleading data to the participants. This was so pronounced that some of the participants reported that they had 'learnt' that air pollution was higher in parks than next to busy roads. For many of the researchers this was not a problem, since the sensors were merely a vehicle for running tests on the participants inside a virtual laboratory. In this case, the low-cost and networking capabilities of the smart sensors were crucial for enabling experimentation on the public. To summarise, in this case study the smart sensors were being used for '*running tests on the public*'.

Air Quality Egg and Smart Citizen Kit

These two commercial devices were based on the same gas sensors and are functionally very similar.[3] They were both promoted to the public by the project coordinators as visualising environmental exposure. Yet, in both cases the devices produced uncalibrated data that users found hard to understand or use. Both devices provided raw electrical resistance values from the gas sensors rather than pollutant concentrations in the parts-per-million or billion ranges, which prevented users from comparing their measurements against official data. The Air Quality Egg coordinators publicly acknowledged 'that any single datapoint that we collect has low value while the breadth, resolution, and update frequency of the network has high value' (Air Quality Egg, 2012). Interestingly, both projects placed the focus not on the environmental data but on the smart networking capabilities that allowed data to be sent to the network. A promotional video for the Smart Citizen Kit described the usage like this: 'I use my kit everyday, normally I take it in the morning just to have a global awareness of what is going on' (Acrobotic Industries, 2013). While another user suggested, 'I check it every day to see how the information is updated and how the data is uploaded for other people to see' (ibid). Rather than focusing on nitrogen dioxide as a gas with health impacts, both devices framed the data as abstract network traffic that demonstrates the integrity of a smart network. The Air Quality Egg was partly funded by a company that was hosting the project's data in order to demonstrate the potential of their IoT platform. While the language of environmental pollution was used to encourage people to back the devices, the focus of the projects was on prototyping globally distributed smart sensing networks. In this case study we see environmental sensors being used as a vehicle for '*deploying smart networks*'.

Meet je stad! (Measure your city)

This project involves residents of the small town of Amersfoort in Holland, exploring data in relation to climate change. The focus is on the process, experience and learning involved in creating one's own sensors rather than trying to measure specific pollutants. The project coordinators highlight the importance of self-discovery and fun without predefined goals and conceptualise the sensors as building blocks:

> if you have a box of Lego, you start building whatever the Lego allows you to build and you get better by doing it and you have to start small and just tinker with it, only then you start to appreciate the capabilities that you have.
>
> *(author interview)*

They suggest this sensor 'tinkering' or 'fiddling' provides the participants with 'sensor literacy' that extends their skills. Participants start with basic temperature and humidity sensors rather than complex air quality sensors since the data are not considered as important as the learning process itself. In the interview, the organisers contrasted their approach with scientific methods that inhibit participants because they do not have specialist knowledge to get started with sensing. They also argue that their approach enables participants to critically assess institutional environmental data and builds social relations that are key to developing climate resilience at the level of a town. In this case study, 'sensing' is used to provide practical skills and politicise the generation of knowledge to '*create community resilience*'.

257

Pacco Test

This project's goal was to create a sensing device that would make water quality understandable and allow communities and stakeholders in Brussels to take care of local water bodies. Unfortunately, the project coordinators were overwhelmed by the technical challenge of building and calibrating the sensing device. They were disappointed because they felt that it could not measure high-enough quality data and after a year of development decided to abandon the device. Nevertheless, the team were pleased that the project process had managed to gather a coalition of actors to manage a local pond as a common resource. The coordinators suggest that the development of the sensor brought local residents, politicians, councillors and technical experts together and agree that maintaining the pond was not just a technical or managerial topic but a 'community thing'. The project prototyped a novel governance system that designated a resident with the role of 'pond master'. 'When there's a change in the water parameters that is alarming people, then there is a roadmap on how to intervene and who needs to intervene' (author interview). The concept of the Pacco Test was that certain data thresholds would dictate actions such as turning on a fountain to increase the oxygen in the pond. The sensor development process meant that competing interest groups such as fishermen and ecologists had to work together to communally agree governance thresholds for the pond. Despite the abandonment of the sensor hardware, the charitable funders perceived the project as a success because it had gathered such a broad range of actors together. In addition, a politician who had taken part in the project suggested that it would support their effort to designate the area as a smart city. In this case study, the smart sensors were used for '*gathering coalitions*'.

WideNoise

WideNoise is a free smartphone app that creates geo-located sound measurements and collects meta data via sliders and textual tags.[4] The device has a complicated history having been designed as a commercial demo and then adopted as an instrument within academic research. The technical ability of device to sense decibel is poor, but it is easy to use. As part of the EveryAware research project, the app was deployed in relation to Heathrow airport in London, where it was framed as measuring aircraft noise as a contentious issue. The official metrics used to legislate noise at Heathrow airport rely on modelled averages that do not take into account the sudden sensory impact of the loud aircraft. This gap meant that local residents and activists welcomed the app as a tool for demonstrating what they saw as the 'real noise' of the aircraft. Most of the local participants used the app to try and capture the sensory disturbance of the loud flights.

> I was going to take some quiet ones [measurements] and but don't want them to just pull down the average. I wanted to stress the loudness. That's what Heathrow already has is averages. I thought we were trying to say in reality the loud noise that we are in.
>
> *(interview quoted in Nold, 2017)*

The way the participants used the app was to capture peak intensities, which they felt better represented the sensory impact of the aircraft rather than the airport's averaging metric. Furthermore, the pressure group in their press release argued that 'the number of people logging readings and the passion of those contributing at community meetings demonstrates how people are worn down by the noise from Heathrow' (HACAN ClearSkies, 2012). The pressure group didn't focus on the decibel content of the data as environmental knowledge. Instead, for both the local participants and the pressure group the physical practice of gathering environmental data itself became a political act. In this case study, smart sensors were not used for their epistemic content but for '*creating political pressure*'.

Discussion

This survey of six case studies has demonstrated that smart cities and the IoT are creating diverse and novel environmental sensing practices that are intertwined with a range of existing agendas. Smart sensors are:

- running tests on the public,
- deploying smart networks,
- creating community resilience,
- gathering coalitions,
- creating political pressure.

This list of practices is not meant as an exclusive taxonomy of environmental sensing, but represents archetypal practices that are recognisable across a variety of other sensing projects. What is striking is the range of radically different practices the environmental sensors support and generate. Yet, all the case studies have one thing in common: the quality of the data from the environmental sensors was not the focus of the projects. In none of the studies do we see the sensor being used in a straightforward way to represent the state of the environment. None of the case studies created meaningful visualisations of environmental exposure that would allow people to reduce their health impact. There was no simple translation from a measured data point towards more knowledge about the environment in time or space. Instead, the sensors functioned as part of larger project goals such as running academic experiments, building commercial networks, creating community resilience, gathering local coalitions or creating political impacts. Nevertheless, all the projects described themselves as using sensors to gather knowledge about the environment.

So, what is going on here? Many people would not recognise experimenting on the public or building technical networks as 'environmental'. Indeed, smart sensing devices offer something radically different from existing environmental sensors and institutional datasets produced by industry and governments. Instead of creating certainty about health effects or locating pollutants, these devices are enacting new forms of smart city environments that I have previously termed 'neo-environmental sensing' (Blok *et al.*, 2017). What matters with neo-environmental devices is their low-price, modularity, networking, tinker-ability, public engagement and high publicity impact. Their unique ability is to enact new kinds of 'environments' that are highly intentioned with a variety of agendas that stretch beyond pollutants. Fundamentally their aims are to enrol people and actors into sensing practices that lead to these different agendas being fulfilled. Across the case studies, the suggestion is not that the sensors or the data were 'useless' but rather that they function as social and institutional agents that drive complex agendas. These findings challenge the epistemic framing of these devices that I examined earlier. What is needed is a new language for articulating the complex and subtle enactments of the environment taking place in these sensing practices that would benefit the broader smart cities and environmental sensing literatures.

Yet, this survey also highlights political and ethical questions about which kinds of environmental practices should be enacted. Is it ethical to involve the public on the promise of measuring pollutants when they are only generating arbitrary data, or treating them as subjects of a behavioural experiment? Yet on the contrary, in the Wide-Noise study, a group of activists managed to transform a technically inferior measuring device into a useful tool for communicating their reality of noise by focusing not on its epistemic content but its political potential. Some of the practices from this survey seem to empower communities while others mislead. This range of practices has the potential to have dramatic impacts on vulnerable pollution affected communities. Yet, these impacts cannot be identified merely by analysing the discourse of smart cities or the possibilities and limitations of the sensor hardware. Instead these differences only come to the fore through an ethnographic focus on people's practices with smart technologies. The issue revolves around the extent to which participants can take ownership of smart devices that are by design: technically complex, networked and institutionally distributed. Further research is needed to identify what allows some people to take control of these devices and develop good practice guidelines for ethical environmental sensing in smart cities. A helpful starting point might be to reject claims that smart sensing devices create neutral knowledge and engage with processes of critical 'tinkering', as suggested by the Meet je stad! study. Ultimately, it is only by focusing on 'sensing practices' that we can understand the ways that smart city technologies have already changed our world.

Acknowledgements

This research has been supported by the European Commission via the DITOs project under the grant agreement No. 709443 as well as EU RD contract IST-265432 and the EPSRC Institutional Sponsorship awards to UCL 2011-12 (Grant References EP/J501396/1 & EP/K503459/1).

Notes

1 These case studies are derived from my PhD thesis (Nold, 2017) as well as two recent studies, Meet je stad! and Pacco Test, that will be described in more detail in forthcoming papers.
2 http://cs.everyaware.eu.
3 The case studies refer to the original Kickstarter versions of the Air Quality Egg and Smart Citizen Kit. Later versions of the devices use different sensors and hardware configurations.
4 This case study refers to version 3.0 of WideNoise later renamed WideNoise Plus as used during the EveryAware project.

References

Acrobotic Industries. 2013. "The Smart Citizen Kit: Crowdsourced Environmental Monitoring." *Kickstarter.* www.kickstarter.com/projects/acrobotic/the-smart-citizen-kit-crowdsourced-environmental-m.
Air Quality Egg. 2012. "Why Sensor Calibration and Precision Is the Wrong Conversation." 2012. www.kickstarter.com/projects/edborden/air-quality-egg/posts/208180?
Bell, Genevieve, and Paul Dourish. 2006. "Yesterday's Tomorrows: Notes on Ubiquitous Computing's Dominant Vision." *Personal and Ubiquitous Computing* 11 (2): 133–143. doi: 10.1007/s00779-006-0071-x.
Blok, Anders, Kelton Minor, Rolien Hoyng, Marquet Clément, Kelton Minor, Christian Nold, and Meg Young. 2017. "Data Platforms and Cities." *Tecnoscienza* 8 (2): 175–219.

Bria, Francesca. 2014. "Digital Social Innovation: Interim Report." https://waag.org/sites/waag/files/public/media/publicaties/dsi-report-complete-lr.pdf.

Captain, Sean. 2016. "How The World Will Transform Once There Are Environmental Sensors Everywhere." *Fast Company*. www.fastcoexist.com/3054781/elasticity/how-the-world-will-transform-once-there-are-environmental-sensors-everywhere.

Carton, Linda, and Peter Ache. 2017. "Citizen-Sensor-Networks to Confront Government Decision-Makers: Two Lessons from the Netherlands." *Journal of Environmental Management* 196 (July). Elsevier Ltd: 234–251. doi: 10.1016/j.jenvman.2017.02.044.

Choi, Sukwon, Nakyoung Kim, Hojung Cha, and Rhan Ha. 2009. "Micro Sensor Node for Air Pollutant Monitoring: Hardware and Software Issues." *Sensors* 9 (10): 7970–7987. doi: 10.3390/s91007970.

HACAN ClearSkies. 2012. "Do-It-Yourself Noise Challenge to BAA: Wherever You Are, Download the Widenoise App and Use It on Planes." *Neighbournet.Com*. http://neighbournet.com/server/common/conhrw158.htm.

Fernandez, Manu. 2013. "Smart Cities of the Future?" In *Smart Citizens*, edited by Drew Hemment and Anthony Townsend, 43–46. Manchester: FutureEverything.

Hemment, Drew, and Anthony Townsend, eds 2013. "Smart Citizens". *Manchester: FutureEverything.* http://futureeverything.org/wp-content/uploads/2014/03/smartcitizens1.pdf.

Hill, Dan. 2013. "On the Smart City; Or, a 'manifesto' for Smart Citizens Instead." *City of Sound*. www.cityofsound.com/blog/2013/02/on-the-smart-city-a-call-for-smart-citizens-instead.html.

Kinsley, Sam. 2012. "Futures in the Making: Practices to Anticipate 'Ubiquitous Computing.'." *Environment and Planning A* 44 (7): 1554–1569. doi: https://doi.org/10.1068/a45168.

Kresin, Frank. 2013. "A Manifesto for Smart Citizens." *Waag Society*. http://waag.org/en/blog/manifesto-smart-citizens.

Kumar, Prashant, Lidia Morawska, Claudio Martani, George Biskos, Marina Neophytou, Silvana Di Sabatino, Margaret Bell, Leslie Norford, and Rex Britter. 2015. "The Rise of Low-Cost Sensing for Managing Air Pollution in Cities." *Environment International* 75 (Febuary). Elsevier Ltd: 199–205. doi: 10.1016/j.envint.2014.11.019.

Latour, Bruno. 1987. *Science in Action: How to Follow Scientists and Engineers through Society.* Cambridge, MA: Harvard University Press.

Law, John, and Evelyn Ruppert. 2013. "The Social Life of Methods: Devices." *Journal of Cultural Economy* 6 (3): 229–240. doi: 10.1080/17530350.2013.812042.

Mead, M.I., O.a.M. Popoola, G.B. Stewart, P. Landshoff, M. Calleja, M. Hayes, J.J. Baldovi et al. 2013. "The Use of Electrochemical Sensors for Monitoring Urban Air Quality in Low-Cost, High-Density Networks." *Atmospheric Environment* 70 (May). Elsevier Ltd: 186–203. doi: 10.1016/j.atmosenv.2012.11.060.

Nold, Christian. 2017. "Device Studies of Participatory Sensing: Ontological Politics and Design Interventions." UCL.

Pritchard, Helen, Jennifer Gabrys, and Lara Houston. 2018. "Re-Calibrating DIY: Testing Digital Participation across Dust Sensors, Fry Pans and Environmental Pollution." *New Media & Society*. doi: 10.1177/1461444818777473.

Safecast. 2011. "Safecast: Open Environmental Data for Everyone." *Safecast.* http://blog.safecast.org.

Townsend, Anthony, Rachel Maguire, Mike Liebhold, and Mathias Crawford. 2010. "A Planet of Civic Laboratories: The Future of Cities, Information, and Inclusion." *Palo Alto.* http://iftf.me/public/SR-1352_Rockefeller_Map_reader.pdf.

Vanolo, Alberto. 2013. "Smartmentality: The Smart City as Disciplinary Strategy." *Urban Studies* July. doi: 10.1177/0042098013494427.

Zandbergen, Dorien. 2017. "'We Are Sensemakers': The (Anti-) Politics of Smart City Co-Creation." *Public Culture* 29 (3_83): 539–562. doi: 10.1215/08992363-3869596.

19

Co-creating sociable smart city futures

Ingrid Mulder and Justien Marseille

Introduction

> The Future is already here, it is not very evenly distributed.
>
> *(Gibson, 2001, p. 61)*

The challenges of tomorrow's society demand new ways of innovation: a shift in thinking, doing, and organizing. Not only are new strategies, ideas, and ways of organization needed to cope with societal challenges, but also co-creative partnerships demonstrating a sustainable relationship to make a transforming society happen. It is not about who drives, but finding a mutual drive (Mulder, 2014, p. 573). The biggest challenge smart cities face is not the technology, but having an open mindset and a participatory attitude to rethink our future is far more challenging (Mulder, 2014, p. 573). This is in keeping with alternative approaches to the smart city to put the citizen perspective central (e.g., Cardullo and Kitchin, 2019; De Lange and de Waal, 2019; Foth et al., 2015; Mulder, 2015b). Differently put, smart cities need smart citizens.

Living labs are generally embraced as the answer to start co-creative city making, deriving that innovative and smart solutions only work when they fit in with and arise from people's daily practices. Elaborating upon living lab experiences from the very beginning, we refer to a living lab as an experiential, creative social space open to collaboratively designing and experiencing future contexts (Schumacher and Niitano, 2008). Policy makers and citizens can use living labs to design, explore, experience, and refine new policies and regulations in real-life scenarios before they are implemented. It can be said that what distinguishes the living lab approach over traditional user-centric methodologies and other cross-disciplinary approaches on innovation is its multi-contextual sphere in which co-creation with users takes place. We therefore, refer to a living lab as a living network of real people with every-day experiences, allowing partners to co-create in context (Mulder and Stappers, 2009).

Within the smart cities methodologies to co-create alternative futures, this chapter emphasizes living labs as a space or platform for participation and co-creation (Mulder and Stappers, 2009). It envisages that both the participatory turn and the digital transformation can shape a more future fit, *sociable* smart city; it adheres to the positioning of living labs addressed above, though elaborates upon bottom-up innovation processes in the context of social innovation (e.g., Schachter et al., 2012). The corresponding living methodologies and the living lab practices need to address these social and dynamic aspects as well as the future context to inform bottom-up participatory innovation paradigms. The iterative character of the used approach is quite similar to the Form-IT approach of Ståhlbröst (2008), though our focus is human-centred rather than user-centred. We take the view that co-creation is a collaborative effort of engaging multiple stakeholders and is not limited to involving users, and nor do we address the industrial research and development processes in related to technology innovation, competitive advantages, as well as commercialization of new products and services. However, our focus is on social innovation, or on urban living labs rather than an industry-led living lab.

The next section introduces Citylab Rotterdam, which is the centre of the living lab facilities, where students, researchers, and teachers, co-create with Rotterdam-based companies, civil agents, and citizens and rethink the future of Rotterdam. The accompanying living labbing approach embedded in the local urban context emphasizes the human scale in a shared process of knowledge production in which participants collaboratively envision desired futures through participatory prototyping (Brodersen et al., 2008; Carayannis and Campbell, 2012; Mulder, 2014; van Waart et al., 2016). The platform for bottom-up participatory innovation has been established through a longstanding collaboration with making and prototyping in education, and exemplifies the freedom of experimentation (Mulder, 2015a). In this chapter we introduce four labs that aimed for capacity building on future thinking skills, and illustrate the value of future-making activities in participatory innovation. We conclude with a discussion and reflect on lessons learnt to accelerate participatory innovation paradigms.

Platform for bottom-up participatory innovation

Living labbing the Rotterdam way, as introduced in Mulder (2012), refers to practicing a participatory innovation approach by making use of co-creative labbing loops and the facilities in Citylab Rotterdam, while taking into account the social infrastructure of Rotterdam. Therefore, students are explicitly trained in understanding the context and value experimentation (Mulder and Stappers, 2009). Local creative industry supports students in the development process, enabling the various small loops into co-creative looping of participatory innovation. The living lab practices vary in scale as well as duration. Next to that, we deliberately distinguish between practices within educational context and within the real-life context, between down-to-earth learning of skills and releasing existing paradigms, between here-and-now and future paradigms, even though these various co-creative loops are related. Herewith the living lab approach is about capacity building for future thinking.

Living labbing the Rotterdam way

Citylab Rotterdam is a creative hotspot open to citizens enabling making and prototyping, a space for participation and co-creation. The lab has an official FabLab motivated by educational needs to prepare students for new trends in digital fabrication and interactive prototyping, which has been extended with an Internet of Things-lab with a strong emphasis on electronic and sensor devices, and an Applab in which meaningful applications are designed making use of open data.

Developing prototyping skills

The lab provides an infrastructure for experimenting with sensor technology (Sensorlab), digital manufacturing (Fablab), and open data (Applab). In order to fully exploit the lab in our design education and our living lab practices, students are prepared to use the machines and materials of the lab on their own, and to create and use (physical) prototypes in a safe and confident way (Mostert-vanderSar et al., 2013). Not only our own students make use of the lab, also other students and professionals have found the lab as work-, play-, and discussion place (Figure 19.1). Working in the lab, with different peers, stimulates the critical debate on material use and conceptual design issues.

Equally important to providing the technical capabilities, we are concerned at providing appropriate methods to employ these capabilities. These methods are fundamentally rooted in the human-centred approach of design thinking, as popularized by Brown (2008). We have termed these methods 'agile rapid prototyping', referring to both design (prototyping) and software engineering (agile) though sufficiently distinct from 'rapid prototyping' that is used in design as a synonym for 'additive manufacturing' technologies. Those methods essentially entail working on projects in short iterations that include designing, physical prototyping, and (end-user) testing. In our design education, we make use of the lab in both structured and unstructured formats— including electives, minors, and applied research projects. Methodically, design education in our lab focuses on the new challenges of interdisciplinary proficiency in rapid prototyping technologies, the mastery of agile design processes, and co-creating meaningful designs together with real stakeholders (Brower et al., 2007).

Co-creative labbing approach

Each labbing activity starts with framing and describing the scope of the respective lab before exploring the context with an open mindset. A diverging activity is followed by a converging activity to bring structure by analyzing the current findings and using methods like clustering and categorization. The next step is again a diverging activity to generate ideas, followed by converging ideas into concepts. The process continues with making and prototyping to facilitate the co-creative labbing activity. After co-creating several prototypes, the co-creative labbing loop starts over again. Students learn to take risks, to experiment, and most importantly, they learn to make mistakes.

Figure 19.1 The empty space for co-creation (left) and the active making and prototyping activities (middle, and right)

Source: Chapter author(s)

Interface to the city

Although Citylab Rotterdam is a prototyping workshop for students in the first place, it is also an active learning environment for practicing making, co-creation, and participatory design skills; though not only for students, the lab is open to everyone. The lab's open identity contributes to an inspiring learning environment. Teachers and researchers from different educational institutes have found the lab to get inspired with new techniques, design, and methods for educational purposes. Though the lab is not just an inspirational zone, making digital fabrication accessible and engaging citizens and professionals through a participatory bottom-up approach of making and prototyping clearly contributes to their capacity building (Pucci and Mulder, 2015).

Although, our practices rely on longstanding experience of training students in making, prototyping, and co-creating with various stakeholders, having students experienced in gathering, filtering, and recognizing patterns of trends and upcoming themes as a part of that particular co-creation and prototyping process is not that straightforward. The living lab practices, therefore, pay significant attention to training, insight-giving, and anticipation capabilities enabling society to change. In this we aim to raise awareness for relevant trends and possible futures before they become part of the mainstream and common practice, to stay ahead of what is already happening.

Co-creating futures

In the current section, we elaborate upon our conceptual framing within future studies, with a particular focus on gathering and selecting weak signals for change and the clustering of the signals into trends.

New for one can be old fashioned for another: the innovation adoption cycle

In order to prepare students and participants in co-creating alternative futures, trend research on possible futures is key. As something totally new for one person can be already old fashioned for another, the future thinking approach includes the understanding and acknowledgement of the different roles in the process of adopting and adapting the new. These adapting roles and characters are described as innovators, early adopters, early and late majority and laggards by Rogers (2003), while Moore (2014) describes these groups as technology enthusiasts, visionaries, pragmatists, conservatives and sceptics. Moore (2014) emphasizes the importance of understanding the (design) chasm among the innovative creators and the (early) mass and pragmatists. This chasm bridges the gap between the more involved and risk-taking innovators and the mass that will only follow the new when it is low in personal, emotional or financial risk. Common in the different classifications are the attitude to change and the general attention given to the new phenomenon. In co-creating processes, challenge is not only to create a common foresight, but also to gain from the different adoption roles that participants and stakeholders have in the process of adopting and adapting trends. During the co-creative sessions on future thinking, participants received lectures detailing these conceptual lenses as a preparation for their assignment. They also received a brief test to assess their own role in the diffusion prices. In this way, also the differences among the participants have been assessed. It was our ambition that the current course on trend research not only emphasizes what is new in mainstream opinions, but ideally enables participants to anticipate on those trends and changes that will be part of the future normal. In other words, through this course we aim to contribute to smart citizenship by training students in future thinking.

Even the biggest change once started with a first signal

We often see and describe trends as an extrapolation of the past: an incline or decrease on the current that has to be taken into account. Next to the continuous trends, future thinking explores the field of research on paradigm change. Kuhn (1962) describes these paradigms as the current normal that might be disrupted by a new normal. The fast and quick technological changes not only need adjustments, they are leading to paradigm shifts in fields like governance, mobility, or design. This forces the future thinkers to not only explore the existing and already acknowledged trends, but also they are challenged to foresee discontinuities in the already established belief in what relevant trends are.

Weak and strong signals as early warnings for change

A well-known and widely used method for creating insights of the future is the gathering of weak signals and filtering these signals into early trends (Ilmola and Kuusi, 2005). Weak signals are those signals and events and occurrences that are still rare and not part of the 'normal' yet, but have the capacity to become part of a new normal. A weak signal might be as small as a single occasion. Weak signals tend to be underestimated and neglected as they deviate from the normal. The interest in this deviation to the current normal is well known in Kuhn's paradigm circle, where the first signals of change are seen as possible new paradigms instead of deviant data (Roubelat, 2006). The importance of weak signals is also subject in Ansoff's theory on the rise of management and marketing trends; in this theory the weak signals are described as strategic surprises (Holopainen and Toivonen, 2012). More recently the weak, deviating signals gained interest in the field of predicative analysis, where these outliers are seen as a valuable early warning for change, for instance in the field of predictive maintenance (Rathburn, 2012). Weak signals are not only rare in their physical occurrence; Festinger's theory of cognitive dissonance (Festinger, 1957) is explicit that coping with the internal conflicts on what is normal often leads to underestimate its impact.

Where to look, sources for weak signals

Weak signals for change can be gathered through all sorts of sources: from inspiring people and artists to studying fundamental research or newspapers. As described by Choo (2007) and adapted from Wygant and Markely (1988 cited in Hiltungen, 2008), when searching for signals in order to inform innovation, or in the phase of idea generation, these signals can best be collected from sources where phenomena appear for the first time. You might think of artists, science fiction, fringe and alternative press, specialized journals or patent application.

The process of gathering weak signals as a way to collect inspiration and data on possible futures also enables participants to be more aware of the signs of the future that are already among us. Themes and examples of trends given by experts are seen more often and earlier, while for most participants the assignment to focus on the arbitrary outlier signals was a new experience, but contributed to a new and meaningful way of looking. During the co-creation sessions, different sources where selected on their accessibility, validity for further research, and responsibility in the process.

Everyday life as a source

Everyday life has proved a good source of inspiration on possible futures. In everyday life as a source, the focus is on identifying those signals that counter the normal and expected; training the mind to be alert to those signals that usually are neglected as 'non-relevant'. It has been used in many co-creative and trend forecasting sessions as an at-hand source for inspiration and weak signals. This process of using everyday life as a source is also known as a 'cool hunt' (Gladwell, 1997). In this, again, the aim is to open participants mind for the unexpected. In this there is no ultimate right or wrong way of observing, it is about collecting signals as exhibits of what is going on. It is all about open minds and discussion on what might become the trend. It can be concluded that everyday life sources can be useful for living labbing in at least three ways:

- Although it may be expected from designers and social scientist to be aware of everyday life as a source for the new and odd, our minds tend to overestimate those events that line with our common beliefs and underestimate signs that are dissonant to that belief. Therefore, we use inspiration from the real-life environment, especially when the focus is on the inside of an organization, like a school or a department, giving a closer look at the real now and here can open eyes and create awareness.
- Training the focus on the outlier signals. Following the idea that today's oddity might be tomorrow's normal, the gathering of real live signals trains the participants to look for outlying signals, signals that are opposed or dissonant to the normal and expected. In this there is no good or false, as normal for one person might be strange to another. The main goal is to get an open mind. The future is already here; it is just not evenly distributed (Gibson, 2001). Most brains are trained to recognize the known, follow the focus, and trivialize the unknown.
- Detecting examples and raising awareness of emerging trends. Experience shows that participants sometimes have to be helped to adopt to new ideas and concepts, as a part of a strategy, or as a starting point for a new line of thinking. It sharpens the awareness for the new concept, set of signals, and therefore enlarges the capability to recognize examples of the trend in the real world. It however, does bring the risk of agenda setting or even Group Think (Choo, 2007).

Weak versus strong signals

To address the effects of the different roles in the adaption process, as well as to provoke participants to discuss the found signals, participants cluster their findings into weak and strong signals. In this the strong signals are examples and exhibits of already identified or acknowledged trends and themes, but are still moulding towards normal. This relates to earlier mentioned roles in the diffusion process. Signals might be new for one person, but already well known for another. In the co-creating futures approach this is seen as an advantage. It benefits the co-creators in understanding the penetration of the trend in the field of discussion. The next step is to label the remaining, unlabelled signals into new trends and themes. This exercise not only helps participants to understand the mentioned difference in their own role in the adoption process, but it also leads to new insights.

To emphasize the importance of early detection of weak signals as well as to evoke discussion on what signals are still weak and what can already be seen as more established exhibits of upcoming trends, participants are actively involved in the gathering and clustering of signals during the process. The emphasis is on the fuzzy front-end of innovation processes and how collaboratively co-creating futures could make participants more sensitive to identifying weak signals for significant social change. The main goal, however, is to advance living lab practices with the capacity to anticipate on future trends in a bottom-up participatory fashion.

Method

Cool hunting, looking for signals of change, the weak signals that futures are made of, is part of the trend research course for second year bachelor students in the course Communication, Media & Design. Learning objectives are to create awareness for micro signals and capability to abstract and define drivers of change formed by the weak signals. As a final assignment, students need to deliver a conceptualization of a scenario using the findings.

In the current section, we introduce four labs in which possible futures were explored in bottom-up participatory innovation processes. Three of them are part of the bachelor course, and one lab has been conducted with professionals. The four labs referred to are:

- *Newspaper Project* (2013). About 50 art students (BSc level) studied today's papers to identify tomorrow's patterns.
- *Social Smart Cities* (2014). About 140 students (BSc level) collected signals on the theme Social Smart City.
- *The Future of Civil Service* (2014). Over 30 professionals of the civil service department of a municipality collected signals for the co-creation of future scenarios, describing the needs for (their) civil service in 2030.
- *Shopping Lab* (2015). About 25 art students (BSc level) explored possible futures in retail.

First, we introduce the method used and briefly report the main results. Afterwards, we discuss the findings and lessons learned, and reflect on how students and professional participants can raise their awareness on identifying weak signals for significant social change.

Newspaper Project (2013)

Reason for choosing the somewhat 'old fashioned' source was for educational as well as for research purposes. Media literacy together with the argument that the newspaper can be seen as a declining institution has motivated the scope and setup of the newspaper project. Participants were instructed to select 21 newspaper articles, three items a day, during one week. The paper had to be read in the real-paper format in order to be sure the collected data could be analyzed afterwards. The instruction emphasized on the search for those signals that did not meet the common expectation. The signal had to be 'out of the comfort zone', in attention of the innovators (and creators) and sceptics (or laggards) as described by – amongst others – Rogers (2003), Gladwell (2008) and visualized in one figure by Hausman (2014). The newspaper assignment was introduced by means of a lecture on theory on adaption of innovations.

Social Smart City (2014)

As a part of a trend watching course, about 140 students divided among five different classes participated in a smart city cool hunt. An inspirational week of key notes and cases that gave them an overview on the still enormous domain of smart cities, was part of the school's introductory program and prepared them on the theme of smart cities. Moreover, students were able to keep track of the work of peers through an active Facebook group on the topic. Also 'real commissioner' partners actively joined this Facebook group to encourage the debate.

The trend research followed the classic pattern of collecting weak signals, clustering them in possible trends – looking for patterns – researching the found patterns with methods like pestle and scenario-building tools. The students were informed and provided with literature on the theory of diffusion, theory on trend research, scenario building, and collecting meta data. Each participating student was assigned to collect ten signals addressing the new and odd on the theme of smart cities. The assignment was introduced with a lecture and accompanied with literature on trend research for organizations (van der Duin, 2012).

To collect, share, and even analyse the gathered signals, open source facilities like the social photo-sharing site Flickr.com or social bookmarking tool Delicious.com provided workable platforms. The openness of these platforms as well as the opportunity for participants to react and tag the gathered signals proved to be appropriate tools for analysing the data. Using Flickr the found signals can be tagged to meta-levels, providing the participants with inspiration based on the material from all the participants in the project.

The course was divided in three phases with alternating research roles.

- *Phase 1: student as cool hunter, data-gatherer*

In the first phase, as explained well informed students collect signs on possible futures of the Smart Social City.

- *Phase 2: from signal to trend; from data-gatherer to data-analyst*

Here, the found signals were structured into new patterns. By means of group discussion the main findings were abstracted to possible trends. What trends did you(r team) see within the theme of Smart Social Cities? Of course, this phase influenced the quality of the selected signals. Therefore, the final translation to the trend to be researched is also discussed and decided on with an expert group (i.e., the tutors and a large group of colleagues).

Trends were distilled from the collected signals for change. In the described labs, three methods have been used, which are: 1) free categorization, 2) categorization with a Delphi team, and 3) categorization on the basis of the pestle categories.

- *Phase 3: the student as a responsible practitioner*

In-depth research on one of the upcoming trends within Smart Social Cities. Students are asked to describe the trend and its context in not more than 500 words, based on five valuable sources.

These phases were also used in the other labs. The Shopping lab (2015) used the same structure as the Smart City Hunt, but took some lessons into account. These are reflected in the results section. The Future of Civil Service (2014) lab worked with professionals, i.e., civil servants, and combined the methodological setup of the Newspaper Project and the Smart City Hunt.

Findings

Below we report the main findings of the four labs, structured along the three phases of research.

The gathering of the signals

In the first Social Smart City 'hunt', the students collected about 300 signs. However, these signs can be assessed as very 'safe and nearby'. For example, smart traffic lights were spotted far more than normal (Figure 19.2). This seemed to be a result of a strong expert briefing in the kick-off lecture.

Therefore, an extra assignment was given. Students were asked to collect another ten examples that included off-street inspiration. This assignment was also inspired by the sudden media attention on the Internet of Things, ownership, and privacy – a topic that was hitting the early majority, and that they, as specialists should be on top of.

Figure 19.2 An illustration of the risk of a strong expert briefing
Source: Chapter author(s)

The newspaper as a source

Although a newspaper may not be the first source to think of when searching for weak signals for future change, it worked well for over 50 of last years' bachelor art students. While working on the newspaper assignment for a week, participants gathered over 1100 news signals that triggered them into thinking (and thoughts) of the future. For collaboration purposes, all collected signals were uploaded in a closed group at the photo-sharing platform Flickr (Flickr, 2013). Interestingly, the students warmly welcomed the somewhat old-fashioned newspaper source; for some students, it even was a first contact with a newspaper. Students reflected as follows: 'Nice to read more than 140 words on a subject', 'Good to have a news gatekeeper', and 'Wow, it is so huge! I had to take to the floor to read it'.

The expert as a source

The value of the expert appeared to be endless. The following expert roles were observed:

- Agenda setter during kick-offs and expert meetings
- Shaper of research question upfront
- Delphi member in directing the process and connecting the dots between the found signals
- Consultant for the participants during the living lab
- Opposition to break the consequences of Group Think (Choo, 2007)

Table 19.1 summarizes the various sources used for gathering signals, and refers to their methodological contribution.

In the co-creation process of The Future Civil Service, 30 civil agents participated in a weak signal hunt on the streets of Rotterdam. They gathered over 500 signals that provoked their thinking on the future of the city and the role of the civil service within the city. The hunt was introduced by an assignment to get used to the idea of seeing the unseen.

Store and share the found signals

As mentioned earlier, to collect, share, and analyze the gathered signals, open source facilities like the social photo-sharing site Flickr.com or social bookmarking tool Delicious.com were provided as workable platforms. In the Newspaper Project, the Social Smart City lab and the Civil Service lab, Flickr groups were used to collect the gathered signals. Participants were able to use over 500 collected signals, tagged by the trends identified by the Delphi team.

Table 19.1 Methods used for gathering signals and their purpose in the process

	Cool hunt	Newspapers	Experts
Accessibility	■	■	
Validity		■	■
Responsibility			■

The open application program interface (API) of the Flickr platform made it possible to export gathered signals and allowed for external analysis of the data. Unfortunately, Yahoo (i.e., the owner of Flickr) has changed the platform settings, by disconnecting existing groups and accounts. This example illustrates the dependency on commercial parties. In order to enable the reuse of gathered signals across living lab activities, a dedicated platform for collecting signals would be needed.

From signal to trend

Free categorization. In the described labs, the process of filtering signals into trends has been done in a qualitative manner. In the Newspaper Project, students were instructed to categorize their findings. Small groups categorized their materials into upcoming trends, in a next step group results were short-listed to ten overall trends. To search for patterns, and to help elaboration on signals and trends, the students had to meta-tag their own material to the found trends. In the Newspaper Project, students were free to choose from the given ten trends or to choose their own tags. Interestingly, the students largely preferred the latter option. More than 100 different meta-tags were used to describe the gathered signals. This broadness of findings in combination with headstrong participants was not helpful in the process of creating a shortlist. The amount of meta-tags hindered the filtering of weak signals to trends. It can be concluded that less freedom needs to be given to creative students (in this case arts students), who have a preference to invent their own tags.

Delphi categorization. The Social Smart City student group started out a bit problematic as students largely copied the examples given during the lecture. After an intervention, this stage setting issue was neutralized. The group continued gathering more than enough material to inspire a Delhi team of experts to connect the gathered and categorized signals into five trends: playful city, planned city, slow city, transparent city, and hyper-me. These five trends informed their prototyping assignment.

Categorization on pestle drivers

The pestle method includes drivers of change into the measurement of the potential of a given trend, in a given situation, within a given group. The pestle working model refers to political, economic, social, technical, legal, and environmental drivers of change, which are used in measuring the potential of a trend. In the Civil Service process the found signals were used in the pestle categorization. A four-hour session started with an instruction on the pestle method followed by a real-time categorization of the pestle subjects. In the current lab, the found signals were printed and available to work in face-to-face settings. The shared experiences, upcoming drivers of change, and the weak signals showing the change were used to create a shortlist of seven to-be-followed trends.

Jumping to speculations

In the third part of the Civil Service future lab, trends were translated into possible futures. Hereto, a more creative and a more rational way was explored. The creative exercise refers to the creation of 'the newspaper of 2030'. Participants designed a newspaper front page with a headline and arguments supporting their reasoning, relevant data visualization, and an advertisement. The more rational activity conducted was an individual plotting of the trends on the axes of uncertainty and impact, which is often used in scenario analysis. The creative assignment of designing the newspaper of 2030 helped to jump from the extrapolation (the continuation of the trends) towards speculation of new and revolutionary futures. Weak signals are the fact drivers for change to be built on. Speculations are the highly complex, uncertain ideas that spring from the facts, and the forces they drive. The steps in scenario building as described by Zurek and Henrichs (2007) were used to deal with the uncertainty and complexity of co- creating futures.

Unfortunately, the effect of using the current technique to plot trend on a scale of uncertainty and impact was corrupted by discussion of the concepts and biased by the process. Group members had informed themselves during the process and they seem to have adjusted the level of oddness accordingly. Obviously, more reliable insights are to be gained from coding the collected material by external participants.

Shaping the future

The process of shaping, elaboration, and generating the context of the future contexts forms a sub-loop in the co-creating process: from searching for signals outside of one's scope, to pattern recognition, to researching the trend itself or even a prototype for the imagined future. It follows the same pattern as described earlier: it starts with a clear scope defining what is, to what has to be co-created and prototyped, followed by the gathering of weak signals for change in the here-and-now, real, living world. Consequently, the careful formulation of the initial quest, as well as the level of involvement to the subject, contributes to the quality of the outcome of the lab.

The scope of a lab can be defined within the timeframe and program of the lab itself. This seems to work well, when participants are open minded, involved, and trained in abstract thinking. In some cases, however, like forming the mindset of students, the introduction of a Delphi team benefits the outcome. This Delphi team is a first ring of co-creating advisors and/or experts. This team is able to abstract the outcomes and facilitate the process of defining of the scope, as well as provide feedback or even management of the outcome in the next phase: the filtering of the outcome. Scoping the object is necessary to pre-filter the next phase, the phase of gathering information that broadens the horizon of the future, shaping its context. Shaping the context and enriching the awareness level at the start of a lab demands 'state of the art' insights and innovations on the scope of the lab. This might vary from international key note speakers to an actively supported Facebook group by the Delphi group on Social Smart City (Emerging, 2014).

Shopping Lab: open source versus dedicated database

In the Shopping Lab experience, a dedicated database was introduced to collect and support participants. In this lab about 30 art students worked on the subject of trends and innovations in retail. The trend research was started and part of an educational program leading to conceptual design. As 'retail' is a broad subject, the search for trends in retail started with defining the subject itself. Six teams mind mapped their ideas on the subject. Afterwards, each student pointed out two 'most relevant' associations. This resulted in a collective definition of the wider concept of retail.

Sources for the signal hunt. In the Shopping Lab hunt for signals, the choice for sources was open but the program did provide a lecture on source value, using the article 'Good Sources for Weak Signals' (Hiltungen, 2008). In this, participants shared their 'best sources' for inspiration. This led to shared insights and a bigger variety of signals.

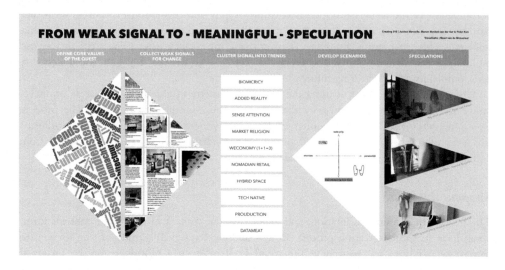

Figure 19.3 Illustration of the process from weak signal to meaningful speculation, as used in the Shoppinglab experience

Source: Chapter author(s)

Signals in a dedicated database. A dedicated database was provided to share and analyse the weak signals. The participants uploaded their signals (pictures or screenshots) into the database, giving location as well as a short description and a url link to the original source.

From 700 signals to 10 trends. The clustering process from signal to trend was a three-step process. At first six groups of students pre-clustered their material to ten trends and gave a short description. The group results were a starting point of an expert group discussion. The full group of students made the final decision on the trends they would use in the project (Figure 19.3).

After, the selection of the trends tags was included in the database. Each participant student tagged their own signals with the ten provided trends. The database enabled a legend on location as well as tags. Herewith the database provided insights in the collection signals tagged with a certain trend. The trend research resulted in design concepts that addressed solutions to address the upcoming need for tactile information in web sites, an experiment with the value of storytelling, and a concept of a stockless store (Marseille, 2016).

Looping forward, prototyping the future context

Although we highlighted the capacity for future thinking, within the co-creative looping process, making and prototyping still play an important role. Making and prototyping not only refers to the designed concepts based on the collected signals, but also to the available support to continue the co-creative process enabling various iterations of several prototypes on many different levels. These prototypes are no longer only a tool for the designer to visualize the intended idea, but also help facilitating the discussion about the idea. In this sense, making and prototyping enable the envisioning of the future. In the current studies, various techniques have been used to generate tangible results; techniques like probing, production of newspaper headings, scenarios, stakeholder maps, customer journeys, business modelling, or even drawing maps for the city of 2030 were used to experience the co-created future context.

The importance of the observant and the observatory

The challenge with gathering 'weak signals' is that things that are 'hardly perceptible' for one, can be totally obvious for somebody else. The reason for this is, amongst situation and location, is the effect cognitive dissonance, selective perception and our own attitude towards change. The effect of cognitive dissonance lead to an underestimation or even denial of the impact of signals that leads to discomfort (Festinger, 1957), selective perception tends to make us overestimate the value of those signals that matches our expectations. This effect is partly illustrated by the shape of Gartner's hype cycle, especially when combined with Rogers' classic theory on roles in the process of diffusion of innovations, shady patterns on the uneven 'predictive value' of the observer emerge (Rogers, 2003). Therefore, understanding one's own role in the spectrum of possible perceptions helps by gathering the right signals. In particular, in the current digital age, it is hard to ignore the noise and not relevant signals.

The main goal of the current study was an education ambition: expose students to alternative design paradigms. Next to that, the living lab approach has been used to ensure that all stakeholders' interests are considered (Anthopoulos et al., 2007; Hartman et al., 2010), and demonstrated that the future making activities can be participatory as well. The active participation of crucial partners has created a sustainable infrastructure to co-create public services and foster further innovation in Rotterdam.

What is the value of open?

In some labbing activities, sharing is required and included in the concept. The open character does, however, not grant the sometimes-needed safe environment. In general, companies and local government valued the open space for experimenting, a place where disruptive ideas are welcome and the in-company habits are left behind. The labs provide an open and safe place for the fuzzy front-end of innovation; gaining insights and sharing experiences seem to happen in ways that are easier and richer when partnerships are actually co-creating. Although in one case the co-creation led to an inconvenient discussion with one of the co-creators who felt disadvantaged by having his contribution part of the shared success of his team. Labs are not the answer to everything and the over-enthusiasm for the approach might jeopardize its future success. In addition, the participatory ideology working with committed companies is not similar to working for them. The labs accommodate a participatory innovation approach to establish co-creative partnerships, and do not allow for client–contractor relationships. It can be concluded that co-creative partnerships are key to co-create a sustainable future that includes new strategies and ways of organization to cope with societal challenges.

Conclusion

Starting from the premise that an open mindset and a participatory attitude to rethink our future is far more challenging than technology, the current chapter elaborated how to empower smart citizens with future thinking capabilities. Taking stock of Citylab Rotterdam, as a platform for participation and co-creation, where students, researchers and teachers, co-create with local entrepreneurs, civil agents, and citizens, to rethink the future of Rotterdam, and its living lab approach, the current chapter explored how to enhance the living lab practices with capacity for training future thinking skills. Four lab activities have been described that deliberately trained students, and professionals as well, in future thinking and anticipation capabilities. The collective definition of subjects and topics was helpful in setting an open mind. Plenary discussions and sharing insights on defining good-quality sources improves the overall results; it also offers a good database for further reading on trends, and enabled participants to use the findings in the further design process, supported in the making and prototyping labs. It can be concluded that participants in the labs have become more sensitive to identifying weak signals for significant change and seem to be prone to act as smarter citizens with future thinking capability, and empowered to contribute to the smart city debate.

References

Anthopoulos, LG, Siozos, P & Tsoukalas, IA 2007, 'Applying participatory design and collaboration in digital public services for discovering and re-designing e-Government services', *Government Information Quarterly*, vol. 24, no. 2, pp. 353–376.

Brodersen, C, Dindler, C & Iversen, OS 2008, 'Staging imaginative places for participatory prototyping', *Co-Design*, vol. 4, no. 1, pp. 19–30.

Brower, J, Dantzler, D, McPolin, L, & Williams, C 2007, 'Agile Rapid Prototyping for Honeycomb', viewed 1 March 2015, www.scribd.com/doc/68977215/14/Agile-Rapid-Prototyping

Brown, T 2008, 'Design thinking', *Harvard Business Review*, vol. 86, no. 6, pp. 84–92.

Carayannis, EG & Campbell, DF 2012, Mode 3 knowledge production in quadruple helix innovation systems: Springer.

Cardullo, P & Kitchin, R 2019, 'Being a 'citizen' in the smart city: up and down the scaffold of smart citizen participation in Dublin, Ireland', *GeoJournal*, vol. 84, no. 1, pp. 1–13.

Choo, CW 2007, 'Social use of information in organizational groups', in A Huizing & EJ de Vries (eds), *Information Management: Setting the Scene*, pp. 111–124. Amsterdam: Elsevier.

De Lange, M & de Waal, M (eds.). 2019, *The Hackable City: Digital Media and Collaborative City-Making in the Network Society*. Singapore: Springer.

Emerging 2014, 'Social Smart City', viewed 1 March 2015, www.facebook.com/emergingsocialsmartcity?fref=ts.

Festinger, L 1957, *A theory of cognitive dissonance*. Stanford, CA: Stanford University Press.

Flickr.com 2013, New Frontiers Findings from a week of weak signal gathering weak, viewed 23 April 2015, https://secure.flickr.com/groups/2287574@N23/

Foth, M, Brynskov, M & Ojala, T (eds.). 2015, *Citizen's right to the digital city: Urban interfaces, activism, and placemaking*. Singapore: Springer.

Gibson, W 2001, 'Broadband Blues', The Economist, 21 June 2001.

Gladwell, M 1997, *The coolhunt*, The New Yorker, 17 March, pp. 78–88.

Gladwell, M 2008, *Outliers: the story of success*. New York: Little, Brown and Company.

Hartman, A, Jain, AN, Ramanathan, J, Ramfos, A, van der Heuvel, W, Zirpinis, C, Tai, S, Charalabidis, Y, Pasic, A, Johannessen, T, Gronsund, T 2010, 'Participatory design of Public Sector Services', In *Electronic Government and the Information Systems Perspective*, KN Andersen et al. (ed), EGOVIS 2010, LNCS 6267, pp. 219–233. Springer-Verlang: Berlin-Heidelberg.

Hausmann, B 2014, 'Innovation: Adoption and Diffusion in the Age of Social Media', viewed 22 April 2015, www.hausmanmarketingletter.com/innovation-adoption-diffusion-age-social-media/

Hiltungen, E 2008, 'Good sources for weak signals. A Global Study of Where Futurist look for Weak Signals', *Journal of Futures Studies*, vol. 12, no. 4, May 2008, pp. 21–44.

Holopainen, M, Toivonen, M 2012, 'Weak signals: Ansoff today', *Futures*, vol. 44, no. 3, April 2012, pp. 198–205.

Ilmola, L & Kuusi, O 2005, 'Filters of weak signals hinder foresight: Monitoring weak signals efficiently in corporate decision-making', *Futures*, vol. 38, no. 8, pp. 908–924.

Kuhn, T 1962, *From the structure of scientific revolutions*. Chicago and London: University of Chicago Press.

Linden, A, Fenn, J 2003, Understanding Garner's hype cycle. Strategic Analysis Report, Gartner. Viewed april 27, 2019, www.bus.umich.edu/KresgePublic/Journals/Gartner/research/115200/115274/115274.pdf

Marseille, JH 2016, "From Weak signal to Speculation", contribution to DRIVE, Design Research & Innovation Festival, Eindhoven, October 26, 2016.

Moore, G 2014, *Crossing the Chasm*, 3rd ed., New York: HarperCollins Publishers.

Mostert-vanderSar, M, Mulder, I, Remijn, L & Troxler, P 2013, 'FabLabs in Design Education' (pp. 629–634), in Proceedings of E&PDE 2013, International conference on engineering and product design education, 5–6 September 2013, Dublin Institute of Technology (DIT), Dublin, Ireland.

Mulder, I 2012, 'Living Labbing the Rotterdam Way: Co-Creation as an Enabler for Urban Innovation', *Technology Innovation Management Review*, 2(9), pp. 39–43.

Mulder, I 2014, 'Sociable Smart City: Rethinking our future through co-creative partnerships' In N Streitz & P Markopoulos (eds), *Proceedings of Distributed, Ambient, and Pervasive Interactions 2014*, pp. 566–574. Switzerland: Springer International Publishing.

Mulder, I 2015a, 'A pedagogical framework and a transdisciplinary design approach to innovate HCI education', *IxD&A*, vol. 27, pp. 115–128.

Mulder, I 2015b, 'Opening up: Towards a Sociable Smart City' In M Foth, M Brynskov & T Ojala (eds), *Citizen's right to the digital city: Urban interfaces, activism, and placemaking*, pp. 161–173. Singapore: Springer. Available online: http://link.springer.com/book/10.1007%2F978-981-287-919-6

Mulder, I & Stappers, PJ 2009, 'Co-creating in practice: results and challenges', in Proceedings of ICE 2009, the 15th International Conference on Concurrent Enterprising, Leiden, the Netherlands, 22-24 June 2009.

Pastoor, R 2013, 'New Frontiers Lab', viewed 13 June 2014, http://rolandpastoor.nl/lectoraat/fronteers/visualisations.php

Pucci, EL & Mulder, I 2015, 'Star(t) to shine: Unlocking hidden talents through sharing and making' In N Streitz & P Markopoulos (eds), *Proceedings of Distributed, Ambient, and Pervasive Interactions* (pp. 85–96). Cham: Springer.

Rathburn, 2012. 'Where Agile Meets Analytics', at the TDWI BI Symposium in Toronto on 25 June 2012.

Rogers, E 2003, *Diffusion of Innovations*, 5th. New York: Free Press.

Roubelat 2006, 'Scenarios to challenge strategic paradigms: Lessons from 2025', *Futures*, vol. 38, no. 5, June 2006, pp. 519–527.

Schachter, ME, Matti, CE, & Alcántara, E 2012, 'Fostering quality of life through social innovation: A living lab methodology study case', *Review of Policy Research*, vol. 29, no. 6, pp. 672–692.

Schumacher J & Niitano V-P. (eds). 2008, *European Living Labs – a new approach for human centric regional innovation*. Berlin: Wissenschaftlicher Verlag Berlin.

Ståhlbröst, A 2008, Forming future IT: the living lab way of user involvement. Doctoral thesis. Luleå University of Technology.

van der Duin, P 2012, Toekomstonderzoek voor organisaties. Plaats: Assen Uitgeverij: Van Gorkum.

van Waart, P, Mulder, I & de Bont, C 2016, 'A Participatory Approach for Envisioning a Smart City', *Social Sciences Computer Review*, vol. 34, no. 6, pp. 708–723.

Zurek, MB & Henrichs, T 2007, 'Linking scenarios across geographical scales in international environmental assessments', *Technological Forecasting & Social Change*, vol. 74, pp. 1282–1295.

Part III
Smart city visions

Section 1

Urban planning, city models and smart storytelling

20

Smart cities as corporate storytelling

Ola Söderström, Till Paasche and Francisco Klauser

Editor's introduction

> On the surface, the dominant smart cities' storyline is about efficient and sustainable cities ... This story should not be taken at face value of course.
>
> *(Söderström et al., this volume)*

One of the core critiques of smart cities is that they represent a techno-utopian fantasy (Datta, 2015; Marvin *et al.*, 2016). Söderström, Paasche and Klauser's investigate how this fantasy is constructed and played out through an analysis of IBM's corporate Smarter Cities programme. Drawing on critical planning theory, they reveal how smart cities rhetoric adopts a storytelling approach in order to construct narratives around a city's past, present and future. Whilst smart city language typically is used to describe how technology can be adopted to address urban challenges, what Söderström et al. put forward is that smartness itself should be understood more in terms of a discursive activity. The future smart city is one which shapes a certain imaginary, where the narrative is a utopian one. The materialization of the smart city is in fact extremely vague, as the concrete vision of what a smart city would be and how it would really work is uncertain, and currently relies on utopian urban planning concepts dominated by technological 'solutions'.

The text draws on urban planning literature that has sought to reveal how planning as a practice is inherently practiced as forms of storytelling about the future (Mandelbaum, 1996; Sandercock, 2003b) (Throgmorton, 1996b) (Van Hulst, 2012b). Through critical planning theory it is possible to understand how planners (as authors) choose to characterize (name and describe) the people who inhabit and activate their stories and how those characters are expected to act and relate to one another. So, the author of the story is the actor who has power. Söderström et al. take reference from actor network theory (ANT) to understand how certain actors try to make themselves indispensable in addressing an identified urban challenge through Callon's (1986, 183) concept of 'obligatory passage points' (OPPs). They argue that the smart city discourse can be interpreted as a way to make certain actors and technologies OPPs or key actors in the implementation of certain types of urban management systems to address a catalogue of problems associated with urbanisation. The chapter authors write that it is through this mechanism that: 'the smart city discourse becomes a story with a plot and can be fruitfully interpreted in the light of research on storytelling in planning'. They outline how the smart city discourse has its roots in two tropes of urban planning: the city described as a system of systems, and a utopian narrative that proposes a magical fix for complex urban problems. In both these tropes, the common feature of utopianism is the creation of a 'new world'. According to Carey (1999) the problem with most utopian theories is that the existing world must be destroyed in order for a new world to take its place. From Ebenezer Howard's Garden Cities of Tomorrow to Le Corbusier's Ville Radieuse and Archigram's Plug in City, spatial imaginaries have provided powerful imagery that promise a radical new solution to a whole host of urban problems by starting from scratch. In fact, as Söderström et al. point out, utopia is closely linked with theories of urban form, and it has played an important role in urban planning (e.g. Choay, 1965). They do not argue for a refutation of utopian rhetoric, but instead point out that it is important to question who tells the story (Throgmorton, 1996b). In their review of the IBM smarter cities programme they highlight the problem that arises in the smart city story is that the author of the story is not an individual urban planner or a city planning office, but a private company (Hollands, 2015). Through claiming the smart city narrative, IBM positions the company as the planner, and instead of a spatialized planning solution they propose an optimization of city systems and infrastructures through code. In this model, data and optimization are what drive a utopian vision, and 'smartness' lies in the algorithm.

Söderström et al. conclude that what is needed is alternative storytelling about smart cities, where the narrative is not one 'of' planning but 'for' planning. McFarlane and Söderström argue that there is a need to 'remobilize critical planning theories of the 1970s and beyond as they similarly sought to present alternatives to modernist-functionalist planning postures' (2017, p. 317). This draws on a body of work from planners such as Healey (2000) and Sandercock (2003b) who have sought to reclaim imaginaries in urban planning by accounting for the diverse kinds of ways of knowing that exist apart from technical knowledge. Healey has called for planners to actively take part in conscious 'city story-writing' (Healey, 2000, pp. 527–528), whilst Sandercock has proposed digital ethnography as a mode of storytelling through multimedia and filmmaking (Sandercock and Atthili, 2010). These storytelling practices adopt different imaginaries and are narrated by diverse sets of actors; they are place based and experiential and the storytellers are those living in the communities for which the future is to be imagined. Datta's work on unpacking the smart city futures in the Global South context, undertaken through studies on the Indian Smart Cities Mission, proposes an approach that involves bringing stakeholders in government, third sector and subaltern groups together into an active discourse which reveals very different meanings across the table (Datta, 2018, p. 4). This spatializes the storytelling imagination and situates it within everyday realities that make 'explicit normative and political positioning within a system of values and aims' (Söderström et al., 2014, p. 318). In their plea for alternative storytelling about smart cities, the chapter authors highlight the importance of recognizing who tells the story in the smart cities narrative.

References

Carey, J. 1999. *The Faber Book of Utopias*. London: Faber.

Choay, F. 1965. *Urbanisme, utopies et réalités. Une anthologie*. Paris: Editions du Seuil.

Datta, A. 2015. "A 100 Smart Cities, a 100 Utopias." *Dialogues in Human Geography*, 5 (1): 49–53.

Datta, A. (2018) "The Digital Turn in Postcolonial Urbanism: Smart Citizenship in the Making of India's 100 Smart Cities." *Transactions of Institute of British Geographers*, pp. 1–15.

Healey, P. 2000. "Planning in Relational Space and Time: Responding to New Urban Realities." In *A Companion to the City*, edited by G. Bridge and S. Watson, 517–530. Oxford: Blackwell.

Hollands, R. G. 2015. "Critical Interventions into the Corporate Smart City." *Cambridge Journal of Regions, Economy and Society*, 8: 61–77.

Mandelbaum, S. 1996. "Telling Stories." *Journal of Planning Education and Research*, 10 (2): 209–214.

Marvin, S., A. Luque-Ayala, and C. Mcfarlane, eds. 2016. *Smart Urbanism—Utopian Vision or False Dawn?* Abingdon: Routledge.

McFarlane, C., and O. Söderström 2017. "On Alternative Smart Cities- From a Technology-Intensive to a Knowledge-Intensive Smart Urbanism." *City* 21 (3–4): 312–328.

Sandercock, L. 2003b. "Out of the Closet: The Importance of Stories and Storytelling in Planning Practice." *Planning Theory & Practice*, 4 (1): 11–28.

Sandercock, L., and G. Atthili (2010) "Digital Ethnography as Planning Praxis: An Experiment with Film as Social Research, Community Engagement and Policy Dialogue." *Planning Theory & Practice*, 11 (1): 23–45.

Söderström, O., T. Paasche, and F. Klauser 2014. "Smart Cities as Corporate Storytelling." *City*, 18 (3): 307–320.

Throgmorton, J. A. 1996b. *Planning as Persuasive Storytelling: The Rhetorical Construction of Chicago's Electric Future*. Chicago: University of Chicago Press.

Van Hulst, M. 2012b. "Storytelling, a Model of and a Model for Planning." *Planning Theory & Practice*, 11 (3): 299–318.

Introduction

Smart cities, like creative cities, sustainable cities or liveable cities are part of contemporary language games around urban management and development. These games involve experts, marketing specialists, consultants, corporations, city officials, etc. and frame how cities are understood, conceptualized and planned. Although we might consider this discursive activity with some skepticism, it often makes a difference. It is performative, because it shapes the imaginaries and practices of a myriad of actors concretely building the city through particular case studies or pilot projects, decisions and everyday action, like creating a new electricity system for a neighborhood. We therefore take discourse seriously in this chapter and focus on two important aspects of contemporary 'smart city talk': we first look at how the term smart city has been popularized in the discourse of municipalities, media and especially private firms and then, at more length, how it has been given a specific content in IBM's global and massive smarter cities campaign: the most developed attempt by a private company to define a smart model of urban management.[1] In doing this, we analyze key episodes in the struggle over the definition of what smart cities are about, claiming that this struggle is an important element in the competition between private companies over authorship, authority and profit in the smart city business.

Drawing on critical planning theory, we conceptualize IBM's smarter city campaign as a specific form of storytelling in the world of planning (Mandelbaum, 1991; Sandercock, 2003a; Throgmorton, 1996a, 2003; Van Hulst, 2012a) and show that it mobilizes and recycles two long-standing tropes: the city conceived as a system of systems, and a utopian discourse exposing urban pathologies and their cure. On this basis, we develop three main arguments related to the purpose, content and effects of the dominant smart city story. The first is that this story is to a large extent propelled by attempts to create an 'obligatory passage point' (Callon, 1986; Latour, 1987) in the transformation of cities into 'smart' ones. In other words, it is conceived to channel urban development strategies through the technological solutions of information technology (IT) companies. Second, we argue that this discourse promotes a conception of urban management that is a technocratic fiction: one where data and software seem to suffice and where, as a consequence, knowledge, interpretation and specific thematic expertise appear as superfluous. Third, we claim that this discourse prioritizes public investments in IT over other domains of spending and thereby introduces a new 'economy of worth' (Boltanski and Thévenot, 2006), which is particularly problematic in resource-scarce cities.

The chapter is embedded in a burgeoning literature on smart cities and which can be divided into two main categories: first, there are studies focusing on the technological side which is concerned with specific questions such as energy efficiency and carbon emissions, etc. (Bakıcı et al., 2013; Fischedick, 2012; Paskaleva, 2011; Rat-Fischer et al., 2012; Streitz, 2011). This literature seeks to develop smart technologies for cities. There is then a second strand of literature approaching smart cities as an object of analysis and attempting to define the smart city as an assemblage of technologies such as advanced information and communications technology (ICT) infrastructure, smart cards in public transport and e-governance functions aimed at increasing competitiveness, administrative efficiency and (in some cases) social inclusion (Allwinkle and Cruickshank, 2011; Caragliu et al., 2011; Deakin, 2014; Deakin and Al Waer, 2011; Kuk and Janssen, 2011; Schaffers et al., 2011). However, little is actually known about the more fundamental principles and ideas underlying the smart city as a model—that is, as a generic solution to the problems of urban development and management[2]—beyond the self-advertisement of IT companies and municipalities. Until recently, the existing literature was also lacking in critical engagement with the exception of an early text by Hollands (2008). However, since 2011, a series of contributions has more critically scrutinized the phenomenon from different viewpoints: political economy, science and technology studies, governmentality studies and ideological critique, moving research away from the self-celebratory climate around smart cities. This critique can be briefly summarized as follows. In Hollands (2008, 305) pioneering paper, the smart city model is interpreted as a contemporary high-tech clothing of urban entrepreneurialism, which 'plays down some of the negative effects the development of new technologies are having on cities' such as growing social polarization. Looking at smart cities as the places where the concentration and interconnection of 'big data' in cities lies, Kitchin (2014) raises questions of technocratic governance, corporatization of city governance or vulnerability and surveillance, while Wyly (2013) combines technology studies and political economy to argue that smart cities are to be interpreted in the context of the shift to 'cognitive-cultural capitalism' boosted by the takeoff of automated data generation and mining, notably through social networks. In a more Foucauldian perspective, Vanolo (2014) shifts the focus from data to citizens and discusses how the smart city model may be a powerful disciplinary tool to shape 'smart citizens', who are compelled to be technologically literate. Finally, different contributions have targeted smart cities as an ideological construct and as a simplistic model of the urban. For Bell (2011, 73), it is a vision of cities 'that frames all urban questions as essentially engineering problems to be analyzed and solved using empirical, preferably quantitative, methods' which give 'pre-eminence to urban phenomena that can be measured and are deemed important enough to measure'. In the same vein, Greenfield (2013, chap. 13) defines the dominant corporate discourse on smart cities as a return to the high modernism of the period 1880–1960: a Le Corbusier redux bound to repeat the worst planning disasters of the 20th century in the 21st century. Based on a more detailed argument, Townsend (2013) sees in IBM's smarter city discourse a resurrection of the urban cybernetics of the 1970s. These recent contributions have begun to provide answers to the agenda of critical questions identified by McFarlane, Marvin, and Luque (2013, 25): the why?, who?, what?, how?, where? and emerging consequences of smart urbanism.

Our chapter connects some 'whys' and 'hows' of smart urbanism and pushes further the critical analysis of smart urbanism by focusing on the discourse of a central actor: IBM. It thus particularly relates to the critique of smart urbanism as an ideological construct focusing on the storytelling activity of the company as a means of securing and strengthening its market position. Thus, the aim of this chapter is to describe the emergence of a prominent 'spokesperson' for the idea of smart cities in the public sphere and to analyze the constitutive elements and imaginaries in IBM's discourse on smarter cities. Our argument unfolds in three different steps corresponding to three aspects of this discourse. We first trace the origin and diffusion of the term 'smart cities' and examine how cities became problematized as smart and by whom. Then, we focus on IBM's smarter cities campaign to show how, on the one hand, the company translates different dimensions of the urban world into a unitary language and how, on the other hand, cities thus translated are inscribed in a narrative of positive transformation. The chapter is based on three different methods and types of empirical material: the first is a media survey, using LexisNexis, to track down the moments and places of origin of the term smart cities; the second is a critical discourse analysis of the abundant online information of IBM's smarter cities campaign; the third is a series of interviews in 2012 and 2013 with IBM and other specialists involved in projects using smart technologies in Switzerland.

Translating and narrating the smart city

In quite general terms, smart cities involve the creation of new relations between technology and society. According to this vision, as we show below, urban infrastructures and everyday life in cities are optimized through technologies provided by IT companies. These companies are the main producers of a discourse about (the benefits of) smart cities that they produce both to describe their activity in the domain and to stage themselves as central actors of this urban management model. This process is resonant with actor network theory's (ANT) focus on the making of socio-technical networks and how certain actors try to create for their interests what Callon (1986) calls in his seminal paper 'obligatory passage points' (OPPs). For Callon (1986, 180–185), the first and crucial step in the creation of socio-technical networks is the problematization of a situation in order to become indispensable actors in a network. It supposes of course the definition of the problem that needs to be solved, but also that of the actors involved and the creation of OPPs, through which these actors will be in a position to solve the problem. Concretely, an OPP is a place (a geographical one or an institution), or a procedure that becomes unavoidable: a vaccine developed by a pharmaceutical firm to avoid a disease, for instance. Here, the discourse about smart cities can be interpreted as a tool to make certain actors and technologies OPPs or key actors in the development and implementation of specific forms of urban management solutions. As we will show, IBM crafts a story that presents their smart technologies as the only solution for various urban problems and hence becomes an OPP. The use of mediations—from small talk to complex machines—to translate phenomena into a manageable language—is a powerful means of creating OPPs (Latour, 1987). We will show here how the translation of the different dimensions of the urban world into the unitary language of urban systems is crucial in IBM's campaign. Finally, this discourse really gains momentum once cities and their problems thus translated are embedded in a narrative of positive transformation. Here, the smart city discourse becomes a story with a plot and can be fruitfully interpreted in the light of research on storytelling in planning.

Since the 1990s, there has been wide recognition within planning theory of the role of storytelling (Van Hulst, 2012a). Stories are important because they provide actors involved in planning with an understanding of what the problem they have to solve is (Van Hulst, 2012a, 314). More specifically, they play a central role in planning because they 'can be powerful agents or aids in the service of change, as shapers of a new imagination of alternatives' (Sandercock, 2003a, 18).

For Throgmorton (1996a,2003), stories are the very stuff of planning, which, fundamentally, is persuasive and constitutive storytelling about the future. Seen in this perspective, planning is about 'emplotment, characterizations, descriptions of settings, and rhythm and imagery of language' (Throgmorton, 2003, 126). It is also about power, as storytelling in planning calls for a critical analysis that asks 'who has the power to give meaning to things, to name others, to construct the character of collective identities, to shape the discussion of urban politics … ?' (Throgmorton, 2003, 132).

This is what this chapter does: it looks at who has the power to define the smartness of cities and what the discussions around this theme should be concerned with. An important difference, as far as previous analyses of storytelling in planning are concerned, is that in our case the author of the story is not an individual planner or a planning office, but a private company, which addresses (mainly) municipalities to persuade them of the central role that IBM can play in a new era of (smart) planning. Our chapter is therefore not based on interviews with planners, but studies the story that a large private company tells about smart cities in a worldwide marketing campaign. It considers the ingredients of this storytelling to be operators of power in an emergent field of thought and action. This will lead us to ask in our conclusion how other stories about smart cities might be told.

Emergence of the smart city discourse

Hollands (2008) and Vanolo (2014) have shown that the idea of the smart city is related to a double lineage in planning literature: the concept of Smart Growth as theorized by the New Urbanism movement in the USA of the 1980s and also the concept of the technology-based intelligent city. However, if we want to grasp the wider efficacy of the term we have to look at a broader public sphere by studying the media. Therefore, we focus here on the term smart city to show how it emerged in the media sphere. Our findings are based on LexisNexis, a large commercial newspaper database, and we have focused our research on international newspapers in English.[3]

This analysis shows that the term 'smart city' first appeared in the mid-1990s. It was mainly used in the 'self-congratulatory ways' Hollands (2008) has criticized. The content of the newspaper articles using the term shows that in that period cities labeled themselves as 'smart' when they introduced functioning ICT infrastructure, e-governance or attracted high-tech industries to foster economic growth. However, there are two examples where the term smart city is used to describe a more complex discursive and technological phenomenon. In 1994, the Multifunction Polis (MFP), an autonomous smart city, was planned near Adelaide, Australia. In 1997, the two cities of Cyberjaya and Putrajaya, Malaysia, were re-planned as intelligent garden cities labeled as 'smart cities'. What made the Australian and Malaysian cases smart was their vision to use ICT infrastructure not simply to attract business, but, as far as this was possible at the time, to let the ICT grid steer the functioning of the city in order to automatize and optimize its processes. While some of the key words used in the newspaper coverage where still 'high speed ICT', 'e-government' and 'attracting investment', they were extended by the reference to other terms: 'sustainability', 'eco and smart homes', 'environmental innovations' or 'public transport using GPS'. Premised on the idea of an optimization and automation of urban infrastructures, which has become central in the discourse and technology of smart cities, these two cases of the late 1990s can in retrospective be seen as pioneering ones for the more recent developments of the concept. Therefore, the term smart city first emerges in the media as a self-definition of a series of municipalities like Adelaide or Cyberjaya.[4]

Then, we have a second moment after 2008 characterized by the intervention of private companies in the IT sector, among which IBM. On 6 November 2008, in the midst of the financial crisis, Sam Palmisano (IBM's CEO at that time) gave a talk entitled 'A Smarter Planet: The Next Leadership Agenda' which had a large impact in the media.[5] In this speech, Palmisano argues that the world and its cities must become smarter to become more sustainable and economically efficient. Timely to this speech, IBM launched an extensive smarter planet advertisement—discussed below—which is running until the present day.

A few months later, on 25 September 2009, to position itself more firmly in the emerging smart city talk, the company officially files the term 'smarter cities' to be registered as a trademark: 'Mark: 79,077,782; Word Mark: SMARTER CITIES; Serial Number: 79,077,782; Registration Number: 4033245'.[6] The trademark was registered two years later on 4 October 2011. With Palmisano's speech and the trademark, we have a problematization of cities as smart cities, the first step in the creation of an OPP. Cities' problems are defined as the need to become smarter and the central actors of the process—

IBM, municipalities—are identified. If IBM is far from being the unique contender in the business,[7] it has since 2008 acquired a very visible position. We now move on to show how the company's discourse unfolds in its very extensive smarter cities campaign initiated in 2008, focusing on the campaign's website.[8]

Crafting the smarter city story

The 1990s and early 2000s were difficult times for IBM. Its annual losses reached US$8 billion in 1993. This led to drastic strategic changes and the announcement in 2002 of the company's move away from hardware design and production to concentrate on consultancy and software. In 2004, the company sold its PC division to the Chinese company Lenovo. The aim of these changes was 'to "move up the value chain" into "more lucrative fields"' (McNeill, 2013, 2). At the time, IBM had realized the importance for a global company like itself of the market presented by urban technologies. The 2008 launch of IBM's smarter planet campaign is to be interpreted within this context. Studies done by senior cadres of the company in the early 2000s had identified cities as a huge untapped market (Townsend, 2013, 64). According to estimates, this market would represent: $39.5 billion[9] by 2016 or $20.2 billion[10] by 2020. In order to try and obtain the largest possible share in this market, IBM has developed a strategy involving two elements: first, a 'full-scale contracting for city governments' (McNeill, 2013, 7) with flagship contracts such as those with Singapore and Rio; and second, its Smarter Cities Challenge where experts provide 100 municipalities over the world with pro bono consultancy in the hope that this initial investment will yield returns. It also allows the company to claim that its expertise is based on its involvement with 2000 cities worldwide. On the whole, as Hollands (2013, 9) notes, 'this strategy has clearly paid off, generating some 3 billion USD' of income and representing 'currently 25% of IBM's operations'. It makes IBM the market leader in the business of smart urban technologies in terms of sales and strategy.[11]

The smarter cities campaign, to which we now turn, has been designed to provide the company's strategy and services with a global visibility. It makes abundant use of video testimonies, pedagogical diagrams and case studies from around the world as its targets are not technological experts but an audience on the management level (municipalities, security, communication or transport companies) which, if convinced by the argument, is able to decide on the implementation of 'smart' urban technologies. Within this ambitious campaign, costing the company $100 million (Townsend, 2013, 31), two aspects can be analytically distinguished: the translation of the city into a unitary language and its inscription into a transformative narrative. Making our way through these two aspects of the company's discursive strategy we will encounter two well-known topoi in urban planning history working as the main rhetorical devices of the campaign: the systems metaphor and utopianism.

IBM's urban theory rests on two main assumptions. First, the city is based on three main pillars:[12] planning and management services; infrastructure services; and human services. Each of these pillars is sub-divided into three sub-pillars: 'planning and management services' into public safety, smarter buildings and urban planning, government and agency administration; 'infrastructure services' into energy and water, environment and transportation; and 'human services' into social programs, health care and education. The sum of these nine pillars makes 'the city'.[13] Ideally, all nine systems would be monitored and regulated in IBM's 'Intelligent Operations Center', the 'central nerve system' of the city, first experienced in the city of Rio de Janeiro.[14] The city is thus seen from the point of view of a municipality: these pillars redefine the main administrative divisions of most cities around the world.

The second assumption is that each of these pillars is an individual system and the city is a 'system of systems'. Systems thinking is not only used as a practical way to schematize a complex phenomenon—the city—but as a tool used in IBM's service provision to municipalities. Justin Cook, manager of IBM's Foundational Research Team, thus explains, referring to collaboration with the city of Portland, that 'we want to help these people become systems thinkers … to help them see relations'.[15] He goes on to a more general declaration: 'As systems thinkers we need to be thinking about the interconnections between these things and the ways to bring this all together so that the whole planet works in a better way'. Faith in systems thinking is also expressed at the very top of the company's hierarchy: Ginni Rometty, IBM's CEO since 2012, explaining smarter cities at a conference in Rio, thus presents data and systems analysis as the very core of how to make cities smarter.[16] Using an enlightenment rhetoric where data and systems theory are the means through which municipalities can move 'from gutfeeling and impressions to knowledge', the new CEO (probably unconsciously) situates herself in the lineage of the social reformists of the previous turn of the century: a Charles Booth, for instance, describing his survey and mapping of London poverty as bringing the light of science on the city (Söderström, 1996).

Systems thinking is of course not a new perspective in urban theory and planning. There is a long genealogy of works within urban theory and planning defining the city as a system. Systems thinking about cities finds its roots in organicism and more precisely in visions of the city informed by William Harvey's theory of blood circulation in the early 17th century. Harvey's research on the functions of the heart, arteries and veins, was critical in secularizing the understanding of the body as a machinic system of circulations (Sennett, 1994). This vision of the body, Sennett (1994, 263–264) argues, progressively informed urban planning of the Baroque period and the Enlightenment:

> Enlightened planners wanted the city in its very design to function like a healthy body … Thus were the words 'arteries' and 'veins' applied to city streets in the eighteenth century by designers who sought to model traffic systems on the blood system of the body.

Thereby, organicism envisaging bodies like a set of (nervous, sanguine, etc.) systems of circulation provided urban design with an alternative to former rational geometric models of spatial organization (Friedmann, 1987, quoted by Mehmood, 2010, 69).

The common denominator of organicist approaches in planning is a holistic view where cities are approached as composed of functionally related parts. Systems thinking in urban theory is a continuation of the organicist tradition in that respect but building on a different metaphor. If the body (and then more broadly living organisms) is the model of traditional organicism, systems theory builds on the computer metaphor. The urban totality is a large calculating system rather than a biological entity.

Systems theory has been one of the most influential and enduring approaches in urban thought since the 1960s, both in planning theory (Healey and Hillier, 2010) and in urban geography (Mandelbaum, 1985). Forty-four years before IBM launched its campaign, Brian Berry (1964) famously defined 'cities as systems within systems of cities' using the same Russian doll idea that is to be found in IBM's urban systems scheme. A few years later, Churchman (1968, quoted by Mehmood, 2010) identified four different approaches in systems thinking: the efficiency, scientific, humanistic and anti-planning approach. IBM's systems rhetoric clearly prioritizes an efficiency approach 'which concentrates on reducing waste (of time, resources, materials, etc.)' (Mehmood, 2010, 77).[17] More precisely, IBM's approach is indebted to Jay Forrester (1969), a computer engineer, and his work on urban dynamics in the late 1960s. Although applied with no success in Pittsburgh in the early 1960s and New York in the early 1970s, urban dynamics have been resurrected by Justin Cook, the IBM smart city strategist who is also a former graduate of MIT's Sloan School of Management where Forrester had been a professor (Townsend, 2013, 83). There is something apparently odd in this resurrection as it gives the audience of the smarter cities campaign a sense of traveling back to the heroic times of post-war cybernetics. If we consider urban dynamics as a translation device used for the purpose of storytelling, this choice becomes less enigmatic. What urban systems theory provides, seen from this perspective, is primarily a powerful metaphor creating a surface of equivalence. It translates very different urban phenomena into data that can be related together according to a classical systemic approach which identifies elements, interconnections, purposes, feedback loops, delays, etc. Thus, the website is packed with schemes and flash animations showing how contemporary cities are constituted by functioning and measurable (but highly perfectible) urban systems and infrastructures. As millimeter paper is used as a surface of equivalence to translate and mathematize different living organisms observed through a microscope in biology (Lynch, 1988), so urban dynamics translate the city into a single language. We have here, after problematization, a second important step in the constitution of an OPP: translation. The city is made to speak the language of IBM.

Two observations can be made on how urban systems are described in the campaign. First, it takes for granted that there are infrastructures: it never considers cities where the lack, breakdown, worsening, centrally unmanaged urban systems and infrastructures is the norm, as it is in most cities of the Global South (Beall and Fox, 2009; Gandy, 2004). Bakker (2011, 63), for instance,

> suggests that the term 'network', and the interconnectedness it evokes is a poor descriptor of water supply systems in many cities. Rather, the metaphor of the archipelago—spatially separated but linked "islands" of networked supply in the urban fabric—is more accurate than the term 'network'.

In other words, this form of systems thinking presuming the existence of a set of urban systems is very largely North-centric. The second observation derives from systems theory as a surface of equivalence. In this approach, cities are no longer made of different—and to a large extent incommensurable—socio-technical worlds (education, business, safety and the like) but as data within systemic processes. This is of course one of the great advantages of data-based systemic analysis. However, the way the campaign presents the nine pillars and their relation tends to reduce the analysis of the city to a machinic vision of cities. As a result, the analysis of these 'urban themes' no longer seem to require thematic experts familiar with the specifics of a 'field' but only data mining, data interconnectedness and software-based analysis. This is particularly clear in IBM's three Is equation where the smarter city is the result of Instrumentation (the transformation of urban phenomena into data) Interconnection of data the Intelligence brought by software. Complexity, multiplicity is simplified, flattened into the uni-city of scaled systems and presents itself as IBM's fiat lux to its clients. This reduction of expertise has political consequences: as Marcuse (2005, 252) observes, the organic or systems metaphor also creates a fictitious entity—'the city' supporting 'a search for consensus politics, in which the claims of the minority or powerless or disenfranchised or non-mainstream groups are considered disturbing factors in the quest for policies benefiting "the whole"'.

This technocratic approach to cities is presented as the key to efficient urban management through a recurrent trope in the smarter city campaign: the 'if … then …': 'If we think about it as a whole, if we integrate the system, we can keep the city's resources from getting trapped between locations'.[18] Ontological transformation is in other words the source of the model's epistemological power. Moreover, in this version of systems thinking this transformation spares us the difficulties of interpretation: translated into data and systems, the city seems to speak by itself, to be self-explanatory. Therefore, at its core, this is what structures IBM's smarter city discourse: an engineering epistemology applied to humans and non-humans. Nature and culture reunited by the engineering mind. Systems thinking of course also transforms the nine pillars (or themes) in terms that can be addressed by the company's technologies and services. The nine pillars become combinable in one large system where all information is being brought together to be processed and then optimized to turn 'the city' into a 'smarter city'. This does not mean that in practice the company does not need or seek collaborations, but in its discourse it nurtures an imaginary of an urban management reduced to systems engineering. Once defined in this way, the city can be embedded in a larger narrative about the city's past, present and future. This narrative, as we will see now in more detail, is a utopian one, in the strict sense of the genre, with its diagnosis of urban ills and its healing therapy.

Transforming the city

In its campaign, IBM constantly emphasizes the problems and shortcomings of the contemporary city. In general terms, the company argues that with 'rising urban populations, ageing infrastructures, and shrinking tax revenues today's cities demand more than traditional solutions'.[19] Across domains, cities, in IBM's urban theory, are facing the same issues: 'growing demands', 'tightening budgets', 'financial deficits', 'volatile markets', 'growing complexities', 'pollution', 'urban growth'. The city is in other words a 'sick city' permeated by a series of pathologies. To confront them, municipalities are hampered by 'inadequate systems to serve basic needs', 'obsolete' or 'broken technologies', 'litigation costs', 'benefit frauds' and 'wasted time'.[20] In short, the picture is grim and cities appear close to a fatal breakdown.

What this narrative conveys is the negative side of the 'utopian mirror image' (Choay, 1997, 261–262). As planning historian Françoise Choay (1997) has argued, this is since Thomas More's *Utopia* (1516) a constitutive trope of the utopian tradition in urban planning.[21] Utopian urban planning is always conceived, she argues, as a therapeutic discourse starting off with a diagnosis of urban problems and pursuing with a set of universally valid solutions. She defines the utopian genre as: a single voice proposing—through a narrative distinguishing between a corrupted past and a perfect and immutable future—an ideal and universally valid model of society constituted by rational spatial form. IBM's smarter city discourse is in many respects in line with this tradition: its core is utopian storytelling.

First, the smarter cities story is a univocal one: nowhere in the campaign are other approaches or solutions to urban problems mentioned (Vodoz, 2013, 52). Utopianism, like the smarter cities model of urban management, is not a collective project assembling different worldviews and interests, but a singular 'emancipatory' vision. Thus, the reply to Throgmorton's (1996a) question 'who tells the story?' is simple: the corporation. Second, the smarter cities campaign hinges on a before–after demonstration closely related to the above mentioned 'if … (data-mining and systems thinking) then … (cities will become smarter)' argument. This is most vividly conveyed on their website by one of the first elements visitors encounter: an invitation to visit the museum of urban problems: 'Before the City got Smart, Exhibition'. The 'exhibition' proposes a travel in time where problems are portrayed as pictures hung on the line and where visitors learn about problems of the past, now—in the fictitious present of an accomplished smart city—resolved by smart technologies. 'With intelligence infused into the way cities worked, urban blights became history', the introduction to the exhibition proudly announces.[22] This before–after—or 'weightwatchers'—rhetoric is repeated throughout the exhibition as well as in the presentation of IBM's smart services and technologies. Throughout, technological solutions are presented as the pharmakon of contemporary urban pathologies through images and short stories.

One canvas for instance is entitled 'the queue' and explains that:

> Before the advent of smart information systems, people actually had to turn up in person to be seen by health centers, passport offices, post offices, embassies, the DVLA and the DMV. Long lines, known as 'queues', quickly formed as people stood around aimlessly for hours. Finally in the early 21st century, electronic declarations cut queues and billions of euros in administration costs.[23]

The third aspect of a utopian rhetoric, the smarter cities story, depicts a model of a perfectly functioning urban society but, in contrast with classical utopianism, it is governed by code rather than spatial form. Problems are specified as we have seen through 'a culture of analytics and systems' and thereby brought to a level where they can be addressed through code. This requires access to data and their interconnection through software. Data, it is argued, are trapped, 'unsmartly' organized in information silos, lost and not available when needed.[24] They are under-used and their potential should therefore be unleashed.

When detailed information on all aspects of the nine systems is systematically collected and connected, IBM's algorithms can process the data and optimize each system. In case usable data do not exist, they have to be produced. Digital electricity meters with a connection to the internet should, for instance, replace analogous ones to allow the implementation of smart electricity grids enabling a more efficient use of electricity. The solution is summarized in the marketing language of IBM as the three Is: to become smarter, the world (or the city) needs to be Instrumented, Interconnected and Intelligent. In other words, the core of 'smartness' lies in the algorithm.

Optimization through code is therefore the utopia promised by the company. As the vice president of the smarter infrastructure division puts it: 'The ultimate smart city, the Shangri La if you will, is one where all the systems communicate'.[25] This 'ultimate smart city' is a transparent one where all flows within the nine systems are quantified, connected and efficiently managed. Take public safety and traffic, for instance. Here the smarter cities program is not designed to suggest more police officers, police cars or roads, but information about them: where officers are, where accidents or traffic jams occur. Therefore, 'smarter cities' is a mild utopianism: it promises efficiency rather than paradise on earth. It is a utopian rhetoric tempered by market realism: it is easier to sell technologies and services than an ad nihilo urban structure, more convincing to tap on the faith in technology and progress than to promise a brave new city. As Vodoz (2013, 71) notes, the ideal of perfection is transposed from material space to virtual space. In other words, the smarter cities model does not suggest a revolution in urban morphology, such as Howard's Garden City model, but a reformist optimization through data, monitoring, interconnectedness and automatic steering mechanisms. In contrast with other utopian models, smarter cities do not require the replacement of existing spaces, but their digital redoubling.

Finally, in the perfect future of classical utopias, historicity is abolished: the arrow of time is bent into a circular repetition. In the bright future promised by IBM's smarter cities, historicity is not abolished because optimization needs to be constantly renewed: novel technologies need to be constantly introduced for that purpose and codes constantly rewritten. If IBM's storytelling rests on a utopian rhetoric, it constantly makes sure that the future it promotes is a realistic one.

In sum then, IBM's storytelling rests on two rhetoric pillars. The first is systems thinking which inscribes it in a techno-scientific imagination and provides it with the legitimacy of science. More concretely, it allows the translation of the city into a common language on which the company's technology can act. The second is a utopian story which recurs to an imaginary of progress, therapy and conversion (if not redemption). Each rhetoric pillar brings different elements to the persuasive power of the smarter cities campaign. They are building blocks of this discourse's authority. This discursive strategy is meant to persuade municipalities to think of the company as an OPP for an efficient and sustainable urban development.

Conclusion: the smarter city discourse as a framing device

On the surface, the dominant smart cities' storyline is about efficient and sustainable cities, but underneath it is primarily a strategic tool for gaining a dominant position in a huge market where, as Townsend (2013, 63) puts it, 'Siemens and Cisco aim to be the electrician and the plumber .... [and IBM] their choreographer, superintendent, and oracle rolled into one'. Our chapter looked at IBM's storytelling in its smarter city campaign in order to grasp some central specificities of the smart city model of urban development as it is usually presented in the public sphere. This story should not be taken at face value of course. What we have proposed is not a description of how smart cities work on the ground but a deconstruction of a communication strategy: what one of our IBM informants calls a market creation strategy.[26] The smarter city discourse is a framing device, as are other discourses on urban development: it makes us consider cities differently to promote new modes of urban management and development. It more specifically develops what Vanolo (2014, 893) calls a 'smartmentality' through which 'citizens are very subtly asked to participate in the construction of smart cities'. This mentality has of course positive aspects: it favors, for instance, efficient solutions to improve urban sustainability in sectors such as energy or transport. However, it also raises a series of critical questions. We briefly address two of them below. The first is that this discourse promotes an informational and technocratic conception of urban management where data and software seem to suffice and where, as a consequence, knowledge, interpretation and specific thematic expertise appear as superfluous. This is a rather dangerous fiction. It leads us back to an epistemology dominant during the 1950s and 1960s, the heyday of spatial analysis and of the belief in the universal power of quantitative models in a discipline like human geography (Billinge et al., 1984). We must admit that the plurality of approaches and languages developed in urban studies since the 1970s does not facilitate their fruitful use in urban management and development strategies. Nonetheless, a return to positivist dreamlands would hardly be progress as problems cannot be reduced to data problems but need to be interpreted in the light of longstanding political and scientific debates. Furthermore, we've been there before: municipalities in the 1960s and 1970s have already experienced the deleterious consequences of taking such stories about large-scale simulations being the ultimate planning solution at face value (Townsend, 2013, 76–82).

The second and correlated point is that such a discourse promotes a mentality where urban affairs are framed as an apolitical matter. In the smarter cities campaign, causes of urban problems are associated with demographic trends, such as an estimated doubling of world urban population by 2050,[27] climate change and tight municipal budgets. Never with politics. The rhetorical means of the campaign also aspire to political neutrality. Systems thinking is neither progressive, nor conservative: it decomposes a phenomenon into related parts. There are of course left- and right-wing utopias, but the horizon of a utopian structure of thinking is apolitical. As Choay (1997, 174) points out, political power in utopias is an 'epiphenomenon'. Similarly, smart technologies can optimize any system, from the surveillance of political opponents to waste management. It can be sold to democratic regimes such as Denmark as well as to much less democratic ones such as Syria. Very much like Le Corbusier saw functionalist urbanism as an apolitical model he was ready to propose to postcolonial India, fascist Rome or Stalinist Russia, smart urban technologies are an omnibus ready to stop wherever customers are to be found.

The apparent political neutrality of the dominant smart city story is reinforced by the production of evaluation criteria and rankings where cities are classified according to their degree of smartness[28] and by supranational funding programs such as the 2012 European Union (EU) program on 'Smart Cities and Communities'.[29] These initiatives introduce a new 'economy of worth' (Boltanski and Thévenot, 2006)—a new way of evaluating the worth of people and things—urging cities at the bottom to climb up the smart city ladder. Such rankings and financial incentives fuelled by smart talk can of course lead to necessary technological developments, but they might also obfuscate more urgent needs. Becoming a smarter city implies giving priority to investments in technology, while technology-poor affordable housing or sewage systems are arguably more urgent in many of the world's cities. Priority-making is of course not an apolitical matter, but the very core of municipal politics. In other words, IBM's storytelling campaign contributes to subtly introducing a new moral imperative where smartness becomes, like creativity, a new necessary asset of cities. The apparent apoliticism of the campaign naturalizes ubiquitous urban technology as an OPP for municipalities' development.

Recent critical work on smart cities has explored alternatives to this dominant corporate vision of urban futures (Hollands, 2013; McFarlane et al., 2013; Townsend, 2013). For Hollands (2013, 13), 'the real smart city has to begin to think with its collective social and political brain, rather than through its "technological tools"'. This alternative smart city exists. It is made up of myriads of initiatives where technology is used to empower community networks, to monitor equal access to urban infrastructures or scale up new forms of sustainable living. However, contrary to corporate storytelling, no straightforward narrative about the smart city emerges from these initiatives as they can be driven by very different and politically variegated motives. It is in this context that an alternative storytelling about smart cities is necessary. Storytelling in planning is not only a possible model of planning but also for planning (Van Hulst, 2012a) and should not only be used as an instrument of critique but also as an instrument to suggest progressive avenues for urban development (Sandercock, 2003a, 26). This is not an easy task because it requires generalization from initiatives that respond to local needs and are therefore usually local in scope. It also requires being explicit about normative and political positioning as smartness only makes sense within a system of values and aims. However, this effort in storytelling is necessary to move beyond critique, and beyond a mere contrast between corporate grand schemes and what easily might be perceived as anecdotal small-scale actions.

Acknowledgments

This chapter is based on the research 'Smarter Cities: New Urban Policy Model in the Making', funded by the EU-COST Action 'Living in Surveillance Societies' (LiSS).

Notes

1 In this chapter, we alternatively use the terms 'smart cities' and 'smarter cities'. Both refer to the same idea and are often used interchangeably in the literature and online publications. The difference between the two is that 'smart city' is legally unprotected and can thus be used more freely or is interpreted and applied more widely while 'smarter cities', as we explain below, legally belongs to IBM and refers to the company's software and campaigns. We therefore use the term smart city to discuss existing literature and when tracing the origins of the smart city idea, while we use smarter cities when discussing IBM's campaign.

2 A 'model', etymologically, is a figure to be reproduced (Söderström and Paquot, 2012, 41). See also developments below on urban models and the utopian planning genre.

3 As exact search parameters we used 'smart! city' allowing a search for all word combinations starting with smart and city (in newspapers only). There were 1952 matches, date of search: 5 September 2012.

4 The Adelaide project failed in 1997 due to a lack of investments leading to the fact that it eventually became an ordinary business park. For analyses of the Malaysian case, see: Bunnell (2002), Lepawsky (2005) and Brooker (2013).

5 www.youtube.com/watch?v=i_j4-Fm_Svs.

6 www.trademarks411.com/marks/79077782-smarter-cities.

7 In 2001, Cisco started, together with the developer Gale International, to build the smart city of Songdo in South Korea and in 2010 founded the 'Smart and Connected Communities Institute' and the 'Connected Urban Development Initiative' where the company's research on urban technologies is being conducted. In 2011, Siemens decided to invest on a huge scale in a new Infrastructure and Cities division and its own version of smart urbanism. Microsoft entered the game in 2013 with its City Next initiative.

8 The site analyzed is the US site of the smarter cities campaign which is richest in information (www.ibm. com/smarterplanet/us/en/smarter_cities/overview/index.html and www-03.ibm.com/innovation/us/thesmartercity/index_flash.html). Country-specific websites are slimmed down and translated versions of the US sites.

9 According to the market research company ABI (www.abiresearch.com/press/395-billionwill-be-spent-on-smart-city-technologi).

10 According to the market research company Navigant (www.navigantresearch.com/research/navigan tresearch-leaderboard-report-smart-city-suppliers).

11 See, for instance, www.navigantresearch. com/research/navigant-research-leaderboardreport-smart-city-suppliers.

12 In IBM's words: 'What makes a city? The answer, of course, is all three [pillars]' (www.ibm.com/smarterplanet/us/en/smarter_cities/overview/index.html).

13 On its interactive site, IBM does not use the hierarchy of main and sub-pillars but uses a list of 11 individual pillars, but there are only minor differences in content (cf. www-03.ibm.com/innov ation/us/thesmartercity/index_flash.html).

14 The project started in 2010 (http://www-03.ibm. com/software/products/us/en/intelligentopera tions-center/). Siemens's City Cockpit, first showcased in Singapore in 2009, is another variation of the same idea.

15 Speech given at USC Price, School of Policy Planning and Development in November 2011 (www. youtube.com/watch?v=IpFyQOW_ldQ).

16 www.forum48.org/portfolio/ginni-rometty/.

17 Interestingly however, Cook (the company's abovementioned systems thinker in chief) refers to Donella H. Meadows, lead author of *The Limits to Growth*, and author of a humanistic/environmentalist plea for systems thinking building on more elaborate complex systems theory (Meadows *et al.*, 1972).

18 www.ibm.com/smarterplanet/global/files/us en_us cities city_leaders_wsj.pdf.

19 http://ibmtvdemo.edgesuite.net/software/industry/intelligent-oper-center/demo/3137_Control-Room_Web.html.
20 See, for instance: www.ibm.com/smarterplanet/us/en/smarter_cities/overview/index.html www.ibm.com/smarterplanet/us/en/healthcare_solutions/ideas/index.html; www.ibm.com/smarterplanet/us/en/government/ideas/index.html.
21 For Choay (1997) there are two main traditions, the utopian one, that she calls the model, and 'the rule' for which urban planning is based on a set of generative principles and which finds its origin in Leon Battista Alberti's De re aedificatoria (1485).
22 www-07.ibm.com/innovation/my/exhibit/index.html.
23 www-07.ibm.com/innovation/my/exhibit/gallery_low.html#queue.
24 See, for instance: www-03.ibm.com/innovation/us/thesmartercity/communications/chapter3.html#!/0; www-03.ibm.com/innovation/us/thesmartercity/traffic/index. html#!/1; www-03.ibm.com/innovation/us/thesmartercity/airports/index.html#!/3.
25 www.ibm.com/smarterplanet/us/en/green_buildings/overview/index.html.
26 IBM manager, interview 13 February 2013.
27 See www.ibm.com/smarterplanet/global/files/us en_us cities city_leaders_wsj.pdf.
28 For instance: www.smart-cities.eu/benchmarking.html.
29 On the EU program, see: http://ec.europa.eu/eip/smartcities/.

References

Allwinkle, S., and P. Cruickshank. 2011. "Creating Smarter Cities: An Overview." *Journal of Urban Technology* 18 (2): 1–16.
Bakıcı, T., E. Almirall, and J. Wareham. 2013. "A Smart City Initiative: The Case of Barcelona." *Journal of the Knowledge Economy* 4 (2): 135–148.
Bakker, K. 2011. "Splintered Urbanisms: Water, Urban Infrastructure, and the Modern Social Imaginary." In *Urban Constellations*, edited by M. Gandy, 62–64. Berlin: Jovis.
Beall, J., and S. Fox. 2009. *Cities and Development*. London: Routledge.
Bell, S. 2011. "System City: Urban Amplification and Inefficient Engineering." In *Urban Constellations*, edited by M. Gandy, 71–74. Berlin: Jovis.
Berry, B. J. 1964. "Cities as Systems within Systems of Cities." *Papers in Regional Science* 13 (1): 147–163.
Billinge, M., D. Gregory, and R. L. Martin, eds. 1984. *Recollections of a Revolution: Geography as Spatial Science*. London: Macmillan.
Boltanski, L., and L. Thévenot. 2006. *On Justification: Economies of Worth*. Princeton: Princeton University Press.
Brooker, D. 2013. "From 'Wannabe'silicon Valley to Global Back Office? Examining the Socio-spatial Consequences of Technopole Planning Practices in Malaysia." *Asia Pacific Viewpoint* 54 (1): 1–14.
Bunnell, T. 2002. "Multimedia Utopia? A Geographical Critique of High-tech Development in Malaysia's Multimedia Super Corridor." *Antipode* 34 (2): 265–295.
Callon, M. 1986. "Éléments pour une sociologie de la traduction: La domestication des coquilles saint-jacques et des marins-pêcheurs dans la baie de saint-brieuc." *L'Année sociologique (1940/1948–)* 36: 169–208.
Caragliu, A., C. Del Bo, and P. Nijkamp. 2011. "Smart Cities in Europe." *Journal of Urban Technology* 18 (2): 65–82.
Choay, F. 1997. *The Rule and the Model: On the Theory of Architecture and Urbanism*. Cambridge: MIT Press.
Churchman, C. W. (1968) *The Systems Approach*. New York: Delacorte Press.
Deakin, M. 2014. *Smart Cities: Governing, Modelling and Analysing the Transition*. London: Routledge.
Deakin, M., and H. Al Waer. 2011. "From Intelligent to Smart Cities." *Intelligent Buildings International* 3 (3): 140–152.
Fischedick, M., and S. Lechtenböhmer (2012) "Smart City – Schritte auf dem Weg zu einer CO_2-armen Stadt." In *Smart Energy*, edited by H.G. Servatius, U. Schneidewind, and D. Rohlfing, 395–414. Berlin, Heidelberg: Springer.
Forrester, J. W. 1969. *Urban Dynamics*. Cambridge: MIT Press.

Friedmann, J. 1987. *Planning in the Public Domain: From Knowledge to Action*. Princeton, NJ: Princeton University Press.

Gandy, M. 2004. "Rethinking Urban Metabolism: Water, Space and the Modern City." *City* 8 (3): 363–379.

Greenfield, A. 2013. *Against the Smart City*. New York: Do Projects.

Healey, P., and J. Hillier. 2010. *The Ashgate Research Companion to Planning Theory: Conceptual Challenges for Spatial Planning*. Farnham: Ashgate.

Hollands, R. G. 2008. "Will the Real Smart City Please Stand Up?" *City* 12 (3): 303–320.

Hollands, R. G. 2013. "Is an 'Alternative' Smart City Possible? Critically Revisiting the Smart City Debate." Paper presentat at the Smart Urbanim: Utopian Vision or False Dawn?, in Durham University.

Kitchin, R. 2014. "The Real-time City? Big Data and Smart Urbanism." *GeoJournal* 79: 1–14.

Kuk, G., and M. Janssen. 2011. "The Business Models and Information Architectures of Smart Cities." *Journal of Urban Technology* 18 (2): 39–52.

Latour, B. 1987. *Science in Action: How to Follow Scientists and Engineers Through Society*. Cambridge: Harvard University Press.

Lepawsky, J. 2005. "Stories of Space and Subjectivity in Planning the Multimedia Super Corridor." *Geoforum* 36 (6): 705–719.

Lynch, M. 1988. "The Externalized Retina: Selection and Mathematization in the Visual Documentation of Objects in the Life Sciences." *Human studies* 11 (2): 201–234.

Mandelbaum, S. J. 1985. "Thinking About Cities as Systems Reflections on the History of an Idea." *Journal of Urban History* 11 (2): 139–150.

Mandelbaum, S. J. 1991. "Telling Stories." *Journal of Planning Education and Research* 10 (3): 209–214.

Marcuse, P. 2005. "The City'as Perverse Metaphor." *City* 9 (2): 247–254.

McFarlane, C., S. Marvin, and A. Luque. 2013. "*Smart Urbanism: Towards a Critical Research Agenda*." Paper presented at the Smart Urbanism: Utopian Vision or False Dawn? At Durham University.

McNeill, D. 2013. "Flat World Cities? A Critical Analysis of Ibm's Smarter Cities Initiative." Paper presented at the Smart Urbanism: Utopian Vision or False Dawn? at Durham University.

Meadows, D. H., D. L. Meadows, J. Randers, and W. Behrens. 1972. *The Limits to Growth*. New York: Universe Books.

Mehmood, A. 2010. "On the History and Potentials of Evolutionary Metaphors in Urban Planning." *Planning Theory* 9 (1): 63–87.

Paskaleva, K. A. 2011. "The Smart City: A Nexus for Open Innovation?" *Intelligent Buildings International* 3 (3): 153–171.

Rat-Fischer, C., F. Rapp, P. Meidl, and N. Lewald. 2012. "Smart City: Energy Efficiency in a New Scope." *In Resilient Cities* 2, 119–124. Springer.

Sandercock, L. 2003a. "Out of the Closet: The Importance of Stories and Storytelling in Planning Practice." *Planning Theory & Practice* 4 (1): 11–28.

Schaffers, H., N. Komninos, M. Pallot, B. Trousse, M. Nilsson, and A. Oliveira. 2011. "Smart Cities and the Future Internet: Towards Cooperation Frameworks for Open Innovation." In *The Future Internet*. FIA 2011. Lecture Notes in Computer Science, vol 6656, edited by J. Domingue et al., 431–446. Berlin, Heidelberg: Springer.

Sennett, R. 1994. *Flesh and Stone: The Body and the City in Western Civilization*. New York: W.W. Norton & Company.

Söderström, O. 1996. "Paper cities: Visual thinking in urban planning." *Cultural Geographies* 3 (3): 249–281.

Söderström, O., and T. Paquot. 2012. "Modéles urbains." *Urbanisme* 383: 41–42.

Streitz, N. A. 2011. "Smart Cities, Ambient Intelligence and Universal Access." In *Universal Access in Human-Computer Interaction. Context Diversity*. UAHCI 2011. Lecture Notes in Computer Science, vol 6767, edited by C. Stephanidis, 425–432. Berlin, Heidelberg: Springer.

Throgmorton, J. A. 1996a. *Planning as Persuasive Storytelling: The Rhetorical Construction of Chicago's Electric Future*. Chicago: University of Chicago Press.

Throgmorton, J. A. 2003. "Planning as Persuasive Storytelling in a Global-Scale Web of Relationships." *Planning Theory* 2 (2): 125–151.

Townsend, A. M. 2013. *Smart Cities: Big Data, Civic Hackers, and the Quest for a New Utopia*. New York: W. W. Norton & Company.

Van Hulst, M. 2012a. "Storytelling, a Model of and a Model for Planning." *Planning Theory* 11 (3): 299–318.

Vanolo, A. 2014. "Smartmentality: The Smart City as Disciplinary Strategy." *Urban Studies* 51 (5): 883–898.

Vodoz, L. 2013. La ville intelligente d'ibm, topique utopique?: Le rêve de la ville idéale porté par les technologies de l'information et de la communication In MA thesis. Université de Neuchâtel: Bibliothéque de l'Université.

Wyly, E. 2013. "The City of Cognitive–Cultural Capitalism." *City* 17 (3): 387–394.

21

Will the real smart city please make itself visible?

Edward Wigley and Gillian Rose

Introduction

Smart cities depend on many kinds of visualisations (usually digital), refracted through multiple media and offering differing kinds of interfaciality, interaction and affectivity. These visuals can be displayed on many kinds of screens, including smartphones, tablets, laptops and desktops. And they are put to many different purposes. There are advertisements for smart city technologies (Rose, 2018a); art and activist projects (Crang and Graham, 2007); the illustrations in city authorities' vision statements and policy documents; the interfaces of smartphone applications and data hubs (Offenhuber, 2015); photographs of smart city events; the logos of companies designing smart products; and many more. These have diverse aesthetics: they can take the form of 'cute' graphics of primary colours and simplified content, futuristic painted fantasies or photorealistic street scenes, and often cite not the histories of urban visualisation techniques but those of computer games, architectural visualisations (Rose, 2017), selfies or science fiction movies (Vanolo, 2016).

Some of these visualisations of the smart city have already gained scholarly attention, and many have been derided as ridiculous fantasies (Rose, 2018a). Promotional videos, for example, are part of wider corporate hype about the potential of smart to transform cities (Söderström et al., 2014; Wiig, 2016). But many smart city visualisations are concerned to express not an overheated fantastical imaginary but the real: the reality of a smart city. Hollands (2008) important paper asked the 'real smart city' to stand up; this chapter proposes that it is less likely to stand up than to appear, literally, on a screen, asking viewers to engage and interact with it. But what does 'real' mean here? What visual rhetorics are deployed to make images of smart cities convincing and persuasive?

Picturing the smart city as real can happen in different ways, as this chapter demonstrates by exploring a handful of examples. The chapter begins by discussing visualisations of data, often big, real-time data. Big data – its harvesting, combination and analysis – is at the heart of many definitions of smart cities, and it is often converted into a wide range of visual formats. The chapter begins with an iconic picture of that picturing: a photograph of the smart city operations centre in Rio de Janeiro. The chapter considers the various claims to reality being made through this photograph, before moving on to further multiply smart realities by considered two more forms of digital visualisation: an interactive three-dimensional map and a virtual reality experience designed to test reactions to autonomous vehicles.

Visualising smart city operations centres visualising smart city data

Cities have always been places where data have been collected and analysed regarding their infrastructures, finances and citizens (Mattern, 2017; McNeill, 2016). Famously, a pioneer of urban data visualisation, nineteenth-century physician John Snow, used an analogue form of what we would now call Geographical Information Systems to map and identify the source of a cholera outbreak in London in 1854 (Schuurman, 2004). Indeed, smart cities are just the latest manifestation of a range of extensive efforts to improve the management of cities through the analysis of data (Batty, 2013; Halpern, 2015). One of the things that makes 'smart cities' somewhat distinct from their cybernetic and intelligent forebears, however, is the visualisation of very many aspects of their activity. So while forms of visualisation were certainly important to its predecessors, in a smart city visualising data about urban processes is foundational. For example, IBM's Operations Centre in Rio de Janeiro is a room with no windows but with two of its three visible walls covered in screens; each workspace at the rows of tables that fill the floor has a computer with a glowing screen (Figure 21.1). The

Figure 21.1   The control room of the IBM Centro de Operações in Rio de Janeiro
Source: Reproduced with permission

room is full of people looking at screens, or looking at the people looking at screens (from the observation deck above the room's third wall), who are also being looked at by the TV audiences via a local TV broadcast from the Centre. The smart city is managed through such screened images.

Urban data dashboards in smart city operations centres like Rio's display a wide range of image types, including CCTV feeds of traffic flows, key performance indicator 'traffic lights' and many kinds of maps showing everything from sentiment analyses of geolocated tweets to rainfall patterns, each of which are distinct kinds of images with different relations to different kinds of data. There are other kinds of dashboards, too: online portals to open access big datasets about particular cities (for examples of urban dashboards see Kitchin et al., 2015, 2016; Kourtit and Nijkamp, 2018; Mattern, 2015). The screens of the smart city control centre are usually understood to be a form of surveillance, and the work of Foucault (1977) on the panopticon is often invoked (see for example McNeill, 2016). Such control centres, it is argued, enact the panoptic gaze of authority as it disciplines urban subjects, compelling them to conform to appropriate forms of behaviour; in this argument, visualisation of urban data is driven by the need for city authorities to control urban spaces and urban populations (McNeill, 2016). The presentation of the city in dashboard formats as a set of neatly sliced visual indicators produces a real in the form of 'a series of essentialized understandings of the city that render it knowable, governable and, ultimately, intervene-able in particular ways' (Shelton, 2017, p. 3; see also Kitchin et al., 2016).

In this critique of the control centre, control is indeed the fundamental relation enacted by the smart city dashboard and its various forms of data visualisation (Luque-Ayala and Marvin, 2016); the enactment of surveillance in which the complexity of urban processes is distilled, and the heterogeneity of urban dwellers is dissolved into flows of data, intermittently signalling the need for intervention. The emphasis on measurable data produces a 'god-trick' vision of the city (Haraway, 1997), which sees a chaotic and disorderly set of problems waiting to be solved through rationalist and technologically enabled interventions (and see Rose, 1993). Luque-Ayala and Marvin (2016) offer a more nuanced account, pointing out that the regular media engagement (including live TV broadcasts from the Operations Room) of the Rio Operations Centre enables a degree of accountability: the city authorities' access to data is broadcast for the local television viewers to observe. However, as Halpern (2015) observes in the case of Songdo's operation room in South Korea, the role of those pictured overseeing the monitors and dashboards is often irrelevant to the working of the dashboard because computer systems and algorithms are continually adjusting or reacting to the unfolding situation. The production and control of the smart city through using a data dashboard is not in fact located only at the screen interface.

Surveillance and control, then, are the outcomes of the analysis and visualisation of urban data in smart city operation centres. The centre as the site where multiple datasets are integrated and visualised assumes that the data on the screens represent at least some parts of the reality of the city: data are taken as evidence of the material world. That its workers wear white coats only seems to underline its objective, even scientific, character. We would also like to suggest that the widespread use of the Rio Operations Centre itself makes a reality claim. Photographs have long been understood as an indexical record of what was there when the camera shutter snapped (Tagg, 1988). By picturing what looks like a real place, where a city is in fact being managed, this photograph itself suggests that the smart city is a real possibility: here, it is being enacted!

This combination of the city being converted into a quantitative data, and the widely circulated photograph of the operations centre having an indexical relation to the centre, alerts us to the ways in which multiple kinds of realities are being constituted. Multiple strategies for picturing the 'actually existing smart city' or 'real smart city' are possible. The next section looks at two more.

The realities of embodied interaction: maps and virtual reality

The visualisation of smart city control centres can make different claims about the reality of a smart city. The screens quantify the city and pictures of the control centre make smart an actually existing thing. Other smart visualisations appear to offer a glimpse of a real kind of smartness because of the forms of embodied interaction they invite. This section explores an interactive, 3D map and the use of virtual reality (VR) technologies in a smart city in the UK. The use of digital 3D visualisations of urban spaces is increasing rapidly, and it can take different forms. One form is the use of interactive 3D models and another is VR. This section looks at both of these in turn, exploring their visualisations of real smart and the increased level of immersion and embodiedness each iteration produces.

The discussion is based on two projects that visualised the UK city of Milton Keynes (MK), one of the leading smart cities in the UK which over the past decade or so has produced a number of successful bids for smart city projects. Examples include an GB£8 million grant for the deployment of charging infrastructure for electric vehicles, a £13 million smart grids trial and £150 million for a transport innovation centre, the Transport Systems Catapult (TSC). Another project, MK:Smart, was funded by the Higher Education Funding Council for England between 2014 and 2017 (Valdez, Cook and Potter, 2018). Both the TSC and MK:Smart facilitated many smaller projects, often in partnership with other funders. As a result, MK is a smart city full of relatively short-term projects with significant amounts of public and civic as well as corporate involvement, and with a wide range of smart stakeholders including charities, voluntary sector organisations and local campaigners, as well as MK:Smart and TSC partners. This chapter looks at just two projects in this complex domain of organisations and technologies. The larger study of which these are a part gathered 58 semi-structured interviews, which were recorded, transcribed and analysed using NVivo software; interviewees included start-up entrepreneurs, local government workers and policy makers, data scientists, community organisers, transport experts and campaigners.

The material presented here focusses on an app commissioned by the city council from a local digital visualisation company, called MK:3D, and the use of VR technologies as part of trials of autonomous vehicles by the TSC in MK. It uses interviews with various policy makers and designers involved in the design and use of the app and the VR to understand what they understood and assumed these visualisations to achieve. As we will see, in both cases there was a strong concern to produce a 'real' picture of MK through the interactive qualities of the visuals; but these 'reals' were also somewhat different.

Embodied interaction I: swiping through three-dimensional interactive maps

Within Milton Keynes, 3D virtual models have played a number of roles in various smart city-related projects. Visualisation company Virtual Viewing is based in Milton Keynes and a leading organisation within the 'smart city' movement in the city, partnering the local authority as well as the TSC. The company creates computer generated images (CGI), animations, smartphone application and animated graphics, including short animated sequences for the LUTZ Pathfinder (autonomous vehicle) project discussed in the following section. Other work includes 3D visualisations for local authorities in England, seeking to present data 'in a visual, easy-to-understand manner' (Bailey, Deshpande and Miller, 2014, p. 427).

One recent project saw Virtual Viewing in partnership with Milton Keynes Council to create an interactive, 3D map app of the town centre. This smartphone app – called the MK:3D Map – displays information about the city in order to encourage inward investment; there was also some hope that eventually its contents would also enable workers, residents and visitors to navigate the city centre better. As part of a lineage where cities commission digital upgrades (see Willis and Aurigi, 2018 for discussion of DDS Amsterdam), during our interview with Virtual Viewing they reported that they had encouraged the local authority to commission a 3D map app with the incentive that if the Council did not, a private sector organisation was likely to; the implication being that MK Council needed to be in control of its digital assets. Such a sentiment draws attention to the ongoing data colonisation of privately owned public spaces in UK towns and cities as well as concerns over the colonisation of virtual spaces (Cheesman et al., 2002). There is indeed a wealth of 3D virtual objects that are available from commercial providers and can be purchased 'off the shelf' and plugged into virtual worlds: Hum3D.com, for example, offers a range of 3D objects including the Arriva Milton Keynes Electric Bus and an architecture library with models of iconic structures such as the London Eye, Taj Mahal and even the humble Asda supermarket petrol station amongst others.

After previous iterations of a 3D Milton Keynes (Bailey, Deshpande and Miller, 2014), the MK:3D Map was launched in 2018. At a business and public engagement event, the MK CityFest (2017) programme stated the 3D map 'will play a key role in place marketing for the city', a sentiment reinforced several times during its launch presentation at CityFest by speakers from MK Council's Economic Development team and Destination: MK (the local tourist information service). Local businesses were invited to be part of this project, to add data about their 'plots' of virtual Milton Keynes for visitors and potential investors to view. For Virtual Viewing, this app is just the start of a data platform that is openly available for exploitation by third party interests – their aspiration was that users of the app would be able to look up any asset on the smartphone and find information about, for example, the available space for rent or energy usage or transport links or car parking spaces. Indeed, Virtual Viewing envisage a time where the user's phone will beep as they walk past certain buildings to offer place- and time- specific data to their attention. The smartphone app here is functioning as a platform, then, for the companies sharing data with it.

Accessing that data, though, is both a visual and a haptic experience. The map around which the app is based is a 3D model of MK, constructed by the app developers from satellite data. The user swipes the screen to 'fly' over the model, and then taps a location to generate a new screen. The new screen holds a 3D model of a particular building or site, with its data also displayed, and by swiping the screen the viewer can move around the site. This is not the first app of this type that Virtual Viewing have designed, and they do not attempt photographic realism; the app is not in indexical image of the city. Instead, rather like the screens in a control centre, it extracts certain kinds of data, collates and visualises it. What is different is the mode of interaction invited with this version of the city. Rather than the surveillant gaze at the screen and, through the screen, the city, this app engages users both visually and haptically. In that sense, it is a more sensory engagement with city, as fingers swipe, the 3D model rolls, and then using more gestures the user can zoom in and out over the model or change the screen image entirely. The mobility of the point of view is quite different from the static gaze out from the control centre.

This app is clearly not picturing the smart city in the same way as the operations centre does then. The data it presents are far less so it cannot make the same claim to comprehensiveness that the control centre can. It does not occupy the same fixed point, and it offers a far more tactile encounter with urban data. Nonetheless, the aim of these forms of visualisation, according to the app designers, is to make data – objective, useful data – more accessible and therefore more useful. This real is partial and selective in terms of the data that it shows, but it aims to offer a more intense, visceral engagement with that data. The 3D model presents an introductory embodied experience of the visual, an embodied experience which is further amplified in the next iteration of the visual to be discussed, VR.

Embodied interaction II: immersion in virtual reality

VR is the latest in a long line of iterations at immersive technologies, from Victorian panoramas to stereoscopes, to 3D cinema (Elsaesser, 2013) and toys like the 1970s' 'View-Master' which uses stereoscopic lens to display photos in 3D. Artists and designers have attempted to imaginatively transport the viewer to other locations by enveloping their visual field and restricting as much as possible the view outside of this field. The latest VR headsets demonstrate a quantum leap at increasing the depth of movement around a virtually constructed visual and audio landscape. Lower cost VR headsets such as the Google Cardboard glasses are available for the home-user and the commercial and industrial client, expanding the audience beyond specialist cinema and industry organisations.

Quite what constitutes 'virtual reality' is not without dispute. Brodlie et al. (2002) state that the most distinctive characterisation of VR is the fully immersive experience, whilst they later seek a diversity of definitions that present VR as a set of trajectories accommodating variable factors of the level of immersion, granularity of scale, interaction and computing power. There have been significant advances in VR, of course, since this framework was established in 2001. However this flexibility in their conceptual understanding recognises the diverse forms of VR currently available, from the cardboard glasses that work with smartphones to display 360 degree videos, to the massive circular 'Omnideck' treadmill discussed later in this chapter. There is a continuing narrative within these technologies, however, of the receding of the materiality of these devices in perception of the user. For example, recent smartphone models from Samsung and Apple remove the physical framing of the screen to induce an 'infinity' effect where the screen edges blend into the background, merging the real and virtual. Weising (2010) observes that the framing of a screen denotes the unrealness of what is displayed. The removal or partial removal of this screen from flat screens replacing bulky cathode-ray tube monitors to new smartphone models or in-depth immersive VR goggles suggest an attempt to authenticate, or to make real, the virtual scene by increasing the feeling of immersion in it.

The near-total visual immersion of the user in VR has significant consequences for their subjective experience of the virtual world (Sturken and Cartwright, 2001). The non-linear and ambiguity of determinism in the VR experience has been argued to liberate the gaze (Mirzoeff, 1999), since the user is able to interact with, and even to manipulate, their experience of the VR (audio-)visual image. Yet this neglects the limitations imposed upon the user by the VR designer. As Weising (2010) asserts, the user is only interacting with the virtual world within the limits imposed by the design of the VR. What is the imagined real, then, that users of the TSC VR experience?

Of the smart city actors in Milton Keynes, VR has been a major component in one of the core activities of the TSC: the autonomous vehicle 'LUTZ Pathfinder' project, which is part of the UK Autodrive consortium. TSC have employed VR in order to research how both pedestrians and passengers react to these 'driverless cars'. This has been an innovative feature of the LUTZ project since the pods are to drive a kilometre long stretch navigating through a series of 'shared spaces' with pedestrians, cyclists and other vehicles from Milton Keynes Central railway station to a location in the business district of the town centre. The development of 'VR trials' enables the TSC to test pedestrian reactions to the vehicle, and how they may interact, within the relative safety of the 'Visualisation Laboratory' environment.

TSC have invested in the VR technology and built a full 360-degree Omnideck (Figure 21.2) that allows a person – wearing VR headgear and other devices discussed below – to walk across moving rollers in the TSC research laboratory, while seeing a version of an MK street through the screen in their headgear. The virtual street shows the wide boulevards of central MK, slightly mechanical-looking moving pedestrians and of course, the LUTZ Pathfinder autonomous vehicle. Tracking how individuals respond to seeing a (virtual) driverless car approach allows the TSC to better understand the attitudes and behaviours of pedestrians and cyclists, or anyone else who comes into contact with the vehicles: for example, when faced with an oncoming vehicle in a shared space pavement, will the pedestrian freeze, dart to the side of the vehicle's path or continue to walk straight? VR scenarios are repeated with multiple users, and their behaviour can be measured in the safe space of the laboratory, producing an efficient and affordable set of findings, say TSC researchers. The TSC has also constructed a virtual version of being a passenger in a driverless vehicle: passenger seat and pedals with large monitors showing the views from the virtual autonomous pod as it journeys from central MK to the railway station.

Figure 21.2 The 'Visualisation Laboratory' at Transport Systems Catapult
Source: Photograph by Edward Wigley

TSC researchers emphasise that these VR experiences create an emotional and physical response in its human users (Pett, 2017). Emotional and physical responses are anticipated and recorded through the preparation for entering the virtual world in the TSC laboratory in their MK offices. Participants in the TSC research were recruited internally to participate in the VR experience. As TSC acknowledged during an interview, this did potentially limit the diversity of the sample in terms of age, gender, ethnicity and (dis)ability). Participants physically involved in a set of processes that transition them from the 'real' to the 'virtual' world. As 'pedestrians', headsets and handheld devices are attached to a ceiling-mounted rig above the central platform of the Omnideck. Using the Omnideck requires some training and preparation. Appropriate footwear must be worn, specifically the soles must be flat with little or no grip else they impede walking on the Omnideck rollers. Headset goggles are worn along with a vest that organises multiple cables, including those that power the hand-held controllers. Participants were also wired up to physiological sensors that record heart-beat, pulse and perspiration data, and the sensor data are later used to analyse the feelings of anxiety or stress when confronted with the virtual interactions. As well, before the partici-pant begins the research experiment, they must first learn to use the Omnideck, and calibrate their walking speed with the software. This takes about 10–15 minutes of walking forward in wide but small steps and then stopping in order to allow the automated rollers to return the user to the central platform. These steps of modifying the body and its movements to transition from the real to the virtual world act to 'theatricalise' the VR experience (Hohl, 2009), to ensure it is not just another activity but an event.

If the experience is dramatic, it is also highly designed. VR promises worlds that are imaginary yet perceptible and shared by different users; imagination is externalised from the creator and presented for the user's consumption (Weising, 2010). The virtual world is con-structed blemish-free. MK has a growing population of homeless sleepers and tents pitched in the city's underpasses (Coughlan, 2018) which can be seen, just, in the background of a YouTube 360 video promoting the LUTZ project (University of Oxford, 2016), showing the LUTZ pod traveling through MK. Despite their physical presence in the real MK, homeless sleepers are omitted from the virtual MK, leaving a world populated by smartly dressed, anatomically idealised CGI characters. Whether this absence is deliberate or an over-sight, this section of the MK population is rendered invisible and excluded from the repre-sentation of MK, even though the presence of their possessions and the tents may prove to be all-too real obstacle that the actual LUTZ pods will have to navigate.

Unlike the dashboards of the smart city operations centre, however, for the smart city researchers in the TSC, it is the immersive nature of VR rather than the closeness of the visual data to real life that is seen as pivotal for producing physiological reactions to the vir-tual objects. Photo-realistic textures are, to a degree, rejected by the developers at both the TSC and Virtual Viewing for a number of reasons: ergonomic (in order for the user to more clearly focus on the 3D models); economic (less paid labour is needed to produce non-photo-realism); and practical (photorealism requires more computing power to generate) (for further discussions of the compromises made in the construction of VR see Brown et al., 2002). The 360-degree experience is predicated on the movement of the head, and the eyes are fully immersed and unable to see outside of the virtual real world created. This is achieved by the headset which fully enclose the user's eyes and control what can be seen – although ultimate control of actual seeing is retained by the user being able to close their eyes and temporarily exit the virtual world. The effects on users of the immersive experience were recorded physiologically as noted above but also behaviourally: passengers of the LUTZ Pathfinder have been observed to instinctually seek an imaginary brake pedal with their right foot when the virtual pods have been driven too close to each other and looked like they might collide. In the case of VR, then, the real is less the visual itself than what the embodied immersion into VR generates: particular forms of bodily data.

In a sense, then, the use of VR to observe human reactions to autonomous vehicles inverts the way in which visuals are related to smart city realities. Rather than human observers surveilling the city, here the human becomes observed, their physiological reactions to a smart technology turned into data. The real here is the data generated by the visuals: it is not displayed but rather prompted by the immersive VR experience.

Conclusion: claiming the 'real' in smart cities

All three forms of visualisation briefly discussed in this chapter – operation centre dashboards, interactive 3D models and VR – all make some kind of claim to show real data about the smart city visually. Both the screens of the smart city operations centre and the 3D model app offer versions of the city through displaying data: the real smart city is shown in the form of datasets. They monitor and visualise data to inform planning, policing, management and investment strategies, and do so in a way that makes the locality visually legible and understandable. The operations centre, however, is more distant from the city it observes. The MK:3D Map app in contrast is designed to be used in the city, kept at hand to generate data whenever needed, in situ as it were. This is the city-as-data made mobile, available on a phone or a tablet, and manipulatable by hand gesture. In the VR case study, the body is also central. Its main requirement was for immersion, not to display data, but to create it through corporeal practice. Immersion would produce data on emotional and physical responses to the city. VR, then, displaces the body from the subject of the visualisation – the subject that sees real data and the smart city as data – to the object of the visualisation – an object that is itself data.

The power relations enacted in interactions with these various smart city realities are complex. VR in the example here has been used to observe the user: the user is offered a viewing experience but finds themselves observed; their behavioural, emotional and physiological states transformed into the visualised data of written observations charts, tables and graphs. Yet their data also act to influence how the driverless vehicles themselves are observed and behave in front of real-life pedestrians and cyclists. There is then an ongoing dynamic here in which the physical responses of the observed to the virtual situations they confront feedback to influence the real-world behaviours and processes of the pod vehicles.

The spatial modelling of MK in both 3D maps and VR also suggests a somewhat disruptive version of the real, in terms of the city's distinctive spatial layout. The need for the virtual model to reflect the architectural shapes of the real-world location it represents ensures that virtual Milton Keynes is a smart city of low density, low-rise blocks and green spaces, instead of the science-fiction vision of high-rise towers and grey spaces of public squares or walkways absent of humans that characterise many depictions of smart cities (Vanolo, 2016). Pedestrians (albeit from a relatively narrow demographic) are found walking around the shared spaces of the VR Milton Keynes. As an urban centre built around the needs of the private motor car (Edwards, 2001), virtual Milton Keynes retains the dominating network of roads from the real-world Milton Keynes. Virtual Milton Keynes does not therefore fully conform nor dissent from popular depictions of smart cities; the actual locality of the place anchors the virtual representation. Yet equally, the virtual is constructed as a 'blemish-free' idealisation in these case studies in which 'undesirable' or unconventional elements are phased out and in their place, market-driven representations of reality are installed. If the digital is coming to dominate the vision of the real smart city, then creators, designers, academics, planners and citizens need to be actively engaged in critiquing the diverse virtual depictions of the real world they claim to represent.

Acknowledgements

This chapter results from a research project funded by the Economic and Social Research Council grant reference ES/N014421/1, *Smart Cities in the Making: Learning from Milton Keynes*. The research team members were Prof Gillian Rose (Principal Investigator), Dr Nick Bingham, Prof Matthew Cook, Prof Parvati Raghuram, Dr Alan-Miguel Valdez, Prof Sophie Watson, Dr Edward Wigley and Dr Oliver Zanetti.

References

Bailey, S., Deshpande, A. and Miller, A. (2014) The use of 3D visualisation for urban development, regeneration, and Smart City demonstration projects: Bath, Buckinghamshire, and Milton Keynes. *Bollettino Del Centro Calza Bini*. 14 (2), pp. 423–435.

Batty, M. (2013) *The New Science of Cities*. Cambridge, MA: MIT Press.

Brodlie, K., Dykes, J., Gillings, M., Haklay, M.E., Kitchin, R. and Kraak, M.-J. (2002) Geography in VR: context. In Peter Fisher and David Unwin (eds.). *Virtual Reality in Geography*. London: Taylor & Francis. pp. 7–16.

Brown, I.M., Kidner, D.B., Lovett, A., Mackaness, W., Miller, D.R., Purves, R., Raper, J., Ware, J.M. and Wood, J. (2002) Virtual landscapes: Introduction. In Peter Fisher and David Unwin (eds.). *Virtual Reality in Geography*. London: Taylor & Francis, pp. 1–4.

Cheesman, J., Dodge, M., Harvey, F., Jacobson, R.D. and Kitchin, R. (2002) 'Other' worlds: Augmented, comprehensible, non-material spaces. In Peter Fisher and David Unwin (eds.). *Virtual Reality in Geography*. London: Taylor & Francis, pp. 295–304.

Coughlan, S. (2018) Rough sleeping hits hard on local high streets [online]. BBC News 25 January Available from: www.bbc.co.uk/news/education-42781377 [Accessed 26 January 2018].

Crang, M. and Graham, S. (2007) Sentient cities: ambient intelligence and the politics of urban space. *Information, Communication & Society*. 10 (6), pp. 789–817.

Edwards, M. (2001) City design: what went wrong at Milton Keynes? *Journal of Urban Design*. 6 (1), pp. 87–96.

Elsaesser, T. (2013) The "return" of 3-D: on some of the logics and genealogies of the image in the twenty-first century. *Critical Inquiry*. 39 (2), pp. 217–246.

Foucault, M. (1977) *Discipline and Punishment*. London: Penguin.

Halpern, O. (2015) *Beautiful Data: A History of Vision and Reason since 1945 [online]*. Durham, NC: Duke University Press.

Haraway, D. (1997) *Modest_Witness@Second_Millenium: FemaleMan(c)_Meets_OncoMouseTM*. London: Routledge.

Hohl, M. (2009) Beyond the screen: visualizing visits to a website as an experience in physical space'. *Visual Communication*. 8 (3), pp. 273–284.

Hollands, R.G. (2008) Will the real smart city please stand up? *City*. 12 (3), pp. 303–320.

Kitchin, R., Lauriault, T.P. and McArdle, G. (2015) Knowing and governing cities through urban indicators, city benchmarking and real-time dashboards. *Regional Studies, Regional Science*. 2 (1), pp. 6–28.

Kitchin, R., Maalsen, S. and McArdle, G. (2016) The praxis and politics of building urban dashboards. *Geoforum*. 77, pp. 93–101.

Kourtit, K. and Nijkamp, P. (2018) Big data dashboards as smart decision support tools for i-cities – An experiment on Stockholm. *Landuse Policy*. 71, pp. 24–35.

Luque-Ayala, A. and Marvin, S. (2016) The maintenance of urban circulation: An operational logic of infrastructural control. *Environment and Planning D: Society and Space*. 34 (2), pp. 191–208.

Mattern, S. (2015) Mission control: a history of the urban dashboard. Available from: https://placesjournal.org/article/mission-control-a-history-of-the-urban-dashboard/ [Accessed 13 March 2015].

Mattern, S. (2017) *Code and Clay, Data and Dirt: Five Thousand Years of Urban Media*. Minneapolis, MN: University of Minnesota Press.

McNeill, D. (2016) IBM and the visual formation of smart cities. In Simon Marvin, Andrés Luque-Ayala, and Colin McFarlane (eds.). *Smart Urbanism: Utopian vision or false dawn?* London: Routledge, pp. 34–51.

Mirzoeff, N. (1999) *An Introduction to Visual Culture*. London: Routledge.

MK CityFest (2017) MK CityFest 2017 Programme [online]. Available from: www.destinationmilton keynes.co.uk/upload/managerFile/Downloads/MK%20CITYFEST%20PROGRAMME%20V9.pdf [Accessed 13 February 2018].

Offenhuber, D. (2015) Infrastructure legibility–a comparative analysis of open311-based citizen feedback systems. *Cambridge Journal of Regions, Economy and Society*. 8 (1), pp. 93–112.

Pett, M. (2017) paper presented at the Imagine conference, 7 June 2017, Milton Keynes.

Rose, G. (1993) *Feminism and Geography: The Limits of Geographical Knowledge*. Cambridge: Polity Press.

Rose, G. (2017) Screening smart cities: managing data, views and vertigo. In Pepita Hesselberth and Maria Poulaki (eds.). *Compact Cinematics: The Moving Image in the Age of Bite-Sized Media*. London: Bloomsbury Academic. pp. 177–184.

Rose, G. (2018) Smart urban: imaginary, interiority, intelligence. In Christoph Lindner and Miriam Meissner (eds.). *The Routledge Handbook of Urban Imaginaries*. Basingstoke: Routledge. pp. 105–112.

Rose, G. (2018a) Look InsideTM: Visualising the smart city. In Karin Fast, André Jansson, Johan Lindell, Linda Ryan Bengtsson, and Mekonnen Tesfahuney (eds.). *An Introduction to Geomedia: Spaces and Mobilities in Mediatized Worlds*. London: Routledge. pp. 97–113.

Schuurman, N (2004). *GIS: A Short Introduction*. Oxford: Blackwell Publishers Ltd.

Shelton, T. (2017) The urban geographical imagination in the age of Big Data. *Big Data & Society*. 4 (1), pp. 2053951716665129.

Söderström, O., Paasche, T. and Klauser, F. (2014) Smart cities as corporate storytelling. *City*. 18 (3), pp. 307–320.

Sturken, M. and Cartwright, L. (2001) *Practices of Looking: An introduction to Visual Culture*. Oxford: Oxford University Press.

Tagg, J. (1988) *The Burden of Representation: Essays on Photographies and Histories*. London: University of Minnesota Press.

TSC LUTZ Pathfinder Demonstration in Milton Keynes (2016) [online]. Directed by University of Oxford. Available from: www.youtube.com/watch?v=oPXDxAMnnd8 [Accessed 26 January 2018].

Valdez, A.-M., Cook, M. and Potter, S. (2018) Roadmaps to utopia: Tales of the smart city. *Urban Studies*. 55 (15), pp. 3385–3403.

Vanolo, A. (2016) Is there anybody out there? The place and role of citizens in tomorrow's smart cities. *Futures*. 82, pp. 26–36.

Weising, L. (2010) *Artificial Presence: Philosophical Studies in Image Theory*. Stanford: Stanford University Press.

Wiig, A. (2016) The empty rhetoric of the smart city: from digital inclusion to economic promotion in Philadelphia. *Urban Geography*. 37 (4), pp. 535–553.

Willis, K. and Aurigi, A. (2018) *Digital and Smart Cities*. London and New York: Routledge.

22

From hybrid spaces to "imagination cities"

A speculative approach to virtual reality

*Johanna Ylipulli, Matti Pouke,
Aale Luusua and Timo Ojala*

Introduction

This chapter departs with the notion that digital and physical spaces are already profoundly entangled in contemporary cities due to the pervasive use of new digital technology (e.g. Luusua *et al.* 2017; Willis 2008). We explore the experiential implications of the potential next step of this development by discussing the consequences of augmenting public urban places with immersive, large-scale digital layers realized with virtual reality (VR) technologies. We utilize a speculative approach: thus, while our discussion is firmly grounded on empirical studies, we will scale up the results, casting a speculative glance into the near future. The empirical context for the chapter is provided by a novel digital design artefact, the *Virtual Library* (VL), whose design process was guided by a design anthropological approach and executed with methods drawn from participatory design and human–computer interaction (HCI). The design of the application was a long-term transdisciplinary process executed collaboratively by the Center for Ubiquitous Computing (the University of Oulu) and the Oulu City Library in 2016–2018. As the final design outcome, the physical library building was modelled into a detailed 3D virtual environment (VE) that was extended with virtual fantasy environments having no direct counterpart in the real world. Essentially, the VL is a VR application that creates a *hybrid space* comprising of physical and digital manifestations of a public library. With the research data collected from the design and use of the VL, we are able to reflect on what urban inhabitants may expect from and how they experience hybrid urban spaces – if they become more commonplace.

The VR application designed for a public library, the Virtual Library, and research results gained through its design process act as our point of departure. We used a *research by design* (e.g. Cross 1982) approach and especially *design anthropology* (Otto & Smith 2013); we produced a digital artefact that was at the same time a vehicle for studying

312

certain social and cultural phenomena. The project was initiated and lead by researchers but the library was an important partner and enabler – without their support and willingness to collaborate, the process would have been impossible to carry out. Early in the process, we chose *participatory design* (PD) (e.g. Bjerknes & Bratteteig 1995; Blomberg & Karasti 2012) as the primary method for developing the Virtual Library prototype as we wanted to proceed in a democratic spirit and gain a rich array of ideas emerging from various stakeholders' viewpoints. The current version of the Virtual Library is an interactive room-scale VR application in which the user is able to freely roam throughout four different VEs and interact with various thematic services. The 3D models and textures used for the VEs are a combination of self-modeled and commercial assets. The technical implementation allows multi-user interaction across all levels, however it has not been part of our evaluation process so far. We analyzed major themes from the research materials collected throughout the PD process and designed the structure of the application and interactive services based on our findings.

We concentrate especially on the *imaginative aspect* of this near-future medium. This framing was chosen primary due to the results of our empirical studies connected to the Virtual Library: we noted that citizens were especially interested in the imaginative capacity of VR. It was clear from the beginning of the design process that the Virtual Library was not considered as a utilitarian application by our participants but as an aesthetic, adventurous and creative experience. The focus on imagination and fantasy is also topical, as the majority of studies in the domain of 3D cities and their applications tend to focus on utilitarian topics, such as visualization of urban planning processes, emergency management, architecture, education, etc. (e.g. Carrozzino et al. 2009; Julin et al. 2018; Lovett et al. 2015; Pouke et al. 2019). Furthermore, it must be emphasized that we are utilizing a *speculative approach* (e.g. Wilkie 2017). This means that we scale up the results of our empirical studies and reflect on the potential implications of VR that would take place *if* the technology itself and similar applications mirroring and extending public urban spaces would become more commonplace. It has been recently suggested that the speculative approach should be used (also) in social sciences in order to provide views on alternative futures and potential challenges and possibilities these differing visions open up. Further, the approach sets light on the contingent nature of futures; future development is not set but dependent on numerous factors and the decisions we make now (Wilkie et al. 2015 2017; Ylipulli et al. 2016). We deem this especially important when studying emerging technologies such as VR.

To summarize, although the direction of near-future development of described systems is always uncertain, we argue that there is an urgent need to create new understandings of urban space and VR as a novel spatial medium that is able to mediate holistic experiences, causing potentially new kinds of hybrid urban spaces. Our exploration is motivated by the notion that in recent years, there has been a clear and increasing global proliferation of commercial VR. Until recent years, VR was available only for few commercial users, and thus mostly remained within non-commercial private spaces, such as test labs and educational facilities. However, due to the development of cheaper and more lightweight technologies, VR technologies have started to expand outside of these constraints. Although *Google Glass* was a commercial failure, many manufacturers continue developing VR, augmented reality (AR) or combined headsets and other wearables, such as contact lenses, for everyday use (on Google Glass, see e.g. Encheva & Pedersen 2014).

The central concepts forming the backbone of the chapter – *media, hybrid space* and *meta-worlds* – are defined in the following section. The overall theoretical background is further expanded with an introduction to the concept of *imagination age*, coined by Rita J. King

(2016), and a discussion of the concept's potential utility in understanding the futures of cities. We continue by presenting the design process of the Virtual Library VR application in detail, and by analyzing our empirical data, collected throughout this process. The analysis framing the Virtual Library as a hybrid space is scrutinized on three levels, following temporal phases of the project: the Virtual Library as imagined, VL as designed and VL as experienced. We concentrate especially on our participants' ideas and experiences concerning physical space, different digital spaces and their entanglements, as well as the role of imagination. Finally, the chapter arrives at the notion of *imagination city*. Is the smart city heading towards becoming an "imagination city" – a hybrid environment where different digitally created spatial, immersive layers intersect with the physical city? In what terms could this transformation happen?

Defining central concepts – and redefining hybrid space

The central concepts for our exploration are *media*, *hybrid space* and *meta-worlds*. We consider VR as a medium and as a part of the ongoing *mediatization* of society. By "mediatization", we mean the "various processes through which culture is influenced by the modus operandi of the media" (Hjarvard & Petersen 2013, 2), these including the media's technological, institutional and aesthetic modes of operation (Hjarvard 2013). As such, media have become central facilitators of cultural experience, as various cultural domains become ever more dependent of media. Many media have also become cultural institutions in their own right. Framing VR as a medium among other media allows us to apply the approaches of media studies, such as *media archaeology*, to the study of VR and understand it as a part of broader media historical continuums rather than as a singular disruption (e.g. Huhtamo & Parikka 2011). In a similar vein, the perspectives offered by the research on the *domestication of new technology*, developed within media studies and science and technology studies, can complement our understanding of VR's potential role in near-future cities. The domestication approach looks at processes in which people negotiate individually and with others on how to adopt and "tame" new technology in contemporary society. Overall, it can be understood as a pragmatic micro-level approach which intends to explain how people make sense of new phenomena (Green & Haddon 2009; Silverstone & Haddon 1996; Tenhunen 2008; Ylipulli *et al.* 2014). The adoption of a new technology as a part of everyday life is seen as consisting of cultural negotiations: the prevailing cultural values and practices meet the values and possibilities embodied by the new technology. These negotiations are somewhat path-dependent and their results often reflect the adoption processes and use patterns associated with previous technology.

The other focal concept for our exploration is hybrid space that can be defined by following de Souza e Silva (2006). According to her conceptualization, in hybrid space it is possible to experience the physical and digital space simultaneously. Currently, this is manifested in the everyday use of different mobile devices, digital public displays and related applications in the cities; thus, digital space is understood here figuratively as a "cyberspace" and not as an actual space with dimensions and orientation. In other words, different mobile and situated technologies enable media to follow us everywhere and color our experience of urban space with differing content, producing the experience of hybrid space. Most people switch from digital to physical and back almost all the time as they move through the urban space; by listening to music, by reading news, by utilizing navigation, by using social network applications. However, the development of VR and also AR[1] technologies can potentially change the experiential nature of this "hybrid urban living" in intriguing ways. VR can take the notion of hybrid urban space to another level where physical space is entangled not only with digital content – or cyberspace – but also with digital space as a spatial experience.

There is a significant experiential difference between hybrid space created by the use of flat screens of our mobile devices and public displays, and hybrid space that comes into being through utilizing (near-future) VR gear. These differences rise from the nature of VR which is inherently spatial, immersive and multisensory. It engages the foundational elements of *place experience*, such as spatiality (orientation and dimension) and the senses, including hearing and even tactility through haptic technology, and renders these into a novel medium. Mobile devices and public displays are able to provide audiovisual content that engages our sight and/or hearing but they do not create an illusion of being spatially in a completely different environment. Some scholars have envisioned that the spreading of this kind of technology could mean that visions of complete digital meta-worlds or *Metaverse* turn into reality (e.g. Gelernter 1992; Ricci *et al.* 2015; Smart *et al.* 2007; Streitz 2010). This would mean that physical places are coupled with a digital counterpart, providing views to the past or introducing future plans, or offering imaginary content. We elaborate these conceptualizations further in the next section.

Imagination age

Imagination can be seen as a central facet of human cognition. We understand imagination here broadly as the capacity of the human mind to turn things into something else – it allows us to think "what if?" and plan our actions accordingly. It is profoundly tied to our capacity of metaphorical translation, i.e. our tendency to transpose, juxtapose and create new concepts and connections through extraordinary combinations of things (on metaphors, see e.g. Lakoff & Johnson 1980; Lakoff 1993; on metaphors in design, see Ylipulli *et al.* 2017). Murphy notes how "the power to imagine one thing as another, makes the human world a virtual one" (2010, 5). Murphy discusses imagination in relation to new technology and depicts how digital technologies are transforming physical world phenomena virtual; books become virtual readers, mail is replaced by email, money turns into electronic transactions and communities flourish on Facebook, Instagram or WhatsApp (ibid; see also Negroponte *et al.* 1997). We would like to add that through the development of VR, also physical space is turned into virtual space. Imagination can be seen as a prerequisite that enables these transformations.

The *imagination age*, in turn, is a concept coined by Rita J. King, a co-director of a strategic consultancy and a futurist (King 2016). King originally defined it as a period of transition between the industrial era, when the focus has been on tangible things, and the intelligence era, when the focus is turned to less tangible – code, analytics etc. It is a "navigation system" that is supposed to help humanity to transfer itself to a new era, where creativity and imagination become the central drivers of economic value. This notion is also connected to the promise of emerging technology, such as VR, which is forecast to change the ways human beings interact with each other and with their environment, and which calls for user-generated content and creativity. Interestingly, King also anticipates that "100 years from now the relationship between imagination and physical space is going to be radically different. There will be a convergence between hardware, software and bioware" (2016, 53). We must note here that unlike King, we are focusing on near future (next 10–20 years). Secondly, the imagination age is not (yet) a particularly academic concept as it was born within the world of business and consultancy. Therefore, it may lack some theoretical depth; however, we have been scrutinizing the concept in the light of other prominent theories to strengthen the discussion. It can be considered very helpful in understanding the current experiential and societal transformations caused by new digital technologies, especially in the context of urban environments.

For us the concept of the imagination age is especially useful because it enables us to perceive how VR (and AR) can transform experiences of public urban spaces by connecting physical spaces with digital, virtual space. In some sense this development is already happening but it is still very fragmented. *Pokémon Go*, an AR game in which gamers hunt imaginary creatures from public spaces, has probably been the most powerful example of this so far (e.g. Dorward *et al.* 2017; Sicart 2017). Immersive VR has not yet caused a similar global phenomenon. However, there are some early examples of combining public spaces and VR; such as virtual factory tours that became popular already in the early days of World Wide Web (Mitchell & Orwig 2002), which have now begun to emerge as on-site VR instances complementing physical factory tours. By following the thoughts of many other authors, we can speculate that through VR technologies, digital urban spaces begin increasingly to overlap with physical urban spaces; it will be possible to experience virtual representations or reproductions of existing physical urban spaces, dubbed for example *digital cities*[2] or *mirror cities* (cf. Couclelis 2004; Gelernter 1992; Ricci *et al.* 2015; Smart *et al.* 2007; Streitz 2010). Alternatively, it will also be possible to experience *imaginative versions* or *fantastic extensions* of physical spaces. These digital spaces can, for example, reflect or enhance – in creative and imaginative ways – some affordances of correspondent physical space (cf. Ylipulli *et al.* 2016). They are linked to imagination not only through the non-realistic content they offer but also because they can support people's creative endeavors; they may allow users to create and imagine individually and collectively.

Furthermore, we can anticipate that if VR technology becomes more common, the amount and quality of different kinds of virtual urban spaces will grow. At the same time, the amount and quality of services and functions these immersive digital spaces offer, increases.

By rooting our speculation in contemporary media culture and leaning on domestication of new technology and media archaeological perspective, we can formulate two broad hypotheses: firstly, when looking at current everyday life in the cities, we can speculate that citizens will probably dive into new VR worlds and back to physical space as effortlessly as they now switch between physical and digital with their mobile devices (e.g. Silverstone & Haddon 1996; Ylipulli *et al.* 2014). This switching may result in prevalence of combinatory AR and VR systems that have various degrees of being more or less immersive or augmented, as the emergence of aviation-like displays for cars indicate (Wu *et al.* 2009). However, our empirical materials consist of immersive VR so this is not evident from our materials, and thus, our focus here is on immersive VR. Secondly, when looking at all the previous media, from books to films and video games, we have absolutely no reason to assume that imagination would not flourish in all imaginable (sic!) ways in the content (and in spaces) enabled by VR. We hypothesize this concerns also VR applications connected to urban spaces (e.g. Huhtamo & Parikka 2011). Curiously, this imaginative aspect has gained much less attention in literature concerning virtual cities. For example, in his extremely influential book, *Mirror Worlds: or the Day Software Puts the Universe in a Shoebox* … David Gelernter (1992) sees computer-generated mirror worlds – and also mirror cities – mainly as a simulation that allows unprecedented control over the reality; they grant us the ability to penetrate inside the reality, to manipulate it and to investigate it from a god-like perspective from a distance. Alternatively, in their famous *Metaverse Roadmap*, Smart *et al.* (2007) make a clear distinction between the mirror worlds and (fantastic) virtual worlds and anticipate differing uses for them.

Our approach differs from these perspectives as we want to emphasize the role and meaning of non-realistic and imaginative VR experiences, and also the dual-existence of realistic simulations of urban spaces and imaginative worlds side by side. Therefore, we introduce the term "imagination city" as it highlights the imaginative and creative capacity of VR as a medium and experiences it enables. At the moment the scenarios depicted in this section are familiar mainly from science fiction; in the remaining part of the chapter we explore, in the light of our empirical findings, whether these kinds of visions are plausible in near future and in what terms. To understand the implications and limitations of the development, we discuss our study participants' experiences about the interlacing of the physical world, the mirror world and the fantasy world(s) in the hybrid space enabled by the Virtual Library.

Empirical context: the design process of the virtual library

The design of the Virtual Library VR application was a long-term collaborative process executed with the Oulu City Library in 2016–2018. The general approach we used can be defined as design anthropology, aiming to combine design and anthropological ways on knowing (e.g. Clarke 2010; Otto & Smith 2013). The goal was to design a concept and a prototype by utilizing methods and principles drawn mainly from PD, but at the same time, we studied people's views and experiences of VR, and developed new PD methods based partially on anthropological theories of magic as a cultural category (Ylipulli et al. 2017). Thus, the intention was to design an artefact and at the same time conduct research that is grounded in anthropological perspectives and theories (cf. research by design). For the purposes of this chapter, we have scrutinized especially the participants' accounts on imaginative nature of VR and their experiences on different spatial layers. We do not focus on scrutinizing the participants' experiences concerning, for example, *presence* or *embodiment*, although they are popular concepts in VR research; rather, we explore the experience more generally through an inductive approach.

We intended to respect the central principles of PD; although we initiated the project, the project goals were not determined by the researchers only but they were set together with the library administration and also with the library patrons. Furthermore, our intention was to facilitate processes of mutual learning (e.g. Bjerknes & Bratteteig 1995; Blomberg & Karasti 2012). The development process included a *preliminary interview* with the library administration (spring 2016) as well as numerous meetings with the researchers and library staff; in these meetings, we discussed both the goals of the project as well as practicalities. However, the most significant part of the design process was the *two workshop sessions* (autumn 2016 and spring 2017) that were set up to create ideas and gather specifications for the prototype. The process continued during autumn 2017 by arranging *user tests* and *interviews* to evaluate the created Virtual Library prototype. The Virtual Library was opened to the general public during the summer 2018 when a public VR access point was deployed to the Oulu Main Library.

The first multi-stakeholder workshop was arranged at the premises of the Oulu Main Library, likewise the rest of the participatory activities connected to the design process. The aim was to bring researchers, library staff and library patrons together to create ideas and preliminary concepts for a *Hybrid Library*, a vision of a future library, where 3D web technologies, AR and VR technologies could be utilized to enhance and extend the library experience. We did not predefine how these technologies could function in the library context. The workshop participants were able to experiment with commercial VR demos as well as the very first Virtual Library prototype which consisted of non-interactive photorealistic 3D models of the real-world library building. The workshop lasted for half a day and it had 35 participants: 14 library patrons, 13 library staff members and 8 university researchers, aged between 20–56 years. The researchers came from slightly different backgrounds, ranging from computer science to design and architecture but all of them were familiar with 3D technologies. The total the number of females was a bit higher, 21 persons, but all the different participant groups had both males and females.[3] All the discussions, presentations and activities included in the workshop were audio and/or video-recorded (for more detailed description of the methods and workshop results, see Holappa *et al.* 2018; Pouke *et al.* 2018; Ylipulli *et al.* 2017). After analyzing the collected research material with thematic analysis and using inductive approach (e.g. Patton 2014), we decided to concentrate primarily on immersive VR.

A second half-day long, multi-stakeholder workshop was arranged six months later. Thematically it was more focused; the idea was to develop forward certain chosen ideas, drawn from the workshop 1, together with the participants. This time we had 17 participants; the majority of them had participated also in the first workshop: eight participants belonged to library staff, five were library patrons and four researchers. We utilized a slightly more advanced Virtual Library prototype, created by following the ideas that came up in the workshop 1. The prototype included some interactive features, but more importantly, it piloted one of the central ideas that came up in the initial workshop: the virtual representation of the actual library was connected with fantasy worlds that exist only in digital reality. This structure intends to reflect how books and other library material can act as a portal transporting people into alternate realities. The first proper concept for the Virtual Library as well as the current interactive prototype with several services was created after all the material from the two workshops had been transcribed and analyzed with thematic analysis; the results were presented to the library staff and the prototype has since then been iteratively developed according to feedback.

In the evaluation stage (autumn 2017), we conducted *user tests* and *semi-structured theme interviews* with 12 participants who had also participated at least one of the workshops. The results of this part are interesting considering, more broadly, the future of technologizing cities and potential breakthrough of VR as a spatial medium connected to different urban spaces. The created prototype was tested at the premises of the city library, which enabled our participants to experience different layers of the space at the same time. Five of the participants were library patrons and the rest belonged to the staff; their age varied between 28 and 55 years. The participants could use the prototype freely, which was followed by a semi-structured theme interview. Notably, the participants spent on average 45 minutes in the Virtual Library.

The Virtual Library as a hybrid space

Virtual Library as imagined

From the start of the PD process, the participants pictured the Virtual Library as an aesthetic, adventurous and creative experience rather than a utilitarian one. This was also the initial idea of the library administration that came up in the preliminary interview with them. In the first, larger and thematically more open workshop (Figure 22.1), an idea that surfaced repeatedly was a possibility to travel to other, alternative realities. Throughout the two workshops, all participants (including the librarians), emphasized aspects such as adventure, discovery, excitement and collaborative activities. However, they also pondered that the Virtual Library should include some aspects that connect it to the physical library and create a feeling of familiarity:

> there should definitely be some place that would nourish and soothe your sense of hearing, for example a woodland where you could listen to the sounds of the forest, so if you can't go to a forest for real, in the virtual library you'd still able to choose a woodland setting or a beach or some place like that, where you can hear the soundscape.[4]
>
> *(Workshop 1, group 8)*

> it [the experience] would always begin with a safe and familiar environment and through that you could head towards adventure and exploration.[5]
>
> *(Workshop 2, group work 1, focusing on "mood")*

Furthermore, the Virtual Library was imagined as a radically new way to explore all the contents provided by the library: books, art, music etc. For these reasons we begun focusing on developing an imaginative VR experience which also contains something very familiar – the

Figure 22.1 The participants of the workshop 1 reflecting on their relationship with the library
Source: © Oulu City Library

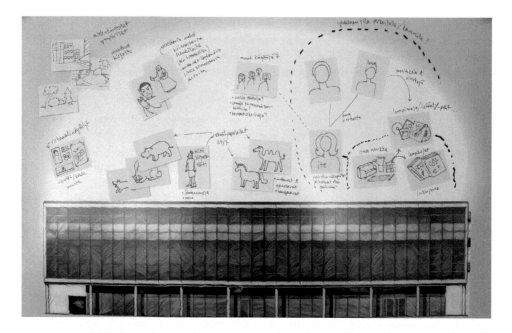

Figure 22.2 A visualization of the Virtual Library's services, created by a participant in workshop 2
Source: Chapter author(s)

photorealistic model of Main Library's lobby area and second floor. Another recurrent suggestion was that the Virtual Library should enable community building and events; it should be a meeting place and a venue for differing happenings, ranging from small meetings revolving around common interests to large thematic cultural events (Figure 22.2).

We must note here that the specific methods used in the workshops resonate with the notions of "imagination age". In the first workshop we used creative metaphors and especially the metaphor of "magic" to foster participants' imagination (see Ylipulli *et al.* 2017). This was done in order to create radically new ideas and to emphasize the vast potential of VR technologies. Thus, we recognize that we purposefully emphasized imaginative and creative possibilities offered by the application under development; however, we do not deem that this was the only reason fantastic and experiential aspects became so prominent in the design. We claim that the affordances of VR as a new medium and library as a specific environment had a strong role in the process.

Virtual Library as designed

The purpose of this sub-section is to introduce the Virtual Library as it was designed. The whole concept was based on the recurrent ideas that came up in the workshops. Naturally, all the suggestions were not feasible due to the limitations of technology or our resources; for example, at this point we could not implement the elements and services that would have enabled community building and events.

As stated earlier, the Virtual Library is an interactive room–scale VR application. It is not designed for gamers only, but instead for all users of the library. Instead of continuous virtual movement similar to video games, the navigation within the environment is based on physically walking combined with teleportation; this has been seen as an effective way to reduce vection-based cybersickness (Llorach *et al.* 2014), making the application more comfortable and accessible for all user groups. Most interaction is point-and-click based, the user indicating an object of interest through pointing with a VR controller device and subsequently confirming a selection by squeezing a controller button. There are also objects that the user can pick up by reaching towards to object and pressing the upper trigger button. Some of the objects contain interactive in-game menus.

The Virtual Library consists of four separate virtual environments, VEs. These VEs are the *City Library, Fantasy Village, Future Alley* and *Study* (Figures 22.3, 22.4, 22.5 and 22.6). The themes of the VEs were chosen according to workshop findings and they loosely correspond to different literary genres. Each VE also contains one or multiple interactive services that were developed according to workshop findings as well.

The starting VE represents the bottom two floors of the actual physical City Library (referred to also as a mirror world). It contains a functionality to search the actual library database from which the user can browse their online reviews (goodreads.org). The user can also view navigation instructions to the virtual corresponding locations of the actual books found in the physical library. In addition, the City Library VE also contains a virtual art exhibition as well as an elevator which can be used to transport users to the fantasy VEs. The Fantasy Village VE contains book recommendations authored by the library staff. The book recommendations in this level are mostly fantasy books for young adults. The recommendations are hidden within the objects in the levels and can be invoked by pointing and clicking the corresponding object. An example can be seen in Figure 22.5 (lower right). The Study VE represents a cozy room and contains interactive objects the user can play with (such as books, bottles, photos, a fireplace and an animated cat). The main interactive service of the level is an interactive book object for collaborative writing (see Figure 22.4), the book is on the table. The writing takes place utilizing a "refrigerator magnet" metaphor; the user has a set of words from which he/she can drag and drop to create stories. The user can save the story so that other users can view it later. The users can also continue stories saved by the others. The purpose of this room is to provide interaction between users since libraries are also seen as public meeting places (Aabø *et al.* 2010; Ylipulli & Luusua 2019). The Future Alley VE is similar to the Fantasy Village level with the exception that the theme of the VE (including the book recommendations) is futuristic instead of medieval fantasy.

Virtual Library as experienced

In the following we analyze the material collected in user tests and interviews, scrutinizing the prototype described above. The tests and following interviews reveal interesting aspects of people's place experiences and VR. Regarding the general experience of the participants, most were stunned by the visual quality and immersive nature of the experience. Many also commented the best part was just "being in there", being and moving around the digital space and interacting with virtual tangible objects. Partially this excitement can be explained by the fact that none of them had used immersive VR before except in the Virtual Library workshops. There were three key findings; the nature of *people's experience of being in the physical and digital place at the same time, the transition between virtual and physical library* and *the positive experience of the combination of mirror world VE and fantasy VEs.*

Figure 22.3 The Virtual library prototype: City Library. Top: Actual library lobby. Bottom: the Virtual Library lobby with the interactive art exhibition to the left. The elevator acts as a portal that transports users into the fantasy VEs.

Source: Chapter author(s)

Figure 22.4   The Virtual Library prototype: Future Alley

Source: Chapter author(s)

Figure 22.5   The Virtual Library prototype: Fantasy Village. The user is interacting with a book recommendation object.

Source: Chapter author(s)

Figure 22.6 The Virtual Library prototype: Study. The cat in the foreground starts purring when in proximity. The particle effect and the floating text in the background indicates the virtual storytelling book object.

Source: Chapter author(s)

People's experience of being in physical and digital place at the same time and transitioning between these was considered mostly "natural" and smooth. We consider especially interesting that being in the mirror world, the City Library VE, and its physical counterpart (the main library building) at the same time did not seem to produce any kind of contradiction or friction; some even described it as a positive experience. Many participants commented they did not think about it at all, or that it felt completely natural. The mirror world included subtle differences compared to the physical one which was considered good: it was perceived as an alternative reality, familiar yet something else than the physical library (Figure 22.7). One participant also said these subtle differences actually made the mirror world more interesting. Nevertheless, many affordances of the physical place were transferred to the digital place resembling it: the participants mentioned during the testing the need to be silent in the library premises, and some did not want to disturb others (human characters situated in the mirror world).

> It was very exciting, maybe because it started from the lobby, and I recognized that it was the same lobby where I just was physically, it was the same but still a virtual world, it was awesome to have that kind of connection. [...] But then the other environments, they were so much fun, it was all so exciting, and I started to think about book recommendations, if they were personalized ...[6]
>
> *(Library patron)*

Returning from the Virtual Library to the physical library was mostly considered "natural" or easy as well, although the participants spent considerable time using VR, on average 45 minutes.

Figure 22.7 Actual lobby exhibition (top) and exhibition in the lobby made interactive
     (bottom)

Source: Chapter author(s)

A few participants commented they were reluctant to return as the virtual world was so
exciting; alternatively, a few experienced returning as a "relief". Interestingly, these latter
participants were less active in VR, they asked a lot of advice and did not want just to
wander around and explore. The digital world we have built requires that the user is
oriented to be active and perhaps has a certain kind of exploratory mindset.

The combination of the mirror world VE representing the physical library and the fantasy VEs was seen very positively by all the participants. Many commented directly that the analogy works well as books, music and other library content act as a portal to alternate realities. Also this part of the experience, the combination of realistic and non-realistic space, was described with the adjective "natural". The realistic representation of the library lobby and first floor, from which the VR experience begins, was seen as putting the whole experience into a context; it had a surprisingly important role. Some participants also mentioned that it is crucial to have a clear and recognizable transition between the mirror world and the fantasy worlds, which in this case was an elevator that takes the user to "imaginary floors", i.e. fantasy VEs.

> I think it [the combination of mirror and fantasy environments] worked especially well because it highlights how the library contains stories and different worlds, so in my opinion this reflects that in a very good way. And I think it is important to have that total resemblance of the physical world at the beginning, because if we would just [directly] go to some fantasy world it would not be part of the library.[7]
>
> *(Member of the library staff)*

We must also note that people's prior experience on different media had a clear impact on their experience: participants who had played video games had different expectations than the ones who had not; people compared the Virtual Library implicitly or explicitly to other media they were more familiar with, from newspapers to games. For example, one participant who had a lot of experience in playing video games, wished the virtual environments had more visual details and referred to sponsored content (ads) in games. However, another participant who had never played any kind of digital games expected more textual content and was quite disappointed that the environment was "just an artwork that you can admire". However, most participants had at least little experience in games, and their opinions were in between these two extremities (cf. Silverstone & Haddon 1996).

Furthermore, one of our central notions concerns the hardware. Based on our empirical findings, the widespread use of VR technology will probably not happen before the technology is more mature. Visual and aural experiences were considered impressive and exciting by our participants, but getting entangled with the cable of the VR glasses, or sweating excessively and getting blurry eye glasses because of the heat that the gear produces, broke down the illusion very effectively. These experiences and problems were prominent throughout the research material. In addition, although our intention is not to present implications for design here, we make one comment on the provided content: the services need to be designed very carefully to enhance the services of the physical world (in our case the library); there needs to be creatively built connections between the physical world and the digital world, especially if VR application is used in the same space that it intends to augment. Our participants, for example, would have wanted to have interesting book suggestions sent to their mobiles in order to find corresponding physical books from the library. This finding supports an overall notion that the design of immersive 3D spaces is not fundamentally different from the design of traditional, physical spaces; the basic notion of contextually driven, place-based design that is central to architecture and urban planning, are crucial to the success of VR spaces as well when they augment physical places. Overall, the services and experiences VR environments offer can alter meanings of physical space and also actions people take. Therefore, also their impact should be considered very carefully and reflected upon in relation to their context.

From hybrid spaces to imagination cities?

Within this chapter, our main focus has been to present the theoretical basis for imagination cities, and support this through the empirical study that we conducted. By analyzing our empirical material in the light of theories connected to imagination, and by taking into account also domestication theories of new technology and perspectives provided by media archaeology, we have arrived to the central arguments of our chapter that can be summarized as follows:

1) Citizens and non-experts of VR in general are very interested in imaginary potential of immersive VR, and if VR becomes more commonplace as an urban medium, imaginary and fantastic content will play a central role. In addition, the development of all the previous media from books to films and games supports this argument.

2) In our research material, the interlacing of the physical world, the mirror world VE and the fantasy VEs, located in public urban space, was experienced mostly as "natural". Although our participants were not experienced in using immersive VR, transitioning between the spatial layers was not experienced as difficult or confusing. This can be compared with the current use of mobile technology in public urban places; people change effortlessly between digital and physical realms. This hints to the possibility that people would utilize also spatial, digital meta-worlds in urban environments with relative ease.

We can scale up these empirical findings by using speculative approach: our study took place in singular urban public space, the public library, but we deem it important to consider what would happen if similar hybrid spaces would become more widespread in cities. Therefore, we arrive to a notion of "imagination city" where spatial, immersive and imaginative digital meta-worlds are interlaced with the physical urban environment, producing new kinds of hybrid urban spaces. According to our findings, it is a potential near-future vision that deserves more research. Imagination cities could be born if VR technology would become more mundane and widespread, and if there would exist (location-based) imaginative digital content that city dwellers would like to experience – in other words, if hybrid spaces described in this chapter would expand. There are many uncertainties connected to this potential development, however it should be scrutinized from different perspectives, also by social scientists and scholars from the arts and humanities, to shed light on the possible negative implications and ethical dilemmas. These include questions related to the access to content and content production, which, in turn, can affect societal power relations; questions of privacy and surveillance; and dilemmas of consumer culture, such as (subliminal) advertising. We do not have a possibility to explore these ethical themes much further in this chapter; however, in what follows we reflect on briefly some broader limitations and future directions for this development.

The first general remark we want to make concerns the nature of all imaginary content of the potential imagination city: it must be connected to people's everyday life practices and perspectives to be meaningful. Urban citizens need *a framework* to interpret and to get interested in these fictive augmentations. A good example of this is Pokémon Go AR game. It has already demonstrated the power of augmented play in urban environments, and citizens find imaginative elements imposed on the physical city enticing (e.g. Sicart 2017). In the case of Pokémon, the cultural framework is formed by the international brand itself and affordances of the original game; in the case of Virtual Library, a similar framework is formed by library's societal role as a portal to other realities. It is also understood as a provider of cultural content and non-commercial space for exploration, as the comments and experiences of our study participants indicated (the societal role of the Finnish library system and new digital technologies is scrutinized further in Ylipulli & Luusua 2019). We acknowledge VR and AR experiences differ from each other substantially, yet we consider them comparable from this perspective.

Secondly, in a similar fashion as the real-world incarnation(s) of *ubiquitous computing* (Dourish & Bell 2011), the reality of an imagination city will probably be messy and fragmented. We do not assume that the imagination city would be one coherent digital city that everybody uses in the same way; rather, it is likely to be something different for different stakeholders; it is a different experience for city officials and for citizens, and different people will use it in differing ways (cf. Ylipulli *et al.* 2016.) Through Pokémon Go, we have already seen how imaginative layers start to impact social behavior and affect how people interpret places and real-word phenomena – and also vice versa (e.g. Dorward *et al.* 2017). More sophisticated and immersive digital meta-worlds can multiply this effect, which at its worst can mean more conflicts over the uses of urban places, or conflicts between different interest groups. In addition, the digital divide and economic factors can place people in unequal positions – experiencing the imagination city requires advanced technology and skills to use it.

Thirdly, it is clear that the VR hardware needs to become significantly more mature, lightweight and portable – otherwise the development envisioned in this chapter will not turn into reality. Nevertheless, in regards of technology, we would like to highlight one developmental strand affecting how people will experience digitally augmented urban places: the potential merging of VR and AR technologies. The appearance of consumer VR devices has seemed to spark a growing interest towards technologies and applications spanning the entire mixed reality continuum (Milgram & Kishino 1994), from true reality into virtual reality. In academic literature, examples include studies regarding the development of proto-type technologies (e.g. Koskela *et al.* 2018; Steed *et al.* 2017) as well as their implications (e.g. Jung *et al.* 2018 regarding gradual transition of reality and its effect on body ownership and presence).

The aim of this chapter has been to initiate a discussion on a potential techno-urban development that can be called "imagination cities". However, this necessarily merely scrapes the surface of the theoretical and empirical possibilities of researching and designing imagination cities. We deem that there is a serious need to consider critically the possibility that urban public spaces in the future can appear as hybrid spaces where imagination and reality are seamlessly interlaced through VR technology.

Acknowledgments

We sincerely thank the Oulu City Library for collaboration. We would also like to express our gratitude to all our study and workshop participants. Dr Luusua gratefully acknowledges the grant from the Academy of Finland (316136 AICity). Dr Ylipulli wishes to thank the City of Helsinki, the City of Espoo and the City of Vantaa for financial support. In addition, this work was partly supported by the COMBAT project (293389) funded by the Strategic Research Council of Academy of Finland, as well as the 6Genesis Flagship (318927) funded the Academy of Finland.

Notes

1 VR creates significantly different experiences than AR; the former refers to experiences of completely artificial, computer-generated environments whereas the latter enables superimposing digital elements on user's view of the physical world. These can be seen, for example, through differing screens. In this chapter we concentrate on experiences created with VR but we note that these two types of experiences can also become mixed in the near future.
2 "Digital city" can be conceptualized also differently, see for example Aurigi (2005b).
3 This trend was present throughout the design process – we always had more female participants. Nevertheless, we always also had both females and males from the different participant groups.
4 ... *niin siellä ehdottomasti ois joku semmonen niinku, just vaikka kuuloaistia ruokkiva paikka, että ois just joku semmonen vaikka metsämaisema jossa sä kuulet metsän ääntä että jos sä et vaikka oikeasti pääse, minnekään metsään niin sä siellä virtuaalikirjastossa voit valita vaikka sen metsäasetuksen, tai joku merenranta tai joku vastaava jossa kuuluu sitä äänimaailmaa.* All quotes translated from original transcription by authors.
5 ... *se lähtis aina siitä tutusta turvallisesta ympäristöstä ja sitten sitä kautta voi päästään sinne seikkailuun ja löytöretkeilyyn.*
6 *Se oli hirveän jännää että, ehkä jo ihan sen takia että kun se alko siitä aulasta ja se oli niin tunnistettavasti se sama aula jossa olin äsken, niin fyysisesti että se aula on siinä mutta sit se oli kuitenkin virtuaalimaailma niin se oli tosi mahtavaa et siinä oli semmonen kiinnekohta ... Mutta sitten ne muut ympäristöt niin nehän oli hirveen hauskoja ja siellä oli kaikkea jännää ja, mä jäin sitä sitten miettimään että kirjasuosituksia sitten että jos ne on personoituja ...*
7 *Se on nimenomaan erittäin toimiva, koska jotenkin just se tulee siinä tavallaan ilmi että miten se kirjasto sisältää niitä tarinoita ja niitä erilaisia maailmoita, et jotenkin se mun mielestä kuvastaa sitä tosi hyvällä tavalla. Ja on mun mielestä tärkeetä se että siinä on nimenomaan se yks yhteen fyysinen se alotusjuttu, koska jos mentäis vaan johonkin tommoseen fantasiamaailmaan niin sit se ei ois kirjastoa.*

References

Aabø, S., Audunson, R., & Vårheim, A. (2010). How do public libraries function as meeting places?. *Library & Information Science Research*, 32(1), 16–26.

Aurigi, A. (2005b). Competing urban visions and the shaping of the digital city. *Knowledge, Technology & Policy*, 18(1), 12–26.

Bjerknes, G., & Bratteteig, T. (1995). User participation and democracy: A discussion of Scandinavian research on system development. *Scandinavian Journal of Information Systems*, 7(1), 73–98.

Blomberg, J., & Karasti, H. (2012). Positioning ethnography within participatory design. *Routledge International Handbook of Participatory Design*, 1, 86–116.

Carrozzino, M., Evangelista, C., & Bergamasco, M. 2009. The immersive time-machine: A virtual exploration of the history of Livorno. In *Proceedings of 3D-Arch Conference*. 198–201.

Clarke, J.A. (ed.) (2010). *Design Anthropology*. New York: Actar.

Couclelis, H. (2004). The construction of the digital city. *Environment and Planning B: Planning and Design*, 31(1), 5–19.

Cross, N. (1982). Designerly ways of knowing. *Design Studies*, 3(4), 221–227.

De Souza e Silva, A. (2006). From cyber to hybrid: Mobile technologies as interfaces of hybrid spaces. *Space and culture*, 9(3), 261–278.

Dorward, L. J., Mittermeier, J. C., Sandbrook, C., & Spooner, F. (2017). Pokémon Go: Benefits, costs, and lessons for the conservation movement. *Conservation Letters*, 10(1), 160–165.

Dourish, P. & Bell, G. (2011). *Divining a digital future: Mess and mythology in ubiquitous computing*. Cambridge MA: MIT Press.

Encheva, L., & Pedersen, I. (2014). 'One Day ... ': Google's project glass, integral reality and predictive advertising. *Continuum*, 28(2), 235–246.

Gelernter, D. (1992). *Mirror worlds: Or the day software puts the universe in a shoebox ... How it will happen and what it will mean*. New York: Oxford University Press.

Green, N., & Haddon, L. (2009). *Mobile communications: An introduction to new media*. Oxford: Berg.

Hjarvard, S., & Petersen, L. N. (2013). Mediatization and cultural change. *MedieKultur: Journal of Media and Communication Research*, 29(54), 1–7.

Hjarvard, S. P. (2013). *The mediatization of culture and society*. London: Routledge.

Holappa, H., Ylipulli, J., Rautiainen, S., Minyaev, I., Pouke, M., & Ojala, T. (2018). VR Application for Technology Education in a Public Library. In *Proceedings of the 17th International Conference on Mobile and Ubiquitous Multimedia*. New York: ACM Press. 521–527.

Huhtamo, E., & Parikka, J. (eds.) (2011). *Media archaeology: Approaches, applications, and implications*. California: University of California Press.

Julin, A., Jaalama, K., Virtanen, J. P., Pouke, M., Ylipulli, J., Vaaja, M., Hyyppä, J. & Hyyppä, H. (2018). Characterizing 3D city modeling projects: Towards a harmonized interoperable system. *ISPRS International Journal of Geo-Information*, 7(2), 55.

Jung, S., Wisniewski, P., Hughes, C. (2018). In Limbo: The effect of gradual visual transition between real and virtual on virtual body ownership illusion and presence. In *Proceedings of 25th IEEE Conference on Virtual Reality and 3D User Interfaces*. IEEE.

King, R.J. (2016). The Future of Science. In Blumenthal, M., Empel, E., Perez Karp, A. & Rogers E. (eds.) *The Future According to Women*. London: Idea Couture. 52–53.

Koskela, T., Mazouzi, M., Alavesa, P., Pakanen, M., Minyaev, I., Paavola, E., & Tuliniemi, J. (2018, February). AVATAREX: Telexistence system based on virtual avatars. In *Proceedings of the 9th Augmented Human International Conference*. New York: ACM Press. Article 13, 8 pages.

Lakoff, G. (1993). The contemporary theory of metaphor. In Andrew Ortony (ed.) (1993) *Metaphor and thought*. Cambridge: Cambridge University Press. 202–251.

Lakoff, G., & Johnson, M. (1980). *Metaphors we live by*. Chicago: University of Chicago press.

Llorach, G., Evans, A., & Blat, J. (2014, November). Simulator sickness and presence using HMDs: Comparing use of a game controller and a position estimation system. In *Proceedings of the 20th ACM Symposium on Virtual Reality Software and Technology*. New York: ACM Press. 137–140.

Lovett, A., Appleton, K., Warren-Kretzschmar, B., & Von Haaren, C. (2015). Using 3D visualization methods in landscape planning: An evaluation of options and practical issues. *Landscape and Urban Planning*, *142*, 85–94.

Luusua, A., Ylipulli, J., Kukka, H., & Ojala, T. (2017). Experiencing the Hybrid City: The role of digital technology in public urban places. In John Hannigan and Greg Richards (eds.) *The SAGE handbook of urban studies*. London: Sage. 535–549.

Milgram, P., & Kishino, F. (1994). A taxonomy of mixed reality visual displays. *IEICE TRANSACTIONS on Information and Systems*, 77(12), 1321–1329.

Mitchell, M. A., & Orwig, R. A. (2002). Consumer experience tourism and brand bonding. *Journal of Product & Brand Management*, *11*(1), 30–41.

Murphy, P. (2010). Introduction. In Murphy, P., Peters, M. A., & Marginson, S. (eds.) (2010) *Imagination: Three models of imagination in the age of the knowledge economy*. New York: Peter Lang. 1–20.

Negroponte, N., Harrington, R., McKay, S. R., & Christian, W. (1997). Being digital. *Computers in Physics*, 11(3), 261–262.

Otto, T., & Smith, R. C. (2013). Design anthropology: A distinct style of knowing. In Gunn, W., Otto, T. & Smith, R.C. (eds.) (2013) *Design anthropology: Theory and practice*. London: Bloomsbury. 1–29.

Patton, M. Q. (2014) *Qualitative Research & Evaluation Methods: Integrating Theory and Practice*, 4th Ed. Thousands Oaks, CA: Sage.

Pouke, M., Ylipulli, J., Minyaev, I., Pakanen, M., Alavesa, P., Alatalo, T., & Ojala, T. (2018). Virtual library: Blending mirror and fantasy layers into a Vr interface for a public library. In *Proceedings of the 17th International Conference on Mobile and Ubiquitous Multimedia*. New York: ACM Press. 227–231.

Pouke, M., Ylipulli, J., Rantala, S., Alavesa, P., Alatalo, T., & Ojala, T. (2019). A qualitative study on the effects of real-world stimuli and place familiarity on presence. In *Proceedings of the 26th IEEE Conference on Virtual Reality and 3D User Interfaces*. 5th Workshop on Everyday Virtual Reality.

Ricci, A., Piunti, M., Tummolini, L., & Castelfranchi, C. (2015). The mirror world: Preparing for mixed-reality living. *IEEE Pervasive Computing*, 14(2), 60–63.

Sicart, M. (2017). Reality has always been augmented: Play and the promises of Pokémon GO. *Mobile Media & Communication*, 5(1), 30–33.

Silverstone, R., & Haddon, L. (1996). Design and the domestication of information and communication technologies: Technical change and everyday life. In Mansell, R. & Silverstone, R. (eds.) *Communication by Design: The Politics of Information and Communication Technologies*. Oxford, UK: Oxford University Press. 44–74.

Smart, J., Cascio, J. & Paffendorf, J. (2007). *Metaverse roadmap: Pathways to the 3D web*. Metaverse: a cross-industry public foresight project.

Steed, A., Adipradana, Y. W., & Friston, S. (2017, March). The AR-Rift 2 Prototype. In *Virtual Reality (VR), 2017 IEEE* IEEE. 231–232.

Streitz, N. (2010, November). Ambient intelligence research landscapes: Introduction and overview. In *Proceedings of International Joint Conference on Ambient Intelligence*. Berlin, Heidelberg: Springer. 300–303.

Tenhunen, S. (2008). Mobile technology in the village: ICTs, culture, and social logistics in India. *Journal of the Royal Anthropological Institute*, 14(3), 515–534.

Wilkie, A., Michael, M., & Plummer-Fernandez, M. (2015). Speculative method and Twitter: Bots, energy and three conceptual characters. *The Sociological Review*, 63(1), 79–101.

Wilkie, A., Savransky, M., & Rosengarten, M. (2017). *Speculative Research: The lure of possible futures*. London: Routledge.

Willis, K. S. (2008). Places, situations and connections. In Aurigi, A., & De Cindio, F. (eds.) (2008) *Augmented urban spaces: articulating the physical and electronic city*. UK: Ashgate Publishing. 9–22.

Wu, W., Blaicher, F., Yang, J., Seder, T., & Cui, D. (2009). A prototype of landmark-based car navigation using a full-windshield head-up display system. In *Proceedings of the 2009 workshop on Ambient media computing*. New York: ACM Press. 21–28.

Ylipulli, J., Kangasvuo, J., Alatalo, T., & Ojala, T. (2016). Chasing Digital Shadows: Exploring Future Hybrid Cities through Anthropological Design Fiction. In *Proceedings of the 9th Nordic Conference on Human-Computer Interaction*. New York: ACM Press. Article 78, 10 pages.

Ylipulli, J. & Luusua, A. (2019). Without Libraries What Have We? Public Libraries as Nodes for Technological Empowerment in the Era of Smart Cities, AI and Big Data. In *Proceedings of the 9th International Conference on Communities and Technologies*. New York: ACM Press. 92–101.

Ylipulli, J., Luusua, A., & Ojala, T. (2017). On creative metaphors in technology design: Case Magic: In *Proceedings of the 8th International Conference on Communities and Technologies*. New York: ACM Press. 280–289.

Ylipulli, J., Suopajärvi, T., Ojala, T., Kostakos, V., & Kukka, H. (2014). Municipal WiFi and interactive displays: Appropriation of new technologies in public urban spaces. *Technological Forecasting and Social Change*, 89, 145–160.

23

The museum in the smart city

The role of cultural institutions in co-creating urban imaginaries

*Carlos Estrada-Grajales, Marcus Foth,
Peta Mitchell and Glenda Amayo Caldwell*

Introduction

From its inception, the concept of the smart city has been characterised by the innovative use of technology and ubiquitous computing. Initial adopters of this tech-driven approach to smart cities include the South Korean government, which decreed a national 'U-city' (ubiquitous city) agenda as early as 2005. Early experiments such as New Songdo were engineered by the national chaebol such as Samsung and SK Telecom, and their deployment was assisted by top-down policy directives (Hwang, 2009). Yet, the 'simplistic imaginary of the smart city,' which sees the 'entangling of neoliberal ideologies with technocratic governance' (Shelton *et al.*, 2015, p. 14) has come under sustained critique (Marvin *et al.*, 2016). Researchers such as Halegoua (2011) also lament a missed opportunity to use the ubiquitous computing at the core of the U-city for community engagement and participation, and more recent contributions to the smart city discourse call for a people-focussed approach to the smart city (Foth *et al.*, 2016; Kitchin, 2016; Mattern, 2017). If a focus on people means community participation and civic engagement, then a legitimate question concerns the role of organisations that have traditionally focussed on people and engagement – that is, cultural institutions such as galleries, libraries, archives, and museums. Libraries have always been at the front of technology adoption, yet with a human-centric mindset. More recently, libraries turned the attack on their raison d'être – books being digitised and available online – into an opportunity to rethink library spaces for novel usages and more and better community engagement (Bilandzic and Foth, 2016).

Museums, too, are a pivotal agent in a city's – whether smart or not – cultural life, as they collect, curate, and communicate their history and heritage, and in so doing, museums reconstruct the past and imagine the future. Progressive approaches to the curation of museum exhibitions do not stop at history and heritage, but focus also on creating urban imaginaries: visions of the future city. This is the case of the Museum of the Future in Dubai, which combines 'elements of exhibition, immersive theatre and themed attraction [to enable visitors] look beyond the present and take your place within possible worlds to come'.[1] Urban imaginaries produced by city museums with the help of technical and data-driven interventions reshape not only how institutions such as museums redefine their narrative practices (Ioannidis *et al.*, 2013), but also how citizens shift their role from passive urban residents to data producers, narrators (Chronis, 2012), and city co-producers (Estrada-Grajales *et al.*, 2018; Foth, 2017a). Under the umbrella of the smart cities discourse, city museums embrace the 'revolutionary principle' (Fleming, 1996) of enabling co-creation as a narrative engagement practice. It appears citizens are critical agents in the process of narrating the city, as they perform as 'data sensors' (Goodchild, 2007) capable of narrating and reshaping their urban spaces and the narrative representations attached to those spaces (Gordon and de Souza E Silva, 2011). In this chapter, we take the notion of co-creation, as Sanders and Stappers (2008) define it, 'refer[s] to any act of collective creativity, i.e. creativity that is shared by two or more people'. The term is, they add 'very broad … with applications ranging from the physical to the metaphysical and from the material to the spiritual' (p. 6).

This chapter is concerned with the role of cultural institutions in co-creating urban imaginaries as narrative constructs that help shaping the future of urban environments (Cinar and Bender, 2007; Silva, 2012; Söderström *et al.*, 2014). We report on a study that examined the museum in the smart city, specifically the *100% Brisbane* exhibition in the Museum of Brisbane in Australia – a data-driven installation that encourages museum visitors to reconsider their image of the city. This case study draws attention to the potential role of the city museum in navigating between smart-city-led and smart-citizen-led visions of the city. We argue that the difference is significant in terms of the approach and adoption of smart technologies, on the one side, and in the levels of agency given to citizens to imagine and shape their city, on the other hand. We first propose the notion of urban imaginaries as a useful construct that addresses some of the shortcomings in conventional smart city discourses. We then introduce our case study, the *100% Brisbane* exhibit. Our discussion examines three themes: (1) the exhibition's link with big data analytics; (2) possibilities of co-creation for visitors of *100% Brisbane*, and; (3) the exhibit's focus on similarity and the lost opportunity to showcase and value a city's diversity. As result of our examination, we found the exhibit hints at a people-focussed approach to the design and development of smart cities, and an interesting opportunity to reflect on alternative forms of engage and understand citizens' visions of the future city. Yet, *100% Brisbane* reveals a limitation by operating with co-creation principles based on homogeneity and sameness, rather than encouraging diversity and difference as fundamental aspects of our current urban environments.

Urban imaginaries

The current penetration of the smart cities discourse in the field of urban governance reveals its dominance as a conceptual and practical referent (Greenfield, 2013). Söderström, for instance, argues that the smart cities discourse 'shapes the imaginaries and practices of a myriad of actors concretely building the city through particular case studies or pilot projects, decisions and everyday action, like creating a new electricity system for a neighborhood' (2014, p. 307). The imaginaries referred to by Söderström are, indeed, 'storytelling ideological constructs' embedded in the smart cities discourse. These utopian visions are often coopted by corporations to persuade society to embrace technology and data-driven interactions as desirable futures (Marvin *et al.*, 2016). The smart cities paradigm conveys a promise of optimisation and automatisation of a number of urban processes for growth and prosperity, and boldly pushes a globalist and entrepreneurism-like agenda (Golubchikov, 2010). The smart city's dominant 'narrative of positive transformations' (Söderström et al., 2014, p. 309) increasingly diminishes possibilities for governments and citizens to develop and engage in forms of political participation outside the smart city's marketing ploy.

City governments around the world (Caragliu et al., 2011) transform their engagement and decision-making practices by incorporating interoperable web-based services, enabling, among other services, ubiquitous connectivity and technological infrastructure for the benefit of citizens and business (Kitchin, 2015; Vanolo, 2013). The international Open & Agile Smart Cities (OASC) network has reached over 100 member cities committed to preventing vendor lock-in by creating 'an open smart city market based on the needs of cities and communities'.[2] In Australia, the federal government follows a smart cities agenda based on 'Real time data and smart technology [which] will lead to better utilisation of infrastructure, clean energy and energy efficiency, improvements in services and better benchmarking of cities performance' (Commonwealth of Australia, 2016, p. 3).

Although the smart cities agenda is presented as the ultimate reference for urban governance, there are also examinations on how such a paradigm enables novel forms of imagining, shaping, and governing the city (Greenfield, 2013; Vanolo, 2013). For many, it is imperative to expand the smart cities discourse and practice towards facilitating the conditions for engaging 'smart citizens' in decision-making (Foth et al., 2015), a notion largely absent in many smart city policy implementations (Kitchin, 2015). The latent risks of implementing a smart cities agenda, such as the one promoted by the Australian government (Commonwealth of Australia, 2016), is that such agendas stand as a form of 'corporate storytelling' apparatus (Söderström et al., 2014) and as a 'disciplinary strategy' (Vanolo, 2013) that reduce citizens, their experiences and visions to the data they produce, rather than leveraging and legitimising those citizen imaginaries in the practices of urban governance and planning. It is important here to clarify that not all citizens are interested or engaged in participatory processes that shape the city. Yet, we argue that planning bodies and other organisations, including cultural institutions such as museums, should promote, facilitate, and make use of other forms of envisioning practices, represented in this case under the concept of urban imaginaries (Estrada-Grajales et al., 2018; Silva, 2012), and reshaped by citizens' increasing use of digital and smart technologies in decision-making scenarios (Foth, 2017a; Foth et al., 2015). Many critical examinations of the smart cities agenda also highlight the need for local governments to adopt collaborative strategies that foster co-creation and peer production of citymaking (Foth, 2017a). We second Odendaal's argument (2006, p. 36) that other forms of knowledge in the shape of 'oral histories, storytelling, and poetry' may benefit the redistribution of power and citizen representation in urban planning and decision-making (Van Hulst, 2012), and that 'smart' policies and practices need to embrace qualitative and creative human input.

The smart cities agenda permeates current local government practices by establishing a sort of 'ideological discipline' (Vanolo, 2013) and a 'communicative rhetoric' (Söderström et al., 2014) that affects how society legitimises certain forms of power, governance, and sociability practices. It is therefore fundamental to examine the influence such an agenda has on scenarios where the discursive production of citizen narratives and urban imaginaries is used to support political structures.

Cities today have some resonance with Anderson's (1991) concept of the nation as 'imagined community', which he first formulated in the 1980s. Despite the obvious differences in scale and sociopolitical ramifications between city and nation – national narratives were the origin of nationalism in the early days of republican projects like those in Latin America – cities can also be regarded as imagined communities. McNeill (2001), for instance, argues that Barcelona, under the Maragall's administration (1982–1997), was ideologically constructed 'as a coherent political community in order to mobilize resources and operate at multiple spatial scales' (p. 344). This narrative portrayed the city as a persona with specific political needs and practices, all susceptible to be transformed for the benefit of its residents. Those 'urban narratives' employed in the ideological construction of the city ensure the creation of a sense of unity and comradeship, generating invisible ties and a sense of belonging among fellow citizens.

Museums as cultural institutions play a role in the generation, diffusion, and legitimisation of culturally cohesive narratives (Castells, 2001; Willim and Gustafsson Reinius, 2015). In his revised edition of *Imagined Communities*, published in 1991, Anderson incorporated a new chapter focusing on how three 'institutions of power' – the census, the map, and the museum – are intricately bound up with national imaginaries, serving as support for political projects favouring the construction of nationalistic meta-narratives. Today, modern *city* museums face the complex task of narrating the multicultural populations that constitute diverse urban environments (Lohman, 2006). In this context, city museums, their curators, artists, and managerial staff constantly examine the institution role as 'social change agent' (Fleming, 1996) as they create mechanisms for representing an everyday changing population, and craft, at the same time, cohesive narratives to legitimise local identities in urban space. The power of city museums is, hence, highly political, because they are a prominent contributor to the curation and dissemination of cultural representations – from marginalised to legitimised – of the diverse citizenry and their cultural identities in the cityscape (Castells, 2001).

Museums can play a crucial role for a smart city in the co-creation of urban imaginaries, that is, narrative depictions of the city according to the visions, experiences, and desires of its citizenry (Cinar and Bender, 2007; Silva, 2012; Strauss, 2006). We claim, building upon Silva (2006, 2012), that urban imaginaries are political instruments, crucial to reveal how citizens desire and expect a smart city to be shaped. In this sense, understanding citizen imaginaries constitutes an opportunity to grasp the cultural and ethical principles for the peer production of the smart city. As urban imaginaries narrate how we shape the city, and how the city simultaneously shapes us (Taylor, 2003), they may be considered a cornerstone for complementary urban planning practices (Collie, 2011) that attempt to empower citizens in decision-making processes in cities.

Furthermore, city museums that embrace and adopt data-driven narratives in their exhibitions understand that, as stated by Castells, 'Much of our imaginary and our political and social practices are conditioned and organised by and through electronic communications systems' (2001, p. 428). If cultural institutions such as city museums are shifting their narrative practices and are highly influenced by the discourse brought about by the smart cities discourse – co-creative narratives, data-driven, and real time exhibitions, and technology-enhanced engagement – then it is relevant to question how such narratives impact the generation of citizen-based urban imaginaries, and how such a narrative shift reframes the political role of urban residents as active agents in decision-making in the city. We propose to examine these questions in relation to *100% Brisbane*, a data-driven exhibition presented by the Museum of Brisbane that attempted to reinterpret 'the statistics of the city' – the type of data that the smart cities paradigm privileges – by 'giv[ing] them a face and voice' (Denham, Manning, and Salter, 2016, p. 34).

100% Brisbane: a walk-through

The smart cities rhetoric has institutional cement in Brisbane's urban policy. As pointed in the official visioning document, Brisbane embraces the challenges that a 'new world city' needs to address by promoting and incentivising the creation and adaptation of emerging technologies, and providing 'tailored digital experiences for Brisbane's residents, business an visitors' (Brisbane City Council, 2019, p. 11). By doing so, Brisbane City Council expects to situate the city in the *avant-garde* of urban development, and finally remove the long-standing stereotype of being just a 'big country town' (Bullivant, 2012, p. 264).

The Museum of Brisbane's *100% Brisbane* exhibition was launched in July 2016, and is currently still active (2019). The exhibition aims to investigate 'who Brisbaners are' by holding a lens to 'how we think and feel about ourselves, our city and our place in the world' (Salter *et al.*, 2016, p. 1). With *100% Brisbane*, the Museum of Brisbane attempts to challenge conventional forms of narrating the cultural diversity and social complexity of the city by incorporating novel forms of urban storytelling supported by interactive facilities in order to provide a collaborative 'real-time snapshot of our city'.[3] The *100% Brisbane* exhibition is the most recent incarnation of Berlin-based theatre company Rimini Protokoll's *100% Stadt* or *100% City* project[4] – a transportable live theatre production format that engages 100 citizens (without acting experience) as 'experts of the everyday' and as broadly representative of the city's population and demography (Garde and Mumford, 2016; Zaiontz, 2014). Each civilian performer on a *100% City* theatre stage represents 1 per cent of the city's population, meeting 'the production's demographic prerequisites for participation: age, marital status, neighbourhood, mother tongue, ethnicity or nationality, and place of birth' – they are, in effect, 'numbers made human' (Zaiontz, 2014, p. 101).

Since Rimini Protokoll's first *100%* production in Berlin in 2008, the format has been reproduced in at least 20 cities across a number of countries, including *100% Vienna* (2011), *100% Melbourne* (2012), *100% London* (2012), *100% Penang* (2015), and *100% Yogyakarta* (2015). In many respects, *100% Brisbane* is a reinvention of the *100%* format rather than a new instantiation of Rimini Protokoll's original one. On the one hand, *100% Brisbane* transports the *100% City* ephemeral, live performance format into a more permanent museum-based exhibition. On the other, it takes Rimini Protokoll's interest in creating artistic interventions based on demographics to another level. The central feature of the *100% Brisbane* exhibition is the portrayal of 100 Brisbane residents' profiles selected according to five criteria available in the 2011 Brisbane census data, namely age group, gender, region of birth, household composition, and residential location within the city metropolitan area. The sample is meant to accurately represent the composition of the Brisbane population. For instance, 50.8% of Brisbane's total population in 2011[5] identified as female, hence the Museum of Brisbane recruited 51 women to be part of the exhibition.

The exhibition invites visitors to engage with 11 interactive sections that provide different visual and sensorial stimulation. For instance, 'Scents of the City', the section that recreates the most popular olfactory associations of the city within the group of 100 participants, displays four different scents – mangroves, frangipani, sunshine, and thunderstorm – encapsulated in Perspex tubes with self-handle mechanisms for visitors to interact with. The exhibition also comprises a number of analogue and digital interactive displays (Figure 23.1) enabling visitors to explore, react, and contribute to the ongoing data collection based on visitors' reaction to the exhibition contents.

Perhaps the most significant example of how *100% Brisbane* enables visitors to interact with the exhibition is the section called 'Brisbane DNA'. In this section, the exhibition recognises narrative co-creation as a useful approach to engage citizens in the cultural identity discourse of the city. The Museum of Brisbane's promotional material for the exhibition states that 'People create cities and give cities shape and meaning. The voices of residents, past and present, provide an insight into who we are' (Denham *et al.*, 2016, p. 12). By giving lay citizens agency as co-narrators of the city, especially the group of 100 original

Figure 23.1 'Data you can touch': interactive and tangible data visualisation on *100% Brisbane* exhibition

Source: Chapter author(s), screenshot taken from video made by Liquid Interactive on Youtube, 28 November 2016, https://youtu.be/861ljKgviRY

contributors, as well as data providers as is the case for the museum visitors, the exhibition challenges conventional top-down forms of producing urban imaginaries as well as the 'one-way conversation' that has often characterised the 'traditional' museum experience (McLean, 1999, p. 89). *100%* Brisbane enables audiences to produce, provide, and experience data generated by other fellow city residents. The 'Brisbane DNA' installation relies on an image of an actual Brisbane resident from the dataset (see Figure 23.2) to become personified in an attempt to make it more relatable to visitors (cf. Reeves *et al.*, 2005). By responding to the questions asked by the installation, the visitor contributes their own information to the dataset, therefore they become a co-producer (Chronis, 2012). This part of the exhibition explores the perceptions of 100 recruited residents of Brisbane. In order to craft 'a dynamic picture of who is Brisbane' (Salter *et al.*, 2016, p. 3), the museum requested the 100 participants to respond to a number of personal questions, which were intended to provide a glimpse of the social, cultural, political, and ethical diversity of Brisbane's population. Documented by the Museum of Brisbane, this exploration is presented to the general public in a 'large scale filmic presentation' projected with the purpose of 'delivering a data-driven insight into the identity of our contemporary community' (Salter *et al.*, 2016, p. 3). The museum also made available a website for visitors to audio-record their memories and stories about the city. The exhibition web facility works as an oral depository open for online visitors.

In addition, 'Brisbane DNA' enables visitors to interact and compare their own answers with those of the 100 original Brisbane residents who represent the city according to the exhibition. Visitors can respond to three different questionnaires in an interactive setting and compare in real-time how they differ from or resemble Brisbane's average responses in topics that define the 'community's identity' (Salter *et al.*, 2016). Visitors are exposed to two different

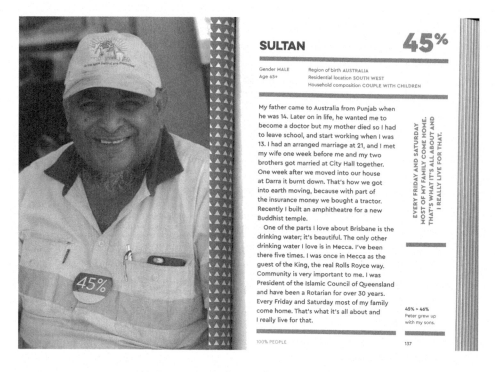

SULTAN   **45%**

Gender MALE   Region of birth AUSTRALIA
Age 65+   Residential location SOUTH WEST
                     Household composition COUPLE WITH CHILDREN

My father came to Australia from Punjab when he was 14. Later on in life, he wanted me to become a doctor but my mother died so I had to leave school, and start working when I was 13. I had an arranged marriage at 21, and I met my wife one week before me and my two brothers got married at City Hall together. One week after we moved into our house at Darra it burnt down. That's how we got into earth moving, because with part of the insurance money we bought a tractor. Recently I built an amphitheatre for a new Buddhist temple.

One of the parts I love about Brisbane is the drinking water; it's beautiful. The only other drinking water I love is in Mecca. I've been there five times. I was once in Mecca as the guest of the King, the real Rolls Royce way. Community is very important to me. I was President of the Islamic Council of Queensland and have been a Rotarian for over 30 years. Every Friday and Saturday most of my family come home. That's what it's all about and I really live for that.

EVERY FRIDAY AND SATURDAY MOST OF MY FAMILY COME HOME. THAT'S WHAT IT'S ALL ABOUT AND I REALLY LIVE FOR THAT.

45% > 46%
Peter grew up with my sons.

100% PEOPLE   137

Figure 23.2 Sultan, participant number 45: *100% Brisbane* uses storytelling to humanise and personify a specific segment of the city's population

Source: Image taken from Denham, Manning, and Salter (2016). *100% Brisbane*. Museum of Brisbane, pp. 136–137.

types of data visualisation for each question. The first visualisation in the upper part of the screen is a basic graphic representation of the number of participants – from the 100 initial participants – who responded that they did not bully anyone in their workplace. The second form of visualisation is presented at the bottom of the screen. The bars visualise the answer in percentages, and also enable visitors to compare their responses against both the 100 participants and previous exhibition visitors. 'Brisbane DNA' collects and visualises in a scoreboard the 'changing views and beliefs we hold as a city and as a community' (Salter *et al.*, 2016, p. 2).

Discussion

In this section, we discuss how *100% Brisbane* represents an opportunity to explore how cultural institutions such as the Museum of Brisbane facilitate engagement among Brisbane residents to narrate and imagine the smart city. In particular, we discuss how the exhibition constitutes an alternative for Brisbane residents to engage, and hence reframe, discursive forms of co-creating the city. The discussion gravitates around three interrelated principles used for the exhibition to construct and illustrate an image of the city. First, we examine how the exhibition weaves different data sources together. Second, we discuss how the *100% Brisbane* exhibition enables co-creative practices. Finally, we assess how the exhibition crafts narratives as part of the visitor experience that are based on similarity more so than difference and diversity, and what potential ramifications this might entail.

A big data exhibition

Byrne and Osborne (2016) question established practices of urban planning, arguing they have been 'failing to meet ordinary citizens' needs'. Such critique is not new. Theorists in the field discuss a paradigmatic shift in planning practices (Goodspeed, 2016; James *et al.*, 2015), suggesting an increased concern for more open and inclusive mechanisms of decision-making processes in urban environments (Van Hulst, 2012), particularly endangered by the emergence and popularisation of big data tools and models associated with smart cities discourse and practice (Kitchin, 2014; Nguyen and Boundy, 2017). As a result of this paradigm shift, urban planning practices have redirected their engagement mechanisms towards a more democratic redistribution of power among the stakeholders involved in decision-making (Odendaal, 2006; Sheedy *et al.*, 2008). Pløger (2004), for instance, advocates for planning models that embrace Mouffe's (2000) concept of 'agonistic pluralism' – that is, the acknowledgement of antagonistic positions in democratic deliberation and the impossibility of achieving logical and inclusive consensus – in order to cater for current multicultural urban societies. Urban planning and civic engagement practices are now required to reframe their scopes, too, to recognise and legitimise citizen expertise (Ehrlich, 2000), once understood as simple residents, and now reinvented as 'critical agents' (de Souza, 2006) and 'urban co-producers' (Estrada-Grajales *et al.*, 2018).

Although *100% Brisbane* is not intended as a planning tool or as a citizen-focussed engagement mechanism for urban planning, we argue that the exhibition, its aims, approach, contents, and interactive tools may contribute for rethinking new approaches to the smart city. The co-creative principle behind the exhibition is revealed, not only in following the Rimini Protokoll to collect a statistically accurate sample comprising the original 100 residents, but also in the display of interactive tools used by the Museum of Brisbane to enable a 'shared understanding and the building of a collective memory' (Denham, Manning, and Salter, 2016, p. 12) of the city. For instance, Rosco, an Everton Hills resident,[6] discloses in the exhibit's ABC Radio Booth how the Ekka (a colloquial hypocorism of Exhibition) – the city's annual agricultural show – helps to remind him of the 'fresh milk provided for free in the early 1950s'. What is a meaningful and 'happy' memory for this person, translates into a powerful element for Brisbane's collective memory, amplified by the use of digital and analogue interactive components. It also reveals the significance of an event such as the Ekka Festival as both an activity (sharing and celebrating fresh local produce) and a social practice (getting together and sharing) for Brisbane residents.

'Brisbane DNA' uses interactive media to enable museum visitors to have a more playful and dynamic interaction with the exhibition contents. The exhibition enables also, as illustrated above with the Radio Booth, opportunities for visitors to engage as 'narrative story-builders and co-creators' (Chronis, 2012). Beyond the debate between critical (Greenfield, 2013; Kitchin, 2015; Leszczynski, 2014; Wong *et al.*, 2009) and more careful-optimistic (Caldwell *et al.*, 2016; Klaebe and Foth, 2007; Schroeter, 2012) approaches related to the use of digital media for facilitating civic engagement and political participation, the exhibition reached more than 55,000 interactions from its opening on 15 July 2016 until 24 April 2017,[7] and in doing so giving 'voice to residents who otherwise would not be easily heard' (Schroeter, 2012, p. 1). In addition, the exhibition organisers emphasise the 'highly collaborative' nature of the process of weaving together different narratives, stories and life-experiences that produce an imaginary representation of the 'collective entity of Brisbane' (Denham, Manning, and Salter, 2016, pp. 13–14).

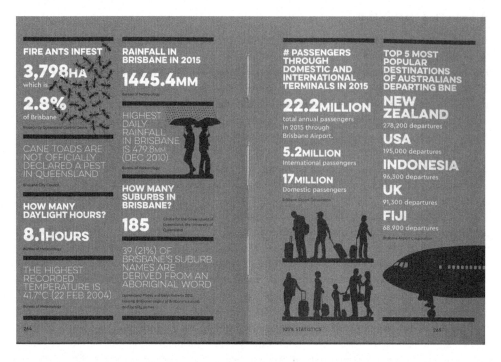

Figure 23.3 Brisbane characterised by statistical data and representations

Source: Image taken from Denham, Manning, and Salter (2016). *100% Brisbane*. Museum of Brisbane, pp. 264–265.

100% Brisbane represents an effort for weaving storytelling elements with statistical data about the city, establishing a form of 'humanising' a big data-inspired dataset (Batty, 2013). In this sense, the exhibition tries to reconcile a dichotomy in narrating the city: the quantitative genre, represented by the interest in statistics about the characteristics of the city and the concern for showcasing a statistically 'accurate' sample of Brisbane's population (Figure 23.3), and the qualitative genre, portrayed by the stories, perceptions, and experiences of the 100 participants and the museum visitors who interact with the exhibit (Figure 23.2). *100% Brisbane*, thus, sets up a 'middle-ground' contribution to rethinking storytelling practices focussed on citizen perspectives (Fredericks *et al.*, 2016; Klaebe *et al.*, 2007; Odendaal, 2006; Van Hulst, 2012), as well as it represents an opportunity for enriching engagement and planning mechanisms, often accused of being highly technocratic and distant of any human sentiment (Odendaal, 2006; Söderström *et al.*, 2014).

Next, we analyse *100% Brisbane* in light of two major operational principles behind the exhibition: co-creation and similarity.

Co-creating the smart city

Based on Sanders and Stappers (2008) definition of co-creation, we discuss the *100% Brisbane* exhibition from the different roles and experiences of co-producers versus that of co-creators.

'Brisbane DNA' intends to increase the understanding of how each citizen relates to other Brisbane citizens; however, what the actual picture of understanding looks like to one another remains hidden.

The experience provided to the participant in the exhibition is individualistic, therefore limiting the ability for co-creation to manifest beyond the participant and their interaction with the exhibit data. Cities are composed of diverse communities, yet the exhibition provides limited possibilities for collective identities of urban communities to be expressed. It is an individualistic picture that is presented, and opportunities to share in a collective creative act are limited. Although participants are asked to record their story of a Brisbane related memory, the only way others can engage with those stories is through the exhibition website. Currently, the website only has one working recording accessible to the public. Additionally, participants are not able to share their ideas or respond in an open-ended manner. There are no mechanisms (either quantitative or qualitative, analogue or digital, interactive or not) available through the exhibition design or set-up that allow participants to openly share their newly created understanding of their position within the city of Brisbane in relation to the selected 100 residents. From a broader citymaking point of view, this is a missed opportunity as there are valuable lessons to be learnt from these shared experiences that could inform urban planning and design, architectural decisions, transport requirements, and a shared vision of the city.

The process of citymaking is creative as it requires the complex composition of many factors such as the distribution of space, infrastructure and navigation, use of materials, technologies, design of structures, the human experience, and cultural expression. Although the exhibition provides a snapshot of the different people who make up a city, it does not promote a collective creativity experience, limiting the co-creation capabilities discussed before (Srinivasan *et al.*, 2009). As participants engage with the exhibition they contribute their own information and increase the size of the dataset, however these data do not then get applied in a meaningful manner (Castells, 2001). The information gathered and the interactive experiences of it do not encourage the development of human relationships, vital to effective collaborative citymaking practices. In this instance the exhibition is similar to most smart city initiatives where the technology is used to entice interaction to collect information and grow datasets, but creative engagement with real issues for effective outcomes is minimal (Maye *et al.*, 2017).

Acknowledging that this exhibition will continue until 2019 and that the dataset will continue to grow over the next two years, we provide some thoughts as to how this information could be applied to co-create a meaningful image of Brisbane. It is our hope that this discussion may attract smart city stakeholders in both urban planning and cultural institutions such as museums and libraries, to consider the co-creation of a diverse and dynamic set of urban imaginaries for co-creative citymaking practices (Foth, 2017a).

Homogeneity versus diversity

Smart city technology is often characterised by its use of big data and algorithmic analysis. Despite only nascent work so far focussing on smart cities (Foth, 2017b; Kitchin, 2016), the repercussions in the context of digital media have been studied in closer detail. Algorithmic curation drives social media news feeds, product recommendations, and search results. It plays a key role in producing sameness and homophily of ideas and opinions, which are partly a result of our connectedness (via digital media) to mostly like-minded people and content (Dvir-Gvirsman, 2016). The compounding aspects of this opinion polarisation in social media have been studied in political science, and media and communication studies, e.g., echo chambers (Aiello *et al.*, 2012; Dvir-Gvirsman, 2016) and filter bubbles (Pariser, 2011).

Cities thrive on diversity (Wood and Landry, 2008). A city's socio-cultural diversity is linked to economic productivity (Ottaviano and Peri, 2006). Cities and their diversity have been described as the engines of innovation and economic growth (Bairoch, 1991), and diversity is seen as a key factor of a city's success (Jacobs, 1969). Global cities, such as London, Paris, New York, and Tokyo, are contributing to world economic growth and innovation largely because of the cultural diversity of their populations (Sassen, 2011). Smart cities are at a critical juncture as social media are infiltrating all aspects of our lives, risking opinion polarisation (Lee *et al.*, 2014). On a global scale this polarisation is rapidly emerging and is manifest in recent political events (Hong and Kim, 2016).

The big data analytics derived content and the interactive features of the *100% Brisbane* exhibit provide touch points between the city and its citizens. These touch points offer opportunities to address issues of the software-sorted city (Foth, 2017b; Graham, 2005) by understanding the role of cultural institutions as innovative avenues for fostering both depolarisation and discovery and value of difference. However, here, it is only realised to a limited extent, partly due to the Rimini Protokoll's original format.

When visitors interact with the exhibition's questionnaires and visualise their results, they are faced with a user experience similar to those enabled by social media platforms, e.g. Tinder, which promote homophily and like-mindedness as design principles (Figure 23.4). The visitor's responses are compared with the *100% Brisbane* dataset looking for similarity and sameness. This not only constitutes socio-ethical issues such as 'over personalisation' (Leszczynski, 2014). It also represents a lost opportunity to discover and value a city's diversity as well as to showcase and highlight the socio-economic disparities faced by different

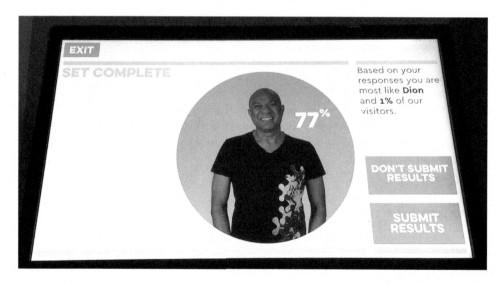

Figure 23.4 Visitors are matched with an existing profile after the completion of 'Brisbane DNA' questionnaires: the results, as shown in the top right corner, are based on similarity of the answers

Source: Image taken by Carlos Estrada-Grajales, 16 November 2017.

groups and communities. Visitors 'discover, often with surprising result, who amongst the community best represents them' (Salter *et al.*, 2016, p. 3). This can foster a sense of belonging based on compatibility and similarity. Yet, it may be useful to complement such experiences with exhibition features that best bestow value difference and those who are different. Giving visitors opportunities to embrace the diversity of the smart city can give rise to urban imaginaries about Brisbane as a 'place for encounters' (Harvey, 2012).

Conclusion

We introduced the *100% Brisbane* exhibition as a case study to highlight challenges and opportunities in the smart city discourse. Many scholars and commentators advocate for a people-focussed approach to the design and development of smart cities. This requires a rethink of current practices in urban planning and civic engagement (Foth and Brynskov, 2016; Foth *et al.*, 2016). Offering a microcosm of the larger smart city macrocosm, *100% Brisbane* illustrates a preponderant preference for quantitative big data (e.g., numbers, charts, and percentages). Yet, it also highlights the potential to use interactive media and urban interfaces to engage citizens in practices of co-creation (e.g., enabling a degree of citizen interaction with the exhibition contents, as if they were sensors). The most critical point, however, is the principle used to conduct the co-creation practices. *100% Brisbane* offers visitors the opportunity to see themselves reflected in a representation of the city, but the image visitors see and interact with is that of a city narrated by those similar to them. Cities and their diversity have been described as the engines of innovation and economic growth (Bairoch, 1991). If a smart city wants to support innovative forms of engaging citizens in co-creating the city, complementary ways to curate and partake in algorithmic culture – as the perpetuation of inequalities via tech-based systems functioning in cities (Foth *et al.*, 2018) – are needed in order to show and treasure 'diversity and depth that underlines the richness of Brisbane's history and identity' (Denham, Manning, and Salter, 2016, p. 14). The questions for future studies are then: what type of city are we imagining? What type of city are we co-creating? And, most importantly, what kind of city can we plan and engage with when interacting with the interfaces and tools of a smart city?

Acknowledgements

The authors would like to gratefully acknowledge the support of the Museum of Brisbane staff, especially the museum curator Phillip Manning, for facilitating access to the exhibit, datasets, discussions, and other resources, without which this research would not have been possible.

Notes

1 www.museumofthefuture.ae/about.
2 www.oascities.org/.
3 www.museumofbrisbane.com.au/whats-on/100-brisbane/.
4 www.rimini-protokoll.de/.
5 http://censusdata.abs.gov.au/census_services/getproduct/census/2016/quickstat/LGA31000.
6 http://museumofbrisbane.com.au/wp-content/uploads/2010/03/Rosco-from-Everton-Hills-EKKA-smells-and-memories-from-childhood-remembers-getting-milk-from-the-cows-before-they-opened-in-1940s.mp3.
7 Date on which the Museum of Brisbane provided a data snapshot for this study.

References

Aiello, L. M., Barrat, A., Schifanella, R., Cattuto, C., Markines, B., & Menczer, F. (2012). Friendship prediction and homophily in social media. *ACM Trans. Web*, *6* (2), 1–33. doi: https://doi.org/10.1145/2180861.2180866.

Anderson, B. (1991). *Imagined communities: Reflections on the origin and spread of nationalism*. Rev. ed. London: Verso Books.

Bairoch, P. (1991). *Cities and Economic Development: From the Dawn of History to the Present*. Chicago: University of Chicago Press. Retrieved from https://market.android.com/details?id=book-Cg7JYZO_nEMC.

Batty, M. (2013). Big data, smart cities and city planning. *Dialogues in Human Geography*, *3* (3), 274–279. doi: https://doi.org/10.1177/2043820613513390.

Bilandzic, M., & Foth, M. (2016). Designing hubs for connected learning: Social, spatial and technological insights from coworking, hackerpaces and meetup Groups. *Place-Based Spaces for Networked Learning*. Retrieved from http://eprints.qut.edu.au/83742/.

Brisbane City Council. (2019). Smart connected Brisbane. Available from: www.brisbane.qld.gov.au/about-council/governance-and-strategy/vision-and-strategy/smart-connected-brisbane.

Bullivant, L. (2012). Smart cities: rethinking the city centre, Brisbane, Queensland, Australia. In Bullivant, L. (ed.) *Master planning futures*. Abingdon: Routledge (pp. 264–274).

Byrne, J., & Osborne, N. (2016). Urban hacktivism: getting creative about involving citizens in city planning. Retrieved July 20, 2017, from http://theconversation.com/urban-hacktivism-getting-creative-about-involving-citizens-in-city-planning-62277.

Caldwell, G., Guaralda, M., Donovan, J., & Rittenbruch, M. (2016). The instabooth: Making common ground for media architectural design. In *Proceedings of the 3rd Conference on Media Architecture Biennale* (p. 3:1–3:8).New York, NY, USA: ACM. https://doi.org/10.1145/2946803.2946806.

Caragliu, A., Del Bo, C., & Nijkamp, P. (2011). Smart cities in Europe. *Journal of Urban Technology*, *18* (2), 65–82. doi: https://doi.org/10.1080/10630732.2011.601117.

Castells, M. (2001). Museums in the information era: Cultural connectors of time and Space. In R. Parry (ed.), *Museums in a Digital Age* (pp. 427–434). London: Routledge. Retrieved from http://archives.icom.museum/pdf/E_news2001/p4_2001-3.pdf.

Chronis, A. (2012). Tourists as story-builders: Narrative construction at a Heritage Museum. *Journal of Travel & Tourism Marketing*, *29* (5), 444–459. doi: https://doi.org/10.1080/10548408.2012.691395.

Cinar, A., & Bender, T. (2007). *Urban imaginaries: Locating the modern city*. Minneapolis, MN: University of Minnesota Press.

Collie, N. (2011). Cities of the imagination: Science fiction, urban space, and community engagement in urban planning. *Futures*, *43* (4), 424–431. doi: https://doi.org/10.1016/j.futures.2011.01.005.

Commonwealth of Australia. (2016). *Smart Cities Plan*. The Department of the Prime Minister and Cabinet, Canberra.

de Souza, M. L. (2006). Social movements as "critical urban planning" agents. *Cityscape*, *10* (3), 327–342. doi: https://doi.org/10.1080/13604810600982347.

Denham, P., Manning, P., & Salter, C. (2016). 100% Brisbane. Museum of Brisbane. Retrieved from https://market.android.com/details?id=book-Z2LbjwEACAAJ.

Dvir-Gvirsman, S. (2016). Media audience homophily: Partisan websites, audience identity and polarization processes. *New Media & Society*, 1461444815625945. doi: https://doi.org/10.1177/1461444815625945.

Ehrlich, T. (2000). Civic Responsibility and Higher Education. Greenwood Publishing Group. Retrieved from https://market.android.com/details?id=book-4gOpzfEqgE4C.

Estrada-Grajales, C., Foth, M., & Mitchell, P. (2018). Urban imaginaries of co-creating the City: Local activism meets citizen peer-production. *Journal of Peer Production*, *11*. Online.

Fleming, D. (1996). Making city histories. In G. Kavanagh (ed.), *Making History in Museums* (pp. 39–43). Leicester: Leicester University Press. Retrieved from https://books.google.com/books?hl=en&id=hoGxAwAAQBAJ&oi=fnd&pg=PA131&dq=Making+city+histories+Fleming&ots=A9SxzSn9m3&sig=lKglotOBNUf0choTbCARiDAeLyo.

Foth, M. (2017a). Lessons from urban guerrilla placemaking for smart city commons. Presented at the 8th International Conference on Communities and Technologies (C&T 2017), Troyes, France. https://doi.org/10.1145/3083671.3083707.

Foth, M. (2017b). The software-sorted city: Big data & algorithms. In N. Odendaal & A. Aurigi (eds.). Presented at the Digital Cities 10: Towards a Localised Socio-Technical Understanding of the "Real" Smart City, Troyes, France: ACM. Retrieved from https://eprints.qut.edu.au/107748/.

Foth, M., & Brynskov, M. (2016). Participatory action research for civic engagement. In E. Gordon & P. Mihailidis (eds.), *Civic Media: Technology, Design, Practice*(pp. 563–580). Cambridge, MA: MIT Press.

Foth, M., Brynskov, M., & Ojala, T. (2015). *Citizen's Right to the Digital City: Urban Interfaces, Activism, and Placemaking*. Berlin: Springer. Retrieved from https://market.android.com/details?id=book-SrtPCwAAQBAJ.

Foth, M., Hudson-Smith, A., & Gifford, D. (2016). Smart cities, social capital, and citizens at play: A critique and a way forward. In F. X. Olleros & M. Zhegu (eds.), *Research Handbook on Digital Trans-formations* (pp. 203–221). Cheltenham, UK: Edward Elgar Publishing. https://doi.org/10.4337/9781784717766.00017.

Foth, M., Mitchell, P., & Estrada-Grajales, C. (2018). Today's internet for tomorrow's cities: On algorithmic culture and urban imaginaries. *Second International Handbook of Internet Research*, 1–22.

Fredericks, J., Caldwell, G. A., Foth, M., & Tomitsch, M. (2017). The city as perpetual beta: Fostering systemic urban acupuncture. In M. de Lange & M. de Waal (eds.). Springer. Retrieved from https://eprints.qut.edu.au/107240/.

Fredericks, J., Caldwell, G. A., & Tomitsch, M. (2016). Middle-out design: collaborative community engagement in urban HCI. In *Proceedings of the 28th Australian Conference on Computer-Human Interaction* (pp. 200–204). ACM. https://doi.org/10.1145/3010915.3010997.

Garde, U., & Mumford, M. (2016). *Theatre of Real People: Diverse Encounters at Berlin's Hebbel am Ufer and Beyond*. London: Bloomsbury Publishing.

Golubchikov, O. (2010). World-city-entrepreneurialism: Globalist imaginaries, neoliberal geographies, and the production of New St Petersburg. *Environment & Planning A*, *42* (3), 626–643. doi: https://doi.org/10.1068/a39367.

Goodchild, M. F. (2007). Citizens as sensors: the world of volunteered geography. *GeoJournal*, *69* (4), 211–221. doi: https://doi.org/10.1007/s10708-007-9111-y.

Goodspeed, R. (2016). The death and life of collaborative planning theory. *Urban Planning*, *1* (4), 1–5. doi: https://doi.org/10.17645/up.v1i4.715.

Gordon, E., & de Souza E Silva, A. (2011). *Net Locality: Why Location Matters in a Networked World*. Malden and Oxford: Wiley-Blackwell. Retrieved from https://market.android.com/details?id=book-v2XKptosHykC.

Graham, S. (2005). Software-sorted geographies. *Progress in Human Geography*, *29* (5), 562–580. Retrieved from http://phg.sagepub.com/content/29/5/562.short.

Greenfield, A. (2013). Against the smart city. *Do Projects*.

Halegoua, G. (2011). The Policy and export of ubiquitous Place: Investigating South Korean U-Cities. In M. Foth, L. Forlano, C. Satchell, & M. Gibbs (eds.), *From Social Butterfly to Engaged Citizen: Urban Informatics, Social Media, Ubiquitous Computing, and Mobile Technology to Support Citizen Engagement*. Cambridge, MA: MIT Press, pp. 315–334. Retrieved from https://eprints.qut.edu.au/13308/.

Harvey, D. (2012). *Rebel Cities: From the Right to the City to the Urban Revolution*. London: Verso Books. Retrieved from https://market.android.com/details?id=book-IKJE02gfP0cC.

Hong, S., & Kim, S. H. (2016). Political polarization on twitter: Implications for the use of social media in digital governments. *Government Information Quarterly*, *33* (4), 777–782.

Hwang, J.-S. (2009). u-City: The next paradigm of urban development. In M. Foth (ed.), *Handbook of Research on Urban Informatics: The Practice and Promise of the Real-Time City* (pp. 367–378). Hershey, PA: Information Science Reference, IGI Global.

Ioannidis, Y., Raheb, K. E., Toli, E., Katifori, A., Boile, M., & Mazura, M. (2013). One object many stories: Introducing ICT in museums and collections through digital storytelling. *2013 Digital Heritage International Congress (DigitalHeritage)* (Vol. 1, pp. 421–424). doi: https://doi.org/10.1109/DigitalHeritage.2013.6743772.

Jacobs, J. (1969). *The Economy of Cities*. New York: Vintage Books.

James, K., Thompson-Fawcett, M., & Hansen, C. J. (2015). Transformations in identity, governance and planning: The case of the small city. *Urban Studies*. doi: https://doi.org/10.1177/0042098015571060.

Kitchin, R. (2014). The real-time city? Big data and smart urbanism. *GeoJournal*, *79* (1), 1–14. doi: https://doi.org/10.1007/s10708-013-9516-8.

Kitchin, R. (2015). Making sense of smart cities: Addressing present shortcomings. *Cambridge Journal of Regions, Economy and Society, 8* (1), 131–136. doi: https://doi.org/10.1093/cjres/rsu027.

Kitchin, R. (2016). The ethics of smart cities and urban science. *Philosophical Transactions. Series A, Mathematical, Physical, and Engineering Sciences, 374* (2083). doi: https://doi.org/10.1098/rsta.2016.0115.

Klaebe, H. G., & Foth, M. (2007). Connecting communities using new media: The sharing stories project. In L. Stillman & G. Johanson (eds.), *Constructing and sharing memory: community informatics, identity and empowerment* (pp. 143–153). Newcastle: Cambridge Scholars Publishing. Retrieved from https://eprints.qut.edu.au/8676/.

Klaebe, H. G., Foth, M., Burgess, J. E., & Bilandzic, M. (2007). Digital storytelling and history lines: Community engagement in a master-planned development. In *Proceedings of the 13th International Conference on Virtual Systems and Multimedia: Exchange and Experience in Space and Place, VSMM 2007.* Brisbane,QLD: Australasian Cooperative Research Centre for Interaction Design Pty, Limited. Retrieved from http://eprints.qut.edu.au/8985.

Lee, J. K., Choi, J., Kim, C., & Kim, Y. (2014). Social media, network heterogeneity, and opinion polarization. *Journal of communication, 64* (4), 702–722.

Leszczynski, A. (2014). Spatial media/tion. *Progress in Human Geography.* doi: https://doi.org/10.1177/0309132514558443.

Lohman, J. (2006). City Museums: do we have a role in shaping the global community? *Museum International, 58* (3), 15–20. doi: https://doi.org/10.1111/j.1468-0033.2006.00562.x.

Marvin, S., Luque-Ayala, A., & McFarlane, C. (2016). *Smart Urbanism: Utopian Vision Or False Dawn?.* London: Routledge. Retrieved from https://market.android.com/details?id=book–mZACwAAQBAJ.

Mattern, S. (2017). A city is not a computer. *Places Journal.* Retrieved from https://placesjournal.org/article/a-city-is-not-a-computer/.

Maye, L. A., Bouchard, D., Avram, G., & Ciolfi, L. (2017). Supporting cultural heritage professionals adopting and shaping interactive technologies in museums. In *Proceedings of the 2017 Conference on Designing Interactive Systems* (pp. 221–232). ACM. https://doi.org/10.1145/3064663.3064753

McLean, K. (1999). Museum exhibitions and the dynamics of dialogue. *Daedalus, 128* (3), 83–107.

McNeill, D. (2001). Barcelona as imagined community: Pasqual Maragall's spaces of engagement. *Transactions of the Institute of British Geographers.* Retrieved from http://onlinelibrary.wiley.com/doi/10.1111/1475-5661.00026/full.

Mouffe, C. (2000). Deliberative democracy or agonistic pluralism. Retrieved from www.ssoar.info/ssoar/handle/document/24654.

Nguyen, M. T., & Boundy, E. (2017). Big Data and Smart (Equitable) Cities. In P. (Vonu) Hakuriah, N. Tilahun, and M. Zellner, (eds.) *Seeing Cities Through Big Data* (pp. 517–542). Springer, Cham. doi: https://doi.org/10.1007/978-3-319-40902-3_28.

Odendaal, N. (2006). Towards the digital city in South Africa: Issues and constraints. *Journal of Urban Technology, 13* (3), 29–48. doi: https://doi.org/10.1080/10630730601145997.

Ottaviano, G. I. P., & Peri, G. (2006). The economic value of cultural diversity: evidence from US cities. *Journal of Economic Geography, 6* (1), 9–44. Retrieved from http://joeg.oxfordjournals.org/content/6/1/9.short.

Pariser, E. (2011). *The filter bubble: what the Internet is hiding from you.* New York: Penguin Press.

Pløger, J. (2004). Strife: Urban Planning and Agonism. *Planning Theory, 3* (1), 71–92. doi: https://doi.org/10.1177/1473095204042318.

Reeves, S., Benford, S., O'Malley, C., & Fraser, M. (2005). Designing the Spectator Experience. In *Proceedings of the SIGCHI Conference on Human Factors in Computing Systems* (pp. 741–750). New York, NY, USA: ACM. https://doi.org/10.1145/1054972.1055074.

Salter, C., Denham, P., & Manning, P. (eds.). (2016). 100% Brisbane at Museum of Brisbane (Vol. 4). International Council of Museums- Museums of Cities Review. Retrieved from http://network.icom.museum/fileadmin/user_upload/minisites/camoc/CAMOC-_REVIEW_No._4_-_Jan_r_02.pdf.

Sanders, E. B.-N., & Stappers, P. J. (2008). Co-creation and the new landscapes of design. *CoDesign, 4* (1), 5–18. doi: https://doi.org/10.1080/15710880701875068.

Sassen, S. (2011). *Cities in a World Economy.* Thousand Oaks, CA: SAGE Publications. Retrieved from https://market.android.com/details?id=book-HdEgAQAAQBAJ.

Schroeter, R. (2012). Discussions in space: interactive urban screens for enhancing citizen engagement. Queensland University of Technology. Retrieved from http://eprints.qut.edu.au/50771.

Sheedy, A., MacKinnon, M. P., Pitre, S., Watling, J.Others. (2008). *Handbook on citizen engagement: Beyond consultation*. Ottawa: Canadian Policy Research Networks. Retrieved from www.sasanet.org/documents/Resources/Handbook%20on%20Citizen%20Engagement_Beyond%20Consultation.pdf.

Shelton, T., Zook, M., & Wiig, A. (2015). The "actually existing smart city." *Cambridge Journal of Regions, Economy and Society, 8* (1), 13–25.

Silva, A. (2006). Imaginarios urbanos. Arango. Retrieved from www.academia.edu/download/39735962/imaginarios_urbanos.pdf.

Silva, A. (2012). Urban imaginaries from Latin America. Documenta II. *Anales Del Instituto de Investigaciones Estéticas, 28* (88), 260–264. doi: https://doi.org/10.22201/iie.18703062e.2006.88.2218.

Söderström, O., Paasche, T., & Klauser, F. (2014). Smart cities as corporate storytelling. *Cityscape, 18* (3), 307–320. doi: https://doi.org/10.1080/13604813.2014.906716.

Srinivasan, R., Boast, R., Furner, J., & Becvar, K. M. (2009). Digital museums and diverse cultural knowledges: Moving past the traditional catalog. *The Information Society, 25* (4), 265–278. doi: https://doi.org/10.1080/01972240903028714.

Strauss, C. (2006). The imaginary. *Anthropological Theory, 6* (3), 322–344. Retrieved from http://journals.sagepub.com/doi/abs/10.1177/1463499606066891.

Taylor, C. (2003). What Is a "Social Imaginary"?. In D. P. Gaonkar, J. Kramer, B. Lee, M. Warner, C. Taylor, D. P. Gaonkar, ... M. Warner (eds.), *Modern Social Imaginaries* (pp. 23–30). London: Duke University Press. doi: https://doi.org/10.1215/9780822385806-003.

Van Hulst, M. (2012). Storytelling, a model of and a model for planning. *Planning Theory, 11* (3), 299–318. Retrieved from http://journals.sagepub.com/doi/abs/10.1177/1473095212440425.

Vanolo, A. (2013). Smartmentality: The smart city as disciplinary strategy. *Urban Studies, 51* (5), 883–898. doi: https://doi.org/10.1177/0042098013494427.

Willim, R., & Gustafsson Reinius, L. (2015). Museum imaginaries: On evocations of possible worlds. Presented at the International SCACA symposium: *Ethnography and Its Audiences*. Halmstad, Sweden: Halmstad University. Retrieved from www.diva-portal.org/smash/record.jsf?pid=diva2:877434.

Wong, Y. C., Law, C. K., Fung, J. Y. C., & Lam, J. C. Y. (2009). Perpetuating old exclusions and producing new ones: Digital exclusion in an information society. *Journal of Technology in Human Services, 27* (1), 57–78. doi: https://doi.org/10.1080/15228830802459135.

Wood, P., & Landry, C. (2008). *The Intercultural City: Planning for Diversity Advantage*. London: Earthscan. Retrieved from https://market.android.com/details?id=book-3VwimAEACAAJ.

Zaiontz, K. (2014). Performing visions of governmentality: Care and capital in 100% vancouver. *Theatre Research International, 39* (2), 101–119. Retrieved from doi:https://doi.org/10.1017/S0307883314000030.

Section 2
Cities and placemaking

24

The hackable city

Exploring collaborative citymaking in a network society

Martijn de Waal, Michiel de Lange and Matthijs Bouw

Introduction

In this chapter, we introduce the hackable city as a heuristic model to explore new modes of collaborative citymaking that have arisen in our current network or platform society (Castells, 2010; van Dijck *et al.*, 2018). The model is closely related to debates about smart cities and smart citizens (Hollands, 2008; Allwinkle and Cruickshank, 2011; Caragliu *et al.*, 2011; Nam and Pardo, 2011; Chourabi *et al.*, 2012; Brynskov *et al.*, 2014; Kitchin, 2014; de Waal and Dignum, 2017; Cardullo and Kitchin, 2017; Hemment and Townsend, 2013; Niederer and Priester, 2016), yet it takes a more relational approach, exploring ethics of citymaking and citizenship, concrete practices for collaborative citymaking as well as affordances of systems for innovation, adaptation and social change. With the lens of the hackable city, we want to highlight a vision of the city as a site of both collaboration as well as struggle and conflicts of interests. In this account, new media technologies enable citizens to organise, mobilise, innovate and collaborate towards commonly defined goals. Yet the hackable city also recognises the messiness of such a process, the conflicts of interest at play and the continuous struggle between the alignment of private goals, collective hacks and public interests. As an alternative imaginary, the hackable city is not a progressive alternative panacea to a neoliberal smart city that will by itself bring out a harmonious, inclusive resilient city, if only citizens would start using the right technological tools and governments would be willing to listen to them. Rather, as a lens, the hackable city aims to bring out the underlying dynamics and (sometimes conflicting) values at stake in citymaking, as well as the concrete practices through which they are enacted. It revolves around using the affordances of digital technologies to find new ways to organise civic initiatives and align these with processes of democratic governance and accountability in a society that is increasingly technologically mediated.

Hacking and citymaking

We – two university researchers and a practicing architect – started exploring this notion of the hackable city in 2012 in a workshop organised by the Dutch Delta Metropolis Association.

Workshop participants were asked to draw up a new vision for the future city: which themes should dominate the urban agenda in the coming decade? What should the next "big visionary project" look like? And how could this be applied to a number of city-regions in the Netherlands?

During the workshop, after discussing various current trends in urban planning, we found ourselves turning those original questions around. When we began exploring what this big vision, masterplan or investment project could look like, we found numerous citymaking initiatives that – in a spirit reminiscent of online hacker cultures – had just started to change things in the city that in their eyes needed improving. We found energy collectives producing and distributing renewable energy; projects that turned empty office space into co-working or housing spaces; and loosely organised citizens using social media to improve the livability of their neighbourhoods through collective action. Sometimes these collectives were organised by informal groups of neighbourhood residents; more often they were initiated by designers, architects or other professionals who were embracing a new approach to their profession. Rather than waiting for new masterplans to appear, they organised collectives of various stakeholders around the issues they deemed urgent.

Their impromptu character does not mean that these initiatives are not visionary. Like many hacker cultures, many of these citymaking initiatives are searching for alternative value systems that could underlie urban planning. Some are actively seeking to catalyse societal transformations, for instance in the domain of renewable energy. Others centre on the notion of the commons: the city as a set of resources that are collaboratively owned and managed. In a world where public services and resources have been increasingly privatised, they seek to bring these resources back into the public domain, reclaiming a citizen-based "ownership" of the city (Franke et al., 2013, 2015a).

Could, we found ourselves arguing, the sum of all these initiatives make up the next visionary project? Taken together, they could potentially make the city more resilient, innovative, livable and social. Rather than approaching the city as a tabula rasa, these initiatives are interested in continuously improving the current city. Local initiatives could quickly identify local issues to address; small-scale initiatives could test out new approaches and technologies and scale them up when proven successful. And their collective and collaborative approach could also bring ownership of the city partly back to the citizenry.

So, at the end of the workshop we formulated our vision, or "urban imaginary", of the hackable city: a city in which new media technologies are employed to open up urban institutions and infrastructures, to improve upon them in the public interest, in practices of collaborative citymaking.

We weren't at that time the first or the last to point out the resemblances between a hacker culture and emerging instances of collaborative citymaking. The term hacking originated in the world of computer science, information and communication technology (ICT) and media technologies. From radio amateurs in the early twentieth century to the US west-coast computer culture that gave rise to first personal computers in the 1970s and the rise of the free/libre and open-source software (FLOSS) movement in the following decades, users have been figured as active creators, shapers and benders of media technologies and the relationships mediated through them (Roszak, 1986; Levy, 2001; von Hippel, 2005; Söderberg, 2010). In general, hacking refers to the process of clever or playful appropriation of existing technologies or infrastructures or bending the logic of a particular system beyond its intended purposes or restrictions to serve one's personal, communal or activism goals.

Where the term was mainly used to refer to practices in the sphere of computer hardware and software, more recently "hacking" has been used to refer to creative practices and ideals of citymaking: spanning across spatial, social, cultural and institutional domains, various practices of "city hacking" can be seen in urban planning, city management and examples of tactical urbanism and DIY/DIWO urban interventions. Various authors have by now described the rise of "civic hackers" (Crabtree, 2007; Townsend, 2013; Schrock, 2016), where citizens are cast in the role of tech-savvy agents of urban change, usually working towards the public good. For instance, in the guise of monitorial citizens (Schudson, 1998) that make use of open data to hold governments accountable (Schrock, 2016); or as coders that take part in programs like Code for America to create apps or websites that can help solve problems posed by local authorities (Townsend, 2013); or alternatively, as participants in hackathons that code more speculative prototypes to spark discussions around issues of concern (Lodato and Disalvo, 2016).

Furthermore, moving beyond the application of technology to civic life, the ethos and spirit of various hacker movements have been invoked to describe new forms of bottom-up, grassroots and collaborative city making. Lydon and Garcia (2015) connect their tactical urbanism paradigm to the iterative, learning-by-doing approach of the hacker movement. Caldwell and Foth (2014) describe the emergence of DIY-placemaking communities around the world, partly inspired by hacking cultures and their ethos of shaping, bending and extending technologies to their needs, often beyond their intended use. In professional circles, Gardner (2015) sees a similar shift in the profession of architecture at large. Architects are moving from the position of "the self-conscious designers of modernism, with its unassailable belief in social engineering" to an ethos of hacking, projecting their imaginations of better futures onto the "full and buzzing activities and structures" of the existing world.

The articulation of civic hacking is especially interesting in this regard. Hacking in these examples refers to the inspiration found in the playful, exploratory, collaborative and sometimes transgressive modes of operation found in various hacker cultures, constructively applied in the context of civics and politics. At the same time, it also connotes the centrality of digital media technologies as tools for mobilisation, communication and civic organisation. As Saad-Sulonen and Horelli (2010) point out, many self-organising civic groups rely on extensive ecologies of digital artefacts, even if their activities themselves are not centred around technology. In addition, the adjective of civic denotes that these activities not only concern societal issues, but should also be understood as taking on a less adversarial position than "regular" activists (or some hacker cultures for that matter) (Hunsinger and Schrock, 2016; Schrock, 2016). Civic hackers are seen as working with – or trying to reform – governments and other institutional actors to address societal issues, such as inequality, community representation, housing affordability and sustainability. The civic hacker, Schrock (2016) writes, seeks "to ease societal suffering by bringing the hidden workings of abstract systems to light and improve their functioning".

The hackable city: a model for collaborative citymaking in a networked society

It were these "civic hacking" initiatives and the role of digital media technologies and their practices that we further sought to understand in a research project we started after the Deltametropool workshop. In a year-long research trajectory situated in the Amsterdam-based urban living-lab and brownfield redevelopment site of Buiksloterham, we further explored the opportunities as well as the challenges presented by the rise of new media technologies for an open, democratic, collaborative citymaking process. We explored these themes through observations and interviews with stakeholders, workshops, the introduction of a number of "design probes", by taking part in various local meetings about the development of the area, as well as through literature study and conceptual explorations. Using a grounded theory approach, we combined the insights gathered through these processes in a "hackable city model". This model is neither purely descriptive nor prescriptive. It acts as a heuristic that allows us to investigate the dynamics at work in processes of collaborative citymaking, and explore these both constructively (how can they be understood as exemplaries, providing directions for design, policy and practices) as well as critically (what new tensions, conflicts and dillema's arise from these new ways of organising citymaking).

The main point we want to underline with our model (Figure 24.1) is that "bottom up" and "top down" should not be regarded as two conflicting modes of citymaking. Nor can the process of citymaking be reduced to just two actors: top-down institutions and bottom-up self-organising citizens. We have found it productive to bring out the level of "collectives" (van Den Berg, 2013). This level consists of groups of citizens and other stakeholders who have organised themselves or are mobilised by a third party that acts on their behalf around a particular theme or issue. It is on the level of the collective that issues are framed, narratives built, agendas set, actions planned and (sometimes) lobbies staged for institutional approval and resources.

Professionals often play a key role at the collective level as initiators and organisers. They have taken on new roles in the process of citymaking as intermediaries, "community orchestrators" (Balestrini et al., 2017) or "urban curators" (Beer et al., 2015). In our mind this involvement doesn't disqualify these collectives as "faux-bottom-up" or "fake citizen initiatives". On the contrary, professional involvement and the development of a sustainable value model at the collective level can be a key factor in the success of these initiatives.

Hackable citymaking revolves around the organisation of individuals into collectives, often through or with the aid of digital media platforms. Individuals contribute resources, such as knowledge, time, information or money, and at the same time reap some form of reward, be it social, economic or political, on an individual or communal level.

Our model also shows that these collectives do not operate in a social or legal vacuum. They generally act within legal and democratic frameworks, and often make use of the resources or infrastructure provided by institutions such as local governments or housing

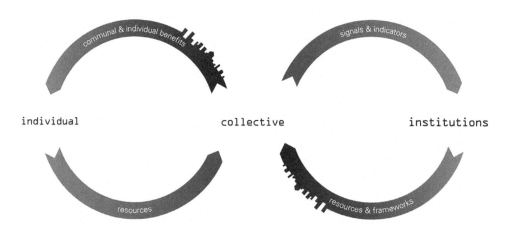

Figure 24.1 The hackable city model for collaborative citymaking
Source: Chapter author(s)

corporations. Often, the goal is to "hack" these infrastructures: to appropriate existing rules or regulations, and to extend or improve upon them.

This raises the issue of the representativeness and legitimacy of these collectives (Tonkens *et al.*, 2015). Collectives will want to argue that their initiatives or "hacks" are justified, for instance by claiming their "rights to the city" or by outlining the public value they produce. Sometimes they will find themselves working in cooperation with these institutions; at other times they may oppose them. Other initiatives may choose to ignore such institutions as much as they can, preferring to build their collectives around alternative social or economic regimes.

From an institutional point of view, within a democratic system the claims of collectives can only be legitimised in a political process in which democratically elected politicians representing the citizenry at large set policy goals and weigh collective and public values against each other. Actors at the institutional level have the responsibility to ensure that hacks serve the public interest, and that essential public services remain inclusive for the public at large. Hackable citymaking makes this relationship between collectives and institutions interactive. How can the city's governing and administrative institutions learn from these collectives' initiatives, and when they contribute to public value, adjust their frameworks accordingly? And how can institutions encourage collectives that underwrite the policy goals they have set?

The perspective of individuals: hacking as an ethos

Citizens play an important role in processes of hackable citymaking. Many initiatives are started by citizens, and many projects revolve around the collaboration of citizens in practices of citymaking. In these projects, citizens contribute their skills, insights and resources. In return they can learn new skills, gain access to collective resources, gain social recognition, or receive financial rewards for their contributions.

It is usually a small core group of active citizens or professionals that propel hackable citymaking projects. They are supported by broader groups and networks who make up the larger collective on whose behalf the core group operates. The core group often includes citizens with professional backgrounds related to the theme itself (e.g. energy) or with a background in design, management or communication. These core organisers usually involve larger groups of citizens, and organise various activities to include them in the process, for instance through co-design sessions or other capacity-building activities. Both in our own research as well as in the literature we have seen many initiatives set up knowledge communities in the form of mailing lists, social media groups, wikis or informal meet-ups. Similarly we have encountered various practices that contribute to capacity-building, including co-creation sessions, workshops, design journeys, storytelling or role-playing. Persuasive or serious games can also play a role here (van Den Berg, 2013).

Individual citizens can also be represented by these collectives. Some citizens may support a particular issue, but do not have the time, resources or energy to actually participate. For instance, a local energy collective may take the initiative to collaboratively develop a solar energy plant on the roof of a local school. Not all members need to actively participate in this process, as it will usually take quite an effort and specific technological, financial and legal skills to get the idea off the ground. Some aspects can be outsourced or taken on by a small group of initiators, backed with the (financial) support of a larger group of stakeholders.

The rise of these collaborative citymaking practices can be understood in relation to shifts in the perception of citizenship, that has been summarised as a shift from "dutified" to "actualising" citizenship (Bennett and Segerberg, 2013; Gordon and Mihailidis, 2016). The former refers to the collective enlistment of citizens in organisations such as churches and unions; the latter can be understood as the organisation of citizens in collectives around issues they are intrinsically motivated for (Levine, 2016). Traditional ways of local community-based organisation of citizens and social capital have given way to the emergence of networked publics (Varnelis, 2008), assemblages of networked individuals (Wellman, 2001) around issues of concern (Foth *et al.*, 2016; de Waal and Dignum, 2017; de Waal *et al.*, 2017). According to Franke *et al.* (2015a), this development should be understood as a reaction to the privatisation of the public domain. As traditional public and civil society organisations have become bureaucratised and more and more market-oriented, citizens try to reclaim the lost ground through commons-based self-organisation around themes such as health, education or public space (Franke *et al.*, 2015a). Faehnle *et al.* (2017) speak of a "self-organisation turn", in which "active citizens adopt new roles and increasingly 'shape and make' their cities through new self-organised forms of action, powered by the internet and social media networking".

The collectives in our model can be understood as vehicles for this process of what Bennett (Bennett and Segerberg, 2013) has called actualisation: citizens and professionals organising themselves around issues which they are intrinsically motivated to address. Many have pointed out that this form of organising also makes it hard to maintain a sharp distinction between citizens and professionals, as in many cases citizens bring their professional knowledge and skills to their cause, or professionals engage laypeople actively in the process (Boyte and Fretz, 2010).

Simon Franke, Bart Lammers and Arnold Reijndorp (2015a) have shown that many examples of collaborative citymaking can be understood in terms of such a framework of shifting citizenship. Many collectives, they argue, aim to recover ground that was lost in the privatisation of the public domain. As such they can be considered a new addition to a civil society that is taking on a new shape. Civil society itself used to consist of collective activities aimed at emancipation or citizen empowerment in the fields of education, housing or health. Many of these organisations have since become privatised or have become less relevant. Many of the collectives in our model can be understood as new attempts to address these issues in the era of the networked society.

Taken together, and seen through our lens of hackable citymaking, we can frame these shifts as the emergence of a "hacker ethos": an attitude that is fuelled by do-it-yourself ethics and professional-amateurism in which citizens exchange knowledge – some bring in their professional skills, others their practical everyday life knowledge. They pool resources, exchange knowledge and collaborate towards a commonly defined goal. These activities are further grounded in the motivation to do something "for the love of it", and/or based on the sensing of an opportunity for public value creation, or a societal need in combination with a sense of responsibility towards an issue of communal concern. This "hacker" ethos can manifest itself at individual and collective levels, and consists of a sense of agency or "ownership" in relation to a particular issue.

The perspective of collectives: hacking as a set of practices

Collaborative citymaking takes shape at the level of the collective. These collectives consist of networked groups of citizens and/or other stakeholders organised around an issue. At the level of the collective, narratives and agendas that propel the project are constructed. Action takes place at this level, and on behalf of the collective, negotiations with institutional parties are carried out. As such the collective could be understood as a new type of actor, or perhaps better defined as a set of roles that need to be coordinated in order to be effective.

These collectives exist in many forms. Many are started by groups of local citizens. As Mariska van Den Berg (2013) found in her study of citymaking collectives in the Netherlands and Germany, once they have identified an opportunity or an urgent issue, these citizens simply start to organise themselves informally, generally without waiting for formal recognition or approval. Over time such a group can take on a more formal character, or develop into a legal body such as a foundation.

In many cases it is professionals such as designers, architects or researchers who start a hackable city project. This corresponds to a shift in the practices of architects and designers that several authors have signalled. For these offices, design is about the identification of pressing urban issues and the organisation of coalitions around these issues. Ole Bouman (2008) has named this shift the emergence of an "unsolicited architecture". In many of these projects, architects and designers have started to include citizens and other stakeholders. In her book *Reactivate!* Indira van't Klooster (2013) has shown how a series of young architectural offices have started to design new procedures of campaigning, crowd-sourcing and crowd-funding to approach citizens as co-creators. According to Edwin Gardner (2015), in such an approach, design practice is focused on programming rather than the design of the built environment. The work of these offices "is almost always proactive and escapes the client-contractor relationship that has traditionally formed the basis of the design practice. They inhabit the overlapping space of the cultural entrepreneur, programmer and designer". Practitioners in this field have called themselves "community orchestrators" (Balestrini *et al.*, 2017) or "urban curators" (Beer *et al.*, 2015). For instance, Saskia Beer, Sabrina Lindemann and Emilie Vlieger have described themselves as "developers without property". They see it as their role to explore the potential of urban transition areas, and to engage communities of local stakeholders around the opportunities they have collaboratively identified.

Design researchers Liesbeth Huybrechts and her co-authors (2017) have argued that this shift is related to broader developments, especially the emergence of a so-called "post-Fordist economy". This term refers to the increasingly flexible and complex relations between various actors and spheres in contemporary society. For instance, work is no longer organised mostly through hierarchical companies that offer jobs for life, with workers organising themselves through unions to act in their interests. Citizens may hold a temporary contract; they may work through an agency; they may find employment as dependent contractors through online platforms; or they may act as freelance entrepreneurs in networks. They may work in an office, but also from home, or in a third place such as local café or co-working space. They will keep their skills updated through courses offered through various organisations, and organise (or fail to organise!) their pensions and health insurance through others. This means that the improvement of working conditions is no longer something that can be promoted simply through the actions of an institution such as a union that can set demands to an employer or to central government. Instead, specific coalitions need to be forged, bringing various stakeholders together around a specific issue in order to address it. In other words, in this vision, many of today's issues can only be addressed in collectives that are (temporarily) organised around issues.

Seen from the perspective of citizens or of the professionals orchestrating such collectives, we have identified a series of "hackable citymaking" practices and roles. Overall, we have started to use the term "dramaturgy" (Hajer, 2005) to describe the design of local settings and stories and the orchestration of events by which collective action is organised in time and place. This concerns the design of a compelling and attractive setting that allows various stakeholders to come together and start collaborating and exchanging knowledge and resources in a meaningful way. It is about building a platform (whether online or offline) on which members of the collective can represent themselves and interact with each other.

In the interaction with citizens and other individual stakeholders (on the left-hand side of the scheme), two main roles are of importance. The first is the activation of these parties around the theme and their inclusion in, for instance, trajectories of co-creation. The second is a role we have called "design integration". This is where professionals can contribute value by adding their specific domain and design expertise; for instance, by integrating ideas that have been brought into a co-creation trajectory in a design that is technologically sound and aesthetically interesting.

The right-hand side of our hackable city model scheme represents interactions between collectives and institutions. On the one hand, collectives will typically engage in activities such as lobbying and assembling evidence for the contributions they are making to the creation of public value. Many hackable city initiatives may engage in activities that aim to bring about a transition in society, for instance in the domain of renewable energy. However, their application may be in a legal grey area, or may even be illegal according to standing legislation. In that sense their activities can literally be understood as hacks: a temporary appropriation of a system beyond its intended use, with the goal of systematic change. This means that their urgency and legitimacy needs to be explained to institutions, as well as being demonstrated by defining indicators and collecting both qualitative and quantitative evidence that their hacks are producing public value.

These consist of practices that Huybrechts *et al.* (2017) have called "institutioning": collective activities take place in various forms of collaboration with, and in opposition to, existing institutional frameworks. Neither the collectives nor these institutions are stable entities, and their purposes and logic may shift or be remade through these interactions. In the course of these interactions, the actions of collectives may actively change the institutions and their frameworks. It should however be noted that not all hackable city initiatives are particularly interested in maintaining relations with traditional institutions. Some of them present themselves as autonomous, and strive to build alternative economic and social models that are not compatible with dominant regimes.

Finally, these collectives need a value or business model in order to thrive. Various options are available: institutional subsidies in exchange for public value creation, crowd-funding, or the design of exchange platforms that administer contributions of various kinds as well as the consumption of collective resources. Some projects are just temporary "hacks" and dissolve after a while. Others would want to adopt a strategy of either "formatising" or "formalising". "Formatising" means that they aim to translate the lessons learned into a "format" or "toolbox" that other organisations could apply to comparable situations in other localities. "Formalising" means that projects will try to find ways to sustain themselves over time by institutionalising themselves, either as independent organisations or as spin-offs that could be adapted by an existing institution. Here, project initiators need to think of themselves as "social" or "civic entrepreneurs" who are inventing new business models around collective action; alternatively, institutions can make funding available to collectives that contribute to public value creation. Both directions have proved hard to get off the ground (Franke *et al.*, 2015a)

For hackable citymaking to be sustainable over time, new financial arrangements for the production of public value are clearly needed. How should civic hacks that contribute to public value be rewarded? If citizens or professionals take initiatives on behalf of a larger group of citizens, and in the interest of the public at large, can we devise new models to remunerate these efforts? A commons-based or public value approach to citymaking does not mean that everyone should work for free, and that there is no business model; rather, in a hackable city, reward systems and value models exist that foster contributions to public value and stimulate stewardship of the commons. The search for sustainable business models that produce both social and economic value is key to the instantiation of hackable citymaking.

The institutional perspective: the "hackability" of the city

Various institutions at national and local levels have – at least on paper – shown an interest in engaging with hackable citymaking collectives, and there are various ways to describe the perspectives of institutions. First of all, institutions can play a role as the initiators of collaborative citymaking projects. So far, most of our attention has been devoted to citizens – often also bringing in their professional backgrounds – who initiate collective projects. But there are also many examples of institutions that have themselves taken the initiative for collective action. Governments around the world have organised open calls around set issues that were opened up to collectives, and cultural institutions in particular have a long track record of organising "dramaturgies" that activate citizens in collectives around particular issues (Knoop and Schwarz, 2017). Likewise, institutions have always functioned as centres of expertise, with professional experts working as civil servants in various departments as well as in specialised institutions such as libraries, archives and institutes for education and research. The expertise and resources they have can be extremely useful in processes of hackable citymaking. Conversely, new opportunities have also emerged for these institutions to capture or further operationalise the knowledge produced in hackable citymaking collectives. When proven successful or urgent, particular collective initiatives can also be incorporated into institutional frameworks.

Over the last few years a number of actors have set out to construct visions and frameworks that allow the further theorisation or shaping of the interaction between traditional institutions and the more volatile, informal, interdisciplinary network-shaped field of collectives we have described here (Foth, 2017a). How can these two fields, each with its own logic, formal responsibilities, rules and ways of doing and thinking, be better aligned? How could collectives play a more active role in public value creation, and how could traditional institutions become more responsive to them?

The frameworks of the "energetic society" (Hajer, 2011) and the "spontaneous city" (Buitelaar *et al.*, 2012), for instance, have explored this direction, arguing for a local or national government that sets the larger policy frameworks, for instance working towards an energy transition. To realise this agenda it is argued, it should make use of the "energy" found in the various collectives already dealing with this theme, and provide frameworks and resources for them to contribute to these goals. Likewise, in a "spontaneous city" urban areas should be developed "spontaneously", where again the government sets up the overall frameworks while various collectives play a role in the actual development of a neighbourhood. It is cities' capacity for innovation that could be extremely useful in the search for solutions to the serious problems, large and small, we are to address, contributing to the resilience of cities. To illustrate this point Potjer *et al.* (2018) refer to a study by Bulkeley and Cástan Broto (2013). Their research showed that at the strategic level of urban policy, they found hardly any comprehensive plans to address climate problems. However, when they explored local initiatives in 100 cities across the world, they found 637 different experiments addressing the challenge. Cities would do well to foster these experiments, find ways to recognise the successful ones and help them to become more successful. In his book *Veranderstad* (2015), Gert-Joost Peek makes a similar argument. In order to increase the capacity of social, economic and ecological systems to adapt to sudden stresses or gradual shifts in trends, it is imperative to involve local communities and their capacities for innovation and self-organisation at the centre of urban development.

In our scheme, this means that governments should set out a vision and translate that vision into infrastructures and frameworks that invite and enable collectives to operate on them. This vision identifies a strong role for national and local institutions as the actors that define societal visions and establish which public values should be put on the agenda. Similarly, these institutions should develop tools to closely monitor and understand the goals, directions and values at stake in the collectives, and use these tools to inform visions, their translation into policies and the execution of policy frameworks. A more participatory society in this vision does not necessarily mean a government that retreats, but rather one that redefines its role in relation to other societal actors.

A related vision is found in *The Responsive City*. In this book, Goldsmith and Crawford (2014) argue that governments need to change the ways they work. The authors challenge them to start making use of digital media and data technologies to become more responsive to the needs of society. For example, governments could make use of all kinds of data to monitor citizens' collectives and anticipate or capitalise on their efforts if they match up with policy goals. Such an approach, the authors claim, could help to bring about an institutional shift away from compliance and towards a focus on results, giving civic servants leeway to deviate from standard procedures when an argument could be made that a particular intervention would contribute to public value creation or to the realisation of policy goals. Governments themselves could also build tools and platforms that allowed citizens and collectives to interact with them in new ways. Governments could also make available various resources that collectives could use in their own projects. The provision of open data could be an example of this, although the authors argue that simply making data available is usually not enough. Data need to be offered in ways that make it understandable and actionable.

To a large extent, a hackable city perspective is congruent with these views, albeit that a responsive city framework could be critiqued for its, at times, overtly rationalist and positivist trust in technology and data as enablers of better relations between citizens, collectives and institutions. Having said that, it can be understood as a call for legitimate institutions to develop a vision for the city or nation at large, and to open up the formation and realisation of these visions to hackable citymaking collectives. On the one hand, the activities of these collectives can inform policy; on the other, institutions should use these frameworks to support the activities of the collectives.

In practice, the transition to such a hackable city model has proven difficult. One of the reasons for this is that institutions and collectives do not always share the same agenda. In their rhetoric surrounding participatory societies, institutions at times express a vision that revolves around activating citizens to carry out their policy goals, or to take over some of their tasks. The real-world dynamic between citizens and institutions is clearly more complicated than this. As we have explained above, citizens do not necessarily act out of a sense of duty, but rather from an intrinsic motivation. This means that citizens are not always interested in taking over the goals and tasks set out by local institutions. Rather, they are looking for ways in which local institutions can help them to realise their own goals (van Den Berg, 2013).

Another issue is the mismatch between the logic of institutions and the modes of operation of collectives. Many civic initiatives spring from what we have called a "hacker's ethos". They just start to address the issues they care about in a DIY approach, and are not very interested in engaging in extensive procurement processes (Beunderman, 2015). Governments often find it difficult to deal with the open-ended approach of many hackable city initiatives. Governments are also looking for processes that are risk averse, with steady and predictable outcomes. There is an obvious rationale for this: governments need to protect their citizens and act as reliable partners. However, the learning-by-doing approach of hackable citymaking is not a good match with the formal working methods of many institutions. This is why Goldsmith and Crawford (2014) argue that procurement procedures need to change. Rather than the high level of detail required in current calls that narrowly describe all facets of a particular product or process, governments should state the desired outcomes and leave the exact execution process more open.

Research by Joost Beunderman (2015) has also shown that regulation around the big society ideal in the United Kingdom has not produced many opportunities for hackable citymaking collectives. Opportunities such as the right to challenge or the right to bid – regulations that in theory allow citizen groups to challenge or bid on government provisions and propose alternative models for the organisation of public services – have mainly been seized by private outsourcing companies rather than civic initiatives. One of the reasons for this is that governments find it easier to deal with large subcontractors rather than with a broad variety of local initiatives.

In a similar vein, Mariska van Den Berg (2013) has shown that the instruments that governments do possess seem to be poorly adjusted to collectives. Firstly, many of these collectives get their finances from funding in arts, culture and design. Budget cuts in these domains make it harder for professionals to apply for this form of funding to sustain hackable citymaking practices. At the same time, the funding that is available for community projects is often targeted at hyper-local short-term interventions such as the organisation of neighbourhood barbecues. It would appear that governments have so far not fully recognised collectives as a new type of actor. Once professionals are involved, however, these initiatives are no longer considered as being "bottom up". What is needed here, according to Van den Berg, is a new approach that could be called "public-collective partnerships". Local governments should recognise collectives for their innovative capacity and potential to create public value, and look for new ways to engage them.

Looking at these issues at the institutional level through our lens of the hackable city, the notion of "hackability" is a useful one. Whereas a hacker's ethos and praxis describe the will, capacity and actual activities through which various actors have engaged in examples of collaborative citymaking, the notion of "hackability" shifts attention to the system that is to be hacked. To what extent are local institutions, their procedures and informal ways of operating, welcoming or even inviting to contributions from (citizen's) collectives? Cities can be ecosystems for innovation, yet the extent to which this potential is actually realised is partly influenced by institutional policy. City governments may or may not set all kinds of legal rules that either facilitate or prohibit the appropriation of urban infrastructures.

Is it possible to experiment with alternative energy systems at a collective level? Is this encouraged through policies and/or resources? Or is it actually very difficult to navigate the legal procedures that protect established order? In effect, as our evaluation above shows, institutions need to find new ways to interact with the open-ended "messiness" of hackable city initiatives in a process of continual exchange, yet at the same time they need to safeguard public values. This has so far proven a difficult proposition. Although there is no lack of vision on how to make cities more "hackable" and on why it might be important to do so, there is still little experience or knowledge about how exactly this could be achieved. As we have seen, to embrace the ideal of the hackable city, much more experimenting and learning are needed at the institutional level and at the interface between the institutional and the collective levels.

Hacking as a critical lense

The hackable city model can thus be used to map dynamics and conflicts between various players and their logics. It can also be used to identify concrete practices – dramaturgies – and new roles and relationships that emerge. At the same time, the notion of the hackable city could also be used to take a critical view on some of the developments described above. Various criticisms have pointed out that there is also a risk involved in the rhetoric of participation at the heart of the hacker's ethos: the risk of "responsibilisation" (Iverson, 2011), befitting a broader neoliberal trend of the dismantlement of the welfare state. Rather than making societies more democratic, it could lead to a situation in which governments step back from their duties to safeguard public values, outsourcing the management and responsibility of essential public provisions to civic initiatives (Thomas *et al.*, 2016), whereas the citizens that are most apt to take on these challenges are those that are highly educated and already well connected with local institutions (Tonkens *et al.*, 2015). Similarly, in a recent publication about precarious labour, well-known critical urban sociologist Sharon Zukin criticises hackathons – a format through which citizens are enlisted for collective action, sometimes by institutions – for being a form of labour extraction and exploitation (Zukin and Papadantonakis, 2017), echoing similar arguments made by others (e.g. Terranova, 2000; Gregg, 2015). Likewise, Evgeny Morozov and Francesca Bria state that "neoliberalism 2.0" casts citizens as "hackers", people who are able to do more with less in the context of austerity of public service expenditure (Morozov and Bria, 2018: 20).

In addition, one could question the legitimacy of these civic initiatives. As Hill (2016) has posed, they may be social, but are they democratic? These collectives may claim their "rights to the city", (Lefebvre, 1996; Mitchell, 2003; Harvey, 2008; Foth *et al.*, 2015) but whose rights are they exercising exactly? After all, Thomas *et al.* argue that the right to the city is a collective one, rather than an individual one, that should be incorporated in "the collective exercising of power in the processes of urbanization" (Thomas *et al.*, 2016). Furthermore, various authors have argued that it would be naïve to expect that self-organisation would automatically lead to positive outcomes. On the contrary, open systems, Rantanen and Faehnle (2007) write, are always vulnerable to misconduct and manipulation.

What these valuable criticisms demonstrate is the conflation of two discussions and fields of study around civic hacking. On the one hand, hacking as we have described it here is both a practice and set of affordances that can be studied empirically and critically as "community of practices". On the other hand, the notion of a hackable city brings out a normative debate about democratic governance and civil society in the network or platform society, producing imaginaries that have become performative in social organisation, political debates and policy.

Research into the hackable city has started to combine these formerly separate domains. As Kitchin (2016) has argued, the risk of normative debates is that academics maintain their ivory tower positions, referring to the perils of dominant smart city imaginaries while these work their ways into society at high speed. "Critical scholars", he argues, "have to become more applied in orientation: to give constructive feedback and guidance and to set out alternatives and to help develop strategies, not just provide critique". That does not mean that critique is not valuable.

On the contrary, as Morozov and Bria (2018) state, constant ideological and intellectual work is needed to think through the application of new technologies in society in relation to power and their implications for democratic governance. Yet, being critical is not enough. The rapid application of technologies in society requires that researchers put their principles into action and contribute to their translation "into practical and political outcomes" (Kitchin, 2016). In this line of thinking, Foth and Brynskov have suggested "participatory action research" as an "indispensable component in the journey to develop new governance infrastructures and practices for civic engagement" (Foth and Brynskov, 2016). The lens of the hackable city can serve as a critical reminder for these methods. It underpins both ethos and praxis: normative discussions about principles and value systems of urban governance, as well as practices to discuss and shape these principles in collaborative ways and take on a learning-by-doing and iterative approach in their implementation, including cycles of critical appraisal to see whether indeed these interventions live up to the goals and expectations.

Acknowledgements

This chapter reflects the insights gained from the research project *The Hackable City: Collaborative City-Making in Urban Living Lab Buiksloterham*, ran from 2015 to 2017 and funded through a Creative Industries Embedded Researcher Grant from the Netherlands Organisation for Scientific Research (NWO). This chapter is based on and made up of parts of our earlier reports: C. Ampatzidou, M. Bouw, F. van de Klundert, M. de Lange, and M. de Waal (2014). *The Hackable City: A Research Manifesto and Design Toolkit*. Amsterdam: Amsterdam Creative Industries Network; M. de Waal, M., M. de Lange, and M. Bouw (2018). *The Hackable City Cahier #1. The Hackable City: a Model for Collaborative Citymaking*. Amsterdam: The Hackable City; M. de Waal and M. de Lange (2019) "Introduction—The Hacker, the City and Their Institutions: From Grassroots Urbanism to Systemic Change." In M. de Lange and M. de Waal eds. *The Hackable City. Digital Media and Collaborative City-Making in the Network Society*. Singapore: Springer; M. de Lange (2019) "Of Hackers and Cities: "How Selfbuilders in the Buiksloterham Are MakingTheir City." In M de Lange and M. de Waal eds. *The Hackable City. Digital Media and Collaborative City-Making in the Network Society*. Singapore: Springer. In addition, it builds further on M. de Waal and M. Dignum (2017) "The citizen in the smart city. How the smart city could transform citizenship." *Information Technology*, 59(6), 263–273. https://doi.org/10.1515/itit-2017-0012.

The following persons were part of the research team and have contributed to the insights delivered in this chapter: Cristina Ampatzidou (Researcher, University of Amsterdam); Bart Aptroot (Architect, One Architecture); Lipika Bansal (Researcher, Pollinize); Matthijs Bouw (Researcher, Director One Architecture); Tara Karpinski (Embedded Researcher, University of Amsterdam); Froukje van de Klundert (Embedded Researcher, University of Amsterdam and One Architecture); Michiel de Lange (Researcher, Utrecht University); Karel Millenaar (Designer, AUAS); Melvin Sidarta (Intern Research); Juliette Sung (Intern Visual Communication); and Martijn de Waal (Project Leader, University of Amsterdam/Amsterdam University of Applied Sciences).

References

Allwinkle, S. and Cruickshank, P. (2011) 'Creating Smarter Cities: An Overview', *Journal of Urban Technology*, 18(2), pp. 1–16.

Balestrini, Mara, Rogers, Yvonne, Hassan, Carolyn, Creus, Javi, King, Martha and Marshall, Paul. (2017) 'A City in Common', in *Proceedings of the 2017 CHI Conference on Human Factors in Computing Systems – CHI '17*, pp. 2282–2294.

Bulkeley, H., & Castán Broto, V. (2013). Government by experiment? Global cities and the governing of climate change. *Transactions of the Institute of British Geographers*, 38(3), 361–375.

Beer, Saskia, Lindemann, Sabrina and Vlieger, Emilie (2015) 'Urban Curators. Ontwikkelen Zonder Eigendom', *Het Nieuwe Stad- maken. Van Gedreven Pionieren Naar Gelijk Speelveld* ed. by Simon Franke, Jeroen Niemans, and Frans Soeter- Broek. Haarlem: Trancity Valiz, pp. 85–97.

Bennett, W. L. and Segerberg, A. (2013) *The Logic of Connective Action: Digital Media and the Personalization of Contentious Politics*. New York: Cambridge University Press.

Beunderman, J. (2015) 'Financiering van de uitkomsteneconomie', Franke, S., Niemans, J., and Soeterbroek, F. (eds) *Het nieuwe stadmaken. Van gedreven pionieren naar gelijk speelveld*. Haarlem and Amsterdam: Trancity; Valiz, pp. 129–142.

Bouman, O. (2008) 'Unsolicited, or: the new autonomy of architecture', *Volume*, 14, 26.

Boyte, H. C. and Fretz, E. (2010) 'Civic professionalism', *Journal of Higher Education Outreach and Engagement*, 14, pp. 67–90.

Brynskov, M. *et al* (2014) *Urban Interaction Design: Towards City Making*. UrbanIxD/Booksprints. Available at http://booksprints-for-ict-research.eu/wp-content/uploads/2014/06/Urban_Interaction_Design_Towards_City_Making.pdf.

Caldwell, G. A. and Foth, M. (2014) 'DIY media architecture: Open and participatory approaches to community engagement', *Proceedings of the 2014 Media Architecture Biennale*.

Caragliu, A., Bo, C. del and Nijkamp, P. (2011) 'Smart Cities in Europe', *Journal of Urban Technology*, 18(2), pp. 65–82.

Cardullo, P. and Kitchin, R. (2019) "Being a 'citizen' in the smart city: up and down the scaffold of smart citizen participation in Dublin, Ireland." *Geo Journal* 84(1): 1–13. https://link.springer.com/article/10.1007/s10708-018-9845-8

Castells, M. (2010) *The information age: economy, society, and culture. The rise of the network society*. Chichester: Wiley-Blackwell.

Chourabi, H. *et al.* (2012) 'Understanding Smart Cities: An Integrative Framework', in *45th Hawaii International Conference on System Sciences (HICSS 2012)*. IEEE, pp. 2289–2297.

Crabtree, J. (2007) 'Civic hacking: a new agenda for e-democracy', www.opendemocracy.net/en/civic_hacking_a_new_agenda_for_e_democracy/, June.

Crawford, S. and Goldsmith, S. (2014) *The Responsive City: Engaging Communities Through Data-Smart Governance*. San Francicso: Jossey-Bass.

de Lange, M. and de Waal, M. (2013) 'Owning the city: new media and citizen engagement in urban design', *First Monday, special issue 'Media & the City'*, 18, 11. doi: 10.5210/fm.v18i11.

de Waal, M., de Lange, M. and Bouw, M. (2017) 'The Hackable City. Citymaking in a Platform Society', *AD Architectural Design*, 87(1), pp. 50–57. doi: 10.1002/ad.2131.

de Waal, M. and Dignum, M. (2017) 'The citizen in the smart city. How the smart city could transform citizenship', *Information Technology*, 59(6), pp. 263–273. doi: 10.1515/itit-2017-0012.

Faehnle, M. *et al* (2017) 'Civic engagement 3.0 – Reconsidering the roles of citizens in city-making'. www.yss.fi/journal/civic-engagement-3-0-reconsidering-the-roles-of-citizens-in-city-making/.

Foth, M. (2017a) 'Some thoughts on digital placemaking', In Haeusler, H. M. *et al.* (ed) *Media Architecture Compendium : Digital Placemaking*, Stuttgart: avedition, pp. 202–213. doi: 10.7238/d.v0i12.915.

Foth, M. and Brynskov, M. (2016) 'Participatory Action Research for Civic Engagement', *Civic Media. Technology | Design | Practice* E. Gordon and P. Mihailidis (eds). Cambridge: The MIT Press, 536–552.

Foth, M., Brynskov, M. and Ojala, T. (2015) *Citizen's right to the digital city urban interfaces, activism, and placemaking, Citizen's Right to the Digital City: Urban Interfaces, Activism, and Placemaking*. Edited by M. Foth, M. Brynskov, and T. Ojala. Singapore: Springer Singapore. doi: 10.1007/978-981-287-919-6.

Foth, M., Hudson-Smith, A. and Gifford, D. (2016) In Olleros, F. X. and Zhegu, M. (eds) 'Smart cities, social capital, and citizens at play: a critique and a way forward', *Research Handbook on Digital Transformations*, Edward Elgar Publishing, pp. 203–222. doi: 10.4337/9781784717766.00017.

Franke, S., Lammers, B. and Reijndorp, A. (2015a) 'De (her)ontdekking van de publieke zaak', Franke, S., Niemans, J., and Soeterbroek, F. (eds) *Het nieuwe stadmaken. Van gedreven pionieren naar gelijk speelveld*. Haarlem and Amsterdam: Trancity; Valiz, pp. 43–58.

Gardner, E. (2015) 'Hack the City!', *Amateur Cities*, 12 October. Available at: amateurcities.com/hack-the-city.

Gordon, E. and Mihailidis, P. (2016) *Civic Media. Technology/Design/Practice*. Cambridge, MA: The MIT Press.

Gregg, M. (2015) 'Hack for good: Speculative labour, app development and the burden of austerity', *The Fibreculture Journal*, 25, pp. 183–201.

Hajer, M. (2011) *The Energetic Society. In Search of a Governance Philosophy for a Clean Economy*. Den Haag: PBL Nether.

Hajer, M. A. (2005) 'Setting the Stage: A Dramaturgy of Policy Deliberation', *Administration & Society*, 36(6), pp. 624–647.

Harvey, D. (2008) 'The right to the city', *New Left Review*, 53, pp. 23–40.

Hemment, D. and Townsend, A. (2013) *Smart Citizens*. Manchester: FutureEverything Publications.

Hill, D. (2016) 'The social and the democratic, in the social democratic European city.', *Medium.com*, 23 May. Available at: https://medium.com/dark-matter-and-trojan-horses/the-social-and-the-demo cratic-in-social-democratic-european-cities-31e0bc169b0b.

Hollands, R. G. (2008) 'Will the real smart city please stand up?', *City*, 12(3), pp. 303–320.

Hunsinger, J. and Schrock, A. (2016) 'The democratization of hacking and making'.

Huybrechts, L., Benesch, H. and Geib, J. (2017) 'Institutioning: Participatory Design, Co-Design and the public realm', *CoDesign*, Taylor & Francis, 13(3), pp. 148–159.

Iverson, K. (2011) 'Mobile Media and the Strategies of Ubran Citizenship: Control, Responsibilization, Politicization', In Foth, M. *et al.* (ed) *From Social Butterfly to Engaged Citizen: Urban Informatics, Social Media, Ubiquitous Computing, and Mobile Technology to Support Citizen Engagement*, Cambridge, MA: The MIT Press, 55–70.

Kitchin, R. (2014) 'Making sense of smart cities: addressing present shortcomings', *Cambridge Journal of Regions, Economy and Society*, p. rsu027. doi: 10.1093/cjres/rsu027.

Kitchin, R. (2016) 'Reframing, reimagining and remaking smart cities', *SocArXiv*, 18 August. doi: 10.17605/OSF.IO/CYJHG.

Klooster, I. van't. (2013) *Reactivate! : vernieuwers van de Nederlandse architectuur*. Amsterdam: trancity valiz.

Knoop, R. and Schwarz, M. (eds) (2017). *Straatwaarden: in het nieuwe speelveld van maatschappelijke erfgoedpraktijken*. Amsterdam: Reinwardt Academie.

Lefebvre, H. (1996) *Writings on cities*. Cambridge: Blackwell Publishers.

Levine, P. (2016) 'Democracy in the Digital Age', Gordon, E. and Mihailidis, P. (eds) *Civic Media. Technology | Design | Practice*. Cambridge: The MIT Press, 44–60.

Levy, S. (2001) *Hackers: Heroes of the computer revolution*. New York: Penguin Books.

Lodato, T. J. and Disalvo, C. (2016) 'Issue-oriented hackathons as material participation', *New Media & Society*, 18(4), pp. 539–557.

Lydon, M. and Garcia, A. (2015) *Tactical urbanism short-term action for long-term change*. Washington, DC: Island Press.

Mitchell, D. (2003) *The right to the city: Social justice and the fight for public space*. Cambridge: The MIT Press.

Morozov, E. and Bria, F. (2018) *Rethinking the smart city*. New York: the Rosa Luxemburg Stiftung.

Nam, T. and Pardo, T. A. (2011) 'Conceptualizing smart city with dimensions of technology, people, and institutions', in *Proceedings of the 12th Annual International Digital Government Research Conference on Digital Government Innovation in Challenging Times - dg.o '11*. New York and USA: ACM Press.

Niederer, S. and Priester, R. (2016). Smart Citizens. Exploring the tools of the urban bottom-up movement. *Computer Supported Cooperative Work*. doi: 10.1007/s10606-016-9249-6.

Peek, G. (2015) *Veranderstad. Stedelijke gebiedsontwikkeling in transitie*. Rotterdam: Rotterdam Hogeschool Uitgeverij.

Potjer, S., Hajer, M. and Pelzer, P. (2018) 'LEARNING TO EXPERIMENT. Realising the potential of the Urban Agenda for the EU'.

Rantanen, A. and Faehnle, M. (2007) 'Self-organisation challenging institutional planning: towards a new urban research and planning paradigm – a Finnish review', *The Finnish Journal of Urban Studies*, 55, 4.

Roszak, T. (1986) *The cult of information: The folklore of computers and the true art of thinking*. New York: Pantheon Books.

Saad-Sulonen, Joanna, and Horelli, Liisa (2010) 'The value of community informatics to participatory urban planning and design: A case-study in Helsinki', The *Journal of Community Informatics*, 6, 2.

Schrock, A. R. (2016) 'Civic hacking as data activism and advocacy: A history from publicity to open government data', *New Media & Society*, 18(4), pp. 581–599.

Schudson, M. (1998) *The Good Citizen: A History of American Civic Life*. New York: Free Press.

Söderberg, Johan (2010) 'Misuser inventions and the invention of the misuser: Hackers, crackers and filesharers', *Science as Culture*, 19(2), pp. 151–179.

Terranova, T. (2000) 'Free labor: Producing culture for the digital economy', *Social Text*, 18(2), pp. 33–58.

Thomas, V. *et al* (2016) 'Where's Wally? In Search of Citizen Perspectives on the Smart City', *Sustainability*, 8(3), p. 207. doi: 10.3390/su8030207.

Tonkens, E. *et al* (2015) *Montessori-democratie. Spanningen tussen burgerparticipatie en de lokale politiek.* Amsterdam: Amsterdam University Press.

Townsend, A. (2013) *Smart Cities: Big Data, Civic Hackers and the Quest for a New Utopia.* New York: W.W. Norton & Company.

van Den Berg, M. (2013) *Stedelingen veranderen de stad. Over nieuwe collectieven publiek domein en transitie.* Amsterdam and Haarlem: Trancity; Valiz.

van Dijck, J., Poell, T. and de Waal, M. (2018) *The Platform Society. Public Values in a Connective World.* Oxford: Oxford University Press.

Varnelis, K. (2008) *Networked publics.* Cambridge, MA: The MIT Press.

von Hippel, E. (2005) *Democratizing innovation.* Cambridge: The MIT Press.

Wellman, B. (2001) 'Physical place and cyberplace. The rise of networked individualism', *Journal of urban and regional research*, 25(2), pp. 227–252.

Zukin, S. and Papadantonakis, M. (2017) 'Hackathons as Co-optation ritual: Socializing workers and institutionalizing innovation in the "new" economy', A. L. Kalleberg, and S. P. Vallas (eds) *Precarious work* Research in the sociology of work Vol. 31, pp. 157–181. Bingley: Emerald Publishing Limited.

25

Designing the city as a place or a product?

How space is marginalised in the smart city

Alessandro Aurigi

Introduction

Smart urbanism is naturally concerned with the role that high technologies, and the shapers of these, play in the way cities are being developed or re-conceived. As Graham and Marvin noted in their seminal 1996 volume *Telecommunications and the City*, one the potential major paradigmatic shifts of the emergence of the digital could be that proximity, and the role of physical space and distances, could be seen as an obsolete urban feature. This chapter's aim is to look at the role of physical urban space and design in "smart" concepts and interventions. It discusses how the role of space seems to have radically shifted with the emergence of smart city visions. Despite in many ways being and having been a key component of what the "city" is, and an essential agent in shaping it and addressing its problems, space is now marginalised within smart city visions as either being a main culprit and generator of problems, or a passive host of technological systems, simply needed for its physicality. The chapter briefly looks at examples of such approaches, and advocates for the need to re-frame smart civic design discourses and processes to include physical urban space as an agent and key component of urban development, rather than a platform for technology. Doing this, the chapter argues, calls for overcoming the idea of the smart city being a product of technology, and indeed articulated through the application of a series of technological products. It is our urban design knowledge and practice that needs to be at the same time updated and included.

The uneasy relationship between digital urbanism and physical space

The relationship between visions inspired by – and inspiring – digital urbanism and the physical space dimension of the city has never been an easy one, often to the detriment and exclusion of the latter from being seen as key to transform urbanity. As the rise of cyberspace was being hailed in the early 1990s, the anti-spatial promise of the benefits of dematerialising functions (Negroponte, 1996) and of re-inventing more fluid forms of architecture and community by exploring virtuality (Beckmann, 1998; Jormakka, 2002; Novak, 1994; Rheingold, 1995) popularised visions based on the obsolescence or greatly diminished importance of physical space and its constraints. The history of the intersection of urban space and digital technology shows how this has been driven by deterministic, tech-first – or even tech-only – perspectives in which cyberspace was the change factor making the difference, whilst people and physical space were at the receiving end of it. Debates in the 1980s and 1990s were dominated by hyperbolic views of digital technology improving an otherwise decaying and disempowered world by making it more environmentally sustainable (see for instance Benedikt, 1991); boosting new forms of human association (Rheingold, 1995); pointing at new ways of settling (or re-settling) in economically and socially viable small towns whilst cities became obsolete (Toffler, 1990); and generally affirming new, prevalently virtualised, revolutionary economic and production models (Negroponte, 1996).

Despite the field having developed greatly ever since, this legacy is still very much skewing visions of urban change involving digital technologies. The proverbial danger of focusing on the finger pointing at the moon rather than admiring the moon itself can be a very real issue, and in "smart city" visions the emphasis is very much on what "smart" is and can do, leaving the city as simply the environment where this happens. As Mattern (2017) has noted referring to the "Y-Combinator" approach to smart cities and civic "effectiveness", "There was hardly any mention of the urban designers, planners, and scholars who have been asking the big questions for centuries: How do cities function, and how can they function better?" In circles of smart city scholars and practitioners it is way too easy to concentrate on everything digital whilst overlooking what we already know about the city. This way, we can end up associating change, solutions and agency to the former whilst looking at space as simply the stage where new technologically induced lifestyles will unfold.

Linear, simplistic views of an all-dominating technology have been met of course with lively reflections and critique from various disciplinary perspectives. Yet, prevalent responses and continuous challenges to technological determinism have mainly engaged the socio-economic and political side of urban management and development, again leaving physical space in the background. Approaches looking at public participation and lack of social inclusion and justice within the digital/smart city have looked at the social dimension of augmented place-making, yet often taking for granted physical space as something already there, that did not need to be seen as a particularly active part of the equation. From early discourses of digital divides, inclusion and engagement (Schuler, 1996, for instance), social constructivist approaches to the shaping of urban civic networks (Aurigi, 2005), to more recent debates on smart citizenship (Hemment and Townsend, 2013; Foth et al., 2015) and community participation, to yet critique engaged with issues of socio-economic development, equality and justice (Hollands, 2008; Rekow, 2013; Datta, 2015; Hollands, 2015), the possible roles of physical space and its potential agency through design and ability to affect relationships could have deserved more attention. This tension between digital and physical was also not helped by the prevalent, cautious view towards the impacts of cyberspace and digital interventions held by mainly architects and urban designers. Norberg-Schulz had issued his early warnings against the loss of meaning stemming from too mobile and fluid conceptions of inhabitation in the era of cybernetic thinking (Norberg-Schulz, 1971; 114). Reiwoldt advocated for architecture to be a refuge and help return to "a real living environment separate from the insubstantial worlds of the computer" (Riewoldt, 1997: 10). Carmona et al. (2003: 110) noted the challenges posed to public space – and the public realm – "through the car and subsequently through the internet". These were an understandable reaction to the often blind enthusiasm underpinning the rise of the digital, but facilitated a widening of the divide between those who were prevalently interested in technology and those who could have affirmed and helped include the agency of physical space and design within "smart" perspectives. More recently, Picon (2015) has more explicitly looked at the smart city as a digital and physical construct, both discussing the relationships between smart urbanism and urban form (110–119), and the limits of "all-digital solutions" (146–148), but still with an overall view of "smart" as infrastructure which gets added to the city.

The divide between scholars and practitioners focusing on the new, digital aspects of place and those warning of the related threats and the need to re-affirm the centrality of physical designs has therefore contributed to perceiving the two sides of the same coin as separate. So what role does urban space have in smart place-making and discourses? Two perspectives are worth highlighting.

Space is the problem: overcoming the context with technology

The widely deterministic and soft-utopian discourse coming from the corporate information and communication technology (ICT) sector (Söderström et al., 2014) and smart city entrepreneurs, implies somehow the need to declare current urban models as terminally ill. This is necessary to then assert the urgent need and fundamental inevitability of a series of high-tech solutions able to save the city from itself. Physical space and things are – within such discourse – the generators and cradle of a plethora of problems, rather than important variables to alter, design and re-design. In the perspective of mainstream, industrial smart city promotion, physical space is often part of the problem, and never part of the solution.

This can be seen as a further evolution of past technocratic approaches to declaring urban space in its present form as critically dysfunctional and in need of a radical re-organisation. For instance, modernist urbanism had challenged the poor life quality conditions of the early 20th century city arguing for a major re-thought of civic planning based on the new "expert" knowledge of functional zoning (Dear, 1995). Architecture in the same period had advocated for the need to replace building practices with more industrialised and internationally de-contextualised and transferable ones, affording more universally healthy lifestyles overcoming the "old hostile framework" (Le Corbusier, 1928: 307) offered by the city and its buildings.

The modernists wanted to fix urban quality with a more "expert" and mechanistic approach to spatial organisation and design, giving space agency though in very different terms and with very different emphases than previously. Grahame notes how "The city theory underlying the City as a Machine involves a simple, mechanistic calculus: four functions, repetition of specialized, standardized building typologies in special zoning enclaves" (Grahame, 2005: 48). The smart urbanism push takes this a – crucial – step further, by replacing a mechanical – hence physical – city with the idea of a digital one. "Smart" firmly takes agency away from space and gives the ability to frame and fix the city to high technology. As Datta notes quoting a TEDx lecture given by Amitabh Kant in Delhi, the importance of digital, smart technology "means you can drive urbanisation through the back of your mobile phone" (quoted in Datta, 2015). The "machine" is therefore no longer urban space or buildings. These alone have failed us – it could be argued – as the hosts and main generators of a series of emergencies. As Cowley and Caprotti (2019) have argued, the generic concept of the "Big Data driven 'control room'" sitting at the heart of the smart city's management structure can be seen on the one hand as a further evolution of positivist, modernist planning, but on the other hand it also disrupts the latter, offering a real-time, data-based way of framing the city. And, it could be argued, an increasingly less "spatial" one, at least as far as the agency of spatial intervention is concerned. Urban redemption now stems from sensing, data and code.

But redemption is needed. Smart city hype and marketing tend to leverage on discourses of critical over-urbanisation, out-of-control densities and the consequent pressures on urban resources, the environment, citizen safety and management practices. These issues are widely discussed and accepted, as part of the United Nation's Sustainable Development Goals. But whilst literature about these describes urban living as an opportunity and in a more balanced way focuses on needs such as tackling urban poverty and making sure physical built environment is affordable and of good quality (Sustainable Development Solutions Network, 2013), correspondent smart city rhetoric shifts this perspective and frames the physical city as terminally ill and in need of a major re-invention through digital technology. Articles and statements from the ICT corporate sector (for instance Menon, 2013; Living PlanIT, 2011; Schneider Electric, n.d.) all have converged on implying that traditional ways of framing, understanding, managing and designing cities have become inadequate and unable to cope with the pace and scale of change, without the redeeming influence of high-tech systems (Aurigi, 2016). This is a revival and commercial leveraging of the utopian hyperbole on the expected predominance of cyberspace over place, which had an acceleration at the dawn of the internet in the early 1990s. The new visions of virtualised lifestyles, seen decades ago as literally replacing many decaying aspects of urbanity and indeed the need for cities itself (Toffler, 1990), are now expected to be the factor that will save an otherwise equally doomed city. Physical space per se might not the object of replacement logics any longer, but its agency largely is.

Space is the platform: neutral stages for social interactions

Industry-dominated soft-utopian perspectives stem from a rather negative stance towards the role of pre-smart urban space, and place virtually all key change agency firmly with what ICTs can do mainly as a control and management infrastructure. However, approaches that are more sensitive to context, particularly social context, do exist. These stem from a less mechanistic and more critical view of the role of smart technologies with, and within, communities and place. But the role of urban space in these visions is often that of a neutral platform, or a passive "host" for the digitally based agency which is envisaged.

Much work both in terms of critique and practice has gone into considering how the development of digitally enhanced places and of situated interactivity could be re-framed in ways that would make it more locally engaged, participated or based on alternative concepts to mainstream e-governmental, city management and place marketing approaches. This work ranges widely. One aspect is the reflection on the non-neutral, contested nature of urban analysis systems and dashboards (Kitchin et al., 2015) and the critical analysis and facilitation of various forms of grassroots or public–private action involving the production of locally relevant smart initiatives and the networking of these (Townsend, 2013). Another is the conception of non-profit systems and digital situated urban gaming associated with the "Hackable City" concept (de Lange and de Waal, 2018) and cognate art installations. Some of the central tenets of these approaches include provoking and encouraging various degrees of community involvement and participation underpinned by a much-needed public space-based, inclusive, interactive and critical dimension to city smartness. Opportunities offered by the hybridity of space and place have been explored in pan-European research networks like Cyberparks (Smaniotto Costa et al., 2019) as well as a variety of initiatives of public interaction design across the globe. Other notable examples are for instance some of the projects associated with the work of the Media Architecture Institute[1] or the Urban Informatics research group[2] as well as international networks and festivals fostering and showcasing urban interactive installations, such as Bristol-originated The Playable City.

This diverse landscape of projects is therefore much more place-conscious and related, and most approaches share an ethos focused on being "located" and socially participated. Agency is seen as not simply embedded within the technological component of a smart intervention, but above all placed within the community, which needs to be able to interact/play with, and even alter if necessary the technological system itself in some instances. It can be argued, however, that even in these more socially oriented cases there remain two potential weaknesses. The first is a degree of self-referentiality, particularly in cases of small-scale or temporary installations. The initiative makers and their colleagues may also end up being the main users, commentators and evaluators, obviously reinforcing a circular logic in reading and interpreting the projects exactly as the creators intended, and confirming their validity. De Lange's set of interviews with Playable City actors in Bristol (2015) highlights the positive shifts from a tech-centric to a people-centric view of the smart city, but also seem to reveal a degree of self-containment, where a relatively close and specialised artist community seems to measure the value and impact of installations through their personal/participant reactions, relying on a direct and implicit understanding and appreciation of the projects' language and intentions.

The second point to reflect upon is a prevalent focus on the project per se, on its technical or interactive character, rather than on the initiative as being conceived, analysed and strictly evaluated as part of any wider place-making strategy. Čakovská et al. (2019) for instance reflect on how to approach the shaping and improving of "blended space" stemming from a series of urban design-related references aimed at highlighting the centrality of people's activities, interactions and participation levels in public spaces. Whilst this is certainly relevant and desirable, the focus ends up being on how ICTs can be added to space, with the latter seen as adequate "physical infrastructure" in which public spaces "have to provide resources for proper functioning of gadgets" (258). The practical methodological approach to achieve these aims shifts to an interaction design-led one.

Adding people and their interactions within civic design to the otherwise purely technocratic smart vision is indeed important. But if spatial design is left out as a gregarious – and most of the times already fixed – component, the shaping of urban digital projects can perhaps inadvertently sum up into the conception of a series of add-on devices or systems. Much of such practice of digitally augmenting place is dominated by projects that are virtual, temporary and/or mobile installations or applications, conceived very much as products able to be seamlessly transferred to or replicated in very different locations with minimal or no change. In a rare and precious effort at systematically tackling reflections and guidelines for the design of urban hybridity, Tomitsch (2018) generally sees "city apps" as artefacts that need to "complement the built environment" which subtly suggest an add-on logic informing smart urban design (126–127).

Interaction design and urban media-based approaches have therefore the very significant merit of framing smart initiatives within a socially involved and contextualised approach, as well as looking at locating projects in real places with a potentially hyper-local dimension. However, it can be argued that whilst the social and the technological spheres are central to these, the spatial dimension remains in a supporting role. Pertinent considerations about the role that "platforms" play in technologically advanced societies, and the need to carefully design and regulate these (Van Dijck et al., 2018) are informing much of the debate on social and urban digital media critique and production. The danger however is that such paradigm – the "platform" – could easily extend to characterise a relatively infrastructural and passive role of urban space, seen as a recipient of initiatives rather than a generator and modifier of relationships itself.

To exemplify this, we can look at a relatively early yet very well-articulated digital augmentation project of a public space, the Sonic Arboretum in Montreal. This scheme, conceived in the mid-2000s, aimed at augmenting the character and functions of a public small park – the Emile Gameline Square in Montreal. A parallel digital environment, accessible both remotely under a 3D/VR modality, as well as at the physical square itself through located devices, augmented the park's physical features – trees and other structures – with digital exchange and interpersonal communication functionalities, and a focus on music files sharing. This meant that users of the system could "meet" and communicate either by physically being present in the park or accessing it through its virtual "mirror" on the internet. Despite such project would indeed manage to interestingly articulate the digital and the physical in potentially synergic re-combinations – the virtual, exchange trees were conceived to match the real ones in the park and afford these to become music information repositories – the focus and design effort was described as:

> the strategy of situating mobile communication activity within the larger framework of urban spaces as ecosystems, in which wireless networks would be more "holistically" incorporated into the environment. This approach allows us to contextualize the flow of information within an expansive stream of other interactions the flow of people, traffic, food, resources, energy, weather, and ideas.
>
> *(CHS UR OWN URBNSM blog, 2007)*

It can be argued that physical space was therefore still subordinate and playing a background, support role. The project's central aim was not understanding, conceiving and re-designing the park as a whole, but more about using the park as a rich physical platform for digitally based interactions. The same initiative could have been exported and replicated potentially anywhere else offering some open space, maybe including pathways and trees, or indeed some other physical elements – lamp posts, rails or even rubbish bins – that could be associated with the idea of hanging/leaving music "objects" there.

Similarly, the already mentioned and well-known Playable City initiative and network, an idea originated in Bristol with increasing transferable, global extensions, aims to put "people and play at the heart of the future city, re-using city infrastructure and re-appropriating smart city technologies to create connections – person to person, person to city" (Playable City Vision, n.d.). This is an interesting and attractive statement, putting an emphasis on some important, social aspects of place-making, yet again fundamentally treating physical space as the "city infrastructure" that can act as a platform or stage for the interactive play facilitating connections. Playable City installations can certainly be provocative, evocative and useful to encourage interaction and increase the range and frequency of use of specific public spaces, but it is debatable whether they really start from place and have a close dialogue with it.

Sidewalk Toronto: high technology as a civic-shaping paradigm

In the two approaches to smart place-making considered so far, the role of civic space shifts from holding a negative connotation as one of the main roots of urban problems, to a somehow neutral one, as the all-needed yet relatively passive context. In a way, these two perspectives about space as something to be overcome by technology or able to accommodate and support it, seem to play against each other, and advocates of the latter approach are very often major critics of the former, particularly on the grounds of the need to encourage community involvement. Yet, in both visions spatial design and agency is marginal at best, leaving technology – and technologists – as the central city-making actor.

The much-debated example of the Sidewalk Toronto prospected regeneration development can be used to briefly reflect on how in a relatively up-scaled and certainly very high-profile smart urbanism scheme, spatial design is present yet extremely marginal in terms of its influence as a change agent. It is well beyond the scope of this chapter to provide a detailed or comprehensive description of Sidewalk Toronto, its genesis and the full range of discussions around it. Much of the critique of the Alphabet/Google proposed urban development has been centred around the many issues that a techno-centric, privately enacted and controlled regeneration scheme can raise in terms of the commodification of smart place and the data it generates. Some have noted how the project is posing itself as an operation of "privatised planning" (Valverde, 2018) allowing Sidewalk to circumvent an otherwise required degree of democratic accountability on its physical interventions and the way public consultations have been carried out (Valverde and Flynn, 2018; Wylie, 2018a, 2018b; Bliss, 2019). Above all much debate has unfolded on the ownership and use of algorithms and data, and the clear potential for non-transparent forms of surveillance (Wylie, 2017; Canon, 2018; Scassa, 2019; Wylie, 2019).

However, it is important here to reflect on how the scheme has approached the relationship between civic space and smart technologies. Some focus is needed on the way debates and commentaries on Sidewalk Toronto seem to highlight how the initiative is informed by a combination of the two approaches towards smart city-making and urban space outlined so far in the chapter.

The city as obsolete: replacing the model

The problematisation and inadequacy of urban space and current civic models seem to have strongly characterised the rationale underpinning Sidewalk. Matti Siemiatycki, an associate professor at the University of Toronto's Department of Geography and Planning, quoted in an article by Josh McConnell (2017), claims that "we aren't just building the city of yesterday, we are really trying to understand how to build the city of tomorrow" somehow characterising current urbanity as obsolete. The same article quotes Dan Doctoroff, CEO of Sidewalk Labs LLC, associating this with a "combination of technologies that are uniquely becoming available at this time, this moment right now, that are capable of addressing some of those big urban challenges" clearly transferring urban design agency, and a "solution" status, to ICTs. Mattern (2017) also refers to Doctoroff's vision of a "revolution" enacted by starting "from scratch in the internet era" and building the city "from the internet up". Canon (2018) refers to a 2016 Google TechTalk video where "Anand Babu of Sidewalk Labs spoke about 'reimagining the city as a digital platform' and using tech to solve the problems big cities face". Valverde (2018) quotes urban affairs expert John Lorinc defining the project as a "built-form version of Facebook". It is therefore quite explicit that Sidewalk's approach is an anti-urban one informed by a vision where civic space and its current organisation and models seem unable to offer convincing answers to a multitude of contemporary challenges. A radical re-framing and total re-thinking of what a city should be like is then needed, and this has its paradigm in the digital machine one. This is a modular, platform-based machine. Urban space as we know it is also there to become a platform supporting this.

City as a platform: urban design by services and buildings as modular add-ons

One of the main tenets of Sidewalk's vision and approach is some generic but powerful advocacy of the importance of flexibility and changeability of the city. Doctoroff has explicitly referred as a major inspiration to Walt Disney's original plan for the experimental EPCOT community in Florida as the blueprint for an ever-evolving city, "a platform that people create on top of" (McConnell, 2017). Whilst the importance of seeing cities as incomplete systems that cannot be "closed" or fully controlled in any way is an important reflection in contemporary urbanism (see for instance Sassen, 2017), this smart scheme uses the concept to assign this change agency, the ability to constantly transform the city, to ICTs and their users. This, and the need to rebuild "from scratch", suggest a plug-in model where modular elements – physical, as well as the prevalent digital ones – populate a place which poses itself as a neutral stage. The urban design of Sidewalk Toronto appears to be a tabula rasa operation, where a context which is basically seen as a wasteland is over-imposed with a new technocratic vision. This vision is populated with constant references to add-on elements, automation and modularity or, in other words, with an urbanism based on the design of products, not place. These include a Carlo Ratti Associati-designed prototype for a "dynamic street" modular intelligent paving (Walsh, 2018), heated pavements and cycle lanes, solar power, energy and waste management systems, and underground tunnels and delivery/service robots (Won, 2018; Woyke, 2018; Lam, 2019; Wilson, 2019), and of course autonomous vehicles.

Choices regarding the buildings themselves, and their relationship with the contiguous public space, are also coherent with the industrial design, highly flexible, modular and plug-in character of the Sidewalk scheme. A contextual concession is made to Canada's economy and local resources through the choice of constructing a series of tower buildings, designed by Snøhetta and Heatherwick Studio, in structural timber. But the whole narrative around these seems to focus on a "digital configurator", computer-aided system to employ a modular kit of parts (Baldwin, 2019; Lam, 2019). The semi-public lower storeys of the buildings, and presumably their outdoor projection onto public space, are dedicated to "'a porous, flexible program we call Stoa that is accessible to everyone', said the team. The Stoa would function as year-round spaces wrapped in transparent and movable facades" (Baldwin, 2018), further reinforcing the open, modular but also fairly non-committal, "blank canvas" vision of a fundamentally neutral and passive built environment offering a generic stage for technological or social plug-ins without taking a stance, or exercising any specific agency.

All of this leads to a specific place, in a specific and complex city, to be treated with an approach that seems to mainly articulate around product design and IT-based projects, which after all are paradigms that the ICT industry is most conversant with. The overall trajectory seems to be stemming not from what the docks area needs, but from borrowing a potentially attractive platform so that a generic technocratic experiment can be run on and through it. The neighbourhood seems to suffer from being dealt with in isolation, as a self-contained close system with Sidewalk Labs recently lamenting the absence of light rail public transport able to connect the area and the rest of the city (The Canadian Press, 2019). Little or no reference to actual context is made in most of the available commentaries or interviews, both spatially and socially. The scheme feels like a *tabula rasa* intervention, relatively independent from the rest of the city, and carrying the potential for becoming the socially homogenous equivalent of a gated community as "Quayside's current plans promise housing for people of all income levels. But the only company so far committed to moving there is Google Canada, suggesting an influx of young, affluent workers" (Austen, 2017).

Towards a holistic view of augmented place

The examples briefly considered so far seem to suggest that jettisoning or anyway marginalising considerations and approaches that value the role of urban space, rejecting spatial design and planning knowledge as somehow obsolete or impotent to make a difference, might have to be re-thought. A more holistic approach could be needed. One able to positively frame smart urban design as something articulating a combination of people, technology and indeed physical space as equally important and inter-connected actants in the making of place.

The importance of looking at digital technology as part of a multi-layered urban place, and the need to imagine and design this as a whole, has of course has been discussed before. Mitchell's trilogy of books on the topic (1995, 1999, 2003) is a seminal contribution, together with work by McCullough (2005, 2013) and Shepard (2011). Following William Mitchell's concept of "recombinant architecture" (1995: 47–105) and trying to expand that perspective with a more operational framework, Thomas Horan (2000) stressed the need to look at a whole place design perspective, rather than at digital add-on solutions.

> At one end of the digital place continuum are "unplugged" designs that manifest little or no digital technology in their appearance and construction. Toward the middle of the continuum are various "adaptive" designs, representing modest attempts to visibly incorporate electronic features into physical spaces. Occupying the far end of the spectrum are "transformative" designs: rooms, buildings, or communities composed of truly interfaced physical and electronic spaces.
>
> *(Horan, 2000: 7)*

Horan's language could still be seen as being affected by a residual dualism – as he talks of physical and electronic spaces as potentially separate layers to be combined. But pervasive computing and the so-called Internet of Things were very much in their infancy then, and so were spaces and objects that could be seen at the same time and in themselves both physical and digital. The concept of "transformative design" was nevertheless powerful in pointing at the fact that successful place-making in the internet era called for a vision where physical space, digital technology and people (themselves conducting re-combined physical/ digital lifestyles) were all contributing to and being active dimensions of any design, and that designers needed to deeply consider their inter-dependencies.

And when projects accept the complex challenges that come from exploring extended, re-imagined ways of using public space and defining useful everyday typologies, rather than being add-on digital art or interaction, physical space becomes part of the equation again. An interesting experience and commentary came from the "Breakout" project aiming at bringing knowledge work into public spaces (Townsend et al., 2011). The research team could observe how the issues of digital living (and working in the specific case) actually combined with those of physical space organisation and design, and greatly depended on contextual factors. The trajectory could not just stick to a deterministic view of high technology impacting and changing the otherwise static platform of space. The other way around was equally true: space actively participated in the equation, affected the "digital" and ultimately the two could not be de-coupled in trying to fulfil the design programme.

Breakout suggests that when physical space is taken into account through a perspective on inhabitation, engagement with the everyday and (relative) persistence rather than a simply performative "installation" mode, this calls for design considerations involving the role of space. This includes looking at private/public thresholds; how different activities help or hamper each other; presence of shelter and seating; filtering with building space and so on. The physical component of place becomes again a very active actor/participant, and sometimes a rather difficult one to deal with, rather than an allegedly docile and passive "host". In other words, and however obvious this might sound, if we intend to design successful augmented places, we need to keep interrogating and articulating all aspects of "place", their relationships and affordances, and in doing this employ all knowledge we have on spatial and urban design, in a holistic way.

Shaping augmented places

What can this mean in practice? How can the process of designing augmented place be enriched? This chapter has argued so far that, whilst attention is being put on the need for more bottom-up social participation in the shaping of smart landscapes, much less thought has been going into re-introducing urban and architectural design principles and knowledge in order to let physical space – and actions involving it – participate themselves as actants.

At the start of the 1990s, in an article on remote communication titled *Being There*, David Brittan mentioned as an example a conversation with Chris Turner, from Olivetti Research Laboratories in Cambridge:

> Do you need to see a video image of someone just to be asked out for a beer? "Well, you don't – Turner admits – but don't you think is rather criminal that you can't?" In his view, the advent of two-way video on computer workstations is a matter of manifest destiny.
>
> *(Brittan, 1992: 43–44)*

This is what we need to move away from: the attitude of deploying technological "answers" just because we can, or we want to, where there are no clear or well-justified place-based questions to address after all. Marteen Hajer, chief curator of the 2016 Rotterdam International Architecture Biennale, questioned whether "If smart technology is the solution, then what was the problem again? It's almost like a solution looking for a problem" (quoted in Frearson, 2016). A good starting point therefore is to move away from a solution – and product – based approach back to an increased awareness of place and the principles and dimensions that can inform its functioning, perception and ultimate shaping.

Asking place-related questions

Place is complex, and that complex overlapping of aspects, issues and opportunities – if an effort to grasp and understand them is sustained – can provide important clues towards its improvement by design. Carmona et al. (2003) for instance identify six interrelated dimensions of our cities as morphological, perceptual, social, visual, functional and temporal layers. Regardless of whether one embraces this specific framework, or a slightly different one, a major mistake here would be thinking that urban digital technology constitutes another, discrete new add-on layer, hence that it can be looked at and designed on its own. High technology however is not a layer the very same way as people and space are not. They are all actants participating in the shaping of place. As Awan et al. put it, referring to Gidden's duality of agency, in the case of architecture and buildings, these

> are not seen as determinants of society (the primacy of the individual) nor as determined by society (the primacy of structure) but rather as in society … Spatial agents are neither impotent nor all powerful: they are negotiators of existing conditions in order to partially reform them.
>
> *(Awan et al., 2011: 31)*

And the interplay of these agents permeates all dimensions. It affects urban form as well as perception and social relationships. It has a bearing on how the city looks like as well as on how it functions, and indeed on how time participates in all these aspects. High technology adds more complexity to them, interplaying with space and people, and participating in a process of constant redefinition of relationships. To understand how, and how to use it within specific urban spaces, we need to "question" place, and proactively incorporate a deep consideration of context, what its layers mean, and what its agents can afford, within any design formulation. Mattern (2017) highlights this need for a holistic approach well, arguing how

> Instead of more gratuitous parametric modeling, we need to think about urban epistemologies that embrace memory and history; that recognize spatial intelligence assensory and experiential; that consider other species' ways of knowing; that appreciate the wisdom of local crowds and communities.

A possible practical example of this line of thought has been discussed by Aurigi (2013) when noticing that the otherwise advanced system of public and interactive terminals in the Finnish city of Oulu was yet deployed as an "ubiquitous" solution, or a product that could be "located" and deployed within urban spaces. As one would expect for a product, questions about usability, information potential or functionality had been raised – together of course with addressing a plethora of technical issues that included the physical design and maintenance of terminals to withstand the harsh climatic conditions of northern Finland. Yet questions about the specific civic character and culture of the locations involved had been overlooked. It has been noted how:

> The terminal/hotspot placed in the market square … could really play a significant – and significantly unique – role in a symbiotic relation with the specific place it is part of. As a market square is eminently a space for exchange and transaction (social as well as financial), this character could be boosted digitally by providing place-based opportunities for digital exchange. The possibilities within such an "augmented market" perspective would still be many and diverse, but they would focus on reinforcing and supporting the place's culture, uniqueness and strengths, rather than providing a "ubiquitous" service. Context would not just be an opportunity, but it would become one of the central generators of the digital intervention.
>
> *(Aurigi, 2013: 138)*

Anybody who has ever participated in a design review – be it of an academic or professional nature – for an urban or architectural scheme, knows how crucial a series of place-probing queries are towards facilitating the formulation of an effective brief and set of design intentions. What relationships, spatial, social and economic, exist there and make the place what it is? What meanings does it have to people? What form(s) does it have and how is it likely to be perceived? How is it used, how do people and things move in it and what happens there? Who lives there or uses it, and why, and what do they think and feel? How does time and environmental conditions affect it? And, more proactively, designers might need to reflect on more complex and choice-embedding questions like: is there anything about the place we need or want to accept or contrast with, through attitudes ranging "from submission, through symbiosis, to domination" (Unwin, 2009: 120)? What potential as well as conflicts and contestations does the context and its history have? Can such space have a role in a wider urban and/or regional strategy? These questions, and more, are essential to start exploring how a new urban element or system – physical, digital or indeed hybrid – could alter the complexity of relationships already characterising a certain place, and even questioning what – if anything – might be needed at all. Yet, these issues are seldom unpacked before a "situated" digital project is conceived. The "city as a computer" paradigm, as Mattern (2017) calls it, "appeals because it frames the messiness of urban life as programmable and subject to rational order", and suggests simplifying but distorting urban design practice into the shaping of a hi-tech product. Product design tends to be in itself much less grounded in context and place than urban design. Whilst the interface and interaction with "users" or some of the conceptual and material aspects of this new wave of hi-tech civic design are often thoroughly looked at, to echo Horan's early concerns, the deep "transformational" and bi-directional interfacing with place is easily overlooked.

Conclusion: extending the place-making design toolbox

This chapter focuses on a gap in research and practice, as the importance and role of physical space and urban design have been looked at only marginally within smart city debates, whilst practice seems to rely on paradigms and visions stemming – unsurprisingly in many ways – from ICT and product design approaches.

A reflection is needed on how on the one hand interrogating and understanding place – and bringing space as a crucial component of it fully back into the picture – could be the first step towards a more sophisticated approach in the design of smart environments. The next, non-trivial challenge however is developing an insight on how the "digital" participates and integrates with the spatial – in a circular relationship rather than a one-way impact trajectory – in making, or re-making place. In other words, once analysis and intentions on how to improve place are clearer and richer, when it comes to actually designing in a hybrid way, how well are we aware of the possibilities (and threats, potentially) of the extended toolbox we are going to use?

Once a rich and place-based brief has been conceived, the issue of updating our design knowledge to reflect the augmented possibilities also arises. On a speculative and intuitive basis it seems clear that issues of spatial relationships and agency, scale, access and mobility, inhabitation, meaning, perception and memory – and more – which we are used to consider carefully when shaping public spaces, are still important, yet often ignored in smart city thinking. On the one hand, they should not be jettisoned in the name of an alleged brand new logic of place only depending on the redeeming and innovating power of ICTs. On the other hand, they do need to be updated and upgraded to inform design in a re-combined world. It is debatable whether this can be achieved simply through repeated and evolving practice. This upgrading probably needs a thorough research effort. In the 20th century much intellectual energy had gone into trying to understand articulations, languages and syntaxes of space, and how designers could harness such principles. From Cullen (1961) and Lynch (1960, 1984), to Norberg-Schulz (1971), Alexander et al. (1977) or Hillier (1996) – just to name a few – ideas on how space, people and things articulate were usefully framed to help designers make sense of the complexity of places, and be more aware of the potential consequences of their own moves. Some work has been carried out trying to bring high technology and urban design knowledge together, though still with a particular emphasis on designing mainly the digital aspects of place and the social interactions they afford (see for instance Paay et al., 2007). It could be important now to intensify those efforts in the light of the emergence of new variables and extended relationships and possibilities.

Should we therefore augment not just spaces, but our questions on place and design thinking too? In those questions, and an increased awareness and mastering of a series of extended principles for understanding how hybrid space works, lies the quantum leap between just designing self-contained interactions, which at best are "located" somewhere, and effectively shaping augmented place, making a significant difference for the way our cities can become smarter.

Notes

1 http://www.mediaarchitecture.org/.
2 https://research.qut.edu.au/designlab/groups/urban-informatics/.

References

Alexander C., Ishikawa S. and Silverstein M. (1977) *A Pattern Language: Towns, Buildings, Construction*. New York: Oxford University Press.

Aurigi A. (2005) *Making the Digital City: The Early Shaping of Urban Internet Space. Design and the Built Environment*. Aldershot, Hants, England; Burlington, VT: Ashgate.

Aurigi A. (2013) "Reflections towards an agenda for urban-designing the digital city", *Urban Design International*, 18 (2), 131–144.

Aurigi A. (2016) "No need to fix: Strategic inclusivity in developing and managing the smart city", in Caldwell G. Smith C. and Clift E. (eds) *Digital Futures and the City of Today - New Technologies and Physical Spaces*, Bristol: Intellect, pp. 9–27.

Austen I. (2017) "City of the Future? Humans, Not Technology, Are the Challenge in Toronto", The New York Times, 29 December 2017, www.nytimes.com/2017/12/29/world/canada/google-toronto-city-future.html, last accessed April 2019.

Awan N., Schneider T. and Till J. (2011) *Spatial Agency: Other Ways of Doing Architecture*. Oxon: Routledge.

Baldwin E. (2018) "Sidewalk Labs Unveils Future City Design for Toronto's Quayside Neighborhood", *ArchDaily* 16 August 2018, www.archdaily.com/900274/sidewalk-labs-unveils-future-city-design-for-torontos-quayside-neighborhood, last accessed May 2019.

Baldwin E. (2019) "Snøhetta and Heatherwick Design a Timber City for Sidewalk Labs", *ArchDaily* 19 January 2019, www.archdaily.com/911805/snohetta-and-heatherwick-design-a-timber-city-for-sidewalk-labs, last accessed May 2019.

Beckmann, J. (ed.) (1998) *The Virtual Dimension: Architecture, Representation, and Crash Culture*. 1st. New York: Princeton Architectural Press.

Benedikt M. (1991) "Introduction", In Benedikt M. (ed) *Cyberspace: First Steps*, Cambridge, MA: MIT Press, 1–25.

Bliss L. (2019) "Critics Vow to Block Sidewalk Labs' Controversial Smart City in Toronto", Citylab 25 February 2019, www.citylab.com/equity/2019/02/block-sidewalk-labs-quayside-toronto-smart-city-resistance/583477/, last accessed May 2019.

Brittan D. (1992) Being There. *Technology Review*, May/June.

Čakovská B., Bihuňová M., Hansen P., Marcheggiani E., and Galli A. (2019) "Methodological Approaches to Reflect on the Relationships Between People, Spaces, Technologies", In

Smaniotto Costa C. et al. (ed) *CyberParks – The Interface Between People, Places and Technology*, LNCS 11380 251–261, Cham: Springer.

The Canadian Press (2019) "Sidewalk Labs could pull out of Quayside project if transit isn't built, CEO says", *CityNews* 6 March 2019, https://toronto.citynews.ca/2019/03/06/sidewalk-labs-could-pull-out-of-quayside-project-if-transit-isnt-built-ceo-says-2, last accessed May 2019.

Canon G. (2018) "'City of surveillance': Privacy expert quits Toronto's smart-city project", *The Guardian*, 23 October 2018, www.theguardian.com/world/2018/oct/23/toronto-smart-city-surveillance-ann-cavoukian-resigns-privacy, last accessed May 2019.

Carmona M., Heath T., Oc T. and Tiesdell S. (2003) *Public Places – Urban Spaces: The Dimensions of Urban Design*. Oxford: Architectural Press.

CHS UR OWN URBNSM blog: http://chseurownurbnsm.blogspot.co.uk/2007/05/sonic-arboretum.html - May 2007 entry 'Sonic Arboretum' – last accessed 9 March 2016.

Le Corbusier (1928) *Toward an Architecture*. London: Frances Lincoln. edition (2008).

Cowley, R. and Caprotti, F. (2019) "Smart city as anti-planning in the UK", *Environment and Planning D: Society and Space*, 37 (3), 428–448.

Cullen G. (1961) *Townscape*. London: Architectural Press.

Datta A. (2015) "New urban utopias of postcolonial India "Entrepreneurial urbanization" in Dholera smart city, Gujarat", *Dialogues in Human Geography*, 5, 3–22.

de Lange M. (2015) The Playful City: Play and games for citizen participation in the smart city. *Short Term Scientific Mission, COST Action TU1306 Cyberparks, mimeo.*

de Lange M. and de Waal M. (Eds) (2018) *The Hackable City*, New York: Springer Berlin Heidelberg.

Dear M. (1995) "Prolegomena to a Postmodern Urbanism", In Healey P., Cameron S., Davoudi S., Graham S., Madani-Pour A. (eds) *Managing Cities: The New Urban Context*, Chichester: Wiley, pp. 27–44.

Foth M., Brynskov M., Ojala T. (Eds) (2015) *Citizen's right to the digital city: Urban interfaces, activism, and placemaking*, Singapore: Springer.

Frearson A. (2016) "'Smart technology is a solution looking for a problem' says Rotterdam Biennale curator", Dezeen 27 April 2016, www.dezeen.com/2016/04/27/smart-technology-driverless-cars-interview-maarten-hajer-rotterdam-biennale-2016-curator-netherlands/, last accessed May 2019.

Graham S. and Marvin S. (1996) *Telecommunications and the City: Electronic Spaces, Urban Places*. London; New York: Routledge.

Grahame S.D. (2005) *Recombinant Urbanism: Conceptual Modeling in Architecture, Urban Design, and City Theory*. Chichester, England; Hoboken, NJ: Wiley.

Hemment D. and Townsend A. (Eds) (2013) *Smart Citizens*, Manchester: FutureEverything.

Hillier B. (1996) *Space is the Machine*. Cambridge: Cambridge University Press.

Hollands R.G. (2008) "Will the real smart city please stand up?" *City: Analysis of urban trends, culture, theory, policy, action*, 12 (3), 303–320.

Hollands R.G. (2015) "Critical interventions into the corporate smart city", *Cambridge Journal of Regions, Economy and Society*, 8, 61–77.

Horan T.A. (2000) *Digital places: Building our city of bits*. Washington, DC: ULI-the Urban Land Institute.

Jormakka, K. (2002) *Flying Dutchmen: Motion in Architecture. The IT Revolution in Architecture*. Basel: Birkhäuser.

Kitchin R., Maalsen S., McArdle G. (2015) "The Praxis and Politics of Building Urban Dashboards" (SSRN Scholarly Paper No. ID 2608988). Social Science Research Network, Rochester, NY.

Lam E. (2019) Viewpoint: Sidewalk Toronto, *Canadian Architect*, www.canadianarchitect.com/features/viewpoint-sidewalk-toronto/, last accessed May 2019.

Lee H. (2008) "Mobile Networks, Urban Places and Emotional Spaces", In Aurigi A. and De Cindio F. (eds) *Augmented Urban Spaces: Articulating the Physical and Electronic City*, Aldershot: Ashgate, pp. 41–59.

Lynch K. (1960) *The Image of the City*. Cambridge, MA: MIT Press.

Lynch K. (1984) *Good City Form*. Cambridge, MA: MIT Press.

Mattern S. (2017) "A City Is Not a Computer", *Places*, February 2017, https://placesjournal.org/article/a-city-is-not-a-computer/, last accessed May 2019.

McConnell J. (2017) "The Android of cities: Alphabet's smartphone-inspired vision for Toronto's waterfront", *Financial Post*, 18 October 2017, https://business.financialpost.com/technology/the-android-of-cities-alphabets-smartphone-inspired-vision-for-torontos-waterfront, last accessed May 2019.

McCullough M. (2005) *Digital Ground: Architecture, Pervasive Computing, and Environmental Knowing*, 1st paperback ed. Cambridge, MA: MIT Press.

McCullough M. (2013) *Ambient Commons: Attention in the Age of Embodied Information*. Cambridge, MA: The MIT Press.

Menon A. (2013) "The Smart City Council – Accelerating an Exciting Growth", in *Cisco Blogs*, https://blogs.cisco.com/news/the-smart-city-council-accelerating-an-exciting-future/, last accessed February 2014.

Mitchell W.J. (1995) *City of bits: Space, place, and the infobahn*. Cambridge, MA: MIT Press.

Mitchell W. J. (1999) *E-Topia: "Urban Life, Jim–but Not as We Know It"*. Cambridge, MA: MIT Press.

Mitchell W. J. (2003) *Me++: The Cyborg Self and the Networked City*. Cambridge, MA: MIT Press.

Negroponte N. (1996) *Being Digital*, Coronet Books, London: Hodder and Stoughton.

Norberg-Schulz C. (1971) *Existence, Space and Architecture*. London: Studio Vista.

Novak M. (1994) "Liquid Architectures in Cyberspace", in Benedikt M. (ed) *Cyberspace: First steps*, Cambridge, MA: MIT Press, 225–254.

Paay J., Dave B. and Howard S. (2007) "Understanding and representing the social prospects of hybrid urban spaces", *Environment and Planning B: Planning and Design*, 34, 446–465.

Picon A. (2015) *Smart Cities: A Spatialised Intelligence*, Ad Primers. Chicester: John Wiley.

Living PlanIT (2011) Cities in the Cloud - A Living PlanIT Introduction to Future City Technology, www.livingplanit.com/resources/Living_PlanIT_SA_Cities_in_the_Cloud_Whitepaper_Website_Edition (2011-09-10-v01).pdf, last accessed 18 November 2013.

Playable City Vision (n.d.) www.playablecity.com/vision/- last accessed 10 August 2016.

Rekow L. (2013) "Including Informality in the Smart Citizen Conversation", in Hemment D. and Townsend A. (eds) *Smart Citizens*, Manchester: FutureEverything, pp. 35–38.

Rheingold H. (1995) *The virtual community: Finding connection in a computerized world*. London: Minerva.

Riewoldt O. (1997) *Intelligent Spaces: Architecture for the Information Age*. London: Laurence King Publishing.

Sassen S. (2017) "The city: A collective good?" *The Brown Journal of World Affairs*, Spring/Summer 2017 volume xxiii(ii), pp. 119–126.

Scassa T. (2019) "As Smart Cities Become Our Norm, We Must Be Smart About a Data Strategy", *Centre for International Governance Innovation*, 15 February 2019, www.cigionline.org/articles/smart-cities-become-our-norm-we-must-be-smart-about-data-strategy, last accessed May 2019.

Schneider Electric (n.d.) Smart Cities, www2.schneider-electric.com/sites/corporate/en/solutions/sustainable_solutions/smart-cities.page, last accessed February 2014.

Schuler D. (1996) *New Community Networks: Wired for Change*. Reading, MA: Addison Wesley.

Shepard M. (ed.) (2011) *Sentient City: Ubiquitous Computing, Architecture, and the Future of Urban Space*, New York City; Cambridge, MA: Architectural League of New York; MIT Press.

Smaniotto Costa, C., Šuklje Erjavec I., Kenna T., de Lange M., Ioannidis K., Maksymiuk G., and de Waal, M. (Eds) (2019) CyberParks – The Interface Between People, Places and Technology: New Approaches and Perspectives, *LNCS* 11380, https://doi.org/10.1007/978-3-030-13417-4.

Söderström O., Paasche T. & Klauser F. (2014) "Smart cities as corporate storytelling", *City: Analysis of urban trends, culture, theory, policy, action*, 18 (3), 307–320.

Sustainable Development Solutions Network (2013) The Urban Opportunity: Enabling Transformative and Sustainable Development, https://sustainabledevelopment.un.org/content/documents/2579Final-052013-SDSN-TG09-The-Urban-Opportunity.pdf, last accessed May 2019.

Toffler A. (1990) *The third wave*. London: Pan Books.

Tomitsch, M. (2018) *Making Cities Smarter: Designing, Interactive, Urban, Applications*. Berlin: Jovis.

Townsend A. (2013) *Smart cities: Big data, civic hackers, and the quest for a new utopia*. New York: W.W. Norton & Company.

Townsend A., Simeti A., Spiegel D., Forlano L., Bacigalupo T., & Shepard M. (eds) (2011) *Sentient City: Ubiquitous computing, architecture, and the future of urban space*, Cambridge, MA: MIT Press.

Unwin S. (2009) *Analysing architecture*, 3. ed London: Routledge.

Valverde M. (2018) "The controversy over Google's futuristic plans for Toronto", *The Conversation* 30 January 2018, http://theconversation.com/the-controversy-over-googles-futuristic-plans-for-toronto-90611, last accessed May 2019.

Valverde M. and Flynn A. (2018) "'More Buzzwords than Answers' To Sidewalk Labs in Toronto", *Landscape Architecture Frontiers/Experiments and Processes*, 6 Issue 2 April 2018, pp. 115–123.

Van Dijck J., Poell T. and de Waal M (2018) *The Platform Society*. New York: Oxford University Press.

Walsh N.P. (2018) Carlo Ratti's Prototype for Sidewalk Labs Shows How the Design of Streets Could Change in Real Time, ArchDaily 18 July 2018, www.archdaily.com/898471/carlo-ratti-associatis-latest-prototype-shows-how-the-design-of-streets-could-change-in-real-time, last accessed May 2019.

Wilson M. (2019) "6 crazy details from Alphabet's leaked plans for its first smart city", *FastCompany*, www.fastcompany.com/90309358/6-crazy-details-from-alphabets-leaked-plans-for-its-first-smart-city, last accessed May 2019.

Won J. (2018) "Smart Cities: Toronto's Google-Infused District and Lessons from Songdo, Korea", *Cornell Real Estate Review*, http://blog.realestate.cornell.edu/2018/11/24/smart-cities-torontos-google-infused-district-and-lessons-from-songdo-korea/, last accessed May 2019.

Woyke E. (2018) "A Smarter Smart City", *MIT Technology Review*, March-April 2018.

Wylie B. (2017) "Think Hard Before Handing Tech Firms The Rights To Our Cities' Data", *The Huffington Post Canada*, 11 August 2017, www.huffingtonpost.ca/bianca-wylie/think-hard-before-handing-tech-firms-the-rights-to-our-cities-data_a_23270793/, last accessed May 2019.

Wylie B. (2018a) "Sidewalk Toronto — We're Consulting on What, Exactly?", *Medium* 23 February 2018, https://medium.com/@biancawylie/sidewalk-toronto-were-consulting-on-what-exactly-f097203b95ed, last accessed May 2019.

Wylie B. (2018b) "Democracy Demands Critical Public Discourse - Sidewalk Toronto Needs More Of It, Not Less", *Medium* 5 March 2018, https://medium.com/@biancawylie/democracy-demands-critical-public-discourse-sidewalk-toronto-needs-more-of-it-not-less-dcaf3ba4dbaf, last accessed May 2019.

Wylie B. (2019) "Why we need data rights: 'Not everything about us should be for sale'", *Financial Post*, 30 January 2019, https://business.financialpost.com/technology/why-we-need-data-rights-not-everything-about-us-should-be-for-sale, last accessed May 2019.

26

Self-monitoring, analysis and reporting technologies

Smart cities and real-time data

Andrew Hudson-Smith, Stephan Hügel and Flora Roumpani

Introduction

Approaching 25 years ago, Putz (1994) in his paper 'Interactive Information Services Using World Wide Web Hypertext' noted that although most World Wide Web (WWW) servers were designed primarily as hypertext file servers, there was an increasing trend towards more dynamic information services where custom documents could be assembled and delivered to a user on request. Since then, the internet has of course transformed into such a system with myriad feeds and data streams from a once largely passive medium for delivering information to one that is interactive in which the user can query, construct and manipulate information on the fly. This level of development can be seen in the emerging fields of smart cities. Networks are beginning to shape themselves into the fabric of cities and in this chapter, we will consider the term 'smart city' to refer to the whole set of opportunities that networked computers offer to cities in terms of their enrichment for greater efficiency and a better quality of life. The smart city is beginning to form into a dynamic information service, but currently, most of the hardware and software which define such elements of the city are based on stand-alone units, not connected to wider networks or linked together in any meaningful way (Hudson-Smith 2014). This is beginning to change but the development is looking to be longer term than the development that Putz (1994) talks about in terms of information services; cities take time to develop and the smart city is one that will develop over the coming decades – these networks are only beginning to form.

The term 'smart' has many other definitions and public perceptions, it has been widely criticised as a meaningless catch-all phrase, as a marketing and public relations ploy, often captured by commercial interests (GLA 2013), indeed as Söderström et al. (2014) note the term can be seen merely as deconstruction of a communication strategy. Therefore, it needs to be defined in context, or it becomes all too easily subject to accusations of marketing and buzzword fodder. Here we use a very self-conscious definition that combines a series of basic functions embracing Self-Monitoring, Analysis and Reporting Technologies (which define the term SMART), and this focuses on our definition of the technologies that are being rolled out to make cities computable in a routine sense. This term originally came from the functions employed in hard disks as a way to internally monitor their own health and performance. Most implementations of SMART, in terms of disk drives, allow users to perform self-tests on the disk and to monitor a number of performance and reliability attributes (Allen 2004). The ability to self-monitor, analyse and report performance and reliability measures is, we argue, a closer definition of how a 'smart city' might function than other looser and wider definitions and one upon which we build in this chapter. We are well aware that many smart city concepts depend on more explicitly social and economic processes, institutions and agencies. We will not review these here for there are some excellent reviews to this area of smart cities work already available (Chin et al. 2011; Debnath et al. 2014). Any complete assessment of smart cities, the technical focus that we have adopted here must be blended with the institutional and organizational and the success of the smart cities movement will depend on a much wider set of forces than those that we describe here (Giffinger et al. 2010).

Such was the momentum 20 years ago when the WWW was first constructed on the back of two generations of networked technologies that Batty (1997) predicted that by the year 2050 everything around us would be some form of computer. In 2019, computerised highways are in prospect, driverless cars and smart buildings which monitor their performance in terms of energy and materials are almost upon us. In 1965, Gordon Moore envisioned in the paper that introduced his famous law 'Cramming More Components onto Integrated Circuits' that integrated circuits would eventually lead to such wonders as home computers – or at least terminals connected to central computers, automatic controls for automobiles and personal communications equipment (Moore 1965). These circuits are increasingly weaving themselves into the city and the fabric of urban form and arguably the prediction by Batty, defined under his term 'Computable City', (Batty 1995) is becoming a reality.

Places and spaces are increasingly becoming internet-connected with mobile communications, and the size of the market for such routine technologies has an addressable market value in terms of transport, utilities and intelligent buildings predicted to amount to some US$70 billion by 2020 (GSMA 2012). The term smart city has become synonymous with how well a city is performing in the current climate of mobile applications (apps), data streams and social networks. As Caragliu et al. (2009) state, it has been introduced as a strategic device to encompass modern urban production factors in a common framework and to highlight the growing importance of information and communication technologies (ICTs) which have become the social and environmental capital in profiling the competitiveness of cities. Yet the smart city is wider than this, for at its heart is a definition of place and space which allows a view from the micro individual up to the macro collective of how a city operates. Indeed even the term smart city itself is perhaps being surpassed as wider contexts develop – the city as a digital twin, for example is becoming a popular term of reference often based on the assumption we are creating a digital duplicate of the physical entity (Datta 2017).

We argue that although phases come and go, the concept of self-monitoring, analysis and reporting technologies is key to the longer-term development. Digital twins can be viewed as a digital representation of the physical world, with the addition of data, collected from anything from building systems through to social and environmental feeds that can be viewed and acted upon. They are like looking at a mirror of our world, a mirror that not only reflects the environment but also displays the invisible: the flows of data (Hudson-Smith et al. 2019). As such they reflect the smart city in mixing the physicality and the flows of data; in this sense the smart city can be viewed as an 'urban hard drive' in which the data can be mined, visualised and modelled for self-monitoring, analysis and reporting which implies a joined-up technology where integration and coordination are key. The challenge of the smart city then is to provide and define the conditions under which this is possible – what is happening now is merely a beginning. We explore challenges that are data stores, data feeds and data visualitions in the following sections, building on work developed at the Bartlett Centre for Advanced Spatial Analysis (CASA), University College London. The Centre has a focus on usability and communication of city-based data with a core around real-time analysis and visualisation.

Data dashboards and data stores

Central to our conception of the smart city is data. Every day, humankind creates over 2.5 quintillion bytes of data – so much that more than 90% of the data in the world today has been created in the last two years alone. These data come from everywhere: sensors used to gather climate information, posts to social media sites, digital pictures and videos, purchases and transaction records, cell phone signals, to name a few (IBM 2013). An increasing amount of this data stream is geo-located, from check-ins via social networking sites like Foursquare, through to tweets and searches via Google Now. The data that cities and individuals emit can be collected and viewed to make it visible, thus aiding our understanding not only of how urban systems operate but opening up the possibility of continuous real-time viewing of the city at large (Hudson-Smith 2014). Cities across the world are at various stages of both releasing and utilising such datasets as both producers and consumers of urban information. It is part of our realisation that smart cities are no longer places where city government acts as the top-down driver of development, but instead, they act as a broker in a much wider ecosystem of data and information (Department for Business Innovation and Skills 2013), though it should be noted that this opening up can itself give rise to interrelated complexities relating to usability, power and marketisation (Kitchin 2013).

Case study: London city data strategy

A key role a city government plays in this emerging ecosystem is as a provider and aggregator of data. The London Data Store in UK is a prime example of how such a system can provide an impetus to the creation of services and added value from data. Developed by the Greater London Authority (GLA), the Data Store has stimulated over 70 mobile applications linking to almost 650 datasets from a combination of the 77 real-time live traffic and transport data feeds it provides. In 2019, the GLA began to actively refine its data strategy, setting out its aims and reasoning in a comprehensive policy document in terms of a roadmap. The roadmap, which is a non-statutory document, builds on the first Smart London Plan, published in 2013. It provides a new approach based on collaborative missions and calls for the city's 33 local authorities and various public services to work and collaborate better with the aid of data and digital technologies (GLA 2019). Recognising that the provision and availability of open data is merely the first step, and drawing upon its expertise and success in this area, the GLA proposes to support the creation of a 'City Data Market'. It hopes this will engage companies who have traditionally viewed data as proprietary and closed to participate in an ecosystem of mixed-permission data products, opening data in order to create demand, and perhaps even trading data with other providers and application developers, in exchange for access or expertise. In so doing, it seeks to promote a sharing culture which can encompass both free and propri-etary data. Furthermore, the GLA recognises that it need not provide the repository for these data itself. The availability of mature, well-tested open platforms such as CKAN (a web-based open-source management system for the storage and distribution of open data) which in fact powers the London Data Store, a variety of federation (that is, sharing and harvesting data) modes are available 'out of the box' as it were. This allows datasets to replicate and propagate across heterogeneous data stores both within the city and beyond, according to granular, well-defined rules. It is expected that this blended approach to 'market making' for data will lead to the simple and wide availability of data from a variety of providers and will facilitate analysis.

We argue that while this is a welcome step forward, it is also not enough; a data 'market' is not a data 'commons', i.e. resources considered public goods, meaning that they are accessible to the public, and also rivalrous, meaning that their use by one precludes their use by another (Beckwith et al. 2019). There are barriers, both obvious and subtle, to participat-ing in this data market, and this necessarily influences the nature of data that will be available within it. Certain contributors of data will be seen as less valuable than others, and some contributions will be less welcome than others. This raises questions as to who participates or contributes to the data market. How, for instance, will the contribution of data from citi-zen science projects be dealt with? Will there be efforts on the part of the marketplace as a whole to encourage participation from demographics who have been historically under-served by the advent of new urban technologies? While the London city data strategy is comprehensive, it fails to note the lack of success of previous attempts to create marketplaces of this kind, such as BuzzData or Freebase to name but two (Dodds 2016). A pilot, currently underway, combines these ideas of a 'hosted' data repository, a market and more nuanced, urgent considerations about the accumulation and use of personal data in a so-called 'data trust' (Wylie and McDonald 2018). Though the precise definition of a data trust has not yet been agreed upon (Hardinges 2018), most trusts conform to one or more of the features distinguished by Hardinges, arguably the most interesting being some form of public over-sight of data access.

London data dashboard

Through the application programming interface (API) provided by the London Data Store, our centre – CASA – has created a city dashboard as a means of viewing a key number of live data feeds. This essentially is a simple interface to the visualisation of these data streams that are updated in real time and is available on the web.[1] In London, the dashboard collates and simplifies over 20 live feeds from air pollution through to energy demand, river flow, the FTSE 100, the number of buses in service, the status of the subway networks and so on, which we illustrate in Figure 26.1.

The London data dashboard is an early example of collating and visualising data feeds to provide a citizen focused view of the city. Not limited to London, the dashboard has also been built for Birmingham, Brighton, Cardiff, Edinburgh, Glasgow, Leeds and Manchester, with a version for Venice also under development. But in these different cities, the types of stream-ing data can be a little different, for the dashboard highlights the variability in the availability of data feeds from city to city. London, at the present time, is the location for a majority of data feeds, with their number updated on a second by second basis. The majority of these data are either collected via an API which is an interface, usually through the web via http(s), where a user can query the status of the system with the live data being delivered to the user (or its

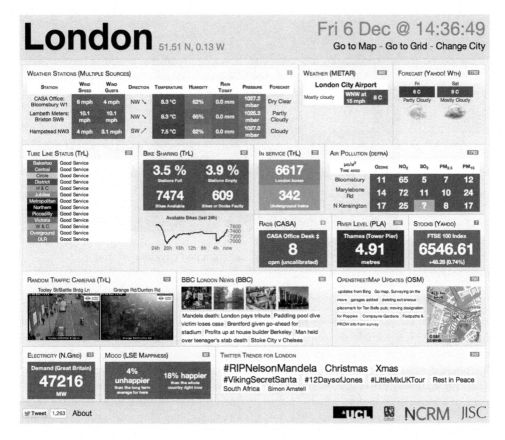

Figure 26.1 Live data feeds into a web-based data dashboard

Source: http://citydashboard.org/london/. Screenshot by chapter author(s)

client) or the data mined in accordance with a data provider's terms and conditions. The ability to tap into these APIs allow the city dashboard to provide a view of the particular city at a glance, with the use of simple colour coding to indicate the positive or negative connotations of the current state of the data. A custom-made version of the London dashboard has been developed for internal use by the London Mayor's Policy Office at the GLA. Designed around 12 iPads and mounted into a single system, the board allows each iPad to display historic and live data relating the city, as we show in Figure 26.2.

The historic nature of the iPad wall version of the dashboard requires an element of data graphing to examine trends in the feeds, and this has led to the next stage of the development of the city dashboard, moving towards an archive system for city data. At its initial conception, the city dashboard was created as a simple viewer for city-related data feeds.

London Data Store

The publishing of data and streams via systems, such as the London Data Store, is creating a new landscape in data availability and arguably the development of a new kind of city-wide information system. The ability to refine and redistribute feeds is perhaps the next step

Figure 26.2   An iPad wall data dashboard to view live streamed city data in London
Source: Chapter author(s)

to their wider use. While we have focused on the distribution of feeds, it should be noted that it is, however, possible to view these streams as a new data archive and plug them directly into systems for building various kinds of urban models ranging from simple 2D and 3D physical visualisations to mathematical abstractions of urban functions such as traffic, rental values, house prices and so on. Such data have the potential to enable the development of a series of indicators which measure the performance of the city in both the short and near term. Indicators covering aspects of the city from urban flow, such as transport and pedestrian movement, through to a city's economic flows and onwards to more social inputs such as wellbeing and happiness, are on the horizon.

The mining and use of APIs from social network data are central to this wider view of city datasets. One of the most popular current social networks is Twitter. First created in 2006, the network now has one billion registered users with over 6000 tweets sent every second (Sayce 2019). Twitter allows users to send a message up to 280 characters in length and a tweet can contain links to other web-based content, user name and a user location. There are a number of emerging 'city toolkits' being developed around collecting a variety of social network feeds with links to geographic location, allowing the data to be understood and mapped at a city scale. One such system is CASA's own Big Data Toolkit[2] which allows the systematic mining of tweets and other social media feeds within a set radius of a location. This allows CASA to map not only the density of tweets but also to collect the text for sentiment analysis, described by Wilson et al. (2005) as the task of identifying positive and negative opinions, emotions and evaluations (Figure 26.3).

Data mining techniques, even via dedicated toolkits, are often limited in terms of the data they can analyse. The 'Twitter Streaming API' is a capability provided by Twitter that allows anyone to retrieve at most a 1% sample of all the data by providing some parameters on the nature of the tweets captured (Morstatter et al. 2013). The Big Data Toolkit used to create the Twitter map was developed to run on multiple cloud-based servers, increasing the percentage of tweets collected. These can be viewed as a sample of the so-called 'twittersphere' and thus potentially usable for analysis of the city. Hill (2008) notes in his article on

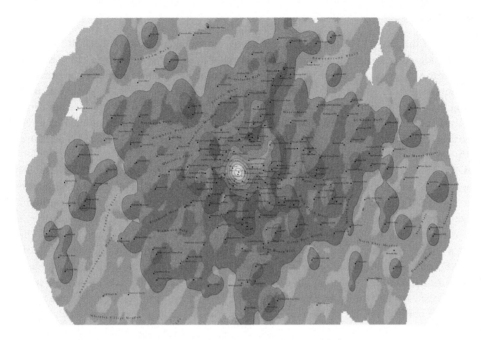

Figure 26.3 Mapping the density of tweets in Central London
Source: www.urbantick.org/

'The Street as a Platform' that we can see how the street is immersed in a twitching, pulsing cloud of data. This is over and above the other well-established sources of data: electromagnetic radiation, crackles of static, radio waves conveying radio and television broadcasts in digital and analogue forms, police voice traffic and so on, which represent the background behavioural data that characterise the city.

This is a new kind of data: both collective and individual, aggregated and discrete, open and closed, constantly logging impossibly detailed patterns of behaviour. The rise of social networks and our ability to communicate via mobile devices is resulting in an ever-growing amount of data being tagged with location. This enables a move towards a real-time view of the city, which at street level in Hill's (2008) view is that 'pulsing cloud of data' travelling outwards to the wider urban datascape or dataverse. Collecting not only more traditional datasets (such as transport, air pollution and building energy use) but also personal data is moving us toward a new view of the city from the urban scale down to the level of the citizen, the group and the crowd. At the present time, crowd-sourced data are arguably more accessible via data mining techniques than more traditional dataset sources. For example, in many cities transport data are still owned by private companies and closed off from public use, which means it is not constituting a part of any public data store or live feed. A similar position exists for energy usage, general utilities and wider city logistics data systems. Public data are increasingly being opened up, but private companies are for logistical or commercial reasons slower to open up feeds and share data. However, the wider public, mainly via mobile technology, are tagging and sharing data to such an extent that participatory sensing allows crowd-sourced datasets to begin to supplement and complement some of the traditional city-based information sources.

Data APIs

The idea of the city dashboard is an integral, but still emergent, part of the shift of the city from the idea of a server of data to a dynamic information system. Today's city dashboards are primarily aggregators of pre-existing data streams e.g. public transport, weather and environmental data, geo-tagged social media data and so on. As we have discussed, in the example of GLA's data market approach, dashboards can lead to a view of this data as merely a resource to be monetised, which as Mattern points out, can and often does impose a particular vision – that of the top-down, technocratic 'management' of city data (Mattern 2015).

Data dashboards are currently designed as 'glanceable' interfaces – ambient, comprehensible screens which quickly convey key information about the city. However, we argue that this concept of the dashboard does not offer the most promising and compelling opportunity for the connected and smart city. Instead it is the rather abstract concept which sees the city as a data server, as we introduced it earlier in the example of retrieving and querying data from a source such as Twitter through its API. This opens up a way for virtual location services, embedded devices (such as urban sensors) and even whole buildings and neighbourhoods to communicate with one another, using ever more elaborate sets of defined actions and messages. In practice, most of these APIs use various http protocols and the representational state transfer (REST) architectural approach (Severance 2015), both of which underlie much of the modern internet's functionality. When you take a photograph with your phone using the Instagram application for example, and a link to the picture can then appear in your Facebook timeline, a series of 'calls' to various APIs have by then been made: your phone makes an http request to the Instagram API, informing it that it wishes to upload a photo and the API responds with a 'go ahead' response. The image is then serialised and uploaded, and an 'OK' response is sent to the app. Conceptually, RESTful APIs can be seen as the lingua franca of today's internet; they are useful abstractions connecting disparate services and applications in a reliable and familiar way. In essence, urban sensor platforms are an attempt to extend the abstraction of an API to the physical world. APIs are not restricted to software; most consumer-grade urban sensors which have become available over the past decade have been capable of communicating with a remote API. It is these APIs which are opening the field of urban analytics within the domain of smart cities.

Visualising and utilising APIs: towards real-time urban analytics

We have highlighted this concept of 'smart' through the use of computers and computation which operate across wide spatial and temporal domains; that is many spatial and temporal scales but from the finest upwards. The focus on moving towards a self-monitoring, analysing and reporting city is on joining up or integrating operations and services and also disseminating the information associated with these activities to users through a variety of computable devices from regular PCs to smartphones. In this, ways of making sense of urban data are key and in the next section we will briefly illustrate the role of visualisation in this process. In addition, these systems, through their embedding into the built environment and their routine use by populations through hand-held devices ranging from smart cards to phones, are delivering large quantities of data about the way cities function. These have the ability to be streamed and archived in real time to either Cloud or local-based storage and analysis systems, thus providing a detailed spatio-temporal record of all that goes on in the functions that are being automated.

In all of this, artificial intelligence, expert systems, routine data mining and the construction of procedures which are automatically invoked are central to developing the concept of the smart city. In the past, many urban services, particularly those dealing with emergencies, have been automated. But now the focus is much more on aiding individuals in the population to make informed decisions with respect to the conditions they encounter, such as those involving the use of energy, transport and the routine purchase of goods. We have not quite reached a world where there is instant access to the internet wherever we are, but this is fast emerging as wireless systems begin to spread out, and phones and other devices are being adapted to searching out such access as the individual moves around the city.

This emergence of a range of advanced and easily accessible digital tools has enabled the realisation of such ideas which are no longer addressed exclusively to the professional, but with a focus on bringing routine information about the city, planning and urbanism closer to the public itself. Characteristic examples include Google Maps, urban themed games (such as Cities: Skylines[3]) and specialised 3D modelling platforms (such as ESRI's CityEngine[4]). These new software applications have facilitated the use of digital environments for testing the consequences of physical planning policies on the future form of cities and are rapidly becoming powerful tools for distributing and communicating spatial information worldwide (Batty, Steadman and Xie 2004).

In the next section we will discuss some examples exploring the use of digital tools and methods which, based on live feeds and open information, will expand city dashboards into a form of dynamic 3D urban analytics. Visualisation lies at the essence of this activity and one thing we examine here is many of the new functions of geographic information systems. One such example is the variability of such information in a 3D spatial context and increasingly these are likely to become available to the non-expert citizens and in the not too distant future digital twins. Finally, we will outline how these systems could be used for understanding the possibilities, problems and consequences of exploring different scenarios relating to the actual form of the future city, thus paving the way for new forms of real-time public participation.

Real-time urban analytics: digital twins

Currently, there are several methods for introducing mined or crowd-sourced data into 3D visual modelling applications. Data, often collected through the use of an API, can be fed into a variety of applications for visualisation purposes, analysis and further editing. As a general requirement, these systems need to have extended editing capabilities such as scripting consoles and specialised libraries. One such example builds on the previously illustrated Twitter map in Figure 26.3 and its visualisation in 3D using CityEngine (Parish and Muller 2001) in order to create a digital twin. This takes the concept of collecting geolocated tweets and placing them into a 3D procedural system for creating data visualisation algorithmically as opposed to manually (Figure 26.4). Tweet City,[5] as the project is known, was developed to create a new 3D urban landscape utilising the Twitter API (Hügel and Roumpani 2013). It allows the user not only to develop different visualisations by editing simple rules but also to develop more sophisticated types of analysis based around city data feeds.

The key advantage of moving city data feeds into a procedural geographic information system is the ability to introduce more advanced spatial analysis functions within real-time data mining. The project aims to include multiple data feeds, such as air quality readings through tracking down enough air quality sensors to form accurate, pinpoint pollution

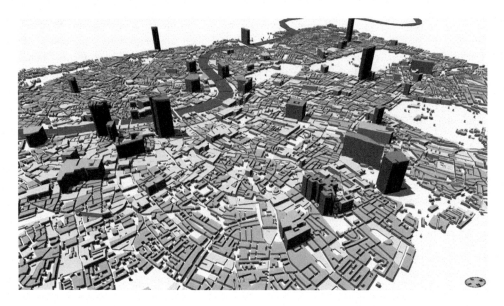

Figure 26.4   Live Twitter feeds (2015) in London visualised as building heights
Source: Chapter author(s)

estimates. The system operates online and, linked to a 3D representation of the city, it extends the concept of a city dashboard towards one integrating wider data feeds into a 3D model visualisation. The real-time data from the city can be visualised on a city-wide or hyper-local urban scale through the 3D interface. The move to a more 'urban' view of the data and its placement in a geolocated reference frame, whilst still maintaining real-time reporting, further develops the concept of a digital twin for London. It is, therefore, possible to produce fully dynamic 3D city scenes, which can be instantly edited from data feeds (as in the case of Tweet City). The interface can offer generic procedures for developing rules that mirror how cities work in terms of their patterns, movement and location.

Next steps: smarter cities, digital twins

These proofs of concept, the London data dashboard and Tweet City, demonstrate the real-time visualisation of urban data, with the latter providing a method for the development of simulations directly in a virtual 3D urban environment. Moreover, they explore the ways in which new means of visualisation and interactivity can aid in the better understanding of the complexity of spatial interaction, flows and patterns of information in cities via what we have defined as self-monitoring, analysis and reporting technologies. In this sense, we view them as being a core part of the emerging smart city. They provide a method in which they facilitate the development of models and introduction of data feeds within urban environments, defining the advantages, limitations and possibilities of the use of models. In relation to planning scenarios, these tools as ways of interacting and visualising data feeds can be used to test the decision-making consequences of urban planning actions on the fly and in real time. This is leading to the introduction of more complex data feeds which will provide a deeper insight into issues such as identifying possible solutions for optimising systems for the allocation of new social infrastructures. These types of method create an intelligent data analytic toolkit: the combination of city dashboard style feeds and the data streams linked to such 3D data models have potential to not only enhance the current state of the city reporting but also self-monitoring.

We are at the beginning of a new era in our understanding of the city, with a shift in the role of government and the citizen in creating, sharing and understanding data which are fed to a variety of stakeholders in real time. It is perhaps also a shift towards the development of a true 'smart city', one that is not only able to self-monitor, analyse and report but also respond. It is a city that ultimately will become self-aware in data terms of its own operations, mirrored via the creation of a digital twin and one that can adapt its infrastructure and services accordingly to be truly 'smart' – or any term it has since been replaced with.

Notes

1 www.citydashboard.org.
2 http://bigdatatoolkit.org.
3 www.paradoxplaza.com/cities-skylines/CSCS00GSK-MASTER.html.
4 www.esri.com/en-us/arcgis/products/esri-cityengine/overview.
5 Tweet City is available for download and online experimentation http://urschrei.github.io/CityEngine-Twitter/.

References

Allen, B. (2004) Monitoring Hard Disks with SMART, *Linux Journal*, 117, 214. www.linuxjournal.com/magazine/monitoring-hard-disks-smart.

Batty, M. (1995) The Computable City, *International Planning Studies*, 2, 155–173.

Batty, M. (1997) The Computable City, *International Planning Studies*, 3, 155–173.

Batty, M., Steadman, P., Xie Y. (2004) Visualization in Spatial Modeling. In J. Portugali (Editor) *Complex Artificial Environments*, Springer, Berlin, 49–70.

Beckwith R., Sherry J., Prendergast D. (2019) Data Flow in the Smart City: Open Data Versus the Commons. In M. de Lange and M. de Waal (Editors) *The Hackable City*, Springer, Singapore, 205–221.

BIS: Department for Business Innovation and Skills. (2013) The Smart City Market: Opportunities for the UK, Research Paper, 136, www.gov.uk/government/uploads/system/uploads/attachment_data/file/249423/bis-13-1217-smart-city-market-opportunties-uk.pdf

Caragliu, A., Del Bo, C., and Nijkamp, P. (2009) *Smart Cities in Europe, Series Research Memoranda 0048*, VU University Amsterdam, Faculty of Economics, Business Administration and Econometrics, Amsterdam.

Chin, H. C., Debnath, A. K., Yuen, B., and Haque, M. M. (2011) Benchmarking Smart And Safe Cities, Research Report, Grant No. R-264-000-251-112, National University of Singapore, Singapore.

Datta, S. P. A. (2017) Digital Twins, *Journal of Innovation Management*, 5, 14–34.

Debnath, A. K., Chin, H. C., Haque, M. M., and Yuen, B. (2014) A Methodological Framework For Benchmarking Smart Transport Cities, *Cities*, 37, 47–56.

Dodds, L. 2016. Data Marketplaces, We Hardly Knew Ye. https://blog.ldodds.com/2016/04/25/data-marketplaces-we-hardly-knew-ye/

Feldt, A. G. (1965) *The Cornell Land Use Game*, Center for Housing and Environmental Studies, Division of Urban Studies, Cornell University, Ithaca, NY.

Giffinger, R., Haindlmaier, G., and Kramar, H. (2010) The Role of Rankings in Growing City Competition, *Urban Research and Practice*, 3, 299–312.

GLA. (2013) The London Datastore, http://data.london.gov.uk/datastore/about

GLA. (2019), Smart London Plan, Smart London Together, www.london.gov.uk/what-we-do/business-and-economy/supporting-londons-sectors/smart-london/smarter-london-together, GLA.

GSMA. (2012) Guide to Smart Cities, www.gsma.com/connectedliving/wp-content/uploads/2013/02/cl_sc_guide_wp_02_131.pdf

Hardinges, J. (2018, July 10). What is a data trust? – The ODI. Retrieved 8 May 2019, from https://theodi.org/article/what-is-a-data-trust/

Hill, D. (2008) The Street as Platform, City of Sound, February 11 2008, www.cityofsound.com/blog/2008/02/the-street-as-p.html

Hudson-Smith, A. (2014) Tracking, Tagging and Scanning the City, *Architectural Design*, 84(1), 40–47.

Hudson-Smith, A., Hay, D., Wilson, D., Gray, S. (2019) *The Little Book of Connected Environments*, ImaginationLab, University of Lancaster. www.petrashub.org/download/little-book-of-connected-environments-and-the-internet-of-things/.

Hügel, Stephan, Roumpani, Flora. (2013). CityEngine-Twitter. Available at: http://dx.doi.org/10.5281/zenodo.9795.

IBM. (2013) Big Data at the Speed of Business, www-01.ibm.com/software/data/bigdata/

Kitchin, R. (2013, November 27). Four Critiques of Open Data Initiatives. Retrieved 2 October 2014, from Impact of Social Sciences website: http://blogs.lse.ac.uk/impactofsocialsciences/2013/11/27/four-critiques-of-open-data-initiatives/

Mattern, S. (2015) Mission Control: A History of the Urban Dashboard, *Places Journal*. doi:10.22269/150309

Mayer I., Bekebrede G., Bilsen A., Zhou Q. (2009) Beyond SimCity: Urban Gaming and Multi-Actor Systems. In E. Stolk and M. Te Brommelstroet (Editors) *Model Town. Using Urban Simulation in New Town Planning*, Sun/INTI, Amsterdam, 168–181.

Moore, G. (1965) Cramming More Components onto Integrated Circuits, *Electronics*, 38(8), 114–117.

Morstatter, F., Pfeffer, J., Liu, H., and Carley, K.M. (2013). Is the Sample Good Enough? Comparing Data from Twitter's Streaming API with Twitter's Firehose. In proceedings of International AAAI Conference on Weblogs and Social Media (ICWSM), Cambridge, MA, 8–11 July 2013, 400–408.

Parish, Y.I.V., and Müller, P. (2001) Procedural modeling of cities. In Proceedings of the 28th annual conference on Computer graphics and interactive techniques (SIGGRAPH '01). ACM, New York, NY, 301–308. DOI: https://doi.org/10.1145/383259.383292.

Putz, S. (1994) Interactive Information Services using World-Wide Web Hypertext, *Computer Networks and ISDN Systems*, 27(2), 273–280.

Sayce, D. (2019) Number of tweets per Day? www.dsayce.com/social-media/tweets-day/

Severance, C. (2015) Roy T. Fielding: Understanding the REST Style, *Computer*, 48(6), 7–9. doi:10.1109/MC.2015.170

Söderström, O., et al. (2014) Smart Cities as Corporate Storytelling, *City*, 18(3), 307–320.

Wilson, T., Wiebe, J. and Hoffmann, P. (2005) Recognizing Contextual Polarity in Phrase-Level Sentiment Analysis. In Proceedings of Human Language Technology Conference and Conference on Empirical Methods in Natural Language Processing.

Wylie, B., and McDonald, S. (2018, October 9). Data Trusts – Defining What, How and Who Can Use Your Data. Retrieved 8 May 2019, from Centre for International Governance Innovation website: www.cigionline.org/multimedia/data-trusts-defining-what-how-and-who-can-use-your-data

27

Reimagining urban infrastructure through design and experimentation

Autonomous boat technology in the canals of Amsterdam

Fábio Duarte, Lenna Johnsen and Carlo Ratti

Introduction

What it is a "smart city"? The term is now so widely used that its meaning has become convoluted and often obtuse. However, its raison d'être is unquestionable: conceptually "smart cities" result from broad technological phenomena that have been unfolding over the last two decades and are now undergoing a dramatic acceleration. Advancements in robotics, the use of machine learning techniques to mine and analyze unprecedented amounts of data, and the infiltration of information technologies into physical space have ushered in a series of unprecedented possibilities in how we can understand, design, and live in a city and make it "smart" (Duarte & Firmino, 2009; Ratti & Claudel, 2016).

Although frequently promoted in a prescriptive way, smart cities entail such a multitude of complex, exciting, and uncertain technological and social challenges that imagining the future of cities in a formulaic fashion is fruitless. The city of the future is not a model, but rather a framework for experimentation. Novel approaches to smart cities arise from the explorations of the integration of the data currently generated in cities and the presence of robots in our daily lives; explorations that create urban experiences that could define the way people, institutions, nature, and infrastructure will interact in the city of the future.

This requires approaches to science, technology, and design where disciplinary boundaries are removed and the future of the city is envisioned through experiment-based proposals. Indeed, the literature on smart cities has stressed such an interdisciplinary approach (Angelidou, 2014; Stratigea et al., 2015). However, what seems to be missing is empirical evidence on how this would actually work. Much of the smart cities literature focuses on a critical reading of the management of urban planning, forging strong links with technological companies, and using a normative approach (Luque-Ayala & Marvin, 2015). Design and experimentation are largely neglected, mostly because the literature is focused on reacting to large projects developed by cities or companies, seldom with researchers participating actively in the conception or deployment of smart cities initiatives. The literature has missed opportunities to develop insights into the iterative and complex process of designing and developing these projects.

On the other side, design projects that endeavor to reshape cities leave details underdeveloped: how might sensors and technology work to create a new urban ecosystem? By cherry picking technologies such as drones to render into promotional photos, but leaving unexplored the logistics of such systems, designers miss the rigor offered by an experimental process and risk superficiality.

In this chapter, we focus on a research project that combines robotics and artificial intelligence, environmental sensing, and design, driven by a clear experiment-based quest: how will autonomy reframe the way we conceptualize urban mobility and urban services? More specifically, how can the development of a fleet of autonomous vehicles in an urban water system unlock new potentials? In this chapter we show how Roboat, an ongoing interdisciplinary design project, addresses both autonomy's technical challenges and imagine its urban applications beyond the form factor of the car dominating the existing self-driving literature (Duarte, 2019). The Roboat project aims to deploy a fleet of autonomous boats in Amsterdam's canals which will eventually be used to provide transportation for people and goods, monitor water quality, and enable the self-assembling of urban infrastructures such as bridges and stages. The project is part of a research collaboration between the Amsterdam-based Advanced Metropolitan Solutions (AMS) Institute and the Massachusetts Institute of Technology (MIT).

The Roboat project integrates many disciplines, from robotics to environmental sensing, from computer-based perception to industrial design. But within the context of this book on smart cities, after a brief description of the Roboat project, here we discuss some urban services that can be provided by Roboat. After all, the future needs to be imagined and built.

Roboat and Amsterdam

Computer science, artificial intelligence, robotics, environmental engineering, urban studies, design: many laboratories at MIT and AMS, involving dozens of researchers, are joining forces to develop a fleet of autonomous boats that will be navigating Amsterdam's canal in a few years, in a project called Roboat. The decision to focus on autonomous boats is not trivial. A rich body of work exists about autonomous vessels and underwater vehicles (Wang & Xie, 2015; Xiang et al., 2017). However, little research has been done on autonomous boats navigating urban waters, such as Amsterdam, where the large network of relatively narrow canals are used daily for a wide range of purposes—but mainly for leisure and tourist boats, moving thousands of people. The challenges involved in this endeavor require an interdisciplinary approach. The complexity increases when we aim to use these autonomous boats to transport people and goods, provide urban services, and create temporary infrastructures, all while continuously sensing the environment (water and air quality, and canal wall infrastructure). Within this context, interdisciplinary researchers feed each other with challenges and solutions.

Amsterdam is uniquely situated as a test site for autonomous boats. Its urban structure is based on rings of canals, which were first built as we know them in the sixteenth century, and are an UNESCO World Heritage site. Although "there is almost nothing on Amsterdam's canals that is not of importance or does not have an interesting history" (Spies, 1991: 15), the main canals—Singel, Herengracht, Keizersgracht, and Prinsengracht—have played the key functions of delivering goods and transporting people for most of their history.

However, since the late nineteenth century and early twentieth century, roads started to be widened, blocks of houses pulled down, and canals backfilled in order to make room for an expanding city. Amsterdam was becoming "dreadfully overcrowded" (Kahn & van der Plas, 1999). Covering and abandoning the canals as a daily infrastructure had two main purposes: to control waterborne diseases, due to the use of the canals as open-air sewers, and to accommodate a higher demand for road traffic (de Haan, 1991). The area of Amsterdam's canals has halved over the years, giving space to an increasing road-based transport system that has resulted in soaring emissions and noise pollution.

According to a Kennisinstituut voor Mobiliteitsbeleid Mobility report, the modal split in Amsterdam for home to work trips is 21% of trips by private motor vehicles, 48% by bicycle, and 16% by public transport, which encompasses buses, 15 trams, and 4 subway lines (a fifth is under construction). Despite being one of the most bike-friendly cities in Europe and having one of the lowest rate of inhabitants per vehicle (3.65, according to Gemeente Amsterdam, 2016a), traffic and related emissions are still a concern. In addition to resident traffic, the city receives 271,000 commuters daily (Eurostat, 2016a), and an average of 45,000 tourists, for a total population of 850,000—all using underground and ground transport (including 2,300 taxis) to move around.

Adding to the movement of people, the transport of goods is a major contributor to traffic. In Amsterdam, there are 20,000 freight trips daily to 40,000 delivery and service points in the city center. Furthermore, 80% of the loading and unloading process happens on Amsterdam's narrow and sinuous streets, impeding the general flow of traffic (van Duin et al., 2014). The accelerating phenomenon of online shopping and home delivery is putting an additional pressure on urban freight. In the Netherlands more than 80% of goods are delivered directly to homes (Weltevreden & Rotem-Mindali, 2009). From 1998 to 2011, in the Netherlands the annual home market for internet shopping grew from EU€41 million to €9 billion, and went from a 2.8% market share, in 2005, to 10% in 2011 (Visser et al., 2014). The Netherlands has the highest waterborne freight transport rate in Europe, transporting 47% of the freight share—similar to the road transport (Eurostat, 2016b). Nevertheless, freight transport was absent from Amsterdam's canals in recent decades until 1997, when DHL began transporting goods by water. Today their boats serve as distribution centers from which bicycles collect and then distribute the parcels (Erdinch & Huang, 2014). In 2010, Mokum Mariteam started operating its first electric freight vessel in the city, using the waterways to deliver goods to shops (Maes et al., 2015).

Although 100 kilometers of canals still cover 25% of the city, their ability to transport people and goods has lost its relevance. Currently, the canals are mostly used by leisure and tourist boats. Since 2017, the city of Amsterdam enforces a maximum speed of 6 kilometers per hour (km/hr) in the canals, and the maximum size of boats in the central canals is 4.25 meters wide and 20 meters long. Since 2018, the use of radio frequency identification (RFID) is mandatory for all boats entering the city, which will allow a better monitoring of the boat traffic. By 2025, canal cruise boats must produce zero emissions, and in the next few years all boats must be electricity powered. These measures would help to address the current exposure to pollutants to the significant population residing close to Amsterdam's canals (van der Zee et al., 2012).

With one of the most extensive canal networks in the world, Amsterdam has the unique opportunity to reclaim the canals, rather than keep saturating its road system and ground transport. In the following sections, we introduce three use cases in which autonomous boats could play a role in Amsterdam: first, we address the use of the canals as a transportation network, comparing it with existing modalities such as cars, bicycles, tramways, and subway. Next, we propose the use of autonomous boats to deliver goods and to tackle one of the critical problems in the historical district of Amsterdam: waste collection, and compare the performance of the proposed systems with the current situation. Finally, we explore the use of Roboat to create temporary urban infrastructures, such as bridges, stages, and floating markets. In discussing both the research methodologies and the physical design of the Roboat units, we hope to show how the interdisciplinary nature of this team allows for unique lines of inquiry and results in novel solutions.

Moving people

As any city, motorization has been increasing in Amsterdam—and despite the pervasive use of bicycles and an extensive transit network including tramways, trains, buses, and subway, traffic has become a problem for the city. While the road network is congested, the canals of the city have been abandoned as a transit infrastructure. We used network analysis to study the possibility of regaining Amsterdam's canals for the movement of people, comparing the performance of autonomous boats with other existing modes, analysis that informed the design of the use cases, which in turn led to further research questions.

In order to assess the feasibility of deploying autonomous boats for the transportation of people and goods using Amsterdam's canals, comparing their performance (coverage area and travel times) with other motorized modes, we divided Amsterdam into a 500 x 500 meter cell grid, totaling 907 cells. Considering only those cells which are traversed by or adjacent to a single connected canal network, 236 cells can be directly served by boats, covering 60 square kilometers, as shown in Figure 27.1.

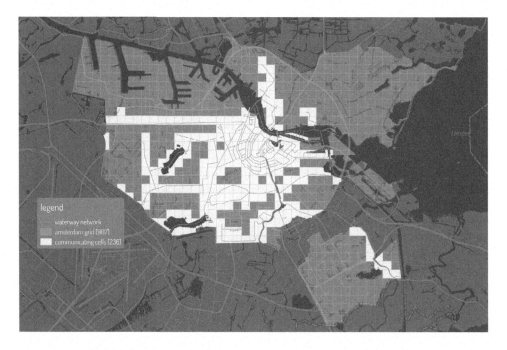

Figure 27.1 Amsterdam's canal network
Source: Chapter author(s)

Cells that can be connected one another through the canals, defining the single largest network of continuous canals in Amsterdam, we call "communicating cells", which include the Ij river and a segment of canal outside city limits. In a city with 850,000 inhabitants, 55% live within the canal network communicating cells. In order to evaluate the use of boats for transport purposes, a test point was established in each cell, using the mean coordinates of the canal segments within each cell, as shown in Figure 27.2. Twenty test points were added manually along the larger bodies of water to ensure realistic points for land transport.

The resulting set of 236 test points were used to calculate travel times across the canal network. For boats, the shortest distance between each pair of test points along the canals was determined using GIS network analysis, specifically an Origin–Destination (OD) Cost Matrix. We used the maximum speed allowed in the canals (6 km/h) to calculate the travel

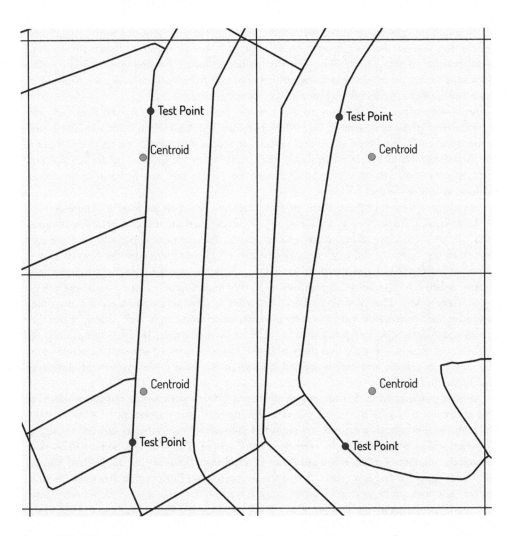

Figure 27.2 To calculate travel times that could be compared across modes, a test point was generated in GIS using the weighted center of the canal segments in each cell

Source: Chapter author(s)

time for each of these possible trips. For the cars and public transport, we developed a script that leverages the Google Distance Matrix web service, which provides travel distance and time for a matrix of origins and destinations. For public transport we used the best combination of buses, trains, trams, and subways.

Combining residents and the 271,000 commuters, daily movements comprise trips to work, shopping, visiting friends and family. We computed travel times from any communicating cell to any other cell in the canal network, for cars, bicycles, public transport, and walking. Not surprisingly for such a bicycle-friendly city, bicycles are the fastest option to almost any trip, any time of the day. Boats, even considering an average speed of 5km/h, can reach a place within a 2 km radius from the origin in less time than the public transport systems in more than half of the trips considered. Boats acquire better results in comparison with cars and public transport if the average speed increases to 10 km/h (Figures 27.3 and 27.4).

Although using boats to move around the city might not be efficient from a travel-time perspective, due to the current low speed limit of 6 km/h, there are other personal and social benefits in using boats. The benefits include reducing on-street parking spaces and decreasing traffic and related emissions, which could be particularly relevant for Amsterdam, where air quality is frequently below European Union standards.

Also, other trips, such as leisure trips, are not sensitive to travel times and thus they consider different parameters. For the 45,000 tourists that visit Amsterdam daily, the city in itself is the attraction; and whereas the routes of existing tourist boats cover, on average, 10 km of canals and only travel on the main canals, the smaller boats we propose in the Roboat project can cover 180 km of canals. Smaller boats could also give more flexibility to tourists wishing to explore farther afield.

Numbers of tourists in Amsterdam are rising rapidly. The resulting issue of congestion and its detrimental impacts on everyday life in the city dominated the mayor's 2016 "Staat van de Stad" (State of the City) address (van der Laan, 2016). A system of boats could also ease congestion in the center of the city by making it easier and more attractive for tourists to visit locations farther away. From anecdotal knowledge, we understand the public transport system in Amsterdam can be confusing for tourists, especially regarding ticket types, costs, and validity on different modes. This confusion serves as a barrier to those seeking to leave the city center —tourists and residents alike are less willing to take multimodal trips, such as a tram and then a bus, and then maybe a regional train to reach their destination. In order to quantify and study this confusion, we measured the number of transfers required to reach attractions from the city center. Here, "transfer" is defined as a switch from one mode or route of transportation to another.

Using OpenStreetMap data for points of interest (POIs), we selected out those related to tourist activities, such as museums, galleries, historic sites, and monuments. Of over 18,000 POIs in the Amsterdam metropolitan region, 4,200 are tourism-related, and over 1,400 of those are within 200 meters of the connected canal system. Then, using Google API, we calculated the number of transfers it would take to travel from the cells with the highest number of hotel beds. On average, it takes 1.1 transfers to reach those POIs within 200 m of the canal system. Autonomous boats could provide an A–B travel system that would allow both tourists and residents to circumvent this problem. While, as discussed earlier, the speed limit in the canals does not allow for fast transport, moving travel back to the canals would add another layer of experience to moving throughout, within, and outside the city of Amsterdam. In the long term, this could aid in the development of museums, galleries, and other cultural facilities outside the core zone, which in turn would diversify the concept of "cultural attraction" in

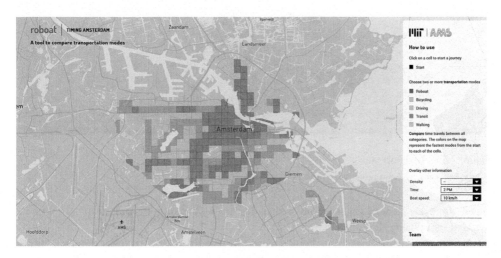

Figure 27.3 and 27.4 Travel times comparison between boats (dark grey) and transit (light grey). The color of the cell shows the best way to reach that location starting from the white cell (Figure 27.3: boat maximum speed 5 km/h. Figure 27.4: boat maximum speed 10 km/h)

Source: Chapter author(s)

Amsterdam. The results from this scientific analysis inform the design of a specific Roboat unit equipped to transport people. An initial design can be seen in Figure 27.5.

Urban services: distributing goods, removing waste

Besides transporting people, autonomous boats could eliminate at least part of the 3,500 trucks and 25,000 vans that drive into Amsterdam daily, consequently decreasing road traffic, and contributing to suit the delivery fleet to the Low Emission Zones enforced in the city center (Teekamp, 2016; Gemeente Amsterdam, 2016b). Moreover, for delivery of goods,

Figure 27.5 Taxi Roboat

Source: Rendering by Pietro Leoni, MIT Senseable City Lab

travel times are generally less sensitive than for people's transport, and rush hour can be avoided by utilizing the canals.

Amsterdam Centraal and the Food Center Amsterdam could serve as potential distribution hubs. Established in 1934 as Centrale Markt, the Food Center Amsterdam initially used canals to distribute goods to the markets and retailers (Gemeente Amsterdam Stadsdeel West, 2014). However, by 1966 those canals were filled in to facilitate the circulation of trucks and other motorized vehicles. Today, there are 70 wholesale companies who operate from the Food Center, selling fruit, vegetables, fish, and meat (Food Center Amsterdam, n.d.). There are plans underway to transform the area into Marktkwartier, a mixed-use district, in which the Food Center Amsterdam continues to be an important distribution hub.

From Amsterdam Centraal and the Food Center, all communicating cells can be reached by boat. As the main destinations, we mapped the supermarkets and restaurants (for larger and smaller deliveries, respectively). Besides these two main hubs, we performed a network analysis to define the most suitable areas along the canals to serve either as intermediate warehouses or boat depots, finding strategic points along the canals that could be developed into distribution centers. Assessing the suitability of cargo boats to supply the demand of restaurants and shops in 21 zones, which cover 2.5 square km in central Amsterdam, van Duin et al. (2014) suggest that just four freight vessels would be enough to supply the total logistic demand in this area during the summer, reducing waiting time for deliveries without interfering with touring boats and pleasure craft.

Waste collection trucks are a particular burden for cities with narrow and sinuous street networks, such as Amsterdam. In the central districts, large trucks collect garbage disposed on the curbside once a week, frequently creating traffic jams while they hoist trash onto the truck. Residents have a 12-hour window to deposit their trash in the designated locations, and face a hefty fine if they place bags in the wrong place and/or at the wrong time. In the peripheral areas of Amsterdam, trucks collect trash from underground refuse containers and bring them to the AEB incinerator. Thus, Amsterdammers have to walk about 100 meters to take their domestic trash to the nearest underground container, as shown in Figures 27.6 to 27.9.

To evaluate the potential for water trash collection we assessed how many buildings in the city were within a convenient walking distance of a canal. Firstly, we have the roads that already run along a canal. Then we took each node in the network of street lines and found the nodes that are on the canals' edge. We then evaluated the shortest walk from each node in the road network to the closest point on the canal. We then redrew these shortest walks from the start point on the canal to the end point in the city, and split the line at the 100 m mark. For the study area chosen, 48% of 37,665 buildings are within 100 m of a canal (Figure 27.10). Therefore, based on the same average walking distances of residents of other neighborhoods, approximately half of the municipal waste in the center of Amsterdam could be collected by boat. This presents some design and logistic challenges, which we have recently addressed focusing on the Centrum district in Amsterdam (Zhang et al., in submission).

Besides deploying autonomous boats as a replacement of trash trucks, the new system could reduce the hazards caused by the plastic trash bags currently left on the curbside, which include being obstacles to pedestrians, attracting pests, and dirtying the streets. Autonomy enables waste collection to operate outside of normal working hours, allows for auto-adjusting based on knowledge of the entire system, and Roboat units might become a key site for data collection on waste and consumption, data which in turn fuel other "smart" aspects of the city. As seen in Figure 27.11, this also has ramifications for the built environment in the design of the garbage modules and their connection to the canal edge. This is one key example of how the diverse experts on the team create feedback loops in the development process, with designers asking questions both formal and functional of the sensors required for autonomy, data scientists identifying optimal locations and re-running models based on feedback from roboticists and designers, and a chorus of voices working together towards the same goal.

Urban services: infrastructure

The goal in developing autonomous boat technology is to realize the potential of Amsterdam's canals to become a responsive infrastructure. As an autonomous system informed by artificial intelligence and machine learning, Roboat can respond in real time to the conditions of the city, such as the ebb and flow of rush hour traffic. Roboat platform units can join together to create temporary bridges, alleviating congestion on Amsterdam's centuries-old bridges and canal-side streets (Figure 27.12). Individual units can also tessellate together to form floating stages and public squares on the canals, a twenty-first century technology enhancing Amsterdam's strong tradition of water-based events. Rather than using autonomy to remove people

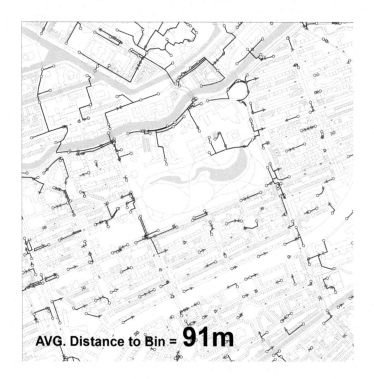

Figure 27.6 Average distance to bin: 91 m
Source: Analysis by Daniel Marshall, MIT Senseable City Lab

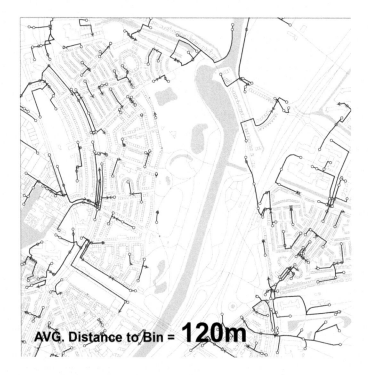

Figure 27.7 Average distance to bin: 120 m
Source: Analysis by Daniel Marshall, MIT Senseable City Lab

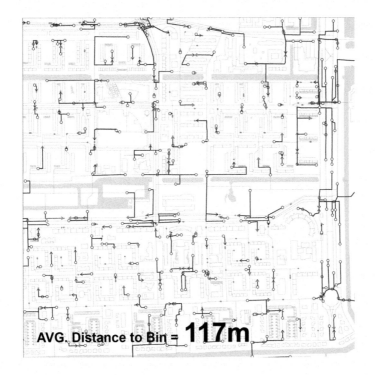

Figure 27.8 Average distance to bin: 117 m

Source: Analysis by Daniel Marshall, MIT Senseable City Lab

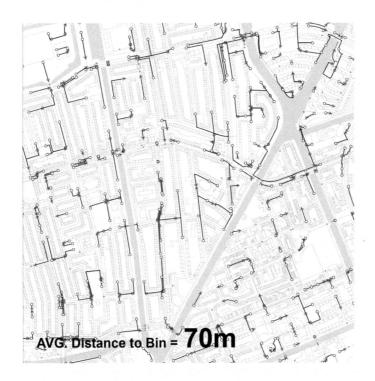

Figure 27.9 Average distance to bin: 91 m

Source: Analysis by Daniel Marshall, MIT Senseable City Lab

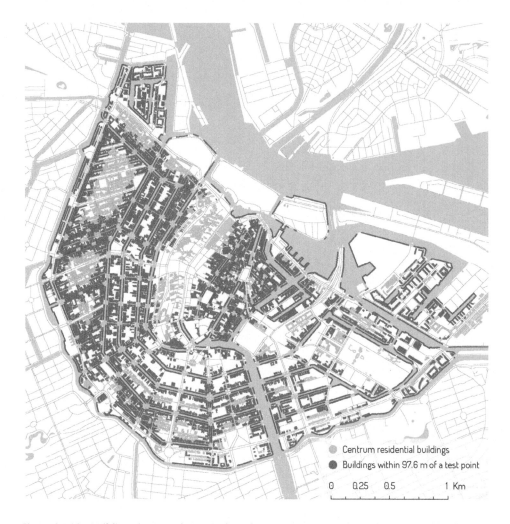

Figure 27.10 Buildings in central Amsterdam that are within a 97.6 m walk of the canal
Source: Map by Snoweria Zhang, MIT Senseable City Lab

from the system, the use of autonomous boats to create novel uses in the important urban space that is Amsterdam's canal network could bring residents and tourists together, creating new shared spaces to exchange ideas, culture, and goods while relieving pressure on roads.

For example, a typical Amsterdam market such as Plein '40–'50 with 150 stalls, it could be estimated that around 150 small vans drive, park, and supply the vendors. This has a huge cost to public space: congestion, as well as noise and CO_2 pollution. With this in mind, a system of floating markets could supplement the robust network of markets already present in Amsterdam, highlighting the potential to tap into the greater region's food production. These markets could function as individual stalls, able to dynamically appear on the canal side for Amsterdammers to collect crates of fresh produce or the Roboat markets could combine together to form larger, more typical markets, on the water, as can be seen in Figure 27.13.

Figure 27.11 Roboat units designed for garbage collection can serve the residents of Amsterdam

Source: Rendering by Pietro Leoni, MIT Senseable City Lab

Figure 27.12 Roboat units can join together to create temporary bridges

Source: Rendering by Pietro Leoni, MIT Senseable City Lab

Figure 27.13 Roboat markets
Source: Rendering by Pietro Leoni, MIT Senseable City Lab

Conclusion

As layers of networks and digital information blanket urban space, new approaches to the study of the built environment are emerging. The way we describe and understand cities is being radically transformed—as are the tools we use to design them. Operating at the intersection of design and science, the Roboat project is developing and deploying autonomous boats to learn about cities. By fostering symbiotic working relationships between roboticists, engineers, data scientists, and urban designers, we seek to avoid the mistakes of smart city experts and instead ask challenging and unexpected questions of our urban environment and technology.

By outlining in this chapter our vision of how autonomous boats can transform the urban fabric of Amsterdam, a centuries-old city, we hope to show that innovation comes from an interdisciplinary approach. Over the centuries, technological innovations have changed how people move in Amsterdam, how goods are delivered and waste removed, and, in general, how the city's canals have been used as urban infrastructure. Although still structuring the urban fabric of the city, today the canals are mainly seen as a tourist attraction and used for leisure activities. Autonomy, arguably the most relevant recent breakthrough in transport technologies, can be used to regain important aspects of Amsterdam's canals.

In this chapter we have assessed the feasibility of redeploying utilitarian boats in Amsterdam, initially from a transport standpoint. We argue that balancing transport efficiency with other co-benefits, such as reducing traffic and related pollutant emissions, boats have great potential to become part of Amsterdam's transport portfolio, for both people and goods. We have discussed how Roboat might radically transform Amsterdam's urban services such as waste collection and food distribution by reclaiming the canals as functional space. And in realizing the ability of autonomous floating platforms to tessellate together, the city could gain an entirely new typology of public space. To imagine that a system of autonomous floating platforms could reinvent Amsterdam's canal system takes roboticists in conversation with historians, engineers collaborating with designers, and data scientists asking questions of planners.

Acknowledgments

We would like to thank the Roboat team, in particular Pietro Leoni, Daniel Marshall, Sebastiano Milardo, and Snoweria Zhang for their contributions to the work discussed in this chapter. We would also like to thank the members of the MIT Senseable City Lab consortium: Dover Corporation, Teck, Lab Campus, Anas S.p.A., Cisco, SNCF Gares & Connexions, Brose, Allianz, UBER, Austrian Institute of Technology, Fraunhofer Institute, Kuwait-MIT Center for Natural Resources, SMART—Singapore-MIT Alliance for Research and Technology, AMS Institute, Shenzhen, Amsterdam, and the Victoria State Government.

References

Angelidou, M. (2014). Smart city policies: A spatial approach. *Cities*, 41 (1), S3–S11. doi: 10.1016/j. cities.2014.06.007.

de Haan, J. (1991). The greatness and difficulty of the work. In Spies, P., Schlüter, L., Camp, D.; de Vries, D. *The canals of Amsterdam*. Amsterdam: D'ARTS,28–45.

Duarte, F. (2019). Self-driving car: A city perspective. *Science Robotics* 4 (28) doi: 10.1126/scirobotics. aav9843.

Duarte, F., & Firmino, R. J. (200910). Infiltrated city, augmented space: Information and communication technologies, and representations of contemporary spatialities. *The Journal of Architecture*, *14* (5), 545–565. doi: 10.1080/13602360903187493.

Erdinch, H. & Huang, C. (2014) City logistics optimization: Gothenburg inner city freight delivery. Master Thesis, University of Gothenburg. At: https://gupea.ub.gu.se/bitstream/2077/37710/1/ gupea_2077_37710_1.pdf

Eurostat (2016a) Transport statistics at regional level. Eurostat Information. At: http://ec.europa.eu/eurostat/ statistics-explained/index.php/Transport_statistics_at_regional_level#Equipment_rates_ for_public_road_transport_passenger_vehicles

Eurostat (2016b) Freight transport statistics. Eurostat Information. At: http://ec.europa.eu/eurostat/statis tics-explained/index.php/Freight_transport_statistics#Inland_waterways_freight

Food Center Amsterdam. (n.d.). *Over ons*. At www.foodcenter.nl/over-ons/wie-we-zijn/

Gemeente Amsterdam. (2016a). De Amsterdamse Thermometer can de Bereikbaarheid 2016. At www. amsterdam.nl/parkeren-verkeer/bereikbaar/thermometer/

Gemeente Amsterdam. (2016b) Infographic milieuzones Amsterdam. At www.amsterdam.nl/publish/ pages/810795/infographic_milieuzone_amsterdam.pdf

Gemeente Amsterdam Stadsdeel West (2014) Bestemmingsplan food center Amsterdam. At www.ruimte lijkeplannen.nl/documents/NL.IMRO.0363.E1204BPSTD-VG01/t_NL.IMRO.0363. E1204BPSTD-VG01.pdf

Kahn, D., van der Plas, K. (1999). City profile: Amsterdam. *Cities*, 16 (5), 371–381.

Luque-Ayala, A., & Marvin, S. (201503). Developing a critical understanding of smart urbanism? *Urban Studies*, 52 (12), 2105–2116.

Maes, J., Sys, C., Vanelslander, T. (2015). City Logistics by Water: Good Practices and Scope for Expansion. In Ocampo-Martinez, C. & Negenborn, R. *Transport of water versus transport over water.* New York: Springer. 413–437.

Ratti, C.; Claudel, M. (2016) *The city of tomorrow: sensors, networks, hackers, and the future of urban life.* New Haven, CT: Yale University Press.

Spies, P. (1991). Jewels in the crown of this city. In Spies, P., Schlüter, L., Camp, D.; de Vries, D. *The canals of Amsterdam.* Amsterdam: D'ARTS,15–27.

Stratigea, A.; Papadopoulou, C-A.; Panagiotopoulou, M. (2015). Tools and technologies for planning the development of smart cities. *Journal of Urban Technology,* 22 (2), 43–62. doi: 10.1080/10630732.2015.1018725.

Teekamp, Raoul (2016) Verkenning goederenvervoer in Amsterdam. At www.amsterdam.nl/parkeren-verkeer/bereikbaar/thermometer/bijlagen/2016/ska-6/

van der Laan, E. (2016, 10). Staat van de Stad 2016. *Gemeente Amsterdam.* Retrieved from www.amsterdam.nl/bestuur-organisatie/college/burgemeester/speeches/toespraak-staat-stad/#_ftn1

van der Zee, S.; Dijkema, M.; van der Laan, J.; Hoek, G. (2012). The impact of inland ships and recreational boats on measured NO x and ultrafine particle concentrations along the waterways. *Atmospheric Environment,* 55 (3), 368–376. doi: 10.1016/j.atmosenv.2012.03.055.

van Duin, J., Kortmann, R., & Boogaard, S. V. (201405). City logistics through the canals? A simulation study on freight waterborne transport in the inner-city of Amsterdam. *International Journal of Urban Sciences,* 18 (2), 186–200. doi: 10.1080/12265934.2014.929021.

Visser, J., Nemoto, T., & Browne, M. (201403). Home delivery and the impacts on urban freight transport: A review. *Procedia - Social and Behavioral Sciences,* 125, 15–27. doi: 10.1016/j.sbspro.2014.01.1452.

Wang, W. & Xie, G. (2014). Online high-precision probabilistic localization of robotic fish using visual and inertial cues. *IEEE Transactions on Industrial Electronics,* 62 (2), 1113–1124.

Weltevreden J. W. J. & Rotem-Mindali, O. (2009). Mobility effects of b2c and c2c e-commerce in the Netherlands: A quantitative assessment. *Journal of Transport Geography,* 17 (2), 83–92.

Xiang, X., Yu, C., Zhang, Q. (2017). On intelligent risk analysis and critical decision of underwater robotic vehicle. *Ocean Engineering,* 140, 453–465. doi: 10.1016/j.oceaneng.2017.06.020.

Zhang, S., Duarte, F., Guo, X., Johnsen, L., van de Ketterij, R., Ratti, C. (in submission 2018). On the feasibility of using waterborne vehicles for waste collection in Amsterdam-Centrum.

28

The death and life of smart cities

Katharine S. Willis

Introduction: the death and life of smart cities

> One of the main things to know is *what kind of problem cities pose*, for all problems cannot be thought about in the same way
>
> *(Jacobs, 1993 [1961], p. 428, italics added)*

At the heart of the rhetoric around smart cities is the idea that cities themselves are the problem. These approaches frame the complexity of urban life as a challenge that can be addressed by the application of a systems solution. A number of authors have shown how the failure of modernist city systems is linked to the legacy of urban planning methods that are perceived to have failed to manage the complexity of cities, and in particular their technological parts (Cowley & Caprotti, 2019; Firmino, Aurigi & Camargo, 2006; Graham, 1997; Van Hulst, 2012). Smart cities tend to position the city as a problem rooted in the crisis of urbanization, and the city is treated as a sick patient, with its urbanity the very core of the problem. The title of this chapter is taken from the challenge posed by Jane Jacobs in her influential book *The Death and Life of Great American Cities* (Jacobs, 1993 [1961]). To explore the way that problems and cities are dealt with, the chapter will look at the urban problems in the context of the field of urban planning through the lens of Jacobs' work. Whilst Jacobs is known as someone who brought a deep understanding of the role of people and communities into the study of the urban, she was also was a systems thinker. For Jacobs, cities were a challenge of organized complexity in which people and community were central. This chapter seeks to reflect on the role of cities as both the source of problems but more importantly as offering solutions. As a contemporary example we draw on the rhetoric and underpinning ideas behind Sidewalk Labs proposals for Waterside Toronto, and reflect on this project through some of Jacobs' thinking. If, as Jacobs states in the quote at the start of this section, it is a matter of knowing what sort of problem cities are, this chapter seeks to address this challenge, by asking what kind of problem a city is. To do this we will look at how systems thinking underpins different planning theories, and the resulting implications for how we might start to thinking differently about the design and planning of smart cities.

The city as a problem to be solved: systems thinking and urban planning

Smart cities are positioned as a new and innovative approach to the failure of urban planning in cities (Cowley & Caprotti, 2019). Yet, this chapter takes the position that smart cities' approaches are themselves a re-marketing and re-making of the very same failed urban planning processes they claim to re-invent. Hall outlines how urban planning 'conventionally refers to planning with a spatial or geographical, component in which the general objective is to provide for a spatial structure of activities (or of land uses) which in some way is better than the pattern that would exist without planning' (Hall, 1992, p. 3). Underlying this is a fundamental rational and often systematic approach, since according to Hall: 'planning as a general activity is the making of an orderly sequence of action that will lead to the achievement of a stated goal' (Hall, 1992, p. 3). This highlights how the primary goal in urban planning is typically depicted as the solving of an urban problem.

This sort of systems thinking underpins many well-accepted urban planning processes, since according to Hall 'planning is a type of management for complex systems' (Hall, 1992, p. 5). Urban planning as a discipline has typically conceptualized the problems of the city as systems in two different approaches: the first was to view cities as systems that could be optimized or made more efficient, and the second saw cities as open-ended processes, which gradually evolved. Both schools of thought treated cities as problems of complexity, although the former was influenced by developments in mathematics and engineering, which took a problem-solving approach, whilst the latter took insights from biology and organicism.

The idea of urban planning as evolutionary was initially shaped by the writings of Geddes, who, as a biologist, saw the city in natural ecological terms and planning as a way to guide its evolution, not determine it (Batty & Marshall, 2009, p. 551). Sennett depicts that systems thinking about cities which saw the city as composed of functionally related parts, developed out of organicist concepts from biology. Here, the problem of the city could be solved by getting the city 'to function like a healthy body' (Sennett, 1994, pp. 263–264), with the city streets as 'arteries' and 'veins' in 'the blood system of the body' (Söderström, Paasche & Klauser, 2014, pp. 312–313). Alexander's book *A City Is Not a Tree* (1965) introduced a contrasting approach to organicism, by proposing a unified science of cities that was a dialogue between the city as a natural phenomenon and other complex systems (Mehaffy, 2019; Willis & Aurigi, 2017). Alexander describes a system as that which focuses on an overall behaviour accomplished through the 'interaction among parts' (Alexander, 1965). One of the first proponents of a systems thinking in planning was Ebenezer Howard, who proposed a new approach to planning in his book Garden Cities of Tomorrow (2010 [1898]). The Garden City envisioned the design of a new socio-economic system marrying city and nature and based on highly planned zones. Howard envisaged further regional systems of garden cities, each focused around a central city. Architect Le Corbusier's Ville Radieuse drew on different inspiration, but similarly proposed a systematic urban model, with a particular focus on the car and highway as an organizing system in the master plan of the city.

However, the most influential branch of systems thinking that pervades the urban planning agenda in smart cities is that which has its roots in cybernetic theories of the late 1960s. Hall describes the impact of cybernetic thinking on urban planning, a development that he terms 'cybernated planning' (Hall, 1992, p. 6). This emerged out of traditional, formalist ideas of planning as production of plans for the future desired state of the area and 'towards the new idea of planning as a continuous series of controls over the development of the area, aided by devices which seek to model or simulate the process of the development so that this control can be applied' (Hall, 1992, p. 5). In this approach the planner develops an information system which is continuously updated as the region develops and changes, much akin to a cybernetic system. The similarities between the organic and cybernetic approaches to urban planning lie in the understanding of the city as a system of parts that could be somehow understood as a dynamic system and be optimized to essentially 'work better'. This departs from urban planning models of the early twentieth century which were mainly a rational and linear process that examined all possible problems and developed solutions based on an optimisation model (Faludi, 1973). Interestingly, Goodspeed (2015) charts an early use of the term 'smart city' to R. E. Hall in a paper informed directly by cybernetic thinking, that described urban centres that are:

> Designed, constructed and maintained making use of advanced, integrated materials, sensors, electronics, and networks which are interfaced with computerized systems comprised of databases, tracking, and decision-making algorithms ... At the simplest level is the basic component and its associated "feedback" or self-monitoring mechanism(s).
>
> *(Hall et al., 2000, p. 1)*

Cybernetics, therefore, had a significant impact on urban planning approaches, and introduced a technologically informed systems thinking to address what were perceived as 'wicked' problems of urban complexity.

One of the key movements that emerged as a challenge to rational and formal urban planning in UK was the 'Non-Plan' manifesto set out by Banham, Price, Cook and Barker, which was informed by cybernetic thinking that Banham et al. argued made 'traditional planning technologically and intellectually obsolete' (1969, p. 422). It positioned itself as an 'experiment in freedom', with the primary emancipatory element being the freedom from the role of the planner having the 'right to say what is right' (Banham et al., 1969, p. 443). Non-Plan was informed primarily by the writings of cyberneticists Pask and Beer and saw the potential of working with vast amounts of information than was hitherto thought possible 'information essentially about the effects of certain defined actions upon the operation of a system' (Banham et al., 1969, p. 442). According to Non-Plan's authors, 'the practical implications are everywhere very large, but nowhere are they greater than in the area we loosely call planning' (Banham et al., 1969, p. 442). Archigram's Plug-In City was a vision of a future city where changes over time, would result in different aspects of cities (bathrooms, workplaces, shops) being plugged out or in at different rates. The Plug-In City could adapt by removing and replacing its parts and most importantly its structures would be computer controlled. This derived from Crompton's Computer City Project (1964): a conceptual proposal where the city would be formed by the channels of information that flow between people. Crompton's design was informed by a cybernetic approach based on 'speculative proposal for a computer system detecting and facilitating patterns of activity amongst a city area of 100,000 people' where the aim was to see 'what happens if the whole urban environment can be programmed and structured for change' (Cook, 1964, p. 33). According to Sadler, each system would have to evolve to manage 'emergent situations', not 'established' ones (Sadler, 2000, pp. 145–146). The problem-solving approach that underpins the cybernetic and systematic planning theory in the 1960s and 1970s, mirrors parts of the rationale for smart cities.

The kind of problem a city is

Taking the rhetoric of smart cities that position the city as a set of problems which technology can solve, it is important to unpack some of the work around the kinds of problems that cities are, that is – to look at the cityness of smart cities. To do this we will look at how systems thinking underpins different planning theories, and the resulting implications for how we might start to thinking differently about the design and planning of smart cities. For Jacobs, cities were a challenge of organized complexity in which people and community were central. The chapter first looks at how the problems of cities was played out in New York, the site of Jacobs' study of neighbourhoods and then will cross over to Toronto, Jacobs' second home and also the location of a more recent experiment in smart city development: Toronto Waterfront. The chapter begins by visiting 555 Hudson Street, Jacobs' home in the Greenwich Village neighbourhood in New York, which was both the site of the perceived problem to city planners and also the source of the solution that Jacobs' set out in *The Death and Life of Great American Cities*. Whilst not directly related to smart cities, it was fundamentally a similar battle to that being played out in the smart city debates: between a rational, top-down approach to urban planning and a social approach to understanding urban complexity. In doing so it reveals some of the kinds of problem a city is, and how competing urban visions can play out in the life (and death) of a city.

555 Hudson street, New York

In her seminal book of 1961, Jane Jacobs set out a comprehensive criticism of modern urban planning: it was, she said, a 'pseudoscience' that was 'almost neurotic in its obsession to imitate empiric failure and ignore empiric success' (Jacobs, 1993 [1961], p. 183). She made a plea to urban planners and city policy makers to understand the real 'kind of problem a city is'. For Jacobs, this was a problem of complexity of interacting factors that are 'interrelated into an organic whole'. Her underlying rationale was to understand and work with the intrinsic order of cities, and to apply the best insights of the new sciences, coupled with the most pragmatic methods.

The context for the book, was that in 1955, New York city planner Robert Moses proposed a highway that would cut through Greenwich Village (Figure 28.1), the district in which Jacobs' home in Hudson Street was located (Figure 28.2). Moses was a key proponent of modernist urban planning, where the city was exposed as sick and the only solution was infrastructural, with the byproduct being the large-scale destruction of their existing features. Jacobs countered this by demonstrating the value of social capital, and how it was linked with the physical form of the city: its streets, sidewalks, parks and mixture of uses. She developed a thesis as to why these elements and character of a small-scale city neighbourhood needed to be valued, and built a body of evidence from observing her own neighbourhood that has been shown to be transferrable to many other urban contexts. One of the central foci of her study was the role of urban everyday life in the street space – a phenomenon she termed 'sidewalk ballet' where:

> Under the seeming disorder of the old city, wherever the old city is working successfully, is a marvelous order for maintaining the safety of the streets and the freedom of the city. It is a complex order. Its essence is intricacy of sidewalk use, bringing with it a constant succession of eyes … The ballet of the good city sidewalk never repeats itself from place to place, and in any one place is always replete with improvisations.
>
> *(Jacobs, 1993 [1961], p. 50)*

By highlighting and championing the value of the chaotic, small-scale and social qualities of a place as well as demonstrating the importance of diversity, she introduced a original approach to urban planning that at the time was revolutionary and is still perceived as relevant today. Many smart cities projects put forward a vision of the city that has many parallels with what

Figure 28.1 Proposed Lower Manhattan Expressway: cross Manhattan Arterials and related improvements

Source: Image courtesy of MTA Bridges and Tunnels Special Archive

Figure 28.2 Jane Jacobs' house (red with white windows) at 555 Hudson Street, New York

Source: Image by Eric Fisher CC BY 2.0

Moses argued in his bid to build a road through New York: the city is broken and can only be fixed by the introduction of top-down, infrastructure-led urban regeneration. In many ways large-scale smart cities projects propose driving a metaphorical motorway through the heart of a city's neighbourhoods. But at a deeper level the technologized infrastructure of the motorway represents the dominance of a controlled and centralized city system against what Jacobs characterized as the much more localized and social complexity of the lived city.

Jacobs challenged conventional modern city planning by arguing that it had consistently mistaken cities as problems of what she termed 'simplicity' and of 'disorganized complexity', and had tried to analyse them and treat them as such. She argued that it was not enough for administrators in most fields to understand specific services and techniques, and that instead they needed to 'understand, and understand thoroughly, specific places' (Jacobs, 1993 [1961], p. 410). Drawing on thinking from the life sciences she put forward a hypothesis that

> cities happen to be problems in organized complexity … they do not exhibit one problem in organized complexity, which if understood explains them all. The variables are many, but they are not helter-skelter, they are 'interrelated in an organic whole'.
>
> *(Jacobs, 1993 [1961], p. 433)*

To illustrate her point, she takes the example of a city neighbourhood park, where

> Any single factor about the park is slippery as an eel; it can potentially mean any number of things, depending on how it is acted upon by other factors, and how it reacts to them … No matter what you try to do to it, a city park behaves like a problem of organised complexity, and that is what it is. The same is true of all other parts or features of a cities. Although the interrelationships of their many factors is complex, there is nothing accidental or irrational about the ways in which these factors affect each other
>
> *(Jacobs, 1993 [1961], p. 433)*

Jacobs acknowledges the park as a problem, but argues that the solution lay in thinking about cities as problems characterized by unexamined but obviously interconnected and understandable relationships that could only be revealed by observing its small-scale processes.

What does it mean for a city to be 'smart'?

If cities are the problem, the logical argument that smart city rhetoric makes is that technology is the only solution. According to Batty, this approach is founded in complex systems theory which characterizes the city as an inherently messy or 'wicked' (Batty et al., 2012) set of problems. Smart city language tends to adopt what Shepard defines as 'a techno determinism that cedes overwhelming agency to new technologies' (2011, p. 14). According to Kitchen et al., smart city developers often adopt an approach where 'their role is to create technologies that solve instrumental problems, such as how to make a process more sustainable, efficient or cost effective,' (Kitchin et al., 2019, p. 3) which fails to acknowledge and address wider social, political and philosophical issues. This perspective sees the city itself as problem, but not one that needs to be understood in a broader holistic sense, but in an instrumental manner. Instead, the city is treated in terms of a problem of technological complexity which Mattern claims has envisioned the smart city from 'from the net up' (Mattern, 2017). Smart city approaches often adopt a computational model of the world, underpinned by a rationale where 'modernist designers and futurists saw morphological parallels between urban forms and circuit boards' (Mattern, 2017; this volume). This optimization model works on the premise that complex social situations can be reduced to quantifiable problems that can be solved through computation (Kitchin, 2014, p.9). The key to this solutionism is the role of data, where the city is instrumented through sensors for data collection and actuators or control devices (Goodspeed, 2015).

In a traditional urban planning approach, data are used as part of a process of analysis, which then informs action. To date, the use of computers for understanding and planning cities have been for purposes of analysis, for example Geographical Information Systems (GIS) (Yeh, 1999), city modeling (Batty, 2001) and for databases. A parallel use has been in through communicating data in different ways and enabling new modes of participation and engagement in planning activities, such as e-planning (Silva, 2010), gamification (Gordon & Mihailidis, 2016) and apps (Innes & Booher, 2004; Wilson, Tewdwr-Jones & Comber, 2017).

However, as early as 1995 Batty identified the changing role of data and computation arguing that 'the notion that computers might be more than simply a means for a better understanding and that computation might be more than simply scientific analysis become significant' (Batty, 1995, p. 158). In this way, the urban planning process is no longer linear, and instead data are informing an ongoing process, rather than informing it or being embedded within it. This means it is important to understand how data become the solution to the problem of cities. In the next section we unpack the role of data in how the city is understood and acted upon.

The end of planning: the predictive and programmable city?

Where the smart cities agenda introduces a new aspect to systems thinking in planning is through a shift in approach from a sequential and cyclical feedback loop to a system undergoing almost constant change. This transforms the model of urban planning from a reflective and responsive process, typically underpinned by periods of analysis and then future planning, to one where the city becomes an almost independent analytical system. Quite literally it is not possible to plan, to make a set of decisions about how to do something in the future, when the nominal future is also the present. This type of city is characterized by Crang and Graham (2007) and by Shepard (2011) as the sentient city, and is made possible by the emergence of computing processes such as artificial intelligence and the interconnected sensors of the Internet of Things (IoT). Crang and Graham describe this as 'a world where we not only think of cities, but cities think of us, where the environment reflexively monitors our behaviour' (Crang & Graham, 2007, p. 789). Although cities have always inherently been reflexive in nature; they are shaped by the people that inhabit them and their practices, cultures and infrastructures (Sassen, 2012), the degree to which smart cities include a level of technological systemization can be seen to shift the balance of this reflexivity. As Sassen highlights 'Bit by bit (or byte by byte), we have been retrofitting various city systems and networks with devices that count, measure, record, and connect' (Sassen, 2012). Kitchin makes the distinction between how these data are used by unpacking how urban science is broadly rooted in a positivistic tradition that has sought to apply scientific principles and methods, drawn from the natural, hard and computing sciences, to social phenomena in order to explain them. He argues that the urban science approach makes it possible to develop, run, regulate and live in the city on the basis of strong, rationale evidence rather weak, selective evidence and political ideology' so that 'the use of such big data provides the basis for a more efficient, sustainable, competitive, productive, open and transparent city.

(Kitchin, 2014, pp. 7–8)

This form of 'real time city' shifts from a city that can be planned to one that is instead programmable, and where the feedback loop is set by parameters based on an almost real time data feed. The question is what happens with the feedback, and how it is integrated into the city? Feedback according to cyberneticist Wiener's definition is 'the property of being able to adjust future conduct by past performance' (1954, pp. 32–33) in order to determine what a machine should do next. In this system the sequential nature of urban planning is challenged by the timescales with which the city is being transformed. Batty argues that real time and complex datasets in smart cities disrupts the timescales that underpin the process of urban planning:

> Out of this comes 'big data' and the idea of the 'smart city' … Real-time data and the embedding of computers into the very fabric of the city itself generates much shorter time horizons than anything planning has dealt with hitherto.
>
> *(Batty, 2014, p. 389)*

More fundamentally, what the changing timescales within the smart city lead to is a way of breaking down the city into a series of quantifiable problems that assumes that all aspects of a city can be measured and monitored and treated as technical problems which can be addressed through technical solutions. This has implications for the linearity of urban planning processes since 'compressing time scales in such a way that longer term planning itself faces the prospect of becoming continuous as data is updated in real time' (Batty et al., 2012, p. 498).

Interestingly, Jacobs addressed the issue of feedback in *The Death and Life of American Cities*, long before smart cities were even imagined. Drawing on the example of organized complexity in cell behaviour she outlines a view of how feedback can be used to modify behaviour in the system. According to Jacobs the feedback process can only be productive when this feedback comes from the social character of the city. She argues that: 'In creating city success, we human beings have created marvels, but we left out feedback. I doubt that we can provide for cities anything equivalent to a true feedback system, working automatically and to perfection' (Jacobs, 1993 [1961], p. 252).

Jacobs counters the optimisation model by arguing that 'routine, ruthless, wasteful, oversimplified solutions for all manner of city physical needs have to be devised by administrative systems which have lost the power to comprehend and to value an infinity of vital, unique, intricate and interlocked details' (Jacobs, 1993 [1961], p. 408). This represents a different approach to city optimization, which takes the feedback from specific places and local knowledge since Jacobs argues that 'intricate minglings of different uses in cities are not a form of chaos' (p. 222). Jacobs rounds on Howard's Garden City project as an example of over simplification, since it sorts and sifts out of the whole certain simple uses, and to 'arrange each of these in relative self containment' (p. 18). Instead of seeing the city as a complex living reality, it treated the city as fundamentally problems in disorganised complexity, susceptible to conversion into problems of simplicity once ranges and averages were worked out, but also to conceive of city traffic, industry, parks, and even cultural facilities as components of disorganised complexity, convertible into problems of simplicity.

(p. 438)

The issues of complexity and simplicity also fundamentally issues of inclusion and participation, since people and the reality of everyday lived experience are either included or excluded from the kind of problems that define a city.

From sidewalks to Sidewalk Labs

After having spent much of her working life in Hudson Street, New York, Jacobs moved to the city of Toronto in 1968. Her activism against large-scale town planning projects continued in her new home and she is credited with stopping a proposed new Spadina Expressway development that would have ripped through downtown the way the Lower Manhattan Expressway would have in New York.[1] Had she been alive in 2019, she may have become an activist to challenge the current Robert Moses of her new home town: Sidewalk Labs. Toronto is the site for a massive urban redevelopment project led by Sidewalk Labs, a company of Alphabet. It is probably no mistake that the name of the company delivering this urban innovation project is Sidewalk Labs, for one of Jacobs' central theses in her book was on the value of sidewalks. Sidewalk Labs appropriates the language of Jacobs throughout its marketing and interview pieces – words like sidewalks and neighbourhoods, and objectives such as having 'a public realm that acts as the city's living room' (Gibson, 2018). But in its master plan for Toronto the project is positioned as one which solves problems based on computational logic since its poses its own question of 'What would a city look like if you started from scratch in the internet era – "a neighborhood built from the internet up?"' (Doctoroff, 2016). It's also a strange irony that Doctoroff's previous defining project before Toronto was Hudson Yards, the first large scale foray into a data-driven urban project (see Shepard's chapter, this volume). Although not in Greenwich Village, it's a few blocks north in a district sharing the same name as Jacobs' street: Hudson Yards. Some of the language may borrow from Jacobs: from the sidewalk ballet Jacobs observed and valued in Hudson Street to the data-driven sidewalks of Sidewalk labs in Toronto, but there is a big gap between the city problems Jacobs cared about and those of Sidewalk Labs.

Underneath the storytelling rhetoric of valuing the public realm and improving the quality of urban life in their marketing material, Sidewalk Labs reveals itself as much more informed by cybernetic-type systemization of urban problems in feedback loops of organization and control. It is a system where the feedback loop creates outcomes that are the product of the system itself not the people or place. According to Doctoroff (2017, p. 18) the quantification of urban processes promises to make the city's growth and scale manageable where:

> Everything from pedestrian traffic and energy use to the fill-height of a public trash bin and the occupancy of an apartment building could be counted, geo-tagged, and put to use by a wifi-connected 'digital layer' undergirding the neighborhood's physical elements. It would sense movement, gather data, and send information back to a centralized map of the neighborhood. With heightened ability to measure the neighborhood comes better ways to manage it.

(Bliss, 2018)

In Sidewalk Labs vision, the city is a problem defined by its inability to be measured, and thus managed or controlled. This represents a new kind of systems complexity problem: one of uncontained urban growth and change that an 'old' or existing city cannot accommodate. In fact, Sidewalk Labs position the very cityness of the city as the problem to be solved by arguing that 'Toronto is straining against its aging infrastructure and the traditionally sluggish pace of urban change … The pace of urban change is too slow' (Sidewalk Labs, 2017, p. 10, 12); also:

> Physical spaces like buildings, streets, and parks can be designed for the opportunities that technology present, rather than forced to retrofit new advances very slowly and at great cost. By merging the physical and the digital into a neighbourhood's foundation, people are empowered with the tools to adapt to future problems no one can anticipate. Such a place quickly becomes a living laboratory for urban innovation.

Sidewalk labs approach is to reduce complexity by measuring behaviour as a set of variables in the system in order to optimize the city infrastructure. It sets out two ways of doing this; the first uses the computational logic of 'if … then', as outlined by Rohit Aggarwala, Sidewalk's head of urban systems. Aggarwala describes how a data-informed system enables the Sidewalk team to make 'fact-based, data-driven planning decisions' for the urban layout based using the following example:

> *If* we knew what are the patterns of how many pedestrians walk down the sidewalk, how many cyclists use the bike lane, how many vehicles go down the street, and how many are private vehicles, how many are delivery vans, how many are street cars, [*then*] we have a much more rational basis *to decide* well the sidewalk ought to be this wide, and yes this route needs a bike lane or not, or the bike lane it has is too narrow, and maybe there should be few travel lanes for vehicles or more travel lanes.
>
> *(Gibson, 2018, author's emphasis)*

Studying human behaviour and analysing it to make informed decisions about future designs of urban space is central to urban planning. In fact, Jacobs was one of the key proponents of the value of understanding a city by observing how people move through it – the sidewalk ballet which she described as a form of complex order:

> Under the seeming disorder of the old city, wherever the old city is working successfully, is a marvelous order for maintaining the safety of the streets and the freedom of the city. It is a complex order. Its essence is intricacy of sidewalk use, bringing with it a constant succession of eyes … The ballet of the good city sidewalk never repeats itself from place to place, and in any one place is always replete with improvisations.
>
> *(Jacobs, 1993 [1961], p. 50)*

This is exemplified in her description of how observation of behaviour, such as the way that 'desire paths' form, could be used to inform future design of pedestrian paths:

I was at a school in Connecticut where the architects watched paths that the children made in the snow all winter, and then when spring came they made those the gravel paths across the green. Why not do the same thing here?

(Gopnik, 2004, p. 34)

The challenge here is not just using data to undertake analysis and propose changes but 'what' data are being gathered and 'who' gets to act upon it. For Sidewalk Labs, the 'what' of data is key and for this reason they have developed their very own urban planning tool: 'Replica' that aims to give 'planning agencies a comprehensive portrait of how, when, and why people travel in urban areas' (Bowden, 2018). Here Sidewalk Labs, a private company, is the gatherer and analyser of urban information, and its benefit is that it can provide 'information that's more accurate, current, and representative than what's typically available' and in so doing 'we can help them respond more quickly to their community's needs today – and prepare for the future' (Bowden, 2018). In this scenario urban planning decisions can be improved because of the volume and completeness of a dataset. Taken to a further level this sort of data-driven systems approach leads to the creation of a predictive digital model of the city that simulates future behaviour where:

we can imagine that with computer simulation we might be able to run a model saying 'we know that this is going on, and where we think the break in the pipe might be' and make a prediction before sending somebody in. That will reduce maintenance costs and improve performance.

(Gibson, 2018)

This is a system that, according to Pask, creates a public image of the world within which all observations are assimilable and in terms of which behavioural predictions are made (1961, p. 22). Its inspiration seems much more informed by cybernetic thinking than the messy and incomplete problems of organized complexity that Jacobs champions. In this respect, it bears more than a passing similarity with Archigram's '-Plug-in City (Cook, 1973), a data-driven architectural machine that would calculate its users' preferences and reconfigure itself to accommodate them, which itself was inspired by Crompton's designs for a Computer City (1965) (Figure 28.3). In fact even the visual language and types of modular configurable building types appear in Sidewalk Labs' visuals for Toronto Waterfront – complete with the cranes that reinforce the idea that the city can be reconfigured (Figure 28.4)
In an interview from 2018, Crompton highlights the parallels of his Computer City project with current smart city projects such as Sidewalk Labs:

In a sense, we are sitting in the network that I drew. The whole world is. It just isn't being applied sufficiently to urban projects. It's very sensitive. It knows well enough where I am, via signals, at any time, to around three meters or whatever it may be. Eventually, all building components will know who they are and where they are. They'll report back to the urban system about their current condition, as you get with certain sensitive structures.

(Anderson, 2018)

The abstracted approach to 'if … then' is further problematic when data about that behaviour are fed back into the environment to alter future human behaviour, since the system starts to reinforce its own biases. According to Mattern 'the data we generate, based on determinist assumptions and imperfect methodologies, could end up shaping populations and building

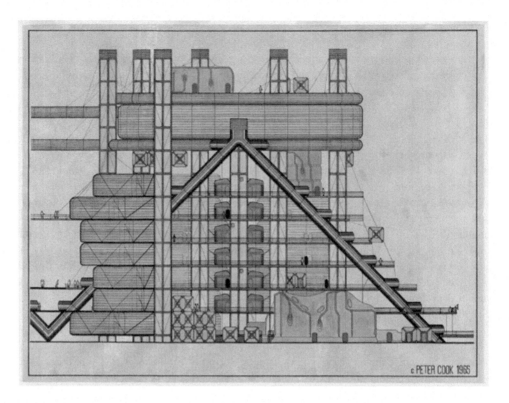

Figure 28.3 Plug-in City – University Node, project (Elevation) 1965 (Peter Cook)
Source: By permission of MOMA archive

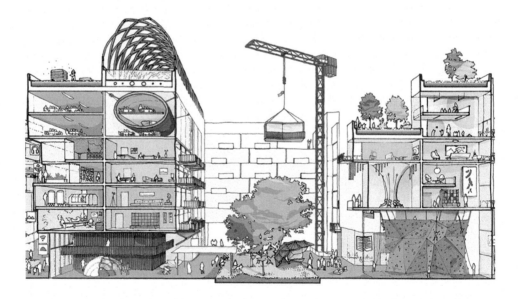

Figure 28.4 Sidewalk Labs' visuals for Toronto Waterfront
Source: Image courtesy of Sidewalk Labs

worlds in their own image' (Mattern, 2016), The smart city as responsive system of feedback systematizes bias and fails to recognize what Jacobs terms 'strips of chaos that have a weird wisdom of their own not yet encompassed in our concept of urban order' (Zipp & Storing, 2016). As Gabrys (2014) points out, these types of 'cyber-netically planned cities' reduce participation to a reflexive 'computational responsiveness' where 'the actions of citizens have less to do with individuals exercising rights and responsibilities, and more to do with operationalizing the cybernetic functions of the smart city' (p. 23). In failing to recognize the informal, everyday and marginalized in the smart city, such feedback systems not only ignore but also exclude people and communities from their right to the smart city (Willis, 2019).

The life of smart cities

> To approach a city, or even a city neighborhood, as if it were a larger architectural problem, capable of being given order by converting it into a disciplined work of art, is to make the mistake of attempting to substitute art for life.
>
> *(Jacobs, 1993 [1961], p. 373)*

Although this chapter has highlighted the issue with treating cities as malign problems, Jacobs' approach too focused on urban problems since she viewed a successful city neighbourhood as 'a place that keeps sufficiently abreast of its problems so it is not destroyed by them. An unsuccessful neighborhood is a place that is overwhelmed by its defects and problems and is progressively more helpless before them' (Jacobs, 1993 [1961], p. 146). The eyes on the street, sidewalk ballet and the local social ecosystems were her data-informed solutions. According to Greenberg,[2] an urban designer and friend of Jacobs in Toronto, the quote at the beginning of this section highlights how Jacobs also saw the neighbourhood as a lab or test-bed since 'everything was a big R&D lab for her; it was trial and error, seeing what succeeded, admitting when there was failure' (Micallef, 2016). The difference is that although it was a data-driven approach, it was based on the observation of what worked in an existing place, with a focus on everyday social processes and how they linked to places. It was this sort of 'data' that informed the set of proposals she made to the design and planning of future neighbourhoods. This is what Kitchin and Lauirault term 'small data', that is characterized as 'generally limited volume, non-continuous collection, narrow variety, and are usually generated to answer specific questions' (2015, p. 463).

This approach can escape what Halpern et al. refer to as the 'nostalgic lament for the beauty of human community before the time when machines, also, were bequeathed sentience' and they argue that 'our networks are often more lively than we can predict. If the test bed is the 'new epistemology, then we should err toward the incalculability of uncertainty rather than the measurable logics of risk' (Halpern et al., 2013). The life of cities lies in responding to specific, place-based and people-centred problems (Aurigi, this volume), and a product of lived experience.

What would Jane Jacobs do?

So, what would Jane Jacobs do (Figure 28.5) about smart cities? Jacobs strength was that she observed everyday, micro-sociality in a particular urban space, and noted what worked (as well as what didn't). It looked for the strengths in the place and sought to give them value and reinforce them, since Jacobs perceived that:

the best way to plan for downtown is to see how people use it today; to look for its strengths and to exploit and reinforce them. There is no logic that can be superimposed on the city; people make it, and it is to them, not buildings, that we must fit our plans.

(Zipp & Storing, 2016)

The same challenge is central to smart cities, since systems thinking requires a problem that technology can solve. As we have shown in this chapter, smart city rhetoric positions the city as problem – but it's the type of problem that is important to keep sight of. Technology needs a problem to solve, so it creates one. Cities aren't a problem, or if they are they always have been. If we continue to position cities as the problem, then we will continue to fall into the trap of systems thinking. The lessons learned from decades of failed urban planning projects is that they cannot be '"planned" in a linear way, from intention, to plan, to outcome as planned' (Healey, 2007, p. 3). Sassen argues for an open source urbanism that allows for incompleteness and allows the city itself to have agency in the process:

Figure 28.5 'What would Jane Jacobs do?' T-shirt created by Spacing, a design store in Toronto

Source: Image courtesy of Spacing Media

But cities talk back. Sometimes it may take decades, and sometimes it is immediate. We can think of the multiple ways in which the city talks back as a type of open-source urbanism: the city as partly made through a myriad of interventions and little changes from the ground up. Each of these multiple small interventions may not look like much, but together they give added meaning to the notion of the incompleteness of cities, the city as somewhat of a mutant.

(Sassen, 2012)

An open source urbanism will require a different approach to the role of data in future cities. Critical to this is that local people understand their own neighbourhoods and decide on the problems they wish to solve. This can be a data-informed approach, but, as MacFarlane and Söderström highlight, questions of 'who produces the data and how, and who interprets the data and how, are vital here' (2017, p. 319). Sandercock argues that 'twentieth century planning turned its back on questions of values, of meaning, and of the art, rather than science of city building' (2003, p. 221) and this requires a refocus on seeing cities as living places of work and as homes, of interactions and of communication. The solution might be to combine the richness of a place-based, social approach to 'small' urban data but combined with some of big data approaches that were highlighted by the Archigram Non-Plan approach. Non-Plan advocated 'providing accurate information to fit into a community investment plan' that championed collective engagement of people and communities in the creation of their own environments (Sadler, 2000, p. 177). This addresses the problematic nature of smart cities' 'invisible citizens', who are 'basically deprived of citizens or, more specifically, the citizens are the quintessence of the subaltern: they are silent, blind, and arguably even "stupid"' (Vanolo, 2016, p. 31). Democratizing practices in smart cities such as citizen science (Haklay, 2013), crowdfunding (Carè et al., 2018) and collaborative or hacking citymaking (de Lange & de Waal, 2019) have been shown to be informed ways in which local and participatory citymaking approaches can drive experimentation in future planning based on real social needs.

Conclusion

This chapter has sought to situate the smart cities agenda within a longer history of the ways we think about problems in cities. It has drawn on the field of urban planning to understand how systems thinking about cities has informed a way of approaching cities as problems. By looking at the challenges of smart cities through the eyes of Jane Jacobs it has revealed that systems thinking was also present in her work in *The Death and Life of Great American Cities*. Jacobs argued for ways of thinking about cities as problems of complexity, that involved understanding the city through studying neigbourhoods and their complex social processes that reveal themselves as a form of socially choreographed street 'ballet'. In contrast, the 'quantified community' approach (Shepard, this volume) of the Sidewalk Labs project for Toronto Waterfront adopts a closed system which defines its own self-directed, contained and quantifiable problems that represent a techno-centred deterministic approach also seen in cybernetic thinking around systems. A data-driven, cybernated planning approach to smart cities relies on internal feedback that reinforces decision-making based on limited sets of inputs that operate in a predictive feedback loop. As Gabrys points out, these types of 'cybernetically planned cities' reduce participation to a reflexive 'computational responsiveness' where 'the actions of citizens have less to do with individuals exercising rights and responsibilities, and more to do with operationalizing the cybernetic functions of the smart city' (2014, p. 23). Whilst traditional urban planning has typically taken a data-informed approach based on linear processes that are informed and then act upon a datafied understanding of the problems to be addressed, a cybernetic approach to planning creates an internalized feedback loop. This closed feedback loop engrains bias, but more importantly it denies people, place and lived experience a role in the way the city develops. It also treats the city as a platform or test bed undergoing continuous experimentation that is performative, but without the ability to define its own problems or to allow for what Jacobs might term messy or incomplete data. Interestingly, what is revealed through her treatise is that Jacobs herself based her ideas on systems thinking, and she too treated her Hudson Street community as a test-bed study of how to understand how a neighbourhood operates. What differentiates Jacobs from Sidewalk Labs' approach to understanding a city's problems is that Jacobs relied on local and place-specific knowledge which was not a closed loop feedback and that might be better described today as 'small data' rather than the big data of the smart city. The chapter finished by asking 'What would Jane Jacobs do?' and looks at approaches such as Sassen's open source urbanism and more participatory and value-based approaches from planners such as Sandercock and Healy. Returning to a cybernetic planning approach it asked whether the democratic ideals of Archigram's Non-Plan when they are 'sufficiently applied to urban projects' (Anderson, 2018) might in fact provide some clues to an approach where not just the people but also the city 'talks back' (Sassen, 2014). This would incorporate and centre approaches such as citizen science and collaborative digital participation in a way that combines some of the potential benefits of knowing through data and the knowing being in the hands of local people in specific places addressing problems that they themselves define. The emergence of the smart cities agenda that is led by private tech companies, such as Sidewalk Labs, represents a failure of systems thinking in the urban planning discipline that sees technology as purely instrumental in the design of cities. This presents the potential for the challenge for urban planning – to address the potential for smart cities to adopt messy, mutant and diverse approaches to technology that build on its strengths rather than solving its problems.

Notes

1 www.tiff.net/the-review/the-town-jane-jacobs-built.
2 Ken Greenberg is currently an advisor on Sidewalk Labs in Toronto.

References

Alexander, C. (1965). The city is not a tree. *Architectural Forum*, 122 (1), pp. 58–62.
Anderson, D. (2018) Archigram's radical architectural legacy. *Citylab*. https://citylab.com/design/2018/12/archigram-the-book-interview-darran-anderson-postmodernism/578389/: Citylab.
Banham, R., Hall, P., Price, C. & Barker, P. (1969). Non-Plan: an experiment in freedom. *New Society*, 338, pp. 435–443.
Batty, M. (1995). The computable city. *International Planning Studies*, 2, pp. 155–173.
Batty, M. (2001). Models in planning: technological imperatives and changing roles. *International Journal of Applied Earth Observation and Geoinformation*, 3 (3), pp. 252–266.
Batty, M. (2014). Can it happen again? planning support, lee's requiem and the rise of the smart cities movement. *Environment and Planning B: Planning and Design*, 41, pp. 381–391.
Batty, M. & Marshall, S. (2009). Centenary paper: the evolution of cities: geddes, abercrombie and the new physicalism. *The Town Planning Review*, 80 (6), pp. 551–574.
Batty, M., et al. (2012). Smart cities of the future. *Eur. Phys. J. Spec. Top*, 214(1), pp. 481–518.

Bliss, L. (2018) How smart should a city be? Toronto is finding out. *Citylab*. https://citylab.com/design/2018/09/how-smart-should-a-city-be-toronto-is-finding-out/569116/

Bowden, N. (2018) Introducing Replica, a next-generation urban planning tool. *Sidewalk Labs* https://sidewalklabs.com/blog/introducing-replica-a-next-generation-urban-planning-tool/.

Carè, S., Trotta, A., Carè, R. & Rizzello, A. (2018). Crowdfunding for the development of smart cities. *Business Horizons*, 61 (4), pp. 501–509.

Cook, P. (1964) Plug-in city. *The Sunday Times Colour Magazine*, 33.

Cook, P. (ed.) (1973). *Archigram*. New York: Praeger.

Cowley, R. & Caprotti, F. (2019). Smart city as anti-planning in the UK. *Environment and Planning D: Society and Space*, 37 (3), pp. 428–448.

Crang, M. & Graham, S. (2007). Sentient cities: ambient intelligence and the politics of urban space. *Information, communication society*, 10 (6), pp. 781–789.

Crompton, D. (1965). Computer city. *Archigram*, no. 5.

de Lange, M. & de Waal, M. (eds.) (2019). *The Hackable City: Digital Media and Collaborative City Making in the Network Society*. Singapore: Springer.

Doctoroff, D. (2016). Reimagining cities from the internet up. In Jaffe, E., Doctoroff, D.L. and Quirk, V. *Sidewalk Talk*. https://medium.com/sidewalk-talk/

Doctoroff, D. (2017) *A Project Vision*. Sidewalk Labs https://sidewalktoronto.ca/wp-content/uploads/2017/10/Sidewalk-Labs-Vision-Sections-of-RFP-Submission.pdf

Faludi, A. (1973). *Planning Theory*. New York: Pergamon.

Firmino, R., Aurigi, A. & Camargo, A. (2006) 'Urban and technological developments: why is it so hard to integrate ICTs into the planning agenda?'. *Sustainable Solutions for the Information Society: 11th International Conference on Urban Planning and Spatial Development for the Information Society (CORP)* Vienna, Austria: 13-16 February 2006.

Gabrys, J. (2014). Programming environments: environmentality and citizen sensing in the smart city. *Environment and Planning D: Society and Space*, 32 (1), pp. 30–48.

Gibson, E. (2018) It takes 40 or 50 years to update an industrial landscape, we want to speed that up, says Sidewalk Labs urban planner. *Dezeen*.

Goodspeed, R. (2015). Smart cities: moving beyond urban cybernetics to tackle wicked problems. *Cambridge Journal Of Regions, Economy And Society*, 8 (1), pp. 79–92.

Gopnik, A. (2004) Cities and Songs. *The New Yorker*.

Gordon, E. & Mihailidis, P. (eds.) (2016). *Civic Media: Technology, Design, Practice*. Cambridge, MA: MIT Press.

Graham, S. (1997). Cities in the real-time age: the paradigm challenge of telecommunications to the conception and planning of urban space. *Environment and Planning A*, 29 (1), pp. 105–127.

Haklay, M. (2013). Beyond quantification: a role for citizen science and community science in a smart city. In Campkin, B. and Ross, R. (eds) *UCL Urban Laboratory Pamphleteer*. London: UCL.

Hall, P. (1992). *Urban and Regional Planning (3rd edition)*. London: Routledge.

Hall, R.E., Bowerman, B., Braverman, J., Taylor, J., Todosow, H. and Von Wimmersperg, U. (2000). The vision of a smart city. Office of Scientific & Technical Information Technical Reports, UNT Library. https://digital.library.unt.edu/ark:/67531/metadc717101/

Halpern, O., LeCavalier, J., Calvillo, N. & Pietsch, W. (2013). Test bed as urban epistemology. In Marvin, S., Luque-Ayala, A. and McFarlane, C. (eds) *Smart Urbanism: utopian vision or false dawn?*. Abingdon: Routledge, pp. 146–168.

Healey, P. (2007). *Urban Complexity and Spatial Strategies: Towards a Relational Planning for Our Times*. Abingdon: Routledge.

Howard, E. (2010 [1898]). *To-Morrow: A peaceful Path to Real Reform'*. Cambridge: Cambridge University Press.

Innes, J. E. & Booher, D. E. (2004). Reframing public participation: strategies for the 21st century. *Planning Theory & Practice*, 5 (4), pp. 419–436.

Jacobs, J. (1993 [1961]). *The Death and Life of American Cities*. New York: Random House.

Kitchin, R. (2014). The real-time city? big data and smart urbanism. *GeoJournal*, 79 (1), pp. 1–14.

Kitchin, R., Coletta, C., Evans, L. & Heaphy, L. (2019). Creating smart cities. In Kitchin, R., Coletta, C., Evans, L. and Heaphy, L. (eds) *Creating Smart Cities*. Abingdon: Routledge, pp. 1–18.

Kitchin, R. & Lauriault, T. P. (2015). Small data in the era of big data. *GeoJournal*, 80 (4), pp. 463–475.

Mattern, S. (2016). Instrumental city: The view from Hudson Yards, circa 2019. *Places Journal*, April 2016.

Mattern, S. (2017). A city is not a computer. *Places Journal*, February 2017.

McFarlane, C. & Söderström, O. (2017). On alternative smart cities- From a technology-intensive to a knowledge-intensive smart urbanism. *City*, 21 (3-4), pp. 312–328.

Mehaffy, M. W. (2019). Assessing Alexander's later contributions to a science of cities. *Urban Science*, 3 (2), 59.

Micallef, S. (2016) Jane up north. *Curbed*.

Pask, G. (1961). *An Approach to Cybernetics*. London: Hutchinson.

Sadler, S. (ed.) (2000). *Non-Plan: Essays on Freedom, Participation and Change in Modern Architecture and Urbanism*. Oxford: The Architectural Press.

Sandercock, L. (2003). *Cosmopolis II: Mongrel Cities of the 21st Century*. London: Continuum International Publishing Group.

Sassen, S. (2012). Urbanising technology. In Burdett, R. and Rode, P. *The Electric City Newspaper*. http://ec2012.lsecities.net/newspaper/

Sassen, S. (2014). Does the City Have Speech?. In Haas, T. and Olsson, K. (eds) *Emergent Urbanism: Urban Planning & Design in Times of Structural and Systemic Change*. Aldershot: Ashgate, pp. 133–140.

Sennett, R. (1994). *Flesh and Stone: The Body and The City in Western Civilization*. London, UK: Faber & Faber.

Shepard, M. (ed.) (2011). *Sentient City: Ubiquitous Computing, Architecture, and the Future of Urban Space*. Cambridge, MA: MIT Press.

Sidewalk Labs (2017) *Vision sections of RFP submission*. http://sidewalk.torontio.ca/wp-content/uploads/2017/10/sidewalk-labs-vision-section-of-rfp-submission-toronto-quayside

Silva, C. N. (2010). The E-Planning paradigm – Theory, Methods and Tools: An Overview. In Silva, C. N. (ed) *Handbook of Research on E-Planning: ICTs for Urban Development and Monitoring*. Hershey, PA: IGI Global, pp. 1–14.

Söderström, O., Paasche, T. & Klauser, F. (2014). Smart cities as corporate storytelling. *City*, 18 (3), pp. 307–320.

Van Hulst, M. (2012). Storytelling, a model of and a model for planning. *Planning Theory & Practice*, 11 (3), pp. 299–318.

Vanolo, A. (2016). Is there anybody out there? the place and role of citizens in tomorrow's smart cities. *Futures*, 82, pp. 26–36.

Wiener, N. (1954). *The Human Use of Human Beings: Cybernetics and Society*. Boston: Houghton Mifflin.

Willis, K. (2019). Whose Right to the Smart City?. In Kitchin, R., Cardullo, P. and Feliciantonio, C.d. (eds) *The Right to The Smart City*. Bingley: Emerald Publishing, pp. 27–41.

Willis, K. S. & Aurigi, A. (2017). *Digital and Smart Cities*. London: Routledge.

Wilson, A., Tewdwr-Jones, M. & Comber, R. (2017). Urban planning, public participation and digital technology: App development as a method of generating citizen involvement in local planning processes. *Environment and Planning B: Urban Analytics and City Science*, 46(2), pp. 286–302.

Yeh, A. G.-O. (1999). *Urban Planning and GIS*. New York: John Wiley.

Zipp, S. & Storing, N. (eds.) (2016). *Vital Little Plans: The short works of Jane Jacobs*. New York: Random House.

Index

Page numbers in *italics* relate to figures; those in **bold** refer to information in tables.